Jonah Field: Case Study of a Tight-Gas Fluvial Reservoir

Edited by

John W. Robinson
Keith W. Shanley

AAPG Studies in Geology 52
Rocky Mountain Association of Geologists 2004 Guidebook

Published by
The American Association of Petroleum Geologists
Tulsa, Oklahoma
and
The Rocky Mountain Association of Geologists
Denver, Colorado

ISBN: 0-89181-059-5

This publication is available from:

The AAPG Bookstore
P.O. Box 979
Tulsa, OK U.S.A. 74101-0979
Phone: 1-918-584-2555
or 1-800-364-AAPG (U.S.A. only)
Fax: 1-918-560-2652
Or 1-800-898-2274 (U.S.A. only)
E-mail: bookstore@aapg.org
www.aapg.org

Rocky Mountain Association of Geologists
820 16th Street
Suite 505
Denver, CO 80202
Phone: 1-303-573-8621
Fax: 1:303-528-0546
E-mail: RMAGdenver@aol.com

Dedication

This book is dedicated to John W. Martin and Mickey McMurry, the two individuals most responsible for the success story that became Jonah field. The story of Jonah field begins in 1991 when John Martin, then President and co-founder of McMurry Oil Company, went to Denver to look through a company's divestiture portfolio. John's background included a term as manager of the Casper airport, but he had also been a logging engineer for Schlumberger after graduating from the Air Force Academy and completing his active duty. "I picked out half a dozen prospects, and Jonah was one. I was particularly looking for gas properties at the time because gas was out of favor and I figured I could get a good price," Martin said. "The Jonah 1-4 stood out as a gas well," Martin recalled. In late 1991, McMurry Oil bought the Jonah properties and quietly amassed a controlling interest in the area.

McMurry Oil had a strong desire to recoup some of their investment money, so they proceeded to build a pipeline to hook the wells up rather than test the wells to the atmosphere. In August 1992, Mick McMurry supervised the installation of 10 miles of surface pipe that was connected to the nearest producing well on the Pinedale anticline. Mick's contribution to the success at Jonah was his construction and facilities background, which was developed after he completed his military service and graduated from the University of Wyoming with an engineering degree. Despite the formation damage from earlier stimulations, the Jonah 1-4 initially flowed 2,000 mcfd. It is thought that the failure in earlier stimulations was related to gel chemistry—the gels didn't break at the proper time to allow flow back and clean up of the well bore. In fact, only 200 of the original 5,000 barrels of fracture fluid injected were recovered from the early stimulation. As John Martin joked later, "It took five years for the gels to break." McMurry Oil and their partners brought the appropriate drilling and completion technology into the Jonah project, an effort that resulted in the "re-discovery" of Jonah field. Jonah field will ultimately have hundreds of producing wells and is poised to become the next multi-tcf gas field in the Rocky Mountain Region.

Through their success at Jonah, John Martin and Mick McMurry have been active in giving back to their communities. The McMurry Foundation has benefited many charitable causes in Wyoming, while the Martin Family Foundation is involved in projects in Colorado, Wyoming, and Montana.

The perseverance and determination that characterizes Mick McMurry and John Martin, especially in an area long abandoned by industry, is a rare quality these days and did not end with the Jonah success. Both gentlemen remain active in the petroleum industry, always on the hunt for the next big opportunity. The re-discovery of Jonah field ignited interest in structurally trapped, tight gas exploration in Upper Cretaceous strata, and stands as an analog for further exploration of vast, sparsely-drilled areas of the Rocky Mountain region.

John W. Robinson
Keith W. Shanley
Co-Editors

About the Editors

John W. Robinson is a consulting geologist and owner of North Ranch Resources, Littleton, Colorado. He received B.S. and M.S. degrees in Geology from San Diego State University and a Ph.D. in Geology from Colorado School of Mines. His career spans 29 years in the energy industry including work for PEMEX, AMOCO, Forest Oil, Snyder Oil, and McMurry Energy. In addition, Robinson has been an adjunct instructor at Colorado School of Mines and a volunteer for the U.S. Geological Survey. In 1993, he was a visiting scientist at Shell Research International, The Netherlands, where he worked with other scientists to create a 3-D reservoir model for braided fluvial strata. As a result of that work, he and Peter McCabe were given the Wallace Pratt Memorial Award in 1999 for the best paper in the AAPG Bulletin.

Robinson is currently Chair of the Headquarters and Business Committee for SEPM and is a Trustee of the SEPM Foundation. He was President of the Rocky Mountain Association of Geologists in 2002 and is a Trustee of the RMAG Foundation. He has served as an Associate Editor of AAPG and was recently elected to Fellow of GSA. He is also a member of SPE, IAS, and the Wyoming Geological Association. He is the Technical Program Chair for the 2004 Rocky Mountain Section AAPG meeting in Denver. His research interests are in fluvial sedimentology and multidisciplinary reservoir studies.

Keith W. Shanley is a consulting geologist in Littleton, Colorado, with more than 22 years of experience in exploration, development, and research. He has worked in a variety of basins around the world for both major and independent oil companies. Shanley received his B.A. degree in Geology from Rice University in Houston, TX and his M.Sc. and Ph.D. degrees in Geology from the Colorado School of Mines in Golden, Colorado. He has served as a volunteer at the U.S. Geological Survey and as interim instructor at the Colorado School of Mines. Shanley has published papers, edited volumes, and organized conferences and seminars dealing with sequence stratigraphy, reservoir architecture, non-marine sedimentology and stratigraphy, and tight-gas. His current research interests include sequence stratigraphy and reservoir architecture, the integration of petrophysics, and risk analysis. Shanley is an Associate Editor of the AAPG. He is a member of the AAPG, SEPM, SPE, RMAG, and SPWLA and is a registered geologist in both Texas and Wyoming.

ACKNOWLEDGMENTS

The American Association of Petroleum Geologists and the Rocky Mountain Association of Geologists would like to thank the following reviewers for their work on this publication:

Steve Bailey
Doug Battin
Michele Bishop
Dave Cramer
Brian Disney
Dean DuBois
Frank Ethridge
Gus Gustason
Bill Hanson
Jane Estes-Jackson
Mark Kirschbaum
Steve Kneller
Barry McBride
Peter McCabe
Phil Nelson
Vito Nuccio
Terrilyn Olson
Lee Shannon
Don Stone
Greg Vigil
Ward Whiteman

CONTENTS

AAPG and RMAG wish to thank the following
for their generous contributions
to

Jonah Field: Case Study of a Light-Gas Fluvial Reservoir

BP America Production Company
Cabot Oil & Gas Corporation
U.S. Geological Survey

Contributions are applied toward the production cost of publication, thus directly reducing the book's purchase price and making the volume available to a larger readership.

CHAPTER 1

Introduction: Jonah Field—Case Study of a Tight-gas Fluvial Reservoir

J. W. ROBINSON
Consultant, Littleton, Colorado, U.S.A.

K. W. SHANLEY
Consultant, Littleton, Colorado, U.S.A.

A core workshop entitled *Tight-gas Fluvial Reservoirs: A Case History from the Lance Formation, Green River Basin, Wyoming*, was held in conjunction with the 2001 AAPG Annual Meeting in Denver, Colorado. As part of that short course, a book containing nine chapters, totaling almost 200 pages, was distributed to participants. In 2002, the editors proposed to AAPG and the Rocky Mountain Association of Geologists (RMAG) an opportunity to copublish a *Studies in Geology* volume with a larger group of papers that would have broad appeal to geoscientists and engineers working in tight-gas fluvial reservoirs.

Jonah field is the largest natural gas discovery in the onshore United States in the last 10–15 yr. Recoverable reserve estimates range from 8 to 15 tcf, depending on the method of analysis. By virtue of being a tight-gas reservoir, it is, in many respects, a nontraditional field. According to many of the recent U.S. and world resource assessments, most of the future gas resources will come from tight, low-permeability sandstones in the deeper portions of basins.

The intent of this volume is twofold: (1) to provide readers with previously unpublished or proprietary data on the field, and (2) to integrate all aspects surrounding the field including geology, geophysics, reservoir engineering, drilling and completion, and regulatory affairs. As such, this volume is a definitive collection that provides a truly integrated perspective of this giant field. The list of topics addressed in this book is by no means complete; however, it does encompass the comprehensive field description that we were striving for. In several cases, operators and/or authors have differing perspectives of the field, and we are fortunate to have captured those differing views in this book.

Chapter 2 by Robinson and Shanley is devoted to several pages of general summary information on Jonah field. As such, this provides a compilation of basic field information useful to those researchers looking for summary statistics.

Chapter 3, "Discovery of Jonah field" by Robinson, is reprinted from an earlier version published by the RMAG in *The Mountain Geologist*. This chapter provides a historical perspective on the field for those not familiar with its unique discovery. The discovery of Jonah is a testament to the insight, perseverance, and vision of John Martin and Mick McMurry, former owners of McMurry Oil Company, Casper, Wyoming.

Chapter 4 by Johnson, Finn, and Roberts presents the regional stratigraphic setting of Maastrichtian strata in the central Rocky Mountain region. The Maastrichtian was a period of mountain building and basin development that ultimately divided the central part of the Rocky Mountain foreland basin into several smaller Laramide basins and uplifts. They show that Jonah field is located in a deep, northwest-trending trough that developed southwest of the Laramide Wind River Range. They speculate that coaly intervals in the underlying Campanian Rock Springs Formation may be the source of the gas at Jonah, and a shale interval near the top of the Lance Formation may be the top seal for the field. These two fundamental aspects are further developed in chapter 7 by Coskey and chapter 8 by Cluff and Cluff.

Chapters 5 and 6 are written by two groups that have been active operators in Jonah field throughout its development. EnCana Oil and Gas (U.S.A.) acquired the McMurry Oil interest in 2000 and has since become the dominant operator in Jonah field. DuBois et al. present

Jonah Field: Case Study of a Giant Tight-Gas Fluvial Reservoir, J. W. Robinson and K. W. Shanley, eds.: AAPG Studies in Geology 52 and Rocky Mountain Association of Geologists 2004 Guidebook

important field maps, including structure, gross and net sandstone isopachs, and net pay isopachs. Their cross sections provide a glimpse of how they correlate units in a thick, fluvial system. Their pressure studies provide us with the first published data from the field. DuBois et al. define Jonah as a combination structural-stratigraphic trap, wherein the bounding fault zones form the lateral trap, and the top seal is formed by the intercalated mudstones of the Lance Formation. BP America (formerly Amoco Production Co.) became active in Jonah field in 1995 and was responsible for acquiring the first three-dimensional (3-D) seismic program across the field. Hanson et al. present an immense amount of information derived from several BP Amoco research programs. Hanson et al. also define Jonah as a structural trap bounded by two wrench faults but state that although structural features control the Jonah trap, regional structural elements shaped the fluvial system that deposited the reservoir sandstones. In addition, they present data on gas isotope composition and thermal maturity of the producing strata, which suggest that the gas was generated in deeper strata underlying the field and migrated upward into the producing interval. This perspective, in combination with the production patterns along the faults, explains the enigmatic charging of these low-permeability sandstone bodies according to structural position.

Chapter 7 by Coskey uses burial history modeling and petroleum systems analysis to evaluate the generative potential of source rocks in the Jonah and Pinedale anticline areas and to determine the timing of generation and expulsion. As such, this study is one of the few published efforts in the Rocky Mountain region to use a quantitative, rigorous approach to address these fundamental issues. Coskey suggests that although several stratigraphic intervals within and below the producing Lance and Mesaverde horizons are capable of generating significant amounts of gas, no single zone is responsible for all the gas trapped at Jonah. He concludes that gas must be generated and migrated from either deeper sources such as the Hilliard Shale or Mowry Shale and/or have migrated to the Jonah trap from a larger fetch area.

Chapter 8 by Cluff and Cluff uses resistivity and sonic logs to identify a log anomaly within and surrounding Jonah. The nature of this log anomaly suggests that at one time, overpressured conditions existed throughout the area surrounding Jonah field up to the base of the Tertiary Fort Union Formation. Late Tertiary relative uplift initiated slow leakage of the overpressure conditions where the system was not tightly sealed. Instead of being a pressure plume or leakage chimney, as some have suggested, Cluff and Cluff suggest that Jonah field represents the remnant of the former regional top of overpressure. If this model is correct, exploration methods should focus on the seal conditions that prevent leakage instead of fracture models that promote leakage.

In chapter 9, Law and Spencer interpret Jonah field to be a gas chimney rooted in a regionally pervasive, direct type, basin-centered gas accumulation. They contend that the field is an excellent example of a structural sweet spot in a basin-centered gas accumulation. They further state that the most likely source rocks are coals and carbonaceous shales in the lower Lance Formation and the upper part of the Mesaverde Group. This differs from the study by Coskey that found these units incapable of generating sufficient amounts of gas to be solely responsible for the field's reserves.

Chapter 10 is a data-rich study by Shanley on the geometry of fluvial sand bodies in the Lance Formation. Shanley has gathered data from fluvial systems around the world and places the Lance sand body dimensions in the context of the larger database. Estimates of sand body dimension are important to understand, in that they can be used to improve reservoir management and well spacing decisions. Up to the time of publication of this book, Shanley had described virtually every core in the Jonah–Pinedale area. Many of those descriptions can be found in Appendix D on the CD-ROM included with this volume.

In chapter 11, Webb et al. describe the petrology and petrophysics of the American Hunter Old Road Unit 1, a well that was drilled and cored prior to the discovery of Jonah field. Although it is outside Jonah field, it is an excellent descriptive study of the Lance Formation and can be used to compare and contrast with core data from inside the field.

Chapter 12 by Cluff and Cluff is a petrophysical study of the Jonah field. This study focuses on 62 wells near the center of the field to characterize porosity, permeability, and water saturation in the Lance reservoirs. The authors estimate that more than 95% of the flow capacity of the Lance is from sandstones with greater than 6% porosity. A small percentage of high-porosity and high-permeability rocks dominate the flow behavior of the reservoir and are probably critical to economic production. They estimate that the average permeability of net pay is 25 md, and the average estimated ultimate recovery per well is approximately 4 bcf gas on 40-ac (0.16-km^2) well spacing.

In chapter 13, Siguaw and Friend used 3-D seismic data in an attempt to extend the updip limit of established production in the Jonah field. Although a reprocessed aeromagnetic survey was initially used to image this area, a 16-mi^2 (41-km^2) high-resolution 3-D survey was acquired to evaluate the southwestern termination of the two major fault systems that form the trap at Jonah.

Chapters 14 and 15 deal with improved fracture stimulation techniques and processes that have been successfully applied to increase the profitability of the drilling and completion programs at Jonah field. Eberhard and Mullen with Halliburton Energy Services use spatial sampling and artificial neural-network analysis to discuss the application of flow-through-fracture stimulation plugs in a multiple-stage completion setting. Using a best practices approach, they provide a list of steps that maximize productivity. Jennings and Ault, with Schlumberger Oilfield Services and Ultra Petroleum, respectively, however, suggest that the use of induced stress diversion, a continuous multistage stimulation operation, can be an effective and cost-efficient method for well completion. With induced stress diversion, completion time was reduced by 4 weeks, and the percentage of producing sandstones in a treated interval increased from 60 to 90%. The net result was a 40% reduction in time and costs without negative impact on reserves. Using this approach, development costs were reduced by 22% while achieving similar production and reserves.

Finally, in chapter 16, Smith provides us with a history of the environmental analyses of Jonah field. The field and the area around it has been the subject of an immense amount of environmental study since the field was discovered. Between 1993 and 2001, the Bureau of Land Management completed one environmental impact statement (EIS) and four environmental assessments. Because of restrictions imposed during the time of these assessments, field activity was reduced, restricting production; an additional EIS is currently underway. This history lesson on Jonah is instructive and represents the type of environmental effort now required to do business in environmentally sensitive areas such as the Rockies.

To allow authors the maximum flexibility in data presentation, this volume includes a CD-ROM filled with additional data.

Appendix A is an Microsoft Excel(TM) file of every well drilled in Jonah field up to June 2000. It is a comprehensive spreadsheet of well information, much of which is being released for the first time.

Appendix B contains two stratigraphic cross sections through Jonah field constructed by Cluff and Murphy. They were prepared to illustrate the distribution of reservoir sandstone bodies and the high level of discontinuity between closely spaced wells. Both sections are hung on the base of the Tertiary Fort Union Formation, a relatively consistent marker. Correlations showing the top of the Lance Formation and top of the Mesaverde Formation are also included; however, there is little agreement among industry geologists as to the exact placement of these stratigraphic horizons.

Appendix C was prepared by Cluff and contains production history plots of the monthly oil, gas, and water production rates for 291 wells in Jonah field.

Appendix D contains detailed core descriptions for three wells in the Jonah–Pinedale area, which is discussed in more detail in chapter 10 by Shanley.

Appendix E contains color versions of the stratigraphic cross sections shown in chapter 8 by Cluff and Cluff. These cross sections are different than the sections in Appendix B.

Appendix F contains supporting documents to chapter 11 by Webb et al., including analytical procedures, digital rock type classification, and a legend for petrographic tables.

We hope you will enjoy reading this volume as much as we enjoyed compiling it. We gathered a prestigious list of contributors from industry, government, consultants, and contractors. We appreciate the diligence of these authors to get their papers completed in a timely fashion.

As the petroleum industry increasingly turns its attention to natural gas, accumulations such as Jonah will broaden our understanding of petroleum systems and allow us to apply and modify lessons from Jonah to the search for hydrocarbons. We hope that this book will serve as a standard reference in the continuing search for new gas accumulations, not just in the Rocky Mountain region, but throughout the world.

CHAPTER 2

Jonah Field: T28–29N, R107–109W, Sublette County, Wyoming

J. W. ROBINSON
Consultant, Littleton, Colorado, U.S.A.

K. W. SHANLEY
Consultant, Littleton, Colorado, U.S.A.

The following information was gathered from various sources and released for publication. Additional information exists among the many operators in Jonah field, but much of that data is considered proprietary. Data on the drilling and completion of individual wells can be found in Appendix A on the CD-ROM included with this volume. Dean DuBois of EnCana Oil and Gas (U.S.A.) Inc. reviewed and revised some of the data.

GENERAL FIELD DATA

Regional setting: Northern Green River basin, west flank of the Farson syncline.

Producing formations: Upper Mesaverde Formation (Campanian); Lance Formation (Maastrichtian); unnamed Tertiary (of Law and Spencer, 1989) (Paleocene). These units comprise the Lance Pool interval (Wyoming Oil and Gas Conservation Commission [WOGCC] Docket 84-2003).

Type of trap: Structural. Updip fault wedge. Fault boundaries on the west and south side. Top seal is the combination of all bounding mudstones in the pay interval. Pay section is a complex of both reservoir and top seal.

Exploration method leading to discovery: Originally tested as a basin-centered gas concept, the first economic completion used new fracture stimulation techniques.

Other significant shows: Blair sandstone, Rock Springs Formation, Ericson Sandstone, Fort Union Formation.

Oldest stratigraphic horizon penetrated: Blair sandstone (Mesaverde Group).

Surface formation: Tertiary.

Spacing: 80-ac (0.32-km^2) spacing, with two wells per spacing unit. Approval of additional wells per spacing unit has been granted in limited areas.

Productive area: 17,000 ac (69 km^2).

Completion practice: Variable through field development. Limited entry and saturation perforation methods used. Three to seventeen fracture stimulation stages with borate gel and KCl water. Proppant has included bauxite, sand, carbolite, econprop, and others.

Logging practice: Open-hole resistivity, neutron density, and gamma ray.

Number of producing wells: 588 (WOGCC, as of March 2004).

Number of abandoned wells: 6 (WOGCC, as of March 2004).

Market for production: Duke Energy Services.

Method of transportation: Gas pipeline; oil is trucked to La Barge.

Major operators: McMurry Oil Company (wholly owned subsidiary of EnCana Oil and Gas [U.S.A.] Inc.), Ultra Petroleum, BP America Production Co., Forest Oil Corp., Devon Corp. (WOGCC).

Cumulative production: 10,152,000 bbl of oil, 1,017,234,762 mmcf gas, 3,591,000 bbl of water (WOGCC, as of March 2004).

DISCOVERY WELL

Operator: Home Petroleum Corp.

Change of operator: McMurry Oil Company, April 1, 1992.

Well name: Jonah 1-4.

Jonah Field: Case Study of a Giant Tight-Gas Fluvial Reservoir, J. W. Robinson and K. W. Shanley, eds.: AAPG Studies in Geology 52 and Rocky Mountain Association of Geologists 2004 Guidebook

Location: SE NE (2149 ft [655 m] from south line, 1980 ft [603 m] from west line), Sec. 4, T28N, R108W, Sublette County, Wyoming.

American Petroleum Institute (API) number: 49035209260000.

Spud date: October 27, 1986.

Total depth date: November 25, 1986.

Mudlog date: October 17, 1986 to November 24, 1986.

Completion date: May 22, 1987.

Date of first production: September 3, 1992.

Initial potential 1987: 145 mcfd.

Recompletion 1992: 874 mcfd.

Recompletion 1997: 4800 mcfd.

Perforations (1987): 8433–10,077.

Perforations (1997): 8433–8791, 8975–9171, 9300–9557, 9901–10,077.

Treatment: 300,000–500,000 lb (136,100–226,800 kg) proppant, 1.1–1.8 MMSCF nitrogen.

Hole size: 12.75 in. (32.4 cm) to 2550 ft (777 m), 8.75 in. (22.2 cm) to 12,593 ft (3838 m).

Casing: 10.75 in. (27.3 cm) at 2550 ft (777 m), 5.5 in. (14 cm) at 12,593 ft (3838 m).

Total depth: 12,594 ft (3839 m), plugged back 11,268 ft (3434 m).

Logged depth: 11,900 ft (3627 m).

Elevation: 7139 ft (2176 m) kelly bushing.

Cumulative production (September 1992 to March 2004): 34,444 bbl of oil, 3,718,000 mmcf gas, 4629 bbl of water.

RESERVOIR DATA

Producing formation: Lance.

Lithology: Sandstone, mudstone, siltstone, and shale.

Gross formation thickness: 2200–3000 ft (670–900 m).

Net sandstone thickness (65 API GR): 500–1000 ft (150–300 m).

Net pay thickness: 150–600 ft (50–180 m).

Geometry of reservoir rock: Lenticular sandstones.

Porosity: 4–12% (core).

Permeability: 0.01–1.0 md (ambient) (core).

Water saturation: 30–47%.

Bottomhole temperature: 210°F (99°C) at 11,000 ft (3400 m).

Producing formation: Upper Mesaverde.

Lithology: Sandstone, mudstone, siltstone, shale, and coal.

Gross formation thickness: 500–1000 ft (150–300 m).

Net sandstone thickness (65 API GR): 50–250 ft (15–80 m).

Net pay thickness: 100–500 ft (30–150 m).

Geometry of reservoir rock: Lenticular sandstones.

Porosity: 4–10% (log).

Permeability: unknown.

GAS ANALYSIS

Well: McMurry Oil Company, Jonah 3-7.

Location: NW SW Sec. 7, T28N, R108W.

Date: December 29, 1994.

Production interval: 10,122–10,380 ft (3085–3164 m).

Formation: Lance.

Gas component analysis: Carbon dioxide, 0.806 mol%; nitrogen, 0.604 mol%; methane, 85.630 mol%; ethane, 7.703 mol%; propane, 3.458 mol%; iso-butane, 0.791 mol%; n-butane, 0.681 mol%; iso-pentane, 0.193 mol%; n-pentane, 0.134 mol%.

***Z* factor:** 0.9972.

Specific gravity: 0.664.

Brtish Thermal Unit (BTU): 1152

RESERVOIR FLUIDS ANALYSIS

Well: McMurry Oil Company, Stud Horse Butte 15-28.

Location: SW SE Sec. 28, T29N, R108W, Sublette County, Wyoming.

Production interval: 10,392–10,646 ft (3167–3245 m).

Formation: Lance.

Date sampled: October 16, 1994.

Flowing tubing pressure: 1000 psig.

Sampling conditions: 500 psig and 76°F (24°C).

Gas-oil ratio: 86.2 mmcf separator gas per 1.0 bbl of separator oil (11.6 bbl oil/mmcf gas).

Dew point: 3500 psia at reservoir temperature of 213°F (100.5°C).

Static bottomhole pressure: 6133 psia.

Gas component analysis: Carbon dioxide, 0.80 mol%; nitrogen, 2.82 mol%; methane, 85.11 mol%; ethane, 6.05 mol%; propane, 2.44 mol%; iso-butane, 0.61 mol%; n-butane, 0.59 mol%; iso-pentane, 0.22 mol%; n-pentane, 0.17 mol%; hexanes, 0.27 mol%; heptanes plus, 0.92 mol%.

***Z* factor:** 0.9973.

Specific gravity: 0.6612.

BTU: 1112.

WATER ANALYSIS

Well: Snyder Oil Corporation, Stud Horse Butte 13-27.

Location: SW SW Sec. 27, T29N, R108W, Sublette County, Wyoming.
Tested interval: 12,411–12,436 ft (3783–3790 m).
Formation: Ericson.
Date sampled: January 1995(?).
R_w: 0.482 at 77°F (25°C).
Chlorides: 5301 mg/L.
Total dissolved solids: 13,321 mg/L.
pH: 8.4.

Well: McMurry Oil Company, Jonah 1-5.
Location: NW NW Sec. 5, T28N, R108W, Sublette County, Wyoming.
Tested interval: 9695–9924 ft (2955–3025 m).
Formation: Lance.
Date sampled: March 22, 1994.
R_w: 8.0 at 68°F (20°C).
Chlorides: 370 mg/L.
Total dissolved solids: 770 mg/L.
pH: 6.2.

Well: McMurry Oil Company, Jonah 1-6.
Location: SE SE Sec. 6, T28N, R108W, Sublette County, Wyoming.
Tested interval: 9520–9729 ft (2902–2965 m).
Formation: Lance.
Date sampled: March 22, 1994.
R_w: 5.05 at 68°F (20°C).
Chlorides: 670 mg/L.
Total dissolved solids: 1370 mg/L.
pH: 5.05.

Well: McMurry Oil Company, Jonah 2-8.
Location: SE SE Sec. 8, T28N, R108W, Sublette County, Wyoming.
Tested interval: 9898–10,044 ft (3017–3061 m).
Formation: Lance.
Date sampled: March 22, 1994.
R_w: 5.18 at 68°F (20°C).
Chlorides: 575 mg/L
Total dissolved solids: 1270 mg/L.
pH: 5.18.

CORED WELLS

1994
McMurry Oil Company, Jonah Federal 2-5
Sec. 5, T28N, R108W
Lance Formation
9577–9631 ft (2902–2918 m)
9773–9813 ft (2961–2973 m)
9969–10,000 ft (3020–3030 m)

1996
Snyder Oil Corporation, Stud Horse Butte 11-27
Sec. 27, T29N, R108W
Lance Formation
8900–8948 ft (2696–2711 m)
10,411–10,448 ft (3154–3166 m)
10,937–10,997 ft (3314–3332 m)

McMurry Oil Company, Corona Unit 7-24
Sec. 24, T29N, R109W
Lance Formation
Rotary sidewall cores

1998
Amoco Production Company, Corona Unit 1-11
Sec. 11, T28N, R107W
Lance Formation
8873–8902 ft (2688–2697 m)
8902–8916 ft (2697–2701 m)
9627–9656 ft (2917–2926 m)
9658–9670 ft (2926–2930 m)

Amoco Production Company, Cabrito Unit 15-25
Sec. 25, T29N, R108W
Lance Formation
9876–9936 ft (2992–3010 m)
10,535–10,595 ft (3192–3210 m)

McMurry Oil Company (EnCana), Stud Horse Butter 16-34R
Sec. 34, T29N, R108W
Unnamed Tertiary
8000–8061 ft (2424–2442 m)
Lance Formation
8734–8760 ft (2646–2654 m)
9121–9181 ft (2763–2782 m)
9721–9781 ft (2945–2963 m)
9962–9984 ft (3018–3025 m)
10,560–10,599 ft (3200–3211 m)

GEOCHEMICAL ANALYSIS

Well: Snyder Oil Corporation, Stud Horse Butte 5-26.
Location: SW NW Sec. 26, T29N, R108W, Sublette County, Wyoming.

Depth (ft)	Total Organic Carbon (wt.%)	T_{max} (°C)	Formation	Sample
7910–7970	0.29	435	Tertiary	cuttings
8170–8200	0.70	433	Tertiary	cuttings
8500–8530	0.59	433	upper Lance	cuttings
8700–8730	2.37	438	upper Lance	cuttings
8920–8950	0.51	441*	upper Lance	cuttings
9240–9270	0.40	436*	upper Lance	cuttings
9810–9840	0.26	443*	upper Lance	cuttings
10,220	0.34	435*	upper Lance	junk basket core
10,400–10,430	0.38	440*	upper Lance	cuttings
10,970–11,000	0.39	444*	middle Lance	cuttings
11,700–11,730	1.40	448	lower Lance	cuttings

*T_{max} value was not reliable because of low S_2 values.

REFERENCES CITED

Law, B. E., and C. W. Spencer, 1989, Geology of tight gas reservoirs in the Pinedale anticline area, Wyoming, and at the multiwell experiment site, Colorado: U.S. Geological Survey Bulletin, v. 1886, 260 p.

Wyoming Oil and Gas Conservation Commission, 2004, http://www.wogcs.state.wy.us (accessed May 30, 2004).

CHAPTER 3

Discovery of Jonah Field, Sublette County, Wyoming[1]

J. W. ROBINSON
Consultant, Littleton, Colorado, U.S.A.

ABSTRACT

Home Petroleum Corporation drilled two wells in 1986 and 1987 to test the concept of a large, basin-centered gas accumulation in the Upper Cretaceous Lance Formation in a remote portion of the northern Green River basin. Although these first wells had excellent shows of gas and calculated log pay, the reservoir sandstones had extremely low permeability. Unfortunately for Home, fracture-stimulation techniques used in the late 1980s were unable to unlock this vast gas resource. In 1991, John Martin at McMurry Oil Company recognized the potential of the play and quietly amassed a controlling interest in the area. McMurry Oil and their partners brought the appropriate drilling and completion technology into the project, an effort that resulted in the "rediscovery" of Jonah field. Jonah field will ultimately have more than 350 producing wells and is poised to become the next multi-tcf gas field in the Rocky Mountain region. The rediscovery of Jonah field ignited interest in structurally trapped, tight-gas exploration in Upper Cretaceous strata and may stand as an analog for further exploration of vast, sparsely drilled areas of the Rocky Mountain region.

INTRODUCTION

This chapter presents information on the discovery and early development of Jonah field, an overpressured, structurally trapped gas accumulation located in T28–29N, R108–109W, Sublette County, Wyoming (Figure 1). The field is located slightly west of the synclinal axis of the Green River basin between the Pinedale anticline and La Barge platform. Gas is produced from the Upper Cretaceous (Maastrichtian) Lance Formation, which is composed of discontinuous, fluvial-channel sandstones and intervening alluvial-plain mudstones. Production from these low-permeability reservoirs ranges from depths of 7500 to 12,500 ft (2280 to 3800 m). Other potential reservoir sandstones in the Fort Union, Ericson, Rock Springs, Blair, Frontier, Muddy, and Dakota formations remain to be tested. The regional geology of the Greater Green River basin and papers dealing with the technical aspects of Jonah field can be found in numerous references, including Law (1984a, b), Law and Spencer (1989), Law et al. (1989), Battin and Clark (1994), Bowker and Robinson (1996), Robinson et al. (1995, 1996), Shaw (1995), Finch et al. (1997), Montgomery and Robinson (1997), and Warner (1997, 1998, 2000).

DAVIS OIL COMPANY WILDCATS

During the 1960s and 1970s, Davis Oil Company of Denver, Colorado, drilled many significant wildcats in the Rocky Mountain region. Marvin Davis was well known as the consummate wildcatter, often drilling wells in remote areas and to previously untested depths. In 1974, the *Oil & Gas Journal* reported on Davis' plans to drill as many as 50 wildcat wells in the Green River and Powder River basins (McCaslin, 1974). The drilling program was funded by Panhandle Eastern Pipeline Company to add gas reserves to support their gas transportation activities. With this infusion of cash, geologists at Davis were charged with finding large, undrilled acreage blocks in areas with gas

[1]This article is reprinted with the permission of the Rocky Mountain Association of Geologists. This article first appeared in *The Mountain Geologist*, v. 37, p. 135–143.

Jonah Field: Case Study of a Giant Tight-Gas Fluvial Reservoir, J. W. Robinson and K. W. Shanley, eds.: AAPG Studies in Geology 52 and Rocky Mountain Association of Geologists 2004 Guidebook

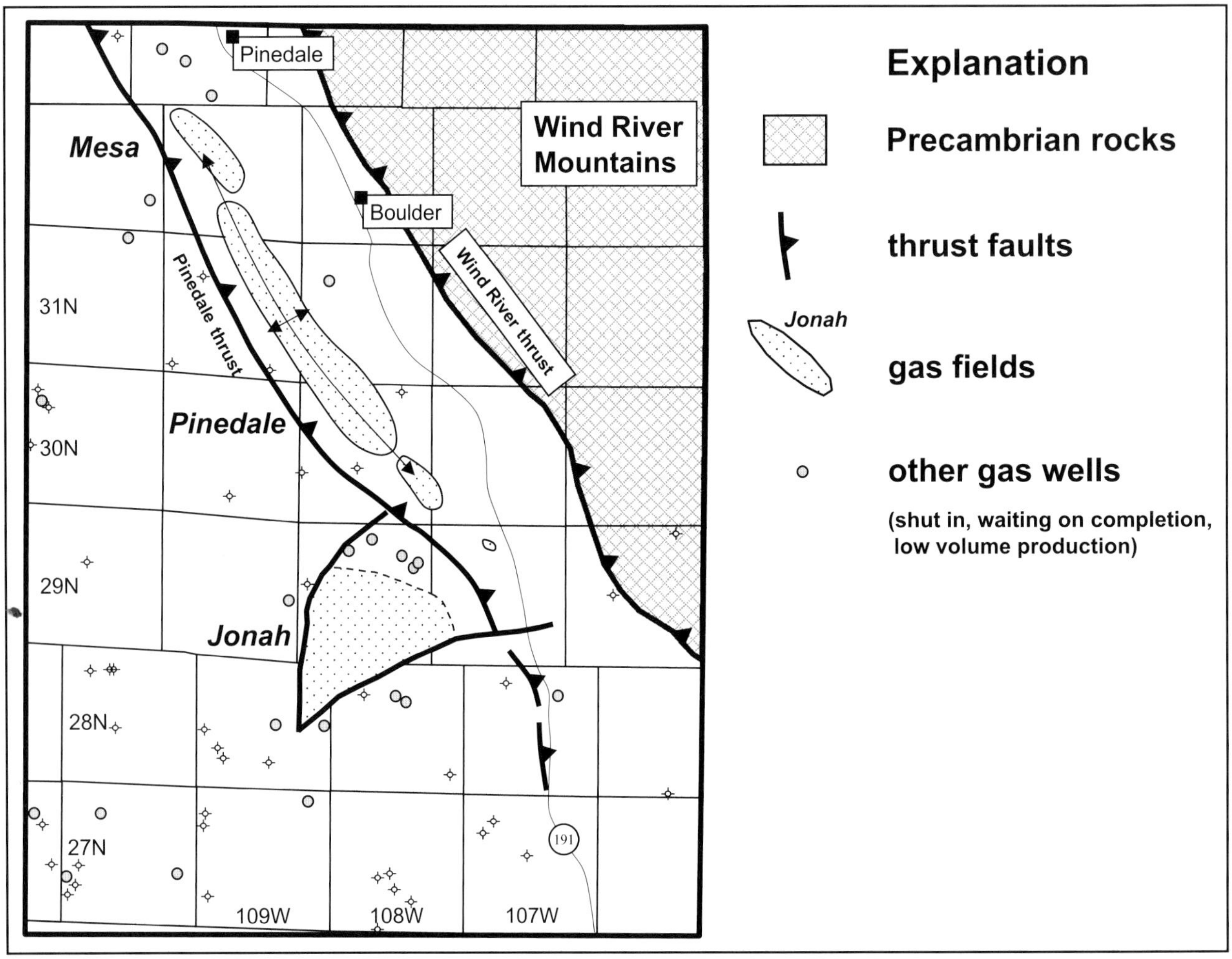

Figure 1. Regional map of the northern Green River basin, Wyoming. Jonah field is located 25 mi (40 km) south of Pinedale, 5 mi (8 km) due west of Highway 191.

potential. One of the areas studied, and later tested, was adjacent to the La Barge platform and Moxa arch. Another area was selected in the northern Green River basin between the La Barge platform and the Pinedale anticline. A few shallow wells drilled in the area had shows of gas, but more importantly, there were large Federal lease blocks held by various companies and individuals that were due to expire soon. Davis eventually negotiated farmouts on about 125,000 ac (50,600 ha) in Sublette and Sweetwater Counties on which they formed two Federal exploratory units and drilled two obligation wells to test the Upper Cretaceous Lance and Mesaverde formations. No specific geologic model or prospect concept was used to justify these wells; they were located in the middle of the unit acreage.

The most significant well was the Davis Oil Company, Wardell Federal Unit 1 (Sec. 9, T28N, R108W) (Figure 2). The well spud July 12, 1975, and reached total depth of 11,324 ft (3452 m) on September 5. Minor coal gas shows were detected in the lower Tertiary Fort Union Formation, and the top of the Lance Formation was encountered at approximately 8100 ft (2460 m). A good gas show from 10,433 to 10,506 ft (3179 to 3194 m) was drillstem tested (DST 1). Flowing and shut-in pressures indicated that the zone was extremely tight [initial flowing pressure (IFP) 114-107, final flowing pressure (FFP) 78-61, shut-in pressure (SIP) 501-735)] and pipe recovery was only 750 ft (230 m) of gas in the pipe and 80 ft (24 m) of mud. The sample chamber recovered 600 cm^3 (37 $in.^3$) of mud and 0.15 ft^3 (0.004 m^3) gas at 35 psi. Conventional wisdom would suggest that the zone would be noncommercial. We now know this interval to be the stratigraphic equivalent of the Jonah zone, the most widespread reservoir in Jonah field.

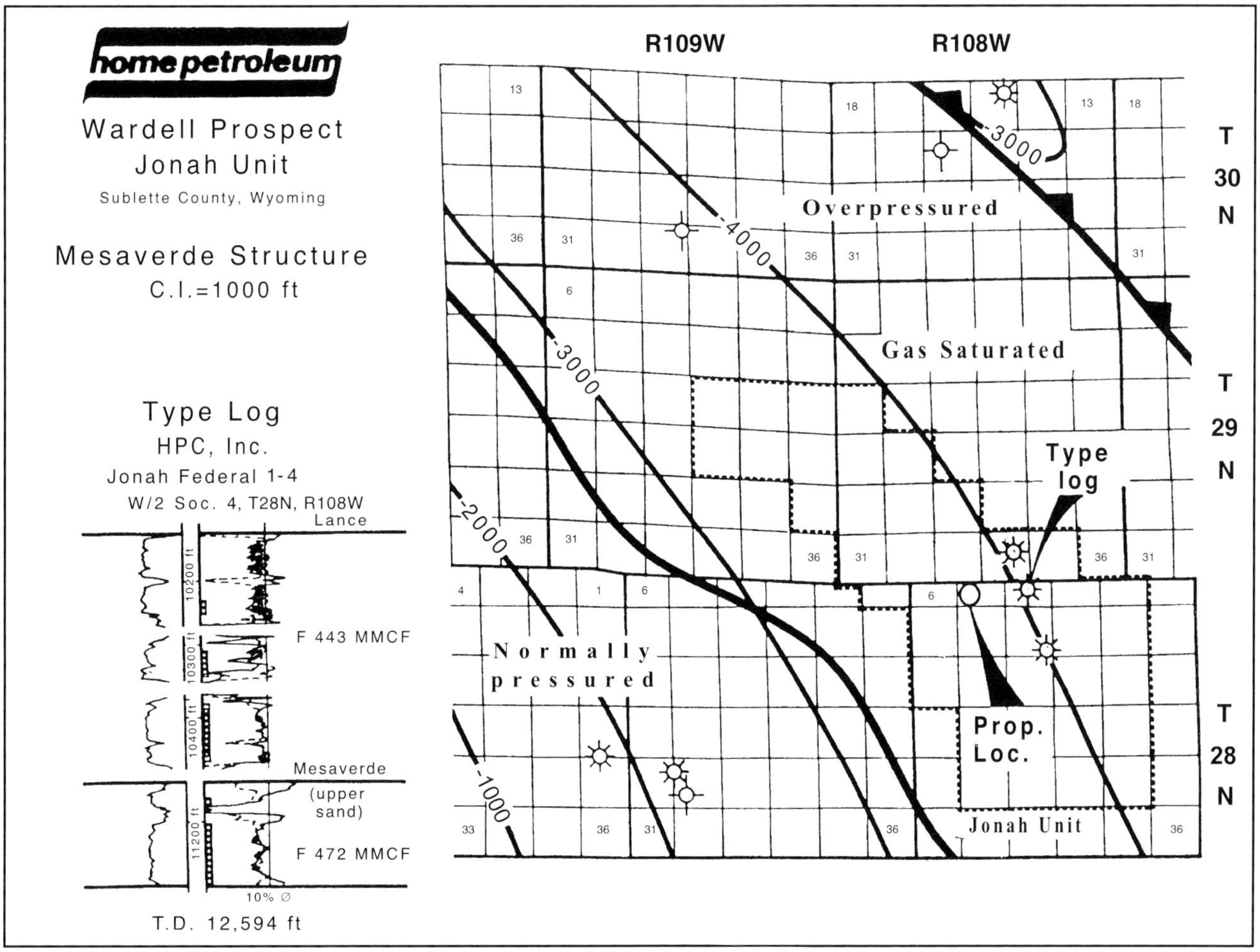

Figure 2. Home Petroleum Corporation prospect map, ca. 1989.

At a depth of 11,100 ft (3380 m), logs were run, but the neutron-density tool became stuck and was fished out of the hole. Two more drillstem tests were attempted over an interval from 11,000 to 11,100 ft (3350 to 3380 m). Gas entry into the borehole cut the drilling mud weight from 9.8 to 7.9 ppg and caused extensive hole sloughing. The mud weight was raised to 11.6 ppg to run DST 3 below the Jonah zone. DST 3 flowed gas at rates from 34 to 39 MCFGD— pipe recovery included the water cushion and 120 ft (36 m) of slightly mud-cut water cushion with a trace of brown oil. Flow pressures were 1013–956 psi/ 960–1013 psi, and shut-in pressures were 4842–5642 psi. The sample chamber recovered 2.5 ft^3 (0.07 m^3) gas and 900 cm^3 (55 $in.^3$) mud at 750 psi. Although the results were better than the first test, the zone appeared to be tight and nonproductive. The well was drilled to a total depth of 11,324 ft (3452 m), and production casing was run.

One zone was selectively perforated from 11,017 to 11,219 ft (3360 to 3420 m), and after being broken down with KCl water, it flowed 17 MCFGD. Extremely cold weather in December 1975 (−38 to −42°F) hindered completion operations. After an acid job, the well flowed 131 MCFGD with a trace of condensate. The well was fracture stimulated with 107,050 gal My-T-Gel, 54,000 lb of 40/60 sand, and 105,000 lb of 20/40 sand, but screened out on the second stage of 20/40 sand. The well flowed for 8 days at rates from 94 to 300 MCFGD and was then shut in for 8 months. When the well was put back on line, it flowed 1002 to 990 MCFGD. An extended test from October 4 to 21, 1976, flowed gas at 566 to 303 MCFGD. The well was then shut in for a pressure buildup test. The final report for the well indicated a production rate of 303 MCFGD and 2 BCPD. Because of the remote location, lack of pipeline, and subcommercial flow rates, the well was shut in. At the time Davis drilled this well, Upper Cretaceous tight-gas sandstone reservoirs were not known to be a viable economic target. In January 1977, the Wyoming Oil and Gas Conservation Commission declared the new field name to be "Jonah," although it was erroneously described in the commission records as a Tertiary Fort Union Formation gas discovery.

In 1981, Davis Oil flow tested the Wardell well, as required by the Bureau of Land Management (BLM) for a shut-in well to hold an held by production (HBP) lease. On a four-point test, the well flowed 680–1220–2000–2675 MCFGD. Extended flow tests of 10 and 24 hr flowed gas at 1900 and 1400 MCFGD, respectively.

ASSETS OF DAVIS OIL SOLD TO HOME PETROLEUM

In 1981, Davis Oil Company sold all of its oil and gas assets to Home Petroleum Corporation (HPC), a subsidiary of Home Oil, Canada. The sale included the Wardell well and surrounding acreage, which at the time was probably not considered one of the prime assets in the transaction. Ben Peterson, who was vice president of HPC, recalled that the seals in the Wardell wellhead were leaking, and the group recommended plugging the well. In an effort to bleed the wellhead pressure down for plugging, a spot rate of 2500 MCFGD was recorded. On an extended 7-day test, the well flowed an average of 416 MCFGD and 8 BOPD, although freezing problems hindered the testing. Although these tests are not spectacular, they were better than the original completion rate of 303 MCFGD. These tests caught the interest of Doug Battin and other explorationists at HPC and prompted a regional study of the northern Green River basin.

Initial reaction to the regional study was met with skepticism within HPC. Fortunately, ongoing studies by Ben Law and Chuck Spencer at the U.S. Geological Survey were able to support the concept of a regional basin-centered gas accumulation in this portion of the basin (Law, 1984a, b; Spencer, 1984; Spencer and Mast, 1986). Law (1984a) addressed the nature of low-permeability reservoirs in basin-centered accumulations and the mechanisms controlling them, specifically the role of active hydrocarbon generation. The Pinedale anticline and northern Green River basin were the focus of much of this work. A schematic cross section that transected Jonah field became instrumental in understanding the play (Law, 1984a, his figure 2). The work of Law (1984a) and Spencer (1984a) postulated that sandstones in the Lance Formation were within the active hydrocarbon generation window and were gas saturated and overpressured as observed in the Wardell well. If a well was drilled into the Lance, it should encounter multiple, stacked, low-permeability sandstones within the active generation window. The limited drainage area from individual units would be overcome by the tremendous cumulative thickness of sandstones observed in the Wardell well. They postulated that emerging fracture technology would overcome the low permeability and prove these sandstones to be commercially productive. Production tests in 1981 and 1985 and articles published in 1984 by the U.S. Geological Survey helped HPC justify drilling the first Jonah wells. The Canadian owners of HPC agreed to pursue the project because of the large reserve potential.

A NEW DRILLING PROGRAM

In 1986, the Denver office of Home Petroleum Corporation received approval to drill another well in the area. A partner could not be found, so HPC had a 100% working interest in the well. The Jonah Federal 1-4 (Sec. 4, T28N, R108W) spud October 27, 1986, and reached a total depth of 12,594 ft (3839 m) in the Ericson Sandstone on November 25 (Figure 2). Unlike the Wardell well, excellent shows of gas were encountered in the Lance Formation and Ericson Sandstone, and a continuous gas flare was established below approximately 9200 ft (2800 m). Excellent sample shows were noted below 9200 ft (2800 m) by Rod Bernasek, the wellsite geologist. Problems associated with gas entry and hole sloughing plagued operations below 11,000 ft (3350 m). No tests or cores were taken, and open-hole logs could not be run below 11,910 ft (3630 m). Production casing was set to total depth, but poor cement bond precluded testing the Ericson and lower Lance. On several occasions, remedial cement jobs were performed with limited success. During completion, gas flowed at erratic rates as much as 3000 MCFGD but would die off or flow slugs of fracture-stimulation fluid. Several zones between 8433 and 10,077 ft (2570 and 3070 m) were subsequently perforated, fracture stimulated, and completed in May 1987, for as much as 472 MCFGD. Ubiquitous drilling shows in the Lance and continuously flared gas below approximately 9200 ft (2800 m) excited the staff. More importantly, a thick, gas-saturated, overpressured envelope of Upper Cretaceous strata was identified.

Discouraged by flow rates and drilling problems in the 1-4, but encouraged by the potential of a significant reserve base, HPC spud a second well in September 1987. The Jonah Federal 32-34 (Sec. 34, T29N, R108W) reached a total depth of 11,476 ft (3500 m) on October 25. Gas flares similar to the Jonah 1-4 were encountered below a depth of 9636 ft (2937 m), and they sustained to total depth. Rod Bernasek meticulously monitored this well and collected condensate and water samples that dripped from the flare line. Cuttings analysis performed by Reservoirs, Inc., Houston, Texas, concluded that porosity in the Lance is occluded by quartz, calcite, and diagenetic clays.

Kaolinite is the most common type of clay, with lesser amounts of illite and chlorite. Fluid retention, fines migration, and acid sensitivity were considered to be potential completion problems. They concluded that the Lance would not be capable of commercial production. The Jonah Federal 32-34 was completed in late 1987 and early 1988. Perforation and fracture stimulation of the Wardell zone between 11,282 and 11,306 ft (3440 and 3450 m) flowed 434 MCFGD and 17 BOPD.

Unfortunately for HPC, fracture-stimulation techniques, which were being performed on low-permeability reservoirs elsewhere in the basin, were in the development stage and not readily available. Production tests were highly variable. Completion rates ranged from 313 to 472 MCFGD per zone and were somewhat discouraging. Nonetheless, the presence of a potentially huge gas reserve in a sparsely drilled area was recognized. It should be noted that HPC's activity occurred during a dismal period for natural gas exploration. Corporate layoffs and budget reductions were common throughout the industry. Only the potential for a vast reserve base kept the project on track.

Eventually, the project lost momentum, largely because of poor market conditions, low gas prices, and lack of a pipeline. In January 1989, the staff at HPC prepared a brochure to sell or farmout the area to an industry partner (Figure 2). Many companies active in the Rocky Mountain region reviewed the deal and rejected it as too risky or too remote to be of value. Finally, in mid-1989, the assets of HPC were sold to Presidio Oil Company for $157 million.

In January 1990, the Jonah project area was reviewed for the first time by Presidio management. The explorationists at Presidio, who were ex-HPC employees, still thought the area had merit and recommended to management to flow test the wells for an extended period of time to determine a stabilized flow rate. Unfortunately, management wanted to focus on exploration and development on the Moxa arch and deemed the Jonah area as a "noncore" asset. In addition, management was deterred by the cost of the flow test because payments would have to be made to royalty owners on the volume of flared gas. The staff was then instructed to bundle up several noncore prospects for a cash sale to industry.

PURCHASE OF ASSETS BY McMURRY OIL

In 1991, John W. Martin, president and cofounder of McMurry Oil Company, went to Denver to look through the Presidio divestiture portfolio. McMurry Oil is based in Casper, Wyoming, and was founded in 1982 by Martin and W. N. McMurry, an owner of Rissler-McMurry Construction Co., a long-time Casper highway construction firm. John Martin's background included a term as manager of the Casper airport, but he had also been a logging engineer for Schlumberger after graduating from the Air Force Academy and completing his active duty. "I picked out half a dozen prospects, and Jonah was one. I was particularly looking for gas properties at the time because gas was out of favor and I figured I could get a good price," Martin said (Klann, 1997, p. 63). Presidio was offering approximately 25,000 ac (10,125 ha) and the associated three wells (Figure 3). "The Jonah 1-4 stood out as a gas well," Martin recalled. "It had all the earmarks— crossover, flares while drilling, logs— and when I looked at the well file, I could see that the completion they had attempted had not been the right type for tight-gas sandstones. They had used hydrochloric acid and some other materials that are death on tight-gas sands" (Klann, 1997, p. 63). In late 1991, McMurry made an offer to buy the Jonah properties that was accepted, but during negotiations, several of the leases in the acreage block expired. As part of the deal terms, Presidio agreed to re-lease key tracts in the December 1991 BLM lease sale. Most of the tracts were leased for the $2.00/acre minimum bonus and the 25,000-ac (10,125-ha) block remained intact. Some months later, the company divested some of the property by offering 25% working interest to Fort Collins Consolidated Royalties Inc., Cheyenne, Wyoming, 25% to Nerd Oil, an oil and gas firm owned by Mick McMurry, son of W. N. McMurry, and 2% to Edward M. Warner, Denver, Colorado. This group is referred to as the McMurry partners and has operated in concert throughout the development of the field.

McMurry had a strong desire to recoup some of their investment money, so they proceeded to build a pipeline to hook the wells up to Williams Field Services rather than test the wells to the atmosphere. In August 1992, Mick McMurry supervised the installation of 10 mi (16 km) of 4.5-in. (11.4-cm) surface pipe that was connected at the New Fork 1 (Sec. 25, T30N, R108W), the nearest producing well on the Pinedale anticline (Figure 3). Despite the formation damage from earlier stimulations, the Jonah 1-4 initially flowed 2000 MCFGD and was able to sustain a long-term rate of 875 MCFGD (Figure 4). It is thought that the failure in earlier stimulations was related to gel chemistry— the gels did not break at the proper time to flow back and clean up the wellbore. In fact, only 200 of the original 5000 bbl of fracture-stimulation fluid injected were recovered from the early stimulation. As John Martin joked later, "It took five years for the gels to break."

Simultaneous with these production operations, the company was quietly nominating and leasing acreage in the area, ultimately acquiring nearly 100,000 ac (40,500 ha)

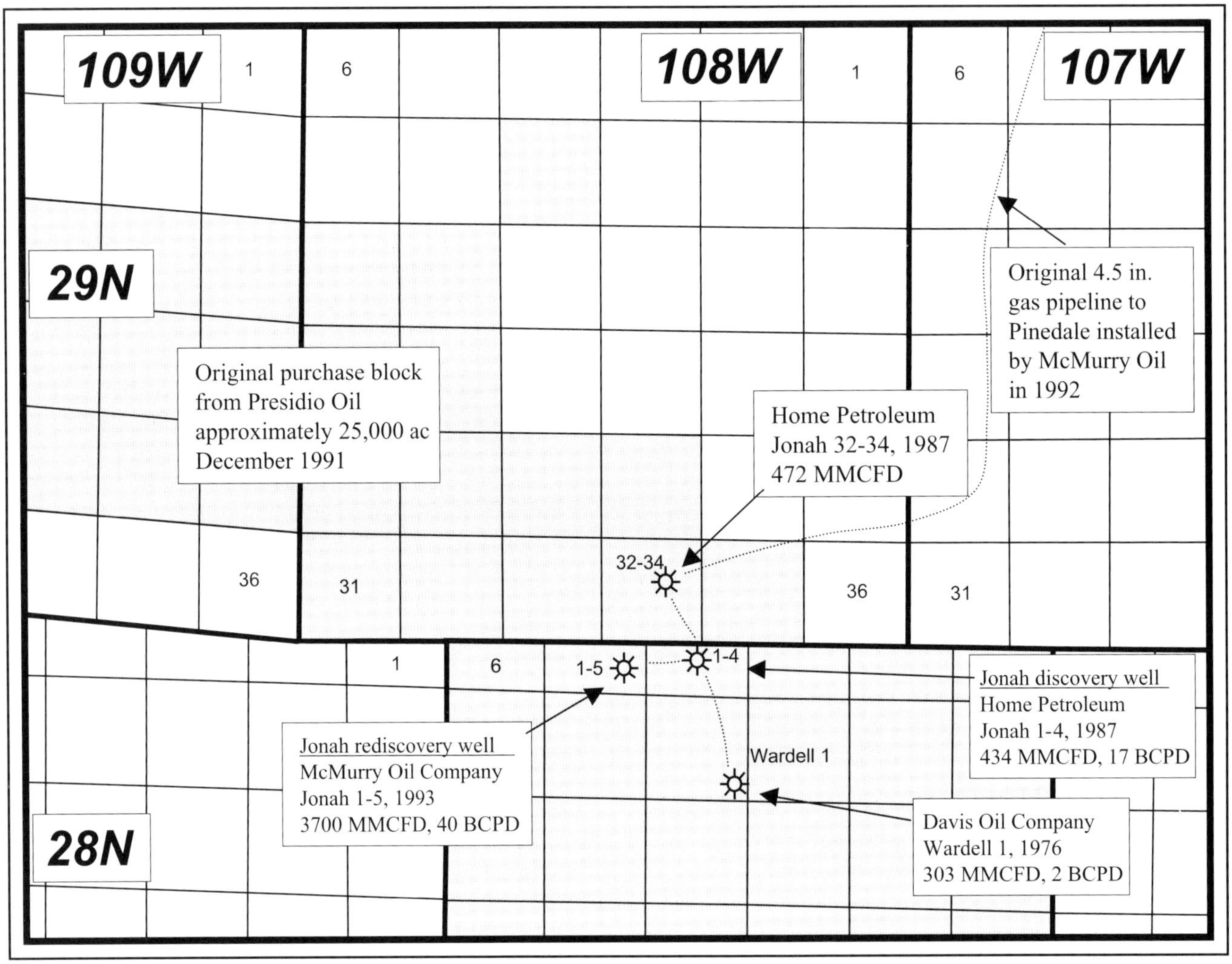

Figure 3. Original purchase block by McMurry Oil from Presidio Oil. The purchase included approximately 25,000 gross ac (10,125 ha) and three shut-in gas wells.

in the play. At this point in the prospect development, there had been little traditional geoscience input. The play was categorized as a "basin-centered gas accumulation," and it was thought to be potentially regional in scope. Acreage was acquired in six townships surrounding the original HPC and Davis wells. Also at this time, some effort was given to looking for a partner. The terms were not onerous, and the intent was to promote a three-well drilling program, but there was no interest in the deal.

THE DISCOVERY WELL—JONAH 1-5

Although the McMurry group was unable to find industry partners, they were able to get some long-time business associates, friends, and family to join in drilling a new well. The Jonah 1-5 location was selected in the SE/4 of Sec. 5, T28N, R108W (Figure 3). When John Martin was asked how the first location was chosen, he stated that there were three factors, (1) it was updip and therefore, shallower, (2) it was on a different lease, and (3) it was conveniently near the road. The Jonah 1-5 spud December 31, 1992, and reached a total depth of 10,793 ft (3290 m) on February 5. Continuous mud log shows were encountered below 8429 ft (2569 m), and samples indicated good porosity and moderate sample stain. Underbalanced drilling techniques were used to minimize formation damage and enhance sample shows. Open-hole logs showed resistivities of 50–100 ohm m and porosities of 8–12% with gas crossover. While the well was drilling, John Martin spent considerable time studying the diversity of fracture-stimulation techniques available from various

Figure 4. Home Petroleum drilled the Jonah 1-4 (Sec. 4, T28N, R108W) in 1986, but was unable complete the well at commercial flow rates. In 1992, McMurry Oil Company purchased the Jonah 1-4 (Sec. 4, T28N, R108W). During a subsequent flow test (shown above), the well produced at rates as much as 2,000 MCFGD into a 4.5-in. (11.4-cm) pipeline and was able to sustain 875 MCFGD. It is thought that the original completions were unable to adequately stimulate the reservoir to establish commercial production (photo by Mick McMurry).

vendors. He was certain that the traditional gel-fracture techniques that Davis Oil and HPC had employed were not appropriate. He surmised that formation damage from drilling and completion operations were the culprit hindering commercial gas production. After considering several alternatives, he decided that a treatment using an inert gas such as carbon dioxide or nitrogen foam might be best suited to the Lance reservoirs, although this type of treatment was more expensive than other techniques. After interviewing several candidates, he selected James Shaw of Stim-Tech, Inc., to design and implement the stimulation treatment. Shaw had been successful treating tight-gas sandstone reservoirs in many areas of the Rockies and agreed with John Martin that a 70-quality foam treatment would minimize formation damage and energize flow back and cleanup after the stimulation (Shaw, 1995). At this period of time, only one zone was selected for completion. A thick package of sandstone units, locally called the Jonah zone of the Lance Formation, was chosen from 9695 to 9924 ft (2960 to 3030 m). On March 29, 1993, John Martin, Mick McMurry, W. N. McMurry, and others watched as crews from Halliburton, under the supervision of James Shaw, pumped 689,000 lb of ceramic prop and 2181 bbl of low-pH, cross-linked fluid energized with 7.99 MMSCF of nitrogen into the perforations (Figure 5). The frac job was successful and put on immediate flow back and cleanup. Although initial rates were higher, a stabilized flow rate of 3.7 MCFGD and 40 BOPD was established about a week after the stimulation. This well now has an estimated ultimate recovery of at least 11 bcf gas.

SUCCESS, AND THEN A DRY HOLE

Encouraged by the success of the Jonah 1-5, the McMurry partners subsequently drilled, in order, the Jonah 1-7, 2-8, 1-6, 2-5, and 3-7 (Figure 6). Mick McMurry was heavily involved in the supervision of these wells, spending most of his time in the field building roads and locations, constructing surface facilities, and installing additional pipelines. The Jonah 2-5 was notable in that three cores were taken from the Lance Formation under a cooperative agreement with the University of Wyoming and the Gas Research Institute (GRI). Every aspect of the rock properties of these cores was analyzed and added greatly to understanding this complex reservoir.

While the drilling program was underway, the McMurry partners continued searching for another industry partner.

Figure 5. McMurry Oil Jonah 1-5 rediscovery well, fracture stimulation treatment, March 29, 1993 (photo by Mick McMurry).

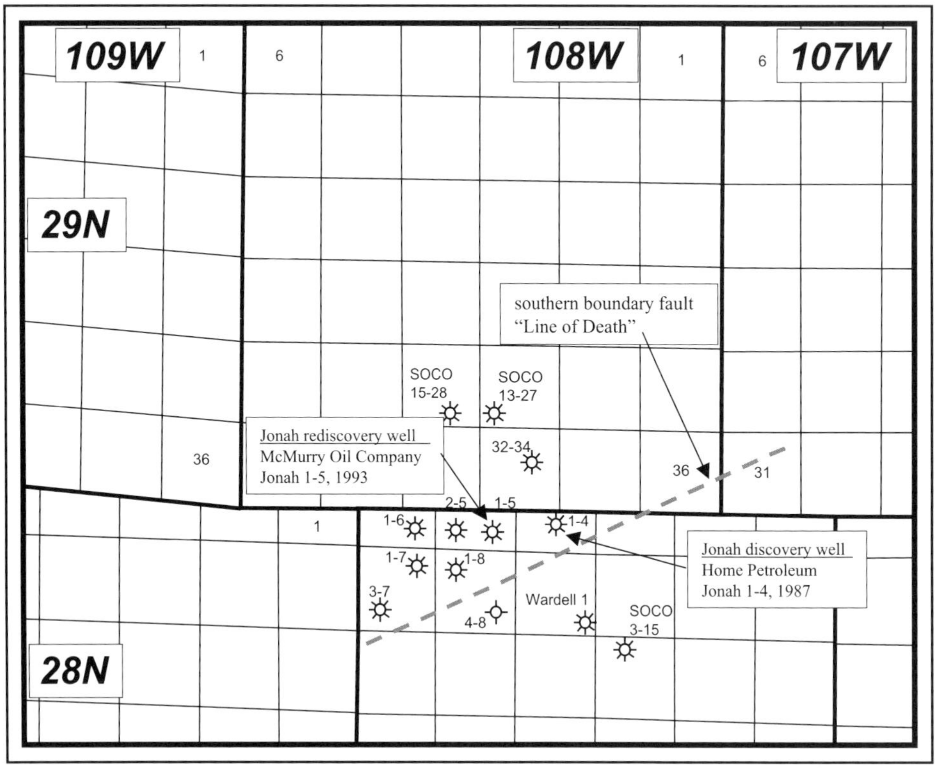

Figure 6. Jonah field, 1995. By this time, the McMurry partners had drilled seven producing wells and one dry hole. SOCO extended the field 1 mi (1.6 km) to the north and confirmed the presence of the fault that delineated the southern boundary of the field.

Because of the million-dollar cost of drilling and completing these wells and the perceived risk of the play, the McMurry partners needed to find someone who could advance the project quickly and find the limits of the play. In 1993 and 1994, they showed the project to more than 30 companies, including some prominent multimillion-dollar exploration and production companies, with no takers. Gas prices were low during this time, and gas supply predictions were not encouraging for Rocky Mountain producers. Most engineers and geoscientists who reviewed the play were dismayed with the low permeability (<1.0 md), the drilling depth (9000–11,000 ft; 2700–3350 m), and the lack of a long-term production profile. Tight-gas sandstone reservoirs in Rocky Mountain basins commonly have steep initial production declines, sometimes exceeding 90% in the first year, that later flatten to 6–8% per year. Early production from the HPC wells was encouraging but was declining rapidly and had not stabilized The nearest production from the Lance Formation was from older wells on the Pinedale anticline and was not encouraging. "I've never been so grateful that so many people said no," commented John Martin (Klann, 1997, p. 63).

In July 1994, McMurry spud the Jonah 4-8, in the SE/4 Sec. 8, T28N, R108W. The 4-8 is 1 mi (1.6 km) south-southeast of the 1-5 discovery well drilled a year earlier. Although the drilling operations were successful, the well did not have any significant shows while drilling to a total depth of 11,000 ft (3350 m), and open-hole logs looked dismal. There was little resistivity greater than 40 ohm m and virtually no porosity greater than 10%. The partners elected to plug the well and reassess the nature of the play. Clearly, there was more variability to the play than originally thought; wells about 1 mi (1.6 km) apart could be drastically different in their apparent pay thickness and gas content.

SALE OF ASSETS TO SOCO

In June of 1994, just prior to drilling the Jonah 4-8, the McMurry partners agreed to sell approximately 50% working interest in the play to Snyder Oil Corporation (SOCO), Denver, Colorado, for acreage considerations and drilling commitments. The Purchase and Sale Agreement with SOCO included several clever twists that allowed McMurry to maintain the heart of the play while SOCO tested surrounding acreage for additional production. In the area adjacent to the producing wells, SOCO could earn as much as nine sections under a farmout provision of drilling one well in each section. In addition, one "deep" well drilled to the Ericson Sandstone would earn to that stratigraphic depth under all nine sections. Drilling and completion operations alternated between SOCO and McMurry, so that technology transfer could occur between the companies. SOCO ultimately earned seven of the nine sections, six in the main part of the field and one outside the field. The remaining two sections are outside the field boundary and were not drilled.

The remaining acreage in the play was checkerboarded into 160-ac (64.8-ha) tracts and bundled into five drilling blocks, A to E. SOCO would earn each block by drilling a well to the Ericson in a specified time frame. The Sagebrush 3-8 (Sec. 8, T27N, R107W), the Jonah State 3-36 (Sec. 36, T28N, R108W), the Stud Horse Butte 7-30 (Sec. 30, T29N, R108W), the Stud Horse Butte 16-20 (Sec. 20, T29N, R108W), and the Cabrito 4-30 (Sec. 30, T29N, R107W) were all earning wells. In all, SOCO drilled 12 earning wells and formed four Federal drilling units (Corona, Cabrito, Cantina, and El Niño). SOCO extended production in the field by about 2 mi (3.2 km) to the west, north, and east. In addition to delineating the field boundaries, SOCO's biggest contribution was the idea of continuous completion operations, in which four to six fracture stimulations were performed at 1-week intervals, until the entire pay section was completed and capable of production. This approach reduced later workover costs, increased the estimated ultimate recovery, and accelerated payout.

THE LINE OF DEATH

SOCO also provided some additional geoscience ideas to the origin and delineation of the field that were first reported by Robinson et al. (1995), at a GRI conference in Denver. SOCO drilled the Jonah 3-15 (Sec. 15, T28N, R108W) in October 1994, which looked identical to the Jonah 4-8. Although there was not a vast amount of seismic data available in this part of the Green River basin, a regional northwest-southeast–trending 2-D seismic line purchased by SOCO showed a fault with a northeast-southwest orientation that separated the nonproducing wells to the south from the producers. Inside SOCO, this fault became known as the "Line of Death" and is now known to be the controlling fault boundary on the south side of the field (Figure 6). Wells north of the fault encounter significant gas shows and sustained flares at drilling depths below 7500–9000 ft (2280–2740 m), whereas south of the fault, continuous gas shows are first found in the regional basin-centered window at about 11,000 ft (3350 m).

In 1996, field production exceeded the capacity of the original 4.5-in. (11.4-cm) surface line going north, so McMurry, now operating with a new subsidiary called Jonah Gas Gathering Company, installed an 8-in. (20-cm) surface line that transports gas approximately 20 mi (32 km) west into gathering and transportation systems in the La Barge area.

OWNERSHIP CHANGES, CASH INFUSION

By the end of 1996, Jonah field was producing 50 MCFGD and had finally overcome the doubts of many skeptics who had dismissed the potential of the area. Several changes in ownership took place during the year in addition to the initiation of an Environmental Impact Study (EIS) that restricted drilling activity until April 1997. Ultra Petroleum Corporation, of Vancouver, British Columbia, Canada, purchased a vast block of acreage in the northern Green River basin from Consolidated Natural Gas of New Orleans, Louisiana. Most of the acreage was east of Jonah field or near the Pinedale anticline, but three sections from the purchase are located in the center of Jonah field. The first well in their drilling program, the Stud Horse Federal 13-23 (Sec. 23, T29N, R108W), was completed in early 1997 for more than 10,000 MCFGD and extended production 0.5 mi (0.8 km) north of the existing production.

In April 1996, McMurry sold an interest in the field to McNIC Oil and Gas Company, Detroit, Michigan. With this financing in place, the McMurry partners were able to undertake a 20-well drilling program that drove production up dramatically. McNIC subsequently sold this interest to Williams Production Company, Tulsa, Oklahoma, in 1998.

In late 1996, SOCO sold a 50% interest in their greater Jonah acreage to Amoco Corporation, Houston, Texas, now BP Amoco. Amoco brought a technological infusion to the field by promoting acquisition of additional core and log data and by undertaking extensive petrophysical investigations. They brought their alliance with Veritas DGC into the area and completed a 51-mi^2 (133-km^2) 3-D seismic survey, which resulted in a revolutionary new image of the subsurface, including the existence of a western bounding fault that forms the opposing boundary of the pressure compartment (Montgomery and Robinson, 1997; Warner, 1997).

In 1998, SOCO traded its remaining producing Jonah interests to BP Amoco for an interest in Beaver Creek field, Fremont County, Wyoming. SOCO retained nonproducing checkerboard interests outside of Jonah field. By the end of 1998, with the EIS completed and active drilling programs underway, Jonah field was producing 170 MCFGD from 106 wells (Figure 7). Additional pipelines were installed

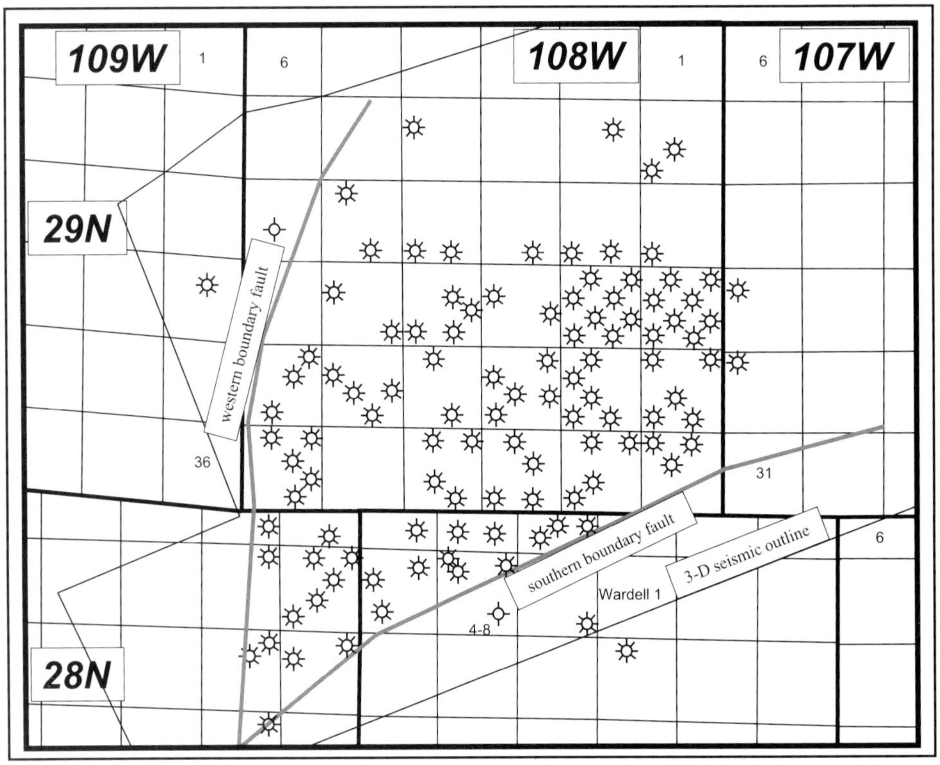

Figure 7. Jonah field, 1998. The field has 106 wells and is producing 170 MCFGD. The location of the both boundary faults is now known, and development drilling is actively underway.

to keep pace with expanding production, and planning for fieldwide compression was begun. An additional 70 mi^2 (183 km^2) of 3-D seismic was acquired to complete the subsurface structural-stratigraphic picture in the area.

In May 2000, the field operators received approval from the Wyoming Oil and Gas Conservation Commission to drill two wells per 80-ac (32.4-ha) spacing unit (40-ac [16.2-ha] spacing). In early June, the Pinedale BLM approved an Environmental Assessment that allows 40-ac (16.2-ha) surface disturbances within the fault boundaries. A new phase of infill drilling activity is underway and will usher in a new producing era for Jonah field.

In June 2000, all of the common stock of McMurry Oil Company, Fort Collins Consolidated Royalties, and Nerd Enterprises was sold to AEC Oil and Gas (U.S.A.), Inc. (now EnCana), a subsidiary of Alberta Energy Company, Ltd., of Calgary, Alberta, for approximately $632 million. In addition, the same companies sold their interests in Jonah Gas Gathering Company to AEC for approximately $167 million.

CURRENT FIELD STATUS

In June 2000, Jonah field was producing approximately 335 MCFGD and 3000 BOPD from 160 wells (Figure 8). Full field development in Jonah field will include several hundred wells, depending on economics of a future 40-ac (16.2-ha) infill drilling program. Production is expected to exceed 400 MCFGD by 2002. Ultimate estimated gas production will exceed 1.5 tcf and could be much greater if market conditions are favorable. Production from deeper and shallower zones, including the Fort Union, Ericson, Rock Springs, Blair, Muddy, and Frontier formations will be tested in the future. Jonah field has now grown beyond the expectations of many skeptics and is poised to become the next multi-tcf gas field in the Rocky Mountain region.

ACKNOWLEDGMENTS

I want to thank John W. Martin and Mick McMurry for filling in the blanks on the McMurry portion of this chapter. John Melby supplied information on Davis Oil Company and events leading up to drilling the Wardell well. Doug Battin provided information on the activity of Home Petroleum and is generally considered to be responsible for the discovery of the field. Greg Anderson provided information on the actions of Presidio Oil. An earlier draft of this manuscript was reviewed by Jane Estes-Jackson, Mark Kirschbaum, and Michele Bishop. I appreciate their comments and suggestions.

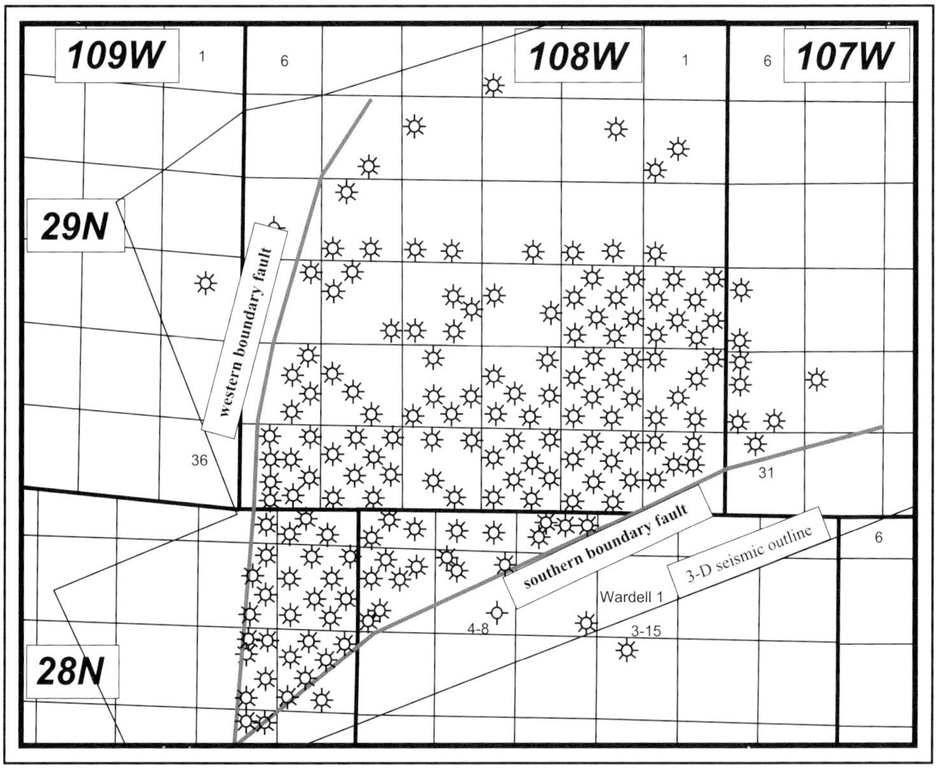

Figure 8. Jonah field, 2000. The field has 160 wells and is producing 335 MCFGD and 3000 BOPD. Infill drilling will double the number of wells on this map.

REFERENCES CITED

Battin, D. P., and S. R. Clark, 1994, Jonah field, a deep basin gas accumulation in the northern Green River basin (abs.): AAPG Annual Meeting Program, v. 3, p. 101.

Bowker, K. A., and J. W. Robinson, 1996, Jonah field: A basin-centered sweetspot in the northern Green River basin, Wyoming, *in* S. Longacre, B. Katz, R. Slatt, and M. Bowman, eds., Compartmentalized reservoirs: Their detection, characterization and management: AAPG/European Association of Geoscientists and Engineers Research Symposium, unpublished proceedings.

Finch, R. W., W. W. Aud, and J. W. Robinson, 1997, Evolution of completion and fracture-stimulation practices in Jonah field, Sublette County, Wyoming, *in* E. B. Coalson, J. C. Osmond, and E. T. Williams, eds., Innovative applications of petroleum technology in the Rocky Mountain area: Rocky Mountain Association of Geologists Guidebook, p. 13–24.

Klann, S, 1997, Jonah's tale: Oil and gas investor: Houston, Texas, Hart Publications, Inc., June, 1997.

Law, B. E., 1984a, Relationships of source-rock, thermal maturity, and overpressuring to gas generation and occurrence in low-permeability Upper Cretaceous and lower Tertiary rocks, Greater Green River basin, Wyoming, Colorado and Utah, *in* J. Woodward, F. F. Meissner, and J. L. Clayton, eds., Hydrocarbon source rocks of the greater Rocky Mountain region: Rocky Mountain Association of Geologists Guidebook, p. 469–490.

Law, B. E., 1984b, Structure and stratigraphy of the Pinedale anticline, Wyoming: U.S. Geological Survey Open-file Report 84-756, p. 1–15.

Law, B. E., and C. W. Spencer, 1989, Geology of tight gas reservoirs in the Pinedale anticline area, Wyoming, and at the multiwell experiment site, Colorado: U.S. Geological Survey Bulletin, v. 1886, 260 p.

Law, B. E., C. W. Spencer, R. R. Charpentier, R. A. Crovelli, R. F. Mast, G. L. Dolton, and C. J. Wandrey, 1989, Estimates of gas resources in overpressured low-permeability Cretaceous and Tertiary sandstone reservoirs, Greater Green River basin, Wyoming, Colorado, and Utah, *in* J. L. Eisart, ed., Gas resources of Wyoming: 40th Field Conference Guidebook, Wyoming Geological Association, p. 39–62.

McCaslin, J. C., 1974, Wyoming wildcat theaters booming: Oil & Gas Journal, (May 20, 1974), p. 135.

Montgomery, S. L., and J. W. Robinson, 1997, Jonah field, Sublette County, Wyoming: Gas production from overpressured Upper Cretaceous Lance sandstones of the Green River basin: AAPG Bulletin, v. 81, p. 1049–1062.

Robinson, J. W., D. Delozier, and R. Finch, 1995, Integrated reservoir description and analysis of the Lance Formation, Jonah field, Sublette County, Wyoming, *in* Geological characterization of the Cretaceous gas plays in the Greater Green River basin: Gas Research Institute/Department of Energy/Independent Petroleum Association of Mountain States Greater Green River Consortium Symposium, p. 297–308.

Robinson, J. W., D. Delozier, and R. Finch, 1996, Integrated reservoir description and analysis of the Lance Formation, Jonah field, Sublette County, Wyoming: AAPG Bulletin, v. 80, p. 980.

Shaw, J., 1995, Jonah field stimulation procedures, *in* Consortium for emerging gas resources in the Green River basin, hydraulic fracturing and completions in the Green River basin: Gas Research Institute/Department of Energy/Independent Petroleum Association of Mountain States Greater Green River Consortium Symposium, Casper, Wyoming, p. 63.

Spencer, C. W., 1984, Overpressured tight gas reservoirs in the Pinedale anticline area, Sublette County, Wyoming: U.S. Geological Survey Open-file Report 84-753, p. 51–59.

Spencer, C. W., and R. F. Mast, 1986, Geology of tight gas reservoirs: AAPG Studies in Geology 24, 299 p.

Warner, E. M., 1997, Geology of Jonah field, a major gas accumulation in the Upper Cretaceous Lance Formation, Sublette County, Wyoming, *in* E. B. Coalson, J. C. Osmond, and E. T. Williams, eds., Innovative applications of petroleum technology in the Rocky Mountain area: Rocky Mountain Association of Geologists Guidebook, p. 1–12.

Warner, E. M., 1998, Structural geology and pressure compartmentalization of Jonah field, Sublette County, Wyoming, *in* R. M. Slatt, ed., Compartmentalized reservoirs in Rocky Mountain basins: Rocky Mountain Association of Geologists Guidebook, p. 29–46.

Warner, E. M., 2000, Structural geology and pressure compartmentalization of Jonah field based on 3-D seismic data and subsurface geology, Sublette County, Wyoming: The Mountain Geologist, v. 37, p. 15–30.

CHAPTER 4

Regional Stratigraphic Setting of the Maastrichtian Rocks in the Central Rocky Mountain Region

R. C. JOHNSON
U.S. Geological Survey, Denver, Colorado, U.S.A.

T. M. FINN
U.S. Geological Survey, Denver, Colorado, U.S.A.

S. B. ROBERTS
U.S. Geological Survey, Denver, Colorado, U.S.A.

ABSTRACT

Jonah field in the northwestern part of the Greater Green River basin produces gas from the Lance Formation of Late Cretaceous (Maastrichtian) age. The Maastrichtian in the central Rocky Mountain region was a complex period of major transitions associated with the onset of the Laramide orogeny. This period of mountain building and basin development, which ultimately divided the central part of the Rocky Mountain foreland basin into much smaller Laramide basins and uplifts, began during latest Campanian and early Maastrichtian time and continued until near the end of the Eocene.

The Maastrichtian section consists largely of continental rocks in the western part of the study area and interbedded continental and nearshore marine rocks in the eastern part. Marine shales deposited in the Maastrichtian seaway are effective seals of regional extent in many parts of the central Rocky Mountain region. Coaly source rocks occur throughout a broad area near the limit of maximum westward marine transgression during the Maastrichtian in this region. Jonah field, however, is located west of the marine shale seals and west of the area with significant Maastrichtian coaly intervals. Coaly intervals in the underlying Campanian Rock Springs Formation at Jonah may be the source of much of the gas, and a shale interval near the top of the Lance Formation may be the top seal for Jonah.

An isopach map of Maastrichtian rocks reveals a complex pattern, including thickening trends toward newly developing Laramide uplifts as well as thickening toward areas of active thrusting along the Sevier orogenic belt. Jonah field is located in a deep, northwest-trending trough that developed southwest of the Laramide Wind River Range during the Maastrichtian. The Maastrichtian interval is overlain by lower Paleocene rocks throughout much of the study area, except near later Laramide uplifts. These thickening trends are the result of both variations in rates of subsidence during the Maastrichtian and differential erosion of Maastrichtian rocks prior to deposition of overlying Paleocene rocks. The Maastrichtian section is thin throughout a broad area in the southwestern part of the Greater Green River basin and Uinta basin. The lack of active thrusting and resultant thrust loading along the adjacent part of the Sevier orogenic belt during the Maastrichtian may be responsible for this thinning. Erosion of the Sevier highlands along this inactive portion of the orogenic belt may have instead produced broad uplift in the adjacent basin areas because of unloading.

Paleocurrent studies in Maastrichtian rocks indicate that eastward-flowing drainage systems off the Sevier orogenic belt were maintained throughout the central Rocky Mountain region despite highly varying subsidence rates. Major Maastrichtian trunk streams tended to remain in place for extended periods of time,

Jonah Field: Case Study of a Giant Tight-Gas Fluvial Reservoir, J. W. Robinson and K. W. Shanley, eds.: AAPG Studies in Geology 52 and Rocky Mountain Association of Geologists 2004 Guidebook creating east–west-trending belts with thick stacks of fluvial sandstones surrounded by much less sandy intervals. Jonah field is located in a major southeast-flowing drainage system that was confined to the rapidly subsiding trough southwest of the Wind River Range. Stream systems that were flowing off rising Laramide uplifts appear to have been relatively short tributary streams to the major east-flowing trunk streams. These tributary streams were commonly mud choked as Cretaceous marine shale sequences were eroded off Laramide uplifts during the initial stages of uplift.

INTRODUCTION

Maastrichtian rocks in the central Rocky Mountain region, including the Lance Formation, contain significant reserves of natural gas in sandstone reservoirs. The central Rocky Mountain region, as the term is used here, extends from southern Montana to southern Colorado (Figure 1). Highly successful recent gas plays in the Lance include Cave Gulch field in the Wind River basin and Jonah field in the Greater Green River basin (Figure 1). Sandstones in the Lewis Shale, a lateral equivalent to the middle part of the Lance Formation, are currently being developed in several fields in the Greater Green River basin east of the Rock Springs uplift. This chapter focuses on the regional depositional setting of the productive Lance Formation at Jonah field, which is a "sweet spot" in an extensive, basin-centered-type, unconventional gas accumulation. The development of this field, which represents one of the great Rocky Mountain exploration successes of the last 10 yr, serves as a model for future exploration in the region.

REGIONAL TECTONIC FRAMEWORK

The Lance Formation and other Maastrichtian-age rocks were deposited in the central Rocky Mountain region during a complex period associated with the onset of the Laramide orogeny. Prior to the orogeny, a single structural and sedimentary basin, the Rocky Mountain foreland basin, extended from the Arctic to the Gulf of Mexico. This continent-long area of downwarping developed as a result of generally east-directed thrusting along the Sevier orogenic belt to the west (Figure 1). Thrusting was in response to active subduction along the west margin of the North American continent. Thrusting along the Sevier orogenic belt is generally thought of as "thin skinned" because of the low angle of the thrust fault planes and because the thrusting generally does not involve the underlying Precambrian crystalline rocks. This is particularly true in the eastern, younger part of the thrust belt; however, Precambrian basement may be involved in some of the older thrusts to the west (DeCelles, 1994). Thrust loading (Price, 1973; Beaumont, 1981; Jordan, 1981; Beck, 1985; Beck et al., 1988) and, possibly, subduction-induced mantle flow (Mitrovica et al., 1989; Gurnis, 1992; Stern and Holt, 1994; Pysklywec and Mitrovica, 1998) created the foreland basin east of the thrust belt. Geologically speaking, the response to thrust loading is generally considered as being nearly instantaneous (Jordan, 1981). Thus, active thrusting and subsidence in the adjacent basin are directly linked. Thrusting began in the western part of the orogenic belt and generally progressed eastward with time (Armstrong and Oriel, 1965). Maximum structural and topographic relief occurred along the older, western part of the thrust belt in large part because of repeated periods of uplift and the further eastward displacement of older thrust sheets riding "piggyback style" on the hanging walls of younger thrust sheets (DeCelles, 1994).

Beginning in the latest Cretaceous and continuing into the Eocene, rising Laramide uplifts segmented the central part of the foreland basin, from Montana to New Mexico, into smaller discrete basins. The Laramide orogeny is characterized by "thick-skinned" deformation manifested by low- to high-angle reverse faults extending deep into basement rocks. Several tectonic models, which were related to changes in subduction along the western margin of the North American continent, have been proposed to explain why compressional forces of the Laramide were concentrated in a fairly restricted area in this region. The style of deformation appears to have occurred during a time of unusually rapid convergence between the Farallon and North American plates (see Beck et al., 1988; Hamilton, 1988, for summaries). The deep troughs of Laramide basins typically developed adjacent to thrust-bounded uplifts, and thrust loading has been proposed as the principal mechanism for inducing subsidence (Beck, 1985; Beck et al., 1988).

DEPOSITIONAL SETTING OF MAASTRICHTIAN STRATA

Figures 1 and 2 depict the depositional setting of Maastrichtian strata in the central Rocky Mountain region. Figure 3 is a generalized stratigraphic chart bracketing the Maastrichtian strata in the basins studied in this report, and Figure 4 shows the ages of the isopach intervals in each basin tied to the North American ammonite zonation (by W. A. Cobban in Molenaar et al., 2002). A generalized approach, which was modified from the work of Roberts and Kirschbaum (1995), was used so that only key elements for the entire Maastrichtian interval are shown. Figures 5 and 6 are schematic cross sections showing stratigraphic relationships and coal-bearing intervals in the western Greater Green River basin. Locations of the cross sections are shown in Figure 7.

In this report, the Maastrichtian interval in the Jonah field area consists of the Almond and Lance formations (Figures 3, 6). However, the Almond is not recognized by all workers here (see, for example, Law and Johnson, 1989), and the interval shown as Almond in Figures 5 and 6 is included by these workers in the overlying Lance. The Almond Formation was originally defined in the eastern part of the Greater Green River basin, where it is a coaly marginal-marine interval between the underlying fluvial Ericson Sandstone of Late Cretaceous (Campanian) age and the overlying marine Lewis Shale of Late Cretaceous (Maastrichtian) age (Figure 3). West of the pinch-out of the Lewis Shale, the Almond, where recognized, is a nonmarine coaly interval that is roughly equivalent to the Almond further to the east (Figures 3, 5, 6). The Lance is about 2800 ft (850 m) thick at Jonah field and is unconformably overlain by the Tertiary Fort Union Formation (Figure 6). This unconformity cuts out the entire Lance and Almond formations toward the La Barge platform (Figure 6).

The maximum westward transgression of the Cretaceous seaway in the early Maastrichtian and the position of the shoreline near the beginning of the late Maastrichtian are shown in Figures 1 and 2. The position of the shoreline at maximum transgression was generally from Roberts and Kirschbaum (1995) but was modified in several areas, including the Bighorn basin, using Keefer et al. (1998); the Wind River basin, using Finn (1993); and in the Crazy Mountains, Greater Green River, and Piceance basins, using unpublished data by the authors. The position of the shoreline veered sharply eastward around the Wind River Range, north of Jonah field, and possibly part of the Granite Mountains area, indicating that activity on these uplifts probably influenced shoreline positions in the early Maastrichtian.

Potential Source Rocks and Seals

Areas where coal-forming mires occurred during the Maastrichtian shown in Figure 2 are from Roberts and Kirschbaum (1995), except in the Greater Green River basin where the area was modified using unpublished data by the authors. Areas of coal-forming mires in the Hanna and Wind River basins were connected at the east end of the Granite Mountains, based on the presence of coal on both sides of the Granite Mountains. Maastrichtian coal-forming mires occur throughout a broad area near the maximum transgression of the seaway during the Maastrichtian (Figure 2). In some cases, the distribution of coal-forming mires can be interpreted to infer the timing of Laramide uplifts. For example, the coal-bearing Maastrichtian Meeteetse Formation occurs in both the Wind River and Bighorn basins (Keefer, 1972), indicating that the intervening Owl Creek Mountains was not a positive feature at that time. Major coal-bearing intervals are absent from the Almond and Lance formations in many areas of the Greater Green River basin west of the Rock Springs uplift, including the Jonah field area, but significant coal occurs in the underlying Rock Springs and overlying Fort Union formations (Figures 5, 6). This supports the hypothesis that much of the gas in the Lance Formation at Jonah migrated vertically from coaly intervals in the underlying Rock Springs Formation (Warner, 1998, 2000).

Marine shale seals of regional extent occur in the lower part of the Maastrichtian interval everywhere east of the line of maximum Maastrichtian transgression shown in Figures 1 and 2. Shale seals in the Maastrichtian-age marine Lewis Shale are important in trapping hydrocarbons in nearshore marine sandstones in the underlying Almond Formation in the eastern part of the Greater Green River basin. Gas sourced from Upper Cretaceous coals and carbonaceous shales is also found in Rocky Mountain reservoirs as young as Eocene, where such seals are not present (Johnson and Rice, 1990, 1993; Johnson et al., 1994b; Johnson and Keighin, 1998). Jonah field is west of the pinch-out of these regional marine shale seals; however, a thick, fine-grained fluvial interval in the upper part of the Lance appears to act as a top seal (Warner, 1998, 2000). Warner (1998, 2000) believed that this seal was critical to the development of Jonah. Seals in fluvial strata, such as the seal that occurs at Jonah field, are of more limited lateral extent than marine shale seals and would require detailed stratigraphic studies to identify. Fine-grained intervals also occur in the overlying nonmarine lower Tertiary section, and some of these may also act as seals. Mapping the extent of these lower Tertiary seals may also prove helpful in understanding the distribution of gas in Maastrichtian rocks.

ISOPACH MAP OF MAASTRICHTIAN ROCKS IN THE CENTRAL ROCKY MOUNTAIN REGION

An isopach map of Maastrichtian rock in the central Rocky Mountain region, constructed using stratigraphic

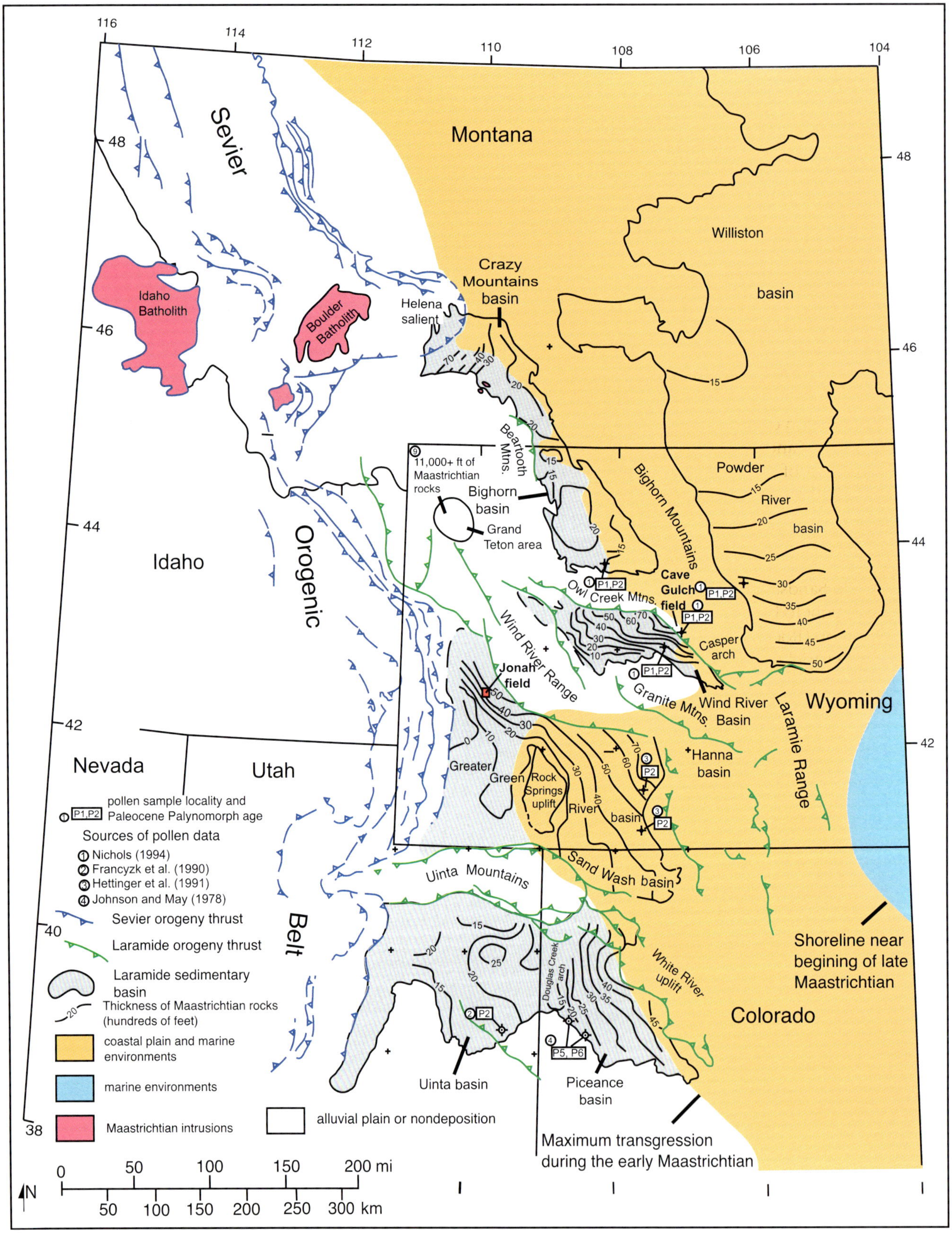
Montana
Sevier
Orogenic
Belt
Idaho
Nevada
Utah
Wyoming
Colorado
Idaho Batholith
Boulder Batholith
Helena salient
Crazy Mountains basin
Williston basin
Beartooth Mtns.
Bighorn basin
Bighorn Mountains
11,000+ ft of Maastrichtian rocks
Grand Teton area
Powder River basin
Owl Creek Mtns.
Cave Gulch field
Casper arch
Wind River Range
Jonah field
Granite Mtns.
Wind River Basin
Laramie Range
Hanna basin
Greater Green River basin
Rock Springs uplift
Sand Wash basin
Uinta Mountains
Uinta basin
Douglas Creek arch
White River uplift
Piceance basin
Shoreline near begining of late Maastrichtian
Maximum transgression during the early Maastrichtian
pollen sample locality and Paleocene Palynomorph age
Sources of pollen data
1 Nichols (1994)
2 Francyzk et al. (1990)
3 Hettinger et al. (1991)
4 Johnson and May (1978)
Sevier orogeny thrust
Laramide orogeny thrust
Laramide sedimentary basin
Thickness of Maastrichtian rocks (hundreds of feet)
coastal plain and marine environments
marine environments
Maastrichtian intrusions
alluvial plain or nondeposition
0 50 100 150 200 mi
50 100 150 200 250 300 km
N

intervals that could be identified on geophysical logs, is shown in Figure 1. Erosion along rising Laramide uplifts during the Paleocene and the Eocene resulted in the partial to complete removal of Maastrichtian strata from the margins of many Laramide structural and sedimentary basins. Thus, the variations in the thickness of Maastrichtian rocks found in Laramide basins today reflect both variations in subsidence rates during the Maastrichtian and differential erosion during the Paleocene and the Eocene. The intervals in each basin extend from the approximate top of the Cretaceous section downward to a lithologic marker horizon that is near the Campanian–Maastrichtian boundary (Figure 3). The approximate time intervals that are represented by the isopach maps in each basin are shown in Figure 4. In all basins shown in Figure 1, except the Uinta and Piceance basins, the base of the isopach interval is the top of the Teapot Sandstone Member of the Mesaverde Formation or its approximate equivalents. The top of the Ericson Sandstone in the Greater Green River basin and the top of the Pine Ridge Sandstone in the Hanna basin are thought to be approximately equivalent to the top of the Teapot (Gill and Cobban, 1966; Gill et al., 1970; Roehler, 1990; Merewether et al., 1997). In the Sand Wash basin of northwestern Colorado (Figure 1), the top of the Twentymile Sandstone Member of the Williams Fork Formation, thought to be approximately equivalent to the top of the Ericson (Roehler, 1987, 1990), was used as the base of the Maastrichtian interval. The Teapot, Ericson, Pine Ridge, and Twentymile sandstones were deposited during and after a time of uplift and beveling through a broad area in western Wyoming, including much of the Greater Green River basin west of the Rock Springs uplift (Gill and Cobban, 1966; Roehler, 1990), and in the Granite Mountains area (Reynolds, 1976). The ammonite zone of *Baculites reesidei* of latest Campanian age has been reported from the lower part of the Lewis Shale and lateral equivalents, and above the Teapot Sandstone Member and lateral equivalents at several localities in the central Rocky Mountain region (Gill and Cobban, 1966, 1973; Gill et al., 1970), indicating that the upper part of the Teapot–Ericson–Pine Ridge–Twentymile interval is late Campanian.

No regional markers are present near the Campanian–Maastrichtian boundary in the Uinta basin in eastern Utah and western Colorado or in the Piceance basin of western Colorado. In the Uinta basin, the isopach interval extends from the marine-flooding surface at the top of the Castlegate Sandstone to the top of the Mesaverde Group (Figure 3), and the isopach map shown is from Johnson and Roberts (2003). The flooding surface is within the *Baculites perplexus* zone of the middle Campanian (Figure 4) (Gill and Hail, 1975). In the Piceance basin, the isopach interval extends from the marine-flooding surface at the top of the Cozzette Member of the Mesaverde or Mount Garfield Formation to the top of the Mesaverde Group (Figure 3), and the isopach map shown is from Johnson (2003). The Cozzette Member was deposited during the *Didymoceras stevensoni* ammonite zone of the late Campanian (Figure 4) (Zapp and Cobban, 1960; Gill and Hail, 1975).

The Maastrichtian section in the central Rocky Mountain region is overlain by Paleocene rocks throughout all but the marginal areas of all Laramide basins. The contact between Maastrichtian rocks and Paleocene rocks is difficult to precisely pick in many areas without palynomorph data. The Fort Union Formation and lateral equivalents overlie the Maastrichtian section in most of the basins, and the base of the Fort Union Formation, as shown on geologic maps, is close to the Cretaceous–Paleocene contact. The age of the basal Fort Union or equivalent strata is shown at several localities in Figure 1, in the marginal areas of basins where it has been dated using palynomorphs. Early Paleocene pollen (P1 and P2 zones of Nichols and Ott, 1978) was collected at most of the localities shown. In some areas, however, the mapped contact for the base of the Fort Union does not correspond closely to the Cretaceous–Paleocene boundary. As an example, Roberts (1972) collected Cretaceous pollen 750 ft (230 m) above the base of the Fort Union in the southwestern part of the Crazy Mountains basin in south-central Montana.

In the Powder River basin (Figure 1), the contact between the Maastrichtian Lance Formation and overlying Fort Union Formation appears to be conformable throughout most, if not all, of the basin (Figure 3). The Cretaceous–Tertiary boundary clay has been identified at three localities in the basin (Bohor et al., 1987; Wolfe and Izett, 1987; Nichols et al., 1992). The contact can be identified

◄

Figure 1. Map of central Rocky Mountain region showing approximate thickness of Maastrichtian rocks, Sevier and Laramide thrust faults, generalized environments of deposition during the Maastrichtian, and age of rocks immediately above the Maastrichtian interval. Thicknesses are in hundreds of feet. P1 and P2 are early Paleocene palynomorph zones (see Nichols, 1994). The Campanian–Maastrichtian boundary is poorly constrained in the Uinta and Piceance basins, and in these basins, the isopach interval includes Campanian rocks. Approximate age of isopach interval in each basin is shown in Figure 3. Location of Sevier thrusts is from Kulik and Schmidt (1988). Location of Laramide thrusts is from Gries (1983, her figure 1). Sources of pollen data are listed on the figure.

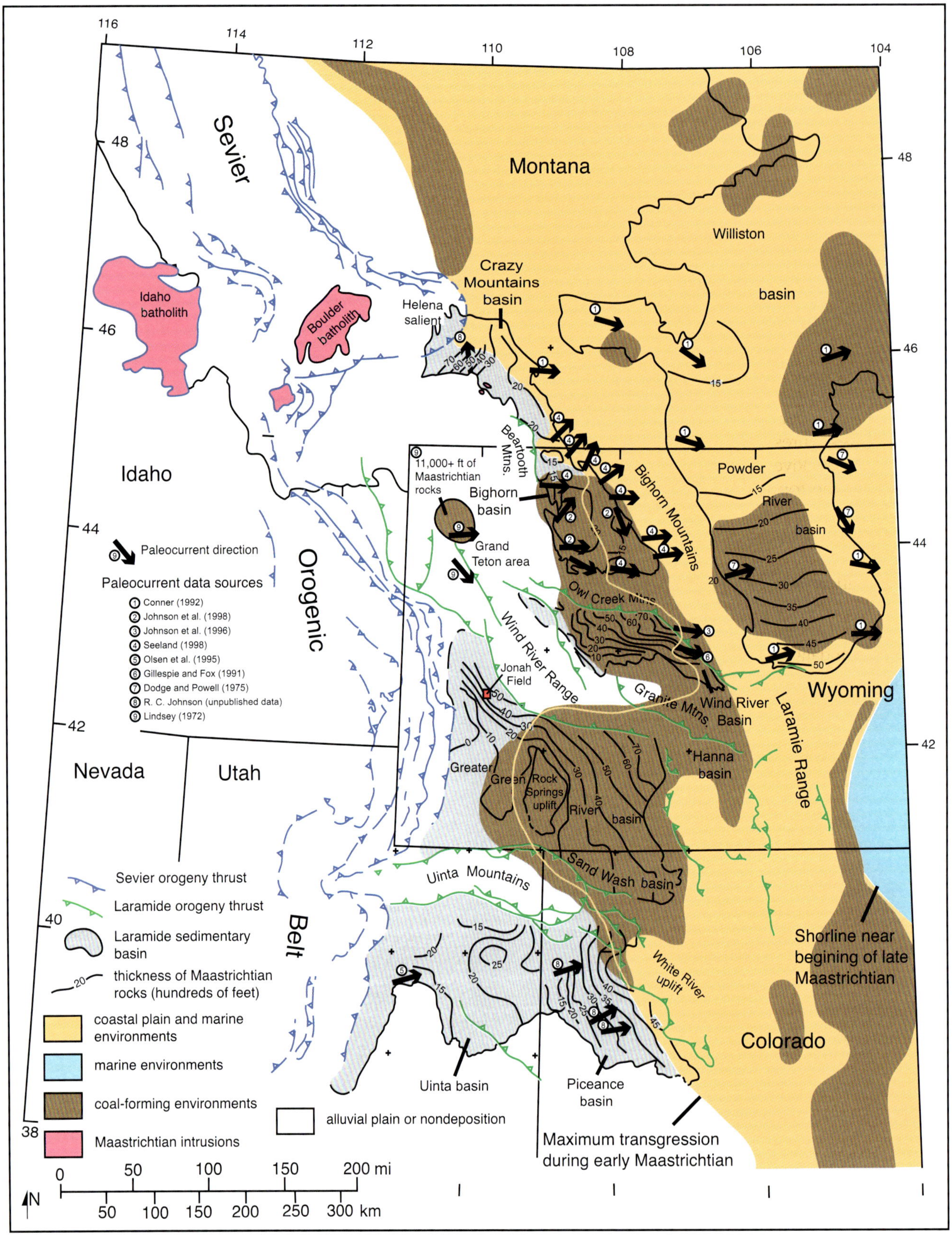
116
114
112
110
108
106
104
48
46
44
42
40
38
Sevier
Orogenic
Belt
Montana
Idaho
Nevada
Utah
Wyoming
Colorado
Idaho batholith
Boulder batholith
Helena salient
Crazy Mountains basin
Williston basin
Beartooth Mtns.
11,000+ ft of Maastrichtian rocks
Bighorn basin
Bighorn Mountains
Powder River basin
Grand Teton area
Owl Creek Mtns.
Wind River Range
Jonah Field
Granite Mtns.
Wind River Basin
Laramie Range
Hanna basin
Greater Green River basin
Rock Springs uplift
Sand Wash basin
Uinta Mountains
Uinta basin
White River uplift
Piceance basin
Shorline near begining of late Maastrichtian
Maximum transgression during early Maastrichtian
Paleocurrent direction
Paleocurrent data sources
1 Conner (1992)
2 Johnson et al. (1998)
3 Johnson et al. (1996)
4 Seeland (1998)
5 Olsen et al. (1995)
6 Gillespie and Fox (1991)
7 Dodge and Powell (1975)
8 R. C. Johnson (unpublished data)
9 Lindsey (1972)
Sevier orogeny thrust
Laramide orogeny thrust
Laramide sedimentary basin
thickness of Maastrichtian rocks (hundreds of feet)
coastal plain and marine environments
marine environments
coal-forming environments
Maastrichtian intrusions
alluvial plain or nondeposition
0 50 100 150 200 mi
50 100 150 200 250 300 km
N

in the subsurface based on the subsurface picks of Connor (1992). The Maastrichtian interval in the Crazy Mountains basin (Figure 1) extends from the base of the Paleocene Fort Union Formation or Fort Union Group to the top of the Teapot equivalent (Figure 3). In the western part of the basin where Cretaceous pollen was collected as much as 750 ft (230 m) above the base of the Fort Union Formation as mapped by Roberts (1972), the isopach interval extends from the top of the Teapot equivalent to the highest occurrence of Cretaceous pollen. In the Bighorn basin, the contact between Maastrichtian and Paleocene rocks is difficult to pick in the subsurface, and lithologic descriptions from surface sections and seismic reflection profiles were used to constrain the contact in the subsurface (Keefer et al., 1998). Similar problems with the Cretaceous–Paleocene contact were encountered in the Wind River basin to the south (Keefer, 1997). The isopach map of the Wind River basin (Figure 1) is from Johnson et al. (1996, their figures 46 and 52). In the eastern part of the Greater Green River basin, the Cretaceous–Tertiary contact is near the top of a thick sandstone interval as identified by Honey and Hettinger (1989), Hettinger and Kirschbaum (1991), and Hettinger et al. (1991). For the rest of the basin, picks on the Cretaceous–Tertiary contact were modified from Tyler (1978, 1979a, b, 1980a, b), Bucurel (1981, 1982), Bucurel-White (1983), Tyler et al. (1981, 1982), and Markochick et al. (1982).

Palynomorph data for the southeastern Uinta basin indicate that little or no Maastrichtian section is preserved there. Francyzk et al. (1990) dated a thin, conglomeratic sandstone unit near the Cretaceous–Tertiary boundary as early Paleocene in age. The sandstone, named the Dark Canyon beds (Figure 3), appears to be bounded by unconformities with upper Paleocene strata above and probable Campanian strata below. Thus, the isopach interval for the Uinta basin shown in Figure 1 may consist almost entirely of Campanian-age rocks. The isopach map shown is from Johnson and Roberts (2003). Along the west margin of the Piceance basin, the Cretaceous–Tertiary contact occurs at the base of a conglomeratic sandstone similar to the lower Paleocene unit in the Uinta basin (Johnson and May, 1978, 1980); thus, the Maastrichtian section is probably very thin or absent there as well. Late Paleocene pollen (P5 and P6 zones of Nichols and Ott, 1978) was collected just above this conglomerate (Figure 1). The isopach interval in the Piceance basin thickens to more than 4500 ft (1700 m) along the east margin of the basin, however, and much of the upper part of this thickened interval appears to be of Maastrichtian age (Johnson, 1989, p. E17–E18).

STRUCTURAL IMPLICATIONS OF THE ISOPACH MAP

Continuous deposition between the Late Cretaceous and the Paleocene has been demonstrated only in the Powder River basin, where the Cretaceous–Tertiary boundary clay has been identified (Nichols et al., 1992). In all other Rocky Mountains basins shown in Figures 1 and 2, there is evidence that some erosion of Maastrichtian rocks occurred prior to deposition of the overlying Paleocene strata, particularly in the marginal areas. Nonetheless, the isopach map can be used to make some general observations about the timing of uplifts in the region.

Proximity to the Sevier orogenic belt exerted some control on the thickness of the Maastrichtian interval preserved in several areas. The belt occurs just west of four areas: the Crazy Mountains basin, the Greater Green River basin, the Uinta basin, and in the Grand Tetons area in northwest Wyoming (Figure 1). In the Crazy Mountains basin, the Maastrichtian interval thickens from less than 1500 ft (460 m) in the eastern and southern areas of the basin to more than 7000 ft (2100 m) adjacent to the Helena salient. This thickening was probably caused by active thrusting and resultant thrust loading along the salient during the Maastrichtian. Active thrusting at this time also occurred along the thrust belt in northwest Wyoming. As much as 11,000 ft (3350 m) of conglomeratic Harebell Formation was deposited in the Grand Tetons area in northwest Wyoming during the Maastrichtian. The Harebell was sourced by the actively rising Targhee uplift along the Sevier orogenic belt to the west (Love, 1973).

Maastrichtian rocks are thin to absent in the western parts of the Greater Green River and Uinta basins adjacent to the thrust belt. Paleocene rocks unconformably overlie the Upper Cretaceous section in both areas; thus, the lack of Maastrichtian strata could be the result of nondeposition and/or later removal. Work by DeCelles (1994) on the thrust belt in southwest Wyoming and northern

◄

Figure 2. Map of the central Rocky Mountain region showing approximate thickness of Maastrichtian rocks, Sevier orogeny and Laramide orogeny thrust faults, environments of deposition including coal-forming environments during the Maastrichtian, and paleocurrent directions in Maastrichtian rocks. Thicknesses are in hundreds of feet. Sources of paleocurrent data are listed on the figure.

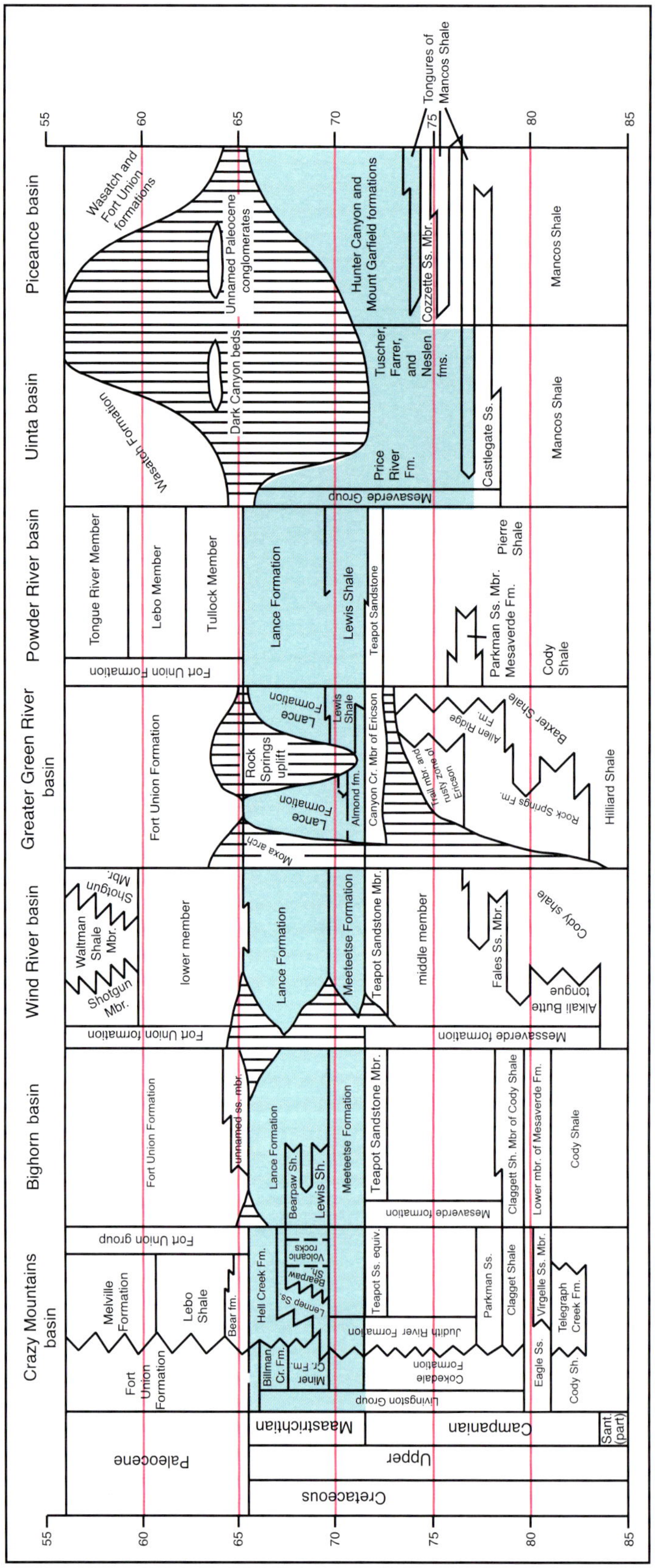

Figure 3. Generalized stratigraphic chart bracketing the Maastrichtian interval for each basin shown in Figures 1, 2. Isopach intervals in each basin are shown in blue.

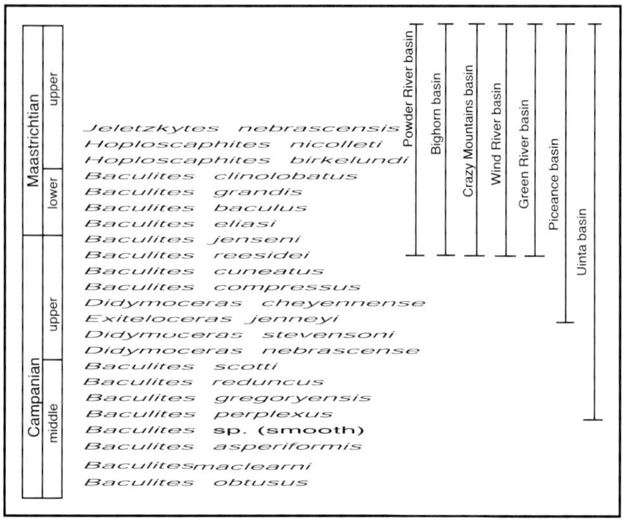

Figure 4. Approximate ages of isopach intervals for each basin shown in Figures 1–3 in relation to the North American ammonite zonation (by W. A. Cobban, from Molenaar et al., 2002).

Utah, however, indicates that little thrusting occurred along this segment of the thrust belt during the late Campanian and early Maastrichtian. The lack of Maastrichtian strata in these areas may, in part, be caused by isostatic rebound in the foreland basin as the Sevier highlands to the west were being eroded during this quiescent tectonic period.

Increases in thicknesses toward Laramide structures are apparent in several areas. In the Greater Green River basin, the Maastrichtian interval thickens from near zero throughout much of the western part of the basin to more than 5000 ft (1500 m) adjacent to the Wind River Range and from less than 2000 ft (610 m) along the Rock Springs uplift to more than 7000 ft (2100 m) adjacent to the Granite Mountains (Figure 1). In the Wind River basin, the interval thickens from a wedge edge along the northwest flank of the Wind River Range to more than 7000 ft (2100 m) adjacent to the Owl Creek Mountains (Figure 1). In the Piceance basin, the isopach interval thickens from less than 1500 ft (460 m) across the Douglas Creek arch to more than 4500 ft (1400 m) toward the White River uplift (Figure 1). The trend in the Wind River basin appears to be the result of both uplift and erosion along the rising Wind River Range and subsidence along the Owl Creek Mountains. The entire Maastrichtian section has been upturned to near vertical along the Casper arch, which forms the eastern terminus of the Owl Creek Mountains. This indicates that most, if not all, of the uplift along the Owl Creek Mountains occurred after the Maastrichtian (Keefer, 1972). Subsidence in the Green River and Wind River basins appears to have started as downwarping along elongate troughs that preceded the onset of faulting and uplift of Laramide uplifts in the area (Love, 1960; Keefer and Love, 1963; Love et al., 1963; Keefer, 1970). The east- to northeast-thickening Laramide trend for Maastrichtian and upper Campanian rocks in the Piceance basin also appears to predate the onset of uplift in the adjacent White River uplift, which may have occurred as late as the Eocene (Tweto, 1975). The thickening trend is not entirely caused by postdeposition; differential truncation of Maastrichtian rocks, however, as the trend can be traced back into the late Campanian (Johnson, 2003).

The isopach map of the Bighorn basin shows little trends, ranging from slightly less than 1500 to more than 2000 ft (460 to more than 610 m) throughout most of the basin. By the end of the Paleocene, however, a deep structural trough had formed in the western part of the Bighorn basin (Johnson, 2003).

In the Powder River basin, the only basin where a complete Maastrichtian section is preserved in outcrop around the basin margins, the interval thickens gradually toward the south from less than 1500 ft (460 m) in the northernmost part of the basin to more than 5000 ft (1500 m) along the south margin, along the north margin of the Laramie Range. The northern Laramie Range was an arch during the Maastrichtian, although some north-directed thrusting occurred along the north end of the range during the Eocene (Gries et al., 1992, their figures 7 and 9). Thus, Maastrichtian subsidence in the Powder River basin was unlikely to have been related to thrust loading. The thickening trend is perpendicular to the Bighorn Mountains, which were not actively rising until the Paleocene (Gries et al., 1992, their figure 8).

PALEOCURRENT DATA

Paleocurrent directions from fluvial sandstones in Maastrichtian strata are shown in Figure 2. The data indicate that general west-to-east regional drainage, away from the Sevier orogenic belt, was maintained during the Maastrichtian throughout much of the study area, despite the onset of uplift on several Laramide structures during this period. Streams originating in these Laramide uplifts were probably relatively short tributary streams flowing into the major east-flowing trunk streams. Some of these east-flowing drainage systems apparently persisted in relatively narrow east–west-trending belts for considerable periods of time. Sandstone percentages along one such drainage system, which flowed southeastward across the southern part of the Bighorn basin in northwestern Wyoming, vary to more than 90% for a greater-than-800-ft (240-m) interval

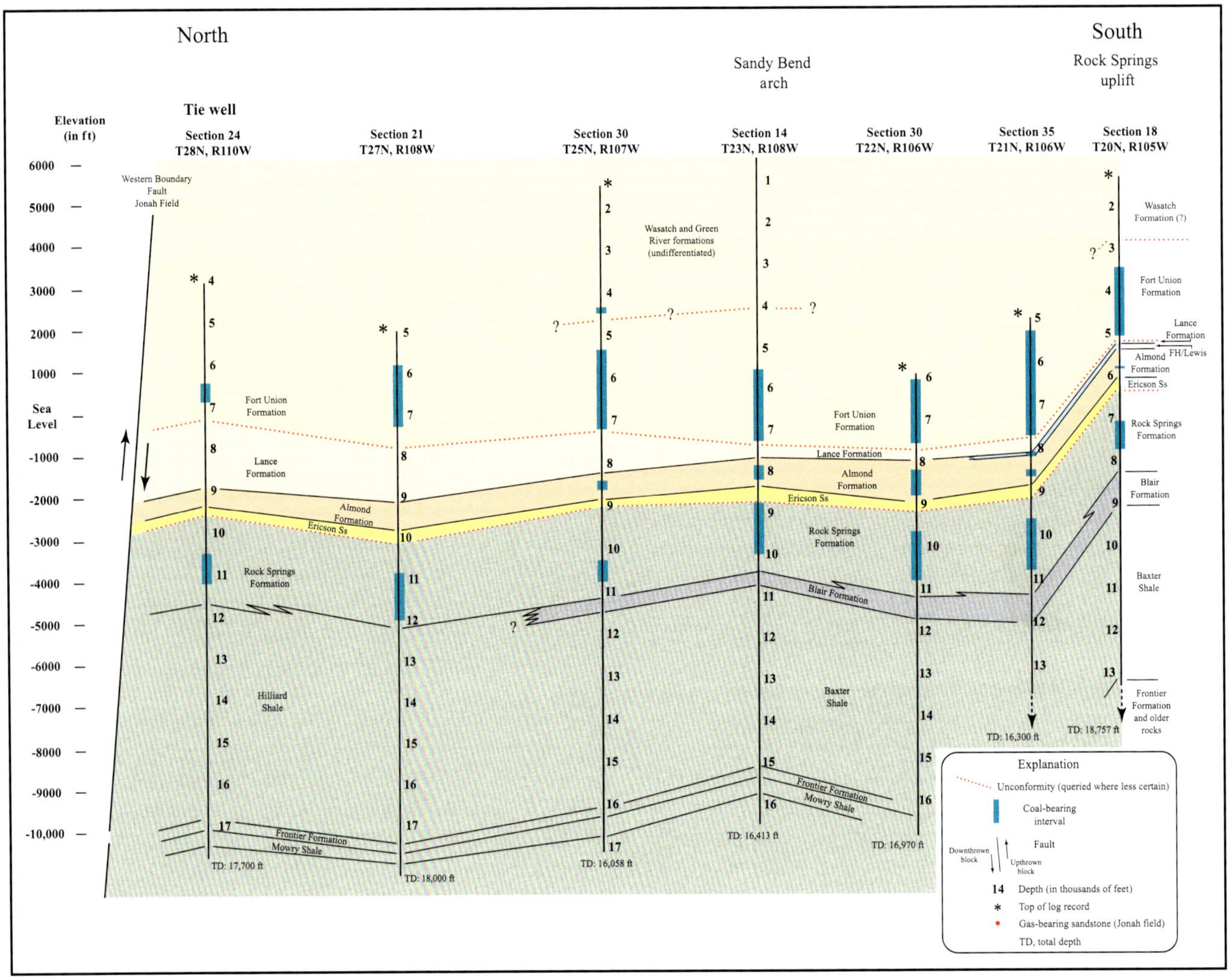

Figure 5. South-to-north cross section from the Rock Springs uplift to Jonah field showing stratigraphic relationships, unconformities, and coal-bearing intervals. At Jonah field, coal-bearing intervals do not occur in the gas-productive Lance Formation but only in the underlying Rock Springs Formation and overlying Fort Union Formation. Location of section is shown in Figure 7.

in the Lance Formation (Johnson et al., 1998). Sandstone percentage in the Lance decreases to less than 50% away from this drainage system. Interestingly, paleocurrent readings for the Powder River basin, where there is a completely preserved Maastrichtian section, indicate that fluvial channels must have run at right angles to subsidence trends (Figure 2). Major variations in subsidence rates there appear to have had surprisingly little impact on drainage systems.

A paleocurrent study by Lindsey (1972) on the Maastrichtian Harebell Formation in the Grand Tetons area (Figure 2, location 9), northwest of Jonah field, indicates a southeast flow direction generally toward the Jonah field area. The Harebell Formation is a conglomeratic interval as much as 11,000 ft (3350 m) thick that contains clasts derived mainly from Paleozoic and Precambrian rocks along the Sevier orogenic belt to the northwest. Love (1973, his figure 15) suggested that a major drainage may have flowed southeast out of the Jackson Hole area along the south flank of the rising Wind River Range during the Maastrichtian.

DISCUSSION

The purpose of this chapter was to gain a better understanding of Jonah field by placing it into a regional depositional context for the Upper Cretaceous (Maastrichtian)

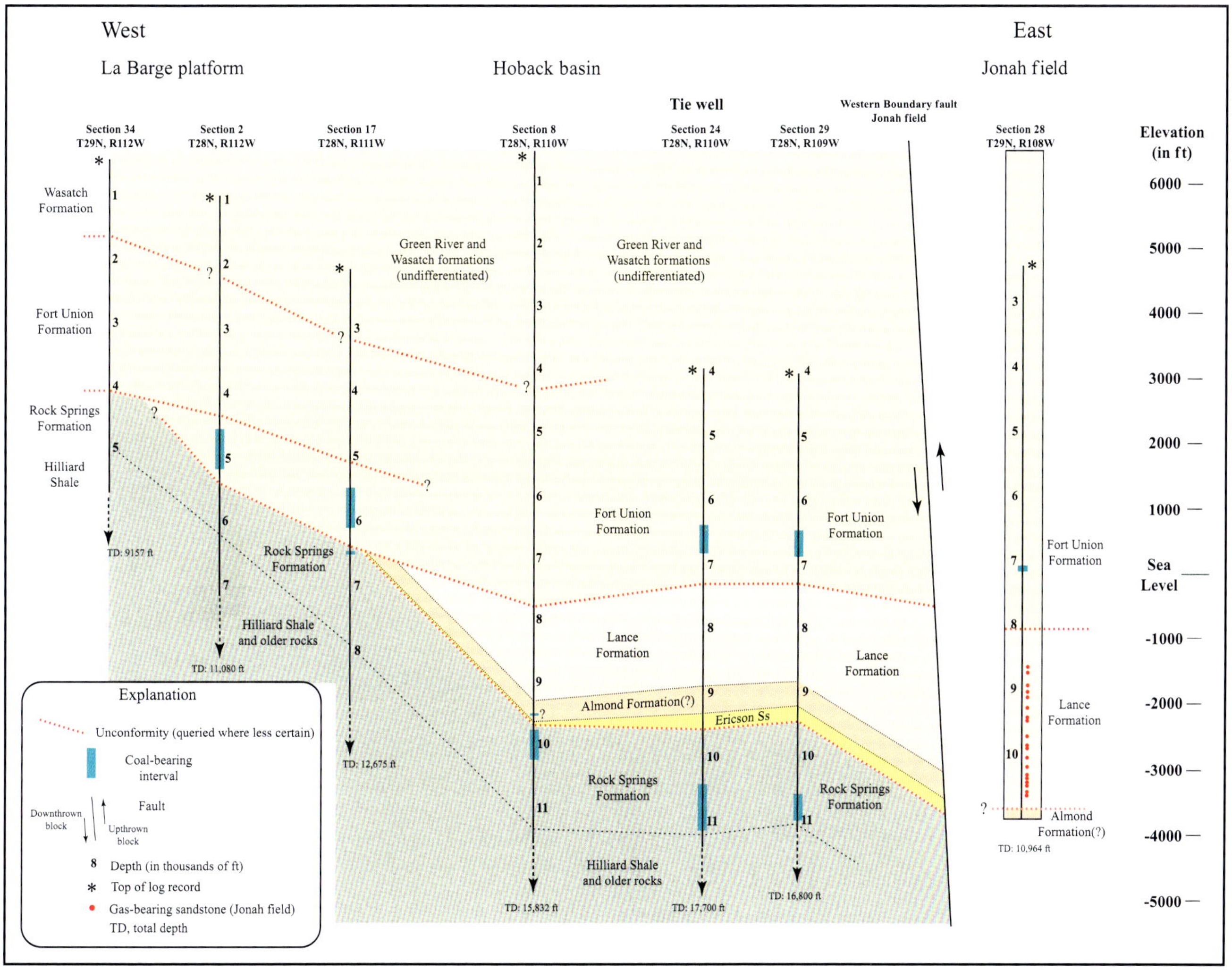

Figure 6. West-to-east cross section from the La Barge to Jonah field showing stratigraphic relationships, unconformities, coal-bearing intervals, and gas-productive interval at Jonah field. Again, note the lack of coal-bearing gas source rocks in the productive Lance Formation at Jonah field. Location of section is shown in Figure 7.

strata in the central Rocky Mountain region and thereby contribute to the future exploration of more Jonah-type accumulations. However, it should be emphasized that although the Maastrichtian Lance Formation is the major producer at Jonah, the source of most of the gas may be the coals and carbonaceous shales in the underlying Campanian Rock Springs section. The shale at the top of the gas accumulation at Jonah appears to have been a critical factor in retarding the vertical migration of gas. The seal rocks are near the Cretaceous–Paleocene boundary and could be Paleocene in age. To find more Jonah-type accumulations may require an extensive knowledge of the distribution of potential seals in the overlying Paleocene section. In addition, faults and fractures have acted as seals in Jonah field, compartmentalizing the gas accumulation. Faults on the south and west side of the field also seal the highly pressured Jonah accumulation from lower-pressured strata surrounding the field. In short, a total petroleum systems approach that involves an extensive knowledge of underlying and overlying strata as well as a detailed analysis of minor structures may be required to explore for more accumulations like Jonah.

Marine shale seals of regional extent in Maastrichtian rocks in the central Rocky Mountain region were important in trapping hydrocarbons in many areas. However, a fine-grained interval in fluvial rocks, which probably has only a limited extent, acts as the seal at Jonah; identifying such seals elsewhere would require more detailed study.

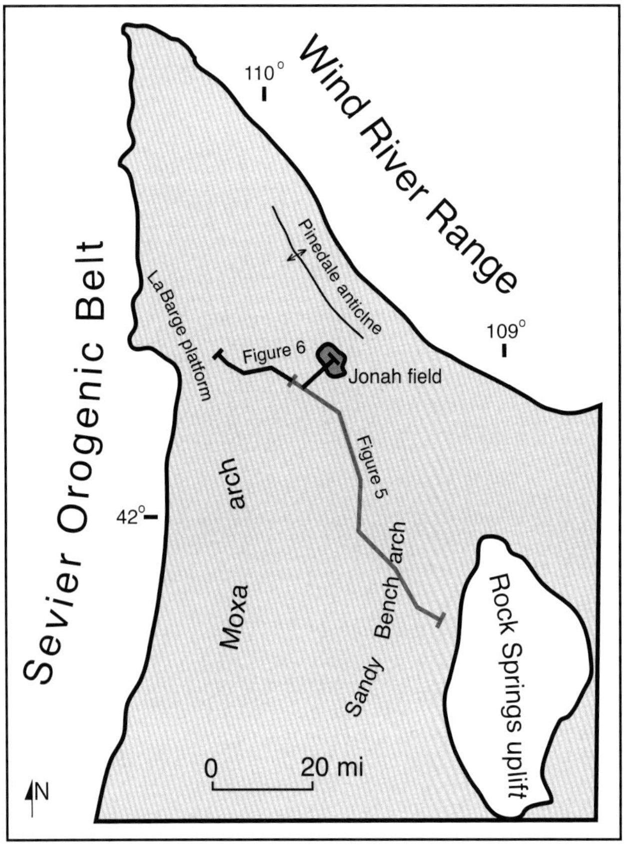

Figure 7. Index map for cross sections shown in Figures 5 and 6.

Major Maastrichtian fluvial systems flowed generally eastward off the Sevier highlands to the west throughout the central Rocky Mountain region; thus, fluvial sandstone reservoirs throughout the area should be oriented generally east–west. Some of these systems remained relatively stationary for extended periods of time, producing sandy east–west-trending belts in the Maastrichtian interval. Limited evidence indicates that a major fluvial system flowed southeast across the northwest part of the Greater Green River basin somewhere near Jonah field during the Maastrichtian. Exceptions would be near rising Laramide uplifts, where drainage systems would have flowed away from these uplifts, but these local drainages may have been quite small and of limited extent. In addition, these local drainages were commonly choked with fine-grained muds that form poor reservoir rocks. The source of these muds was the thick Cretaceous shale intervals that were being eroded off Laramide uplifts during the initial stages of uplift (Flores et al., 1994; Flores, 2003). In the Wind River basin, a mud-dominated fluvial facies occurs in the Maastrichtian Lance and Paleocene Fort Union formations just south of the Owl Creek Mountains (Johnson et al., 1994a). The source of these muds may have been reworked Cretaceous marine shales eroded off of the Owl Creek Mountains. A similar mud facies may be present in the Greater Green River basin and adjacent parts of the Wind River Range.

The thickness of Maastrichtian strata in the central Rocky Mountain region is variable and is related to a combination of variations in rates of subsidence during the Maastrichtian and later uplift and erosion along rising Laramide uplifts. Maastrichtian strata thicken toward thrust fault-bounded Laramide uplifts in many basins, including the Greater Green River basin. The Maastrichtian interval is about 2500–3000 ft (760–900 m) thick in the Jonah field area and thickens to more than 5000 ft (1500 m) to the northeast, adjacent to the Wind River Range. Southwest of Jonah, the Maastrichtian interval thins to a wedge edge on the La Barge platform and is thin throughout much of the western part of the Greater Green River basin.

Coaly source rocks are widespread in Maastrichtian-age rocks in the central Rocky Mountain region but are largely absent from Maastrichtian strata at Jonah field, where vertical migration of gas from deeper source rocks probably largely charged the reservoirs at Jonah field. An understanding of the distribution of coaly source rocks in the underlying Campanian Rock Springs section might contribute to the finding of sweet spots like Jonah in Maastrichtian rocks, not only in the Greater Green River basin, but elsewhere in the central Rocky Mountain region. In summary, an extensive knowledge of underlying and overlying strata will be required to find more Jonah-type fields.

REFERENCES CITED

Armstrong, F. C., and S. Oriel, 1965, Tectonic development of Idaho–Wyoming thrust belt in Nevada and Utah: AAPG Bulletin, v. 49, p. 1847–1866.

Beaumont, C., 1981, Foreland basins: Geophysical Journal of the Royal Astronomical Society, v. 55, p. 291–329.

Beck, R. A., 1985, Syntectonic sedimentation adjacent to Laramide basement thrusts, Rocky Mountain foreland; timing of deformation: M.S. thesis, Ames, Iowa State University, 88 p.

Beck, R. A., C. F. Vondra, J. E. Filkins, and J. D. Olander, 1988, Syntectonic sedimentation and Laramide basement thrusting, Cordilleran foreland; timing of deformation, *in* C. J. Schmidt and W. J. Perry, Jr., eds., Interaction of Rocky Mountain foreland and the Cordilleran thrust belt: Geological Society of America Memoir 171, p. 465–487.

Bohor, B. F., D. M. Triplehorn, D. J. Nichols, and H. T. Millard, Jr., 1987, Dinosaurs, spherules, and the "magic" layer— A new K-T boundary clay site in Wyoming: Geology, v. 15, p. 896–899.

Bucurel, H. G., 1981, Subsurface correlations of some Upper Cretaceous and Tertiary rocks, Great Divide basin, Wyoming, section A–A′: U.S. Geological Survey Open-file Report 81-981, 2 sheets.

Bucurel, H. G., 1982, Subsurface correlations of some Upper Cretaceous and Tertiary rocks, Great Divide basin, Wyoming, section B–B′: U.S. Geological Survey Open-file Report 82-456, 2 sheets.

Bucurel-White, H., 1983, Surface and subsurface correlations of some Upper Cretaceous and Tertiary rocks from the northeast flank of the Rock Springs uplift to the Rawlins uplift, Great Divide basin, Wyoming, section C–C′: U.S. Geological Survey Open-file Report 83-418, 2 sheets.

Connor, C. W., 1992, The Lance Formation— Petrology and stratigraphy, Powder River basin and nearby basins, Wyoming and Montana: U.S. Geological Survey Bulletin, v. 1917-I, 17 p.

DeCelles, P. G., 1994, Late Cretaceous–Paleogene synorogenic sedimentation and kinematic history of the Sevier thrust belt, northeast Utah and southwest Wyoming: Geological Society of America Bulletin, v. 106, p. 32–56.

Dodge, H. W., and J. D. Powell, 1975, Stratigraphic and paleoenvironment data for the uranium-bearing Lance and Fox Hills formations, Crook and northern Weston Counties, northeast Wyoming: U.S. Geological Survey Open-file Report 75-502, 21 p.

Finn, T. M., 1993, Subsurface stratigraphic cross section of Lower and Upper Cretaceous rocks in the southeastern Wind River basin, Wyoming, *in* W. R. Keefer, W. J. Metzger, and L. H. Godwin, eds., Oil and gas and other resources of the Wind River basin, Wyoming: Wyoming Geological Association Special Symposium, Casper, Wyoming, p. 91–94.

Flores, R. M., 2003, Paleocene paleogeographic, paleotectonic, and paleoclimatic patterns of the northern Rocky Mountains and Great Plains region, *in* R. G. Reynolds and R. M. Flores, eds., Cenozoic systems of the Rocky Mountain region: SEPM Rocky Mountain Section, p. 63–106.

Flores, R. M., S. B. Roberts, and W. J. Perry, 1994, Paleocene paleogeography of the Wind River, Bighorn, and Powder River basins, Wyoming, *in* R. M. Flores, K. T. Mehring, R. W. Jones, and T. L. Beck, eds., Organics and the Rockies field guide: Wyoming Geological Survey Public Information Circular 33, p. 1–16.

Francyzk, K. J., J. K. Pitman, and D. J. Nichols, 1990, Sedimentology, mineralogy, palynology, and depositional history of some uppermost Cretaceous and lowermost Tertiary rocks along the Utah Book and Roan Cliffs east of the Green River: U.S. Geological Survey Bulletin, v. 1787-N, 27 p.

Gill, J. R., and W. A. Cobban, 1966, Regional unconformity in Late Cretaceous, Wyoming: U.S. Geological Survey Professional Paper 550-B, p. B20–B27.

Gill, J. R., and W. A. Cobban, 1973, Stratigraphy and geologic history of the Montana Group and equivalent rocks, Montana, Wyoming, and North and South Dakota: U.S. Geological Survey Professional Paper 776, 37 p.

Gill, J. R., and W. J. Hail, Jr., 1975, Stratigraphic sections across Upper Cretaceous Mancos Shale–Mesaverde Group boundary, eastern Utah and western Colorado: U.S. Geological Survey Oil and Gas Investigation Chart OC-68.

Gill, J. R., E. A. Merewether, and W. A. Cobban, 1970, Stratigraphy and nomenclature of some Upper Cretaceous and lower Tertiary rocks in south-central Wyoming: U.S. Geological Survey Professional Paper 667, 53 p.

Gillespie, J. M., and J. E. Fox, 1991, Tectonically influenced sedimentation in the Lance Formation, eastern Wind River basin: The Mountain Geologist, v. 28, no. 2–3, p. 53–66.

Gries, R., 1983, Oil and gas prospecting beneath Precambrian of foreland thrust plates in the Rocky Mountains: AAPG Bulletin, v. 67, p. 1–28.

Gries, R., J. C. Dolson, and R. G. Raynolds, 1992, Structural and stratigraphic evolution and hydrocarbon distribution, Rocky Mountains foreland, *in* R. W. Macqueen and D. A. Leckie, eds., Foreland basins and fold belts: AAPG Memoir 55, p. 395–425.

Gurnis, M., 1992, Rapid continental subsidence following the initiation and evolution of subduction: Science, v. 255, p. 1556–1558.

Hamilton, W. B., 1988, Laramide crustal shortening: Geological Society of America Memoir 171, p. 27–39.

Hettinger, R. D., and M. A. Kirschbaum, 1991, Chart showing correlations of some Upper Cretaceous and lower Tertiary rocks, from the east flank of the Washakie basin to the east flank of the Rock Springs Uplift, Wyoming: U.S. Geological Survey Miscellaneous Investigations Map I-2152.

Hettinger, R. D., J. G. Honey, and D. J. Nichols, 1991, Chart showing correlations of Upper Cretaceous Fox Hills Sandstone and Lance Formation, and lower Tertiary Fort Union, Wasatch, and Green River formations, from the eastern flank of the Washakie basin to the southeastern part of the Great Divide basin, Wyoming: U.S. Geological Survey Miscellaneous Field Investigations Map MF-2151.

Honey, J. G., and R. D. Hettinger, 1989, Cross section showing correlations of Upper Cretaceous Fox Hills Sandstone and Lance formations, and lower Tertiary Fort Union and Wasatch formations, southeastern Washakie basin, Wyoming, and eastern Sand Wash basin, Colorado: U.S. Geological Survey Miscellaneous Investigations Map I-1964.

Johnson, R. C., 1989, Geologic history and hydrocarbon potential of Late Cretaceous–age, low-permeability reservoirs, Piceance basin, western Colorado: U.S. Geological Survey Bulletin, v. 1787-E, 51 p.

Johnson R. C., 2003, Depositional framework of the Upper Cretaceous Mancos shale and lower part of the Upper Cretaceous Mesaverde Group, western Colorado and eastern Utah: U.S. Geological Survey Digital Data Series 69-B, chapter 10, 24 p.

Johnson, R. C., and C. W. Keighin, 1998, Origins of natural gases from Upper Cretaceous reservoirs, Bighorn basin, Wyoming and Montana, and comparison with gases from the Wind River basin, Wyoming, *in* W. R. Keefer and J. E. Goolsby, eds., Cretaceous and lower Tertiary rocks of the Bighorn basin, Wyoming and Montana: Wyoming Geological Association Forty-ninth Guidebook, p. 233–249.

Johnson, R. C., and F. May, 1978, Preliminary stratigraphic studies of the Mesaverde Group, the Wasatch Formation, and the lower part of the Green River Formation, DeBeque area, Colorado, including environments of deposition and investigations of palymomorphs assemblages: U.S. Geological Survey Miscellaneous Field Investigations Map MF 1050.

Johnson, R. C., and F. May, 1980, A study of the Cretaceous–Tertiary unconformity in the Piceance Creek basin, Colorado: The underlying Ohio Creek Formation (Upper Cretaceous) redefined as a member of the Hunter Canyon or Mesaverde Formation: U.S. Geological Survey Bulletin, v. 1482-B, 27 p.

Johnson, R. C., and D. D. Rice, 1990, Occurrence and geochemistry of natural gases, Piceance basin, northwest Colorado: AAPG Bulletin, v. 74, p. 805–829.

Johnson, R. C., and D. D. Rice, 1993, Variations in composition and origins of gases from coal bed and conventional reservoirs, Wind River basin, Wyoming, *in* W. R. Keefer, W. J. Metzger, and L. H. Godwin, eds., Oil and gas and other resources of the Wind River basin, Wyoming: Wyoming Geological Association Special Symposium Volume, p. 319–335.

Johnson R. C., and L. N. Roberts, 2003, Depths to selected stratigraphic horizons in oil and gas wells for Upper Cretaceous and lower Tertiary strata of the Uinta basin, Utah: U.S. Geological Survey Digital Data Series 69-B, chapter 13, 30 p.

Johnson, R. C., R. M. Flores, R. J. Szmajter, and T. M. Finn, 1994a, A preliminary study of coal-forming environments during deposition of the Paleocene Fort Union Formation, Wind River basin, Wyoming, *in* R. M. Flores, K. T. Mehring, R. W. Jones, and L. T. Beck, eds., Organics and the Rockies field guide: Wyoming Geological Survey Public Information Circular 23, p. 69–81.

Johnson, R. C., R. M. Flores, R. J. Szmajter, and T. M. Finn, 1994b, A preliminary study of coal-forming environments during deposition of the Paleocene Fort Union Formation, Wind River basin, Wyoming, *in* R. M. Flores, K. T. Mehring, R. W. Jones, and T. L. Beck, eds., Organics and the Rockies field guide: Wyoming Geological Survey Public Information Circular 23, p. 69–81.

Johnson, R. C., T. M. Finn, W. R. Keefer, and R. J. Szmajter, 1996, Geology of Upper Cretaceous and Paleocene gas-bearing rocks, Wind River basin, Wyoming: U.S. Geological Survey Open-file Report 96-090, 120 p.

Johnson, R. C., W. R. Keefer, C. W. Keighin, and T. M. Finn, 1998, Detailed outcrop studies of the upper part of the Upper Cretaceous Cody Shale and the Upper Cretaceous Mesaverde, Meeteetse, and Lance formations, Bighorn basin, Wyoming, *in* W. R. Keefer and J. E. Goolsby, eds., Cretaceous and lower Tertiary rocks of the Bighorn basin, Wyoming and Montana: Wyoming Geological Association Forty-Ninth Guidebook, p. 59–78.

Jordan, T. E., 1981, Thrust loads and foreland basin evolution, Cretaceous, western United States: AAPG Bulletin, v. 65, p. 2506–2520.

Keefer, W. R., 1970, Structural geology of the Wind River basin, Wyoming: U.S. Geological Survey Professional Paper 495-D, 35 p.

Keefer, W. R., 1972, Frontier, Cody, and Mesaverde formations in the Wind River and southern Bighorn basins: U.S. Geological Survey Professional Paper 495-E, 23 p.

Keefer, W. R., 1997, Stratigraphy and correlation of Cretaceous and Paleocene rocks, northern Wind River basin, Wyoming: U.S. Geological Survey Oil and Gas Investigations Chart OC-146-A.

Keefer, W. R., and J. D. Love, 1963, Laramide vertical movements in central Wyoming: Wyoming University Contributions to Geology, v. 2, no. 1, p. 47–54.

Keefer, W. R., T. M. Finn, R. C. Johnson, and C. W. Keighin, 1998, Regional stratigraphy and correlation of Cretaceous and Paleocene rocks, Bighorn basin, Wyoming and Montana, *in* W. R. Keefer and J. E. Goolsby, eds., Cretaceous and lower Tertiary rocks of the Bighorn basin, Wyoming and Montana: Wyoming Geological Association Forty-Ninth Guidebook, p. 1–30.

Kulik, D. M., and C. J. Schmidt, 1988, Region of overlap and styles of interaction of Cordilleran thrust belt and Rocky Mountain foreland, *in* C. J. Schmidt and W. J. Perry, Jr., eds., Interaction of the Rocky Mountain foreland and the Cordilleran thrust belt: Geological Society of America Memoir 171, p. 75–98.

Law, B. E., and R. C. Johnson, 1989, Structural and stratigraphic framework of the Pinedale anticline, Wyoming, and the multiwell experiment site, Colorado, *in* B. E. Law and C. W. Spencer, eds., Geology of tight gas reservoirs in the Pinedale anticline area, Wyoming, and at the multiwell experiment site, Colorado: U.S. Geological Survey Bulletin, v. 1886, p. B1–B11.

Lindsey, D. A., 1972, Sedimentary petrology and paleocurrents of the Harebell Formation, Pinyon Conglomerate, and associated coarse clastic deposits, northwestern Wyoming: U.S. Geological Survey Professional Paper 734-B, 68 p.

Love, J. D., 1960, Cenozoic sedimentation and crustal movement in Wyoming: American Journal of Science, Bradley Volume, v. 258-A, p. 204–214.

Love, J. D., 1973, Harebell Formation (Upper Cretaceous) and Pinyon conglomerate (uppermost Cretaceous and Paleocene), northwestern Wyoming: U.S. Geological Survey Professional Paper 734-A, 54 p.

Love, J. D., P. O. McGrew, and H. D. Thomas, 1963, Relationship of latest Cretaceous and Tertiary deposition and deformation to oil and gas in Wyoming, *in* The backbone of the Americas— Tectonic history from pole to pole: AAPG Memoir 2, p. p. 196–208.

Markochick, D. J., B. E. Law, and C. W. Spencer, 1982, Preliminary subsurface correlations of some Cretaceous and Tertiary rocks from Moxa arch to Rock Springs uplift, Green River basin, Wyoming, section E–E′: U.S. Geological Survey Open-file Report 82-455, 2 sheets.

Merewether, E. A., J. C. Dolson, W. B. Hanson, W. R. Keefer, B. E. Law, R. E. Mueller, T. A. Ryer, A. C. Smith, D. P. Stilwell, and D. M. Wheeler, 1997, Cretaceous stratigraphy in a northeast-trending transect, northern Utah to south-central South Dakota: U.S. Geological Survey Geologic Investigations Map I-2609.

Mitrovica, J. X., C. Beaumont, and G. T. Jarvis, 1989, Tilting of continental interiors by the dynamic effects of subduction, Tectonics, v. 8, p. 1079–1094.

Molenaar, C. M., W. A. Cobban, E. A. Merewether, C. L. Pillmore, D. G. Wolfe, and J. M. Holbrook, 2002, Regional stratigraphic cross sections of Cretaceous rocks from east-central Arizona to the Oklahoma panhandle: U.S. Geological Survey Miscellaneous Field Studies Map MF-2382, 3 figures.

Nichols, D. J., 1994, Palynostratigraphic correlation of Paleocene rocks in the Wind River, Bighorn, and Powder River basins, *in* R. M. Flores, R. W. Mehring, and T. L. Beck, eds., Organics and the Rockies field guide: Wyoming Geological Survey Public Information Circular 33, p. 17–29.

Nichols, D. J., and H. L. Ott, 1978, Biostratigraphy and evolution of the *Mimipities-Caryapollenites* lineage in the early Tertiary in the Wind River basin, Wyoming: Palynology, v. 2, p. 93–112.

Nichols, D. J., J. L. Brown, M. Attrep, Jr., and C. J. Orth, 1992, A new Cretaceous–Tertiary boundary locality in the western Powder River basin, Wyoming: Biological and geological implications: Cretaceous Research, v. 13, p. 3–30.

Olsen, T., R. Steel, K. Hogseth, T. Skar, and S.-L. Roe, 1995, Sequential architecture in a fluvial succession: Sequence stratigraphy in the Upper Cretaceous Mesaverde Group, Price Canyon, Utah: Journal of Sedimentary Research, v. B65, p. 265–280.

Price, R. A., 1973, Large-scale gravitational flow of supracrustal rocks, southern Canadian Rockies, *in* K. A. Delong and R. Scholten, eds., Gravity and tectonics: New York, John Wiley and Sons, p. 491–502.

Pysklywec, R. N., and J. X. Mitrovica, 1998. Mantle flow mechanisms for the large-scale subsidence of continental interiors: Geology, v. 26, no. 8, p. 687–690.

Reynolds, M. W., 1976, Influence of recurrent Laramide structural growth on sedimentation and petroleum accumulation, Lost Soldier area, Wyoming: AAPG Bulletin, v. 60, p. 12–33.

Roberts, A. E., 1972, Cretaceous and early Tertiary depositional and tectonic history of the Livingston area, southwestern Montana: U.S. Geological Survey Professional Paper 526-C, 120 p.

Roberts, L. N. R., and M. A. Kirschbaum, 1995, Paleogeography of the Late Cretaceous of the western interior of middle North America— Coal distribution and sediment accumulation: U.S. Geological Survey Professional Paper 1561, 115 p.

Roehler, H. W., 1987, Surface-subsurface correlations of the Mesaverde Group and associated Upper Cretaceous formations, Rock Springs, Wyoming, to Mount Harris, Colorado: U.S. Geological Survey Miscellaneous Field Studies Map MF-1937.

Roehler, H. W., 1990, Stratigraphy of the Mesaverde Group in the central and eastern Greater Green River basin, Wyoming, Colorado, and Utah: U.S. Geological Survey Professional Paper 1508, 52 p.

Seeland, D, 1998, Late Cretaceous, Paleocene, and early Eocene paleogeography of the Bighorn basin and northwestern Wyoming, *in* W. R. Keefer and J. E. Goolsby, eds., Cretaceous and lower Tertiary rocks of the Bighorn basin, Wyoming and Montana: Wyoming Geological Association Forty-Ninth Guidebook, p. 137–165.

Stern, T. A., and W. E. Holt, 1994, Platform subsidence behind an active subduction zone: Nature, v. 368, p. 233–236.

Tweto, O., 1975, Laramide (Late Cretaceous–early Tertiary) orogeny in the southern Rocky Mountains, *in* B. F. Curtis, ed., Cenozoic History of the southern Rocky Mountains: Geological Society of America Memoir 144, p. 1–44.

Tyler, T. F., 1978, Preliminary chart showing electric log correlation section A–A′ of some Upper Cretaceous and Tertiary rocks, Washakie basin, Wyoming: U.S. Geological Survey Open-file Report 78-703, 4 sheets.

Tyler, T. F., 1979a, Preliminary chart showing electric log correlation section D–D′ of some Upper Cretaceous and Tertiary rocks, Rock Springs uplift, Wyoming: U.S. Geological Survey Open-file Report 79-1217, 4 sheets.

Tyler, T. F., 1979b, Preliminary chart showing electric log correlation section F–F′ of some Upper Cretaceous and Tertiary rocks, Rock Springs uplift, Wyoming: U.S. Geological Survey Open-file Report 79-1573, 4 sheets.

Tyler, T. F., 1980a, Preliminary chart showing electric log correlation section D–D′ of some Upper Cretaceous and Tertiary rocks, Cherokee Ridge, Wyoming: U.S. Geological Survey Open-file Report 80-191, 4 sheets.

Tyler, T. F., 1980b, Preliminary chart showing electric log correlation section H–H′ of some Upper Cretaceous and Tertiary rocks, south end, Rock Springs Uplift, Wyoming: U.S. Geological Survey Open-file Report 80-1248, 3 sheets.

Tyler, T. F., H. Bucurel-White, and J. R. Peterson, 1981, Preliminary chart showing electric log correlation section J–J′ of some Upper Cretaceous and Tertiary rocks, Washakie basin, Wyoming: U.S. Geological Survey Open-file Report 81-47, 2 sheets.

Tyler, T. F., H. Bucurel-White, and J. R. Peterson, 1982, Preliminary chart showing electric log correlation section I–I′ of some Upper Cretaceous and Tertiary rocks, Washakie basin, Wyoming: U.S. Geological Survey Open-file Report 82-720, 2 sheets.

Warner, E. M., 1998, Structural geology and pressure compartmentalization of Jonah field, Sublette County, Wyoming, *in* R. M. Slatt, ed., Compartmentalized Reservoirs in Rocky Mountain Basins: Rocky Mountain Association of Geologists Symposium, p. 29–46.

Warner, E. M., 2000, Structural geology and pressure compartmentalization of Jonah field based on 3-D seismic data and subsurface geology, Sublette County, Wyoming: The Mountain Geologist, v. 37, p. 15–30.

Wolfe, J. A., and G. A. Izett, 1987, A new Cretaceous–Tertiary boundary section near Teapot dome, western Powder River basin, Wyoming: Eos, Transactions, American Geophysical Union, v. 68, no. 44, p. 1344.

Zapp, A. D., and W. A. Cobban, 1960, Some Late Cretaceous strand lines in northwestern Colorado and northeastern Utah: U.S. Geological Survey Professional Paper 400-B, p. B246–B249.

CHAPTER 5

Geology of Jonah Field, Sublette County, Wyoming

D. P. DuBOIS, P. J. WYNNE, T. M. SMAGALA, J. L. JOHNSON, AND K. D. ENGLER
EnCana Oil & Gas (U.S.A.) Inc., Denver, Colorado, U.S.A.

B. C. McBRIDE
Medicine Bow Energy, Denver, Colorado, U.S.A.

ABSTRACT

Jonah field is located in Sublette County, Wyoming, and lies in the southeastern portion of the Hoback basin, a northwestern extension of the Greater Green River basin. The field is confined by the intersection of two subvertical shear fault zones that form a wedge-shaped structural block. The updip termination at the southwest end of the field is the apex of the block. The downdip limit is somewhat arbitrarily defined as occurring along the synclinal axis separating the basin flank from the Pinedale anticline to the northeast. Within the wedge-shaped block, overpressure conditions exist near the top of the Upper Cretaceous (Maastrichtian) Lance Formation, 2000–3000 ft (610–915 m) above regional occurrence. Immediately to the west and south of the field, overpressure conditions are present near the top of the Upper Cretaceous (Campanian) Mesaverde Group. The trap at Jonah is described as combination structural-stratigraphic. The bounding fault zones form the lateral trap, and the top seal is comprised of the mudstones that are intercalated with the reservoir sandstones of the Lance.

Sandstones in the Lance Formation are the principal reservoir at Jonah field. The Lance Formation is comprised of braided to meandering fluvial sandstones intercalated with overbank siltstones and mudstones. Similar sandstone facies in the upper Mesaverde Group are locally productive. The gross thickness of the Lance Formation increases toward the downdip limit of the field. Near the updip termination, the Lance is 2000 ft (610 m) thick, whereas at the northeastern side of the field, it attains a thickness in excess of 3000 ft (915 m).

Overpressure increases storage capacity and gas saturation in the reservoir and allows for subtle preservation of better porosity relative to sandstones outside the field boundary. Original gas in place in the Lance Formation is estimated to be more than 8.3 tcf. Subcompartments formed by faults inside the field exhibit better per-well recovery near their updip edge; poorer performance is present in downdip regions of each compartment. Pore pressure in each compartment increases by about 1 psi/ft of depth or more than twice the normal hydrostatic gradient. Pressure data suggest that migration of hydrocarbons into the Jonah field compartment is occurring currently. Liquid condensate yield from the gas production increases with depth. Despite high pressures, production from the deepest sandstones tends to be poor because of low permeability and the impact of hydrocarbon liquids on relative permeability.

The lenticular nature of the fluvial sandstones in the Lance has created highly complex reservoir architecture and is a significant challenge to the gas-recovery process. Connectivity is poor, as indicated by the difficulty in correlation of individual sand bodies between wellbores positioned on 40-ac (16-ha) spacing.

INTRODUCTION

The exploitation of Jonah field is a prime example of the successful integration of geology, geophysics, and engineering. From the initial two wells drilled by Home Petroleum in the field boundaries in the late 1980s to present-day production of more than 500 MMCFGD, technical advances have facilitated the economic development of the field. Geophysical work has defined the field limits and internal boundaries, petrophysical analysis

Jonah Field: Case Study of a Giant Tight-Gas Fluvial Reservoir, J. W. Robinson and K. W. Shanley, eds.: AAPG Studies in Geology 52 and Rocky Mountain Association of Geologists 2004 Guidebook

has uncovered parameters that control productivity, and rigorous engineering evaluation has provided the completion practices to optimize the gas-recovery process. The essential ingredient that makes Jonah one of the premier fields in the Rocky Mountain region is the development of effective fracture-stimulation technology to unlock the hydrocarbons from this low-permeability reservoir.

Jonah field covers about 24,000 ac (37 mi^2; 9720 ha) in T28–29N, R107–109W in Sublette County, Wyoming (Figure 1). The field lies between two hydrocarbon-producing areas: the Pinedale anticline to the northeast and the La Barge platform to the southwest. The field is situated in the Hoback basin, the northwestern subbasin of the Greater Green River basin of southwestern Wyoming (Figure 1). This use of the term Hoback basin expands the definition by Dorr et al. (1977) that was based solely on morphology of surface drainage by the Hoback River at the northwestern tip of the Greater Green River basin. In the subsurface, there is good continuity between the Hoback basin of Dorr et al. (1977) and the region to the south of Jonah field.

STRUCTURAL SETTING

Jonah field is bounded by a set of intersecting, subvertical, shear fault zones that delineate a wedge-shaped structural block. The southern limit is called the south-bounding fault (SBF), and the western edge is called the west-bounding fault (WBF). The block widens in the downdip direction to the northeast (Figure. 2). The apex of this wedge-shaped block is positioned at the updip, southwest side of the field. The downdip limit of the field is defined as the synclinal axis that separates the field from the Pinedale anticline to the northeast.

The field-bounding shear faults are complex zones of deformation instead of simple, discrete planes and exhibit complex offset history. Commonly, the current vertical component of offset is opposite that indicated for latest Cretaceous and early Tertiary time. Upper Cretaceous and lower Tertiary strata are thinner outside the Jonah compartment than they are inside (Figure 3), indicating that the Jonah block was a negative feature during the latest Cretaceous and the early Tertiary. Additionally, depositional energy, as represented by thicker sand bodies deposited during the latest Cretaceous, decreases outside the fault-bounded compartment. The present offset across the SBF is down to the south, whereas stratigraphic thickness changes suggest up to the south in the latest Cretaceous and the early Tertiary. Structural inversion of the Jonah compartment during this time is based on changes in thickness inside and outside the compartment, as well as changes in fluvial sandstone architecture (sandstone-stacking patterns and net sandstone content versus gross interval thickness) inside and outside the compartment.

Within the Jonah structural block, several northeast-trending, arcuate shear faults subdivide the field into smaller compartments. Several of the faults exhibit a scissored offset, i.e., the sense of vertical offset changes along its length. The subcompartments exhibit different productivity characteristics that suggest that each fault block is isolated from the adjacent block.

STRATIGRAPHY

Sandstones of the Upper Cretaceous Lance Formation are the principal reservoir at Jonah field. The Upper Cretaceous Mesaverde Group underlies the Lance Formation. The Unnamed Tertiary, Fort Union, and Wasatch formations, in ascending order, overlie the Lance. Figure 4 shows the stratigraphic column from the lower Fort Union through the Blair Formation of the Mesaverde Group.

Mesaverde Group

The Mesaverde is comprised of four recognized units that include, from top to bottom, the upper Mesaverde, the Ericson Sandstone, the Rock Springs Formation, and the Blair sandstone (Figure 4). The overall thickness of the Mesaverde Group in the Yellow Point 11-13 well (NE SW Sec. 13, T28N, R109W) is about 3200 ft (970 m).

The upper Mesaverde consists of mostly siltstone and mudstone with a few intercalated sandstones and is about 600 ft (180 m) thick. The upper Mesaverde at Jonah was deposited in a fluvial, nonmarine setting and is considered to be a lateral equivalent of the Almond sandstones of the eastern Green River basin. The top of this unit is marked by a subtle increase in the conductivity of the mudstones, and drill samples indicate an increase in carbonaceous material when compared to the overlying Lance. This part of the section was considered to be the lowest portion of the Lance Formation by Law and Johnson (1989) in the Wagon Wheel 1 well on the Pinedale anticline. Sandstones of the upper Mesaverde are locally productive in Jonah field.

The Ericson Sandstone is about 250 ft (80 m) thick and consists of well-sorted, subangular to subrounded, upper very fine to medium chert litharenite. The sandstones are cemented by quartz, chlorite kaolinite, illite, and sparse ferroan calcite (Webb, 1999, personal communication).

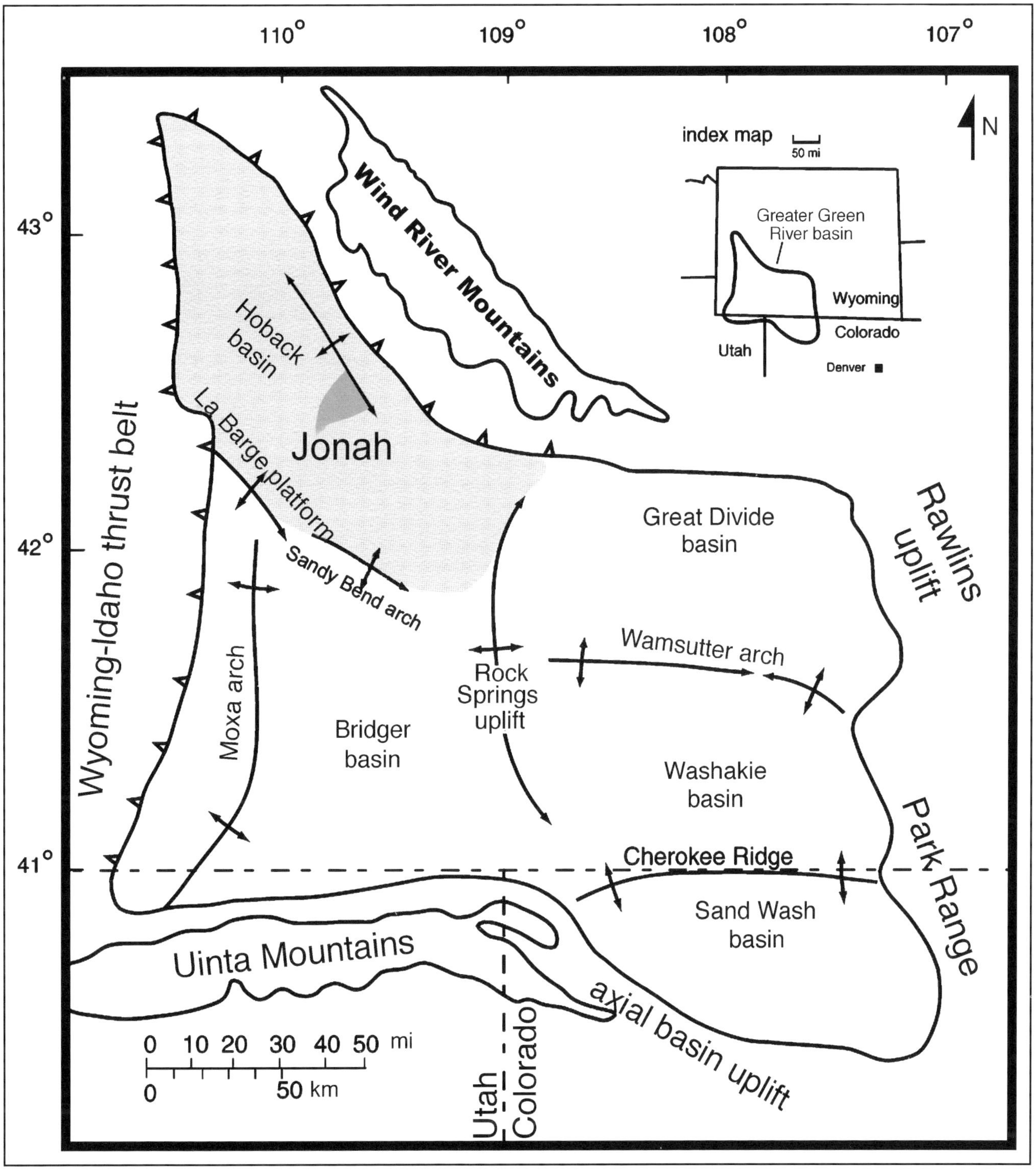

Figure 1. Location map showing the position of the Jonah field in the Greater Green River basin of southwestern Wyoming. Dark shading in the northwestern portion of the Greater Green River basin indicates the area defined as the Hoback basin.

(A)

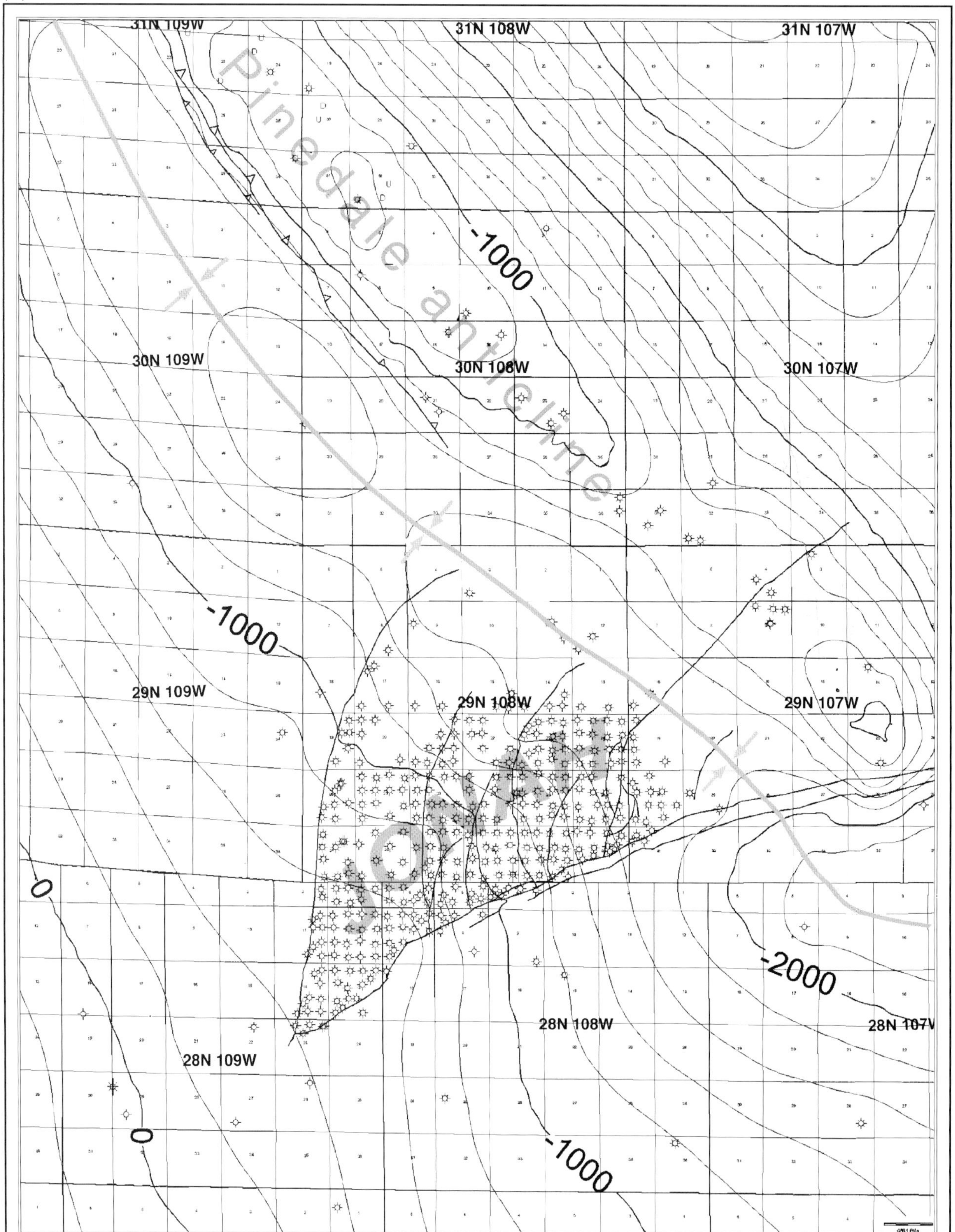
31N 109W
31N 108W
31N 107W
Pinedale anticline
-1000
30N 109W
30N 108W
30N 107W
-1000
29N 109W
29N 108W
29N 107W
JONAH
0
-2000
28N 108W
28N 107W
28N 109W
0
-1000

(B)

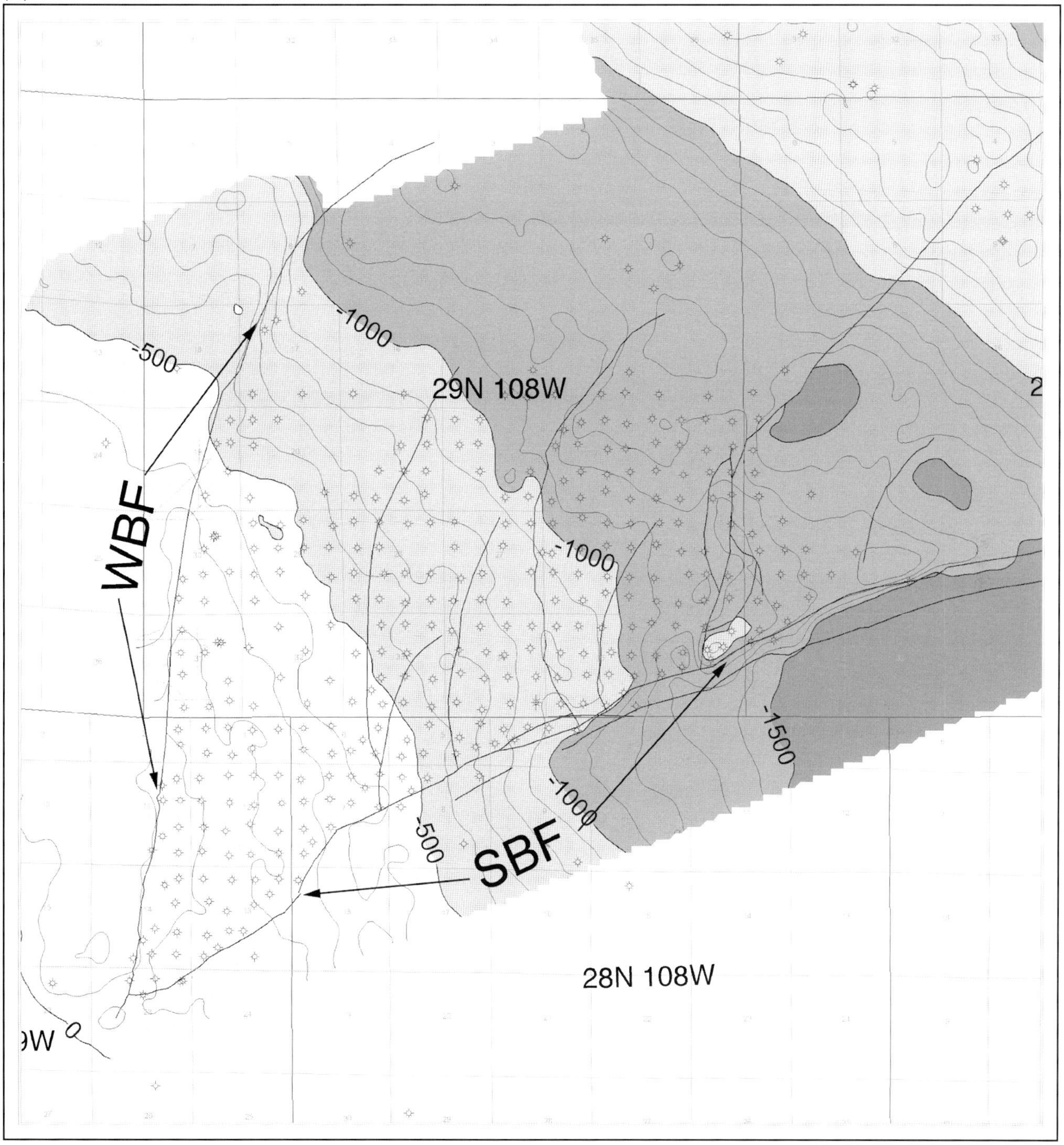

Figure 2. (cont.).

◀

Figure 2. (A) Generalized regional structural contour map on the top of the Lance Formation for an area around Jonah field and a portion of the Pinedale anticline. Synclinal axis that separates the Pinedale anticline from the west flank of the Hoback basin is shown. Contour interval is 200 ft (60 m). (B) Structural contour map on the base of the Tertiary Fort Union Formation. Map generated using a combination of well control and seismic data. WBF = west-bounding fault. SBF = south-bounding fault. Contour interval is 100 ft (30 m).

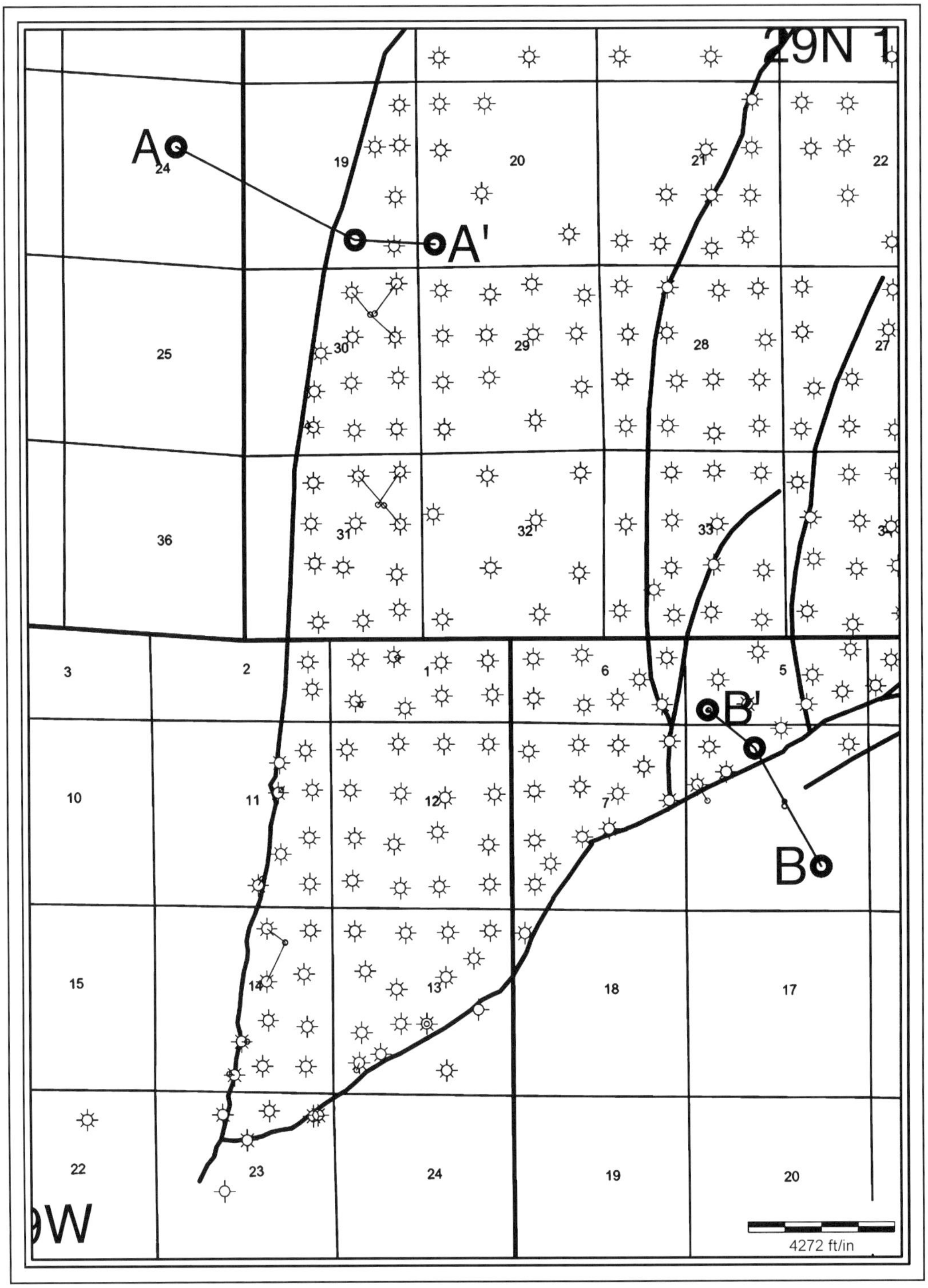

Figure 3. Location map and cross sections across the Jonah field bounding faults. In each cross section, the well on the left is located outside the field, and the well on the right is located inside the field. Datum is the base of the Fort Union Formation. The upper Lance bentonite marker (dashed line below top Lance) is present only in section AA′; its presence verifies the position of the top of the Lance Formation. The top Lance picks in BB′ are less determinate. Logs shown are gamma ray (GR) (left) and deep resistivity (RD) (right). Coals are identified by a combination of bulk density less than 2.0 gm/cm^3, RD greater than 17 ohm-m and GR less than 60 API. Where coals are identified by these log criteria, black horizontal fill between the log curves is shown. Thicknesses of the intervals shown in AA′ and BB′ are shown in Table 1.

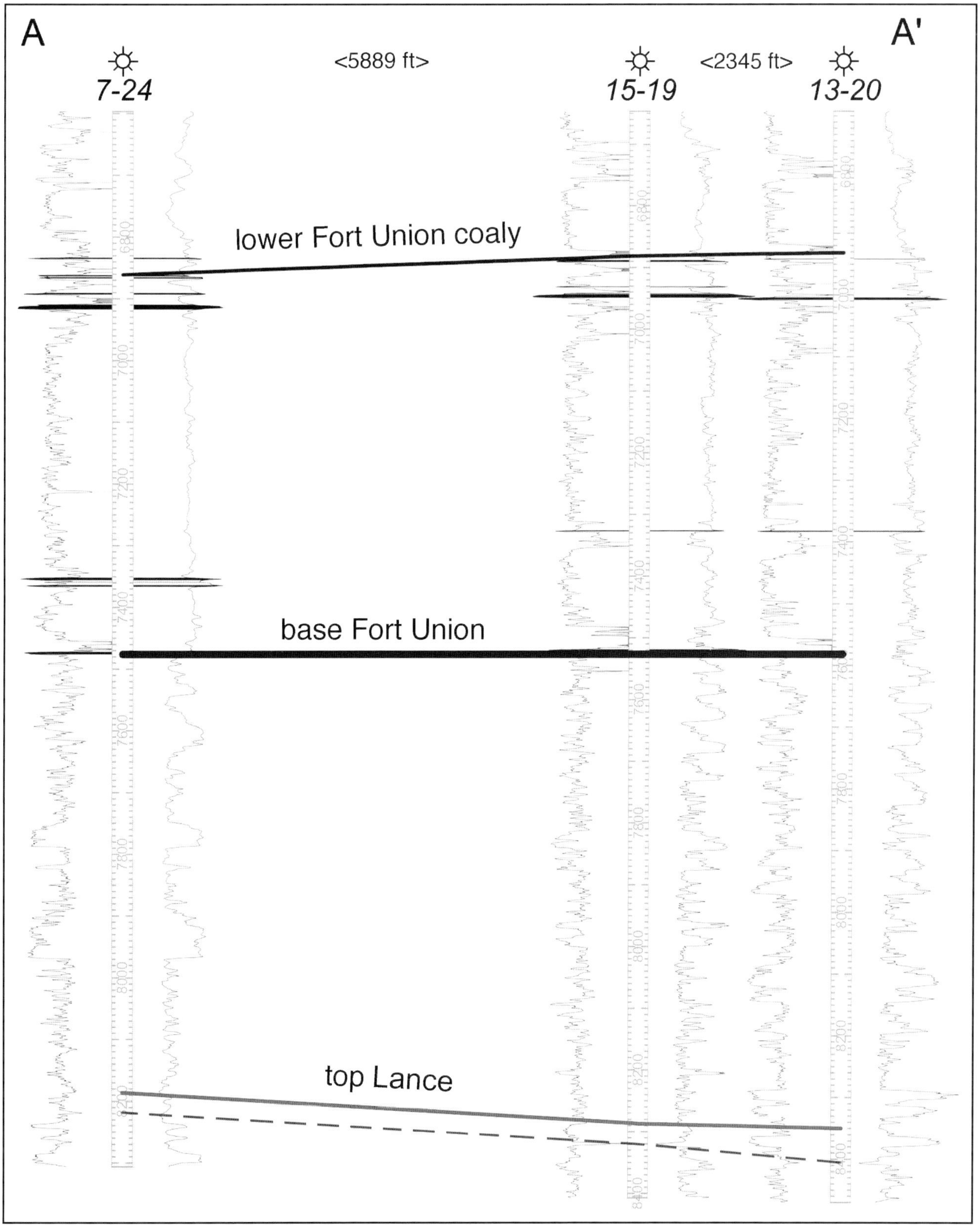

Figure 3. (cont.).

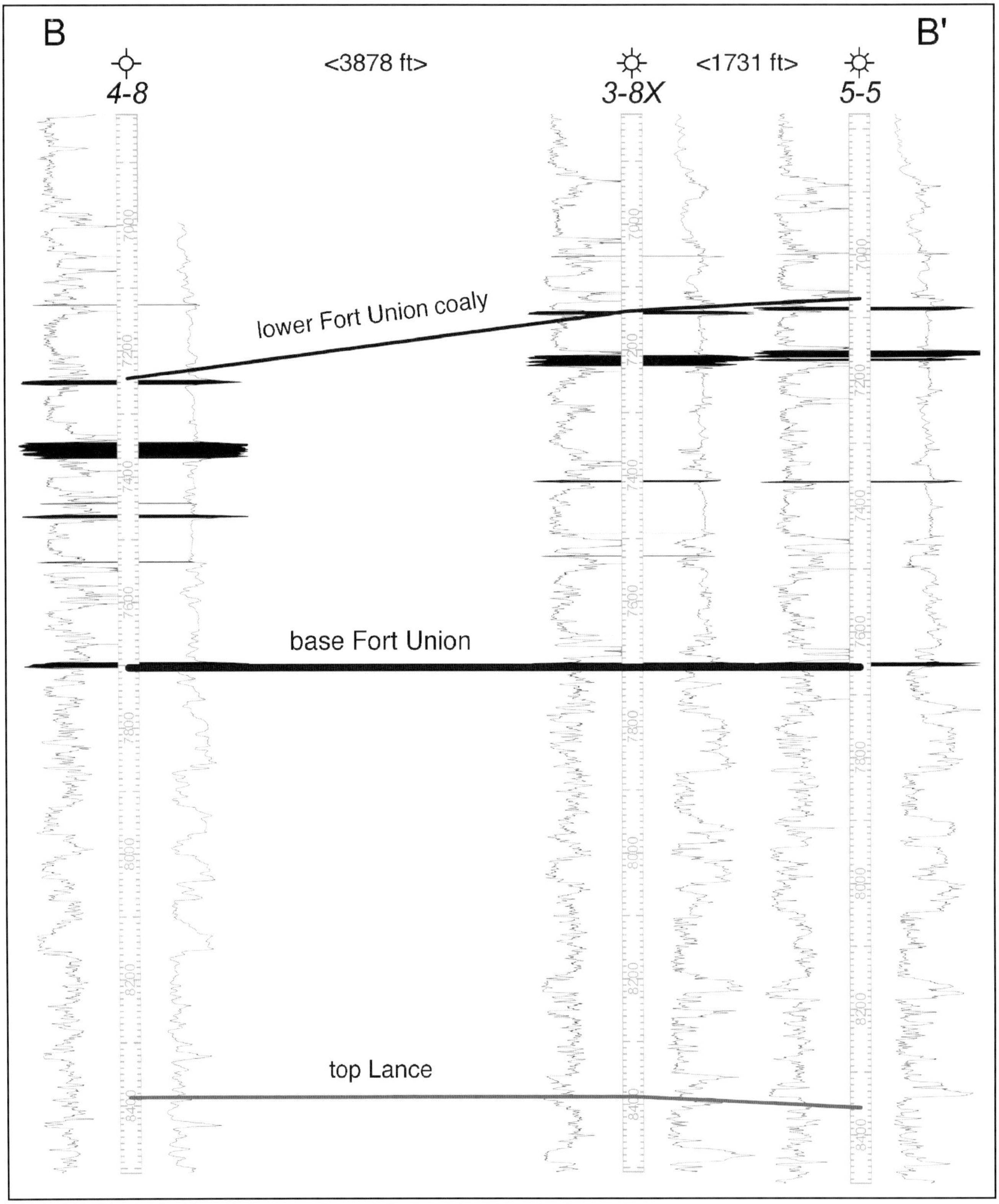

Figure 3. (cont.).

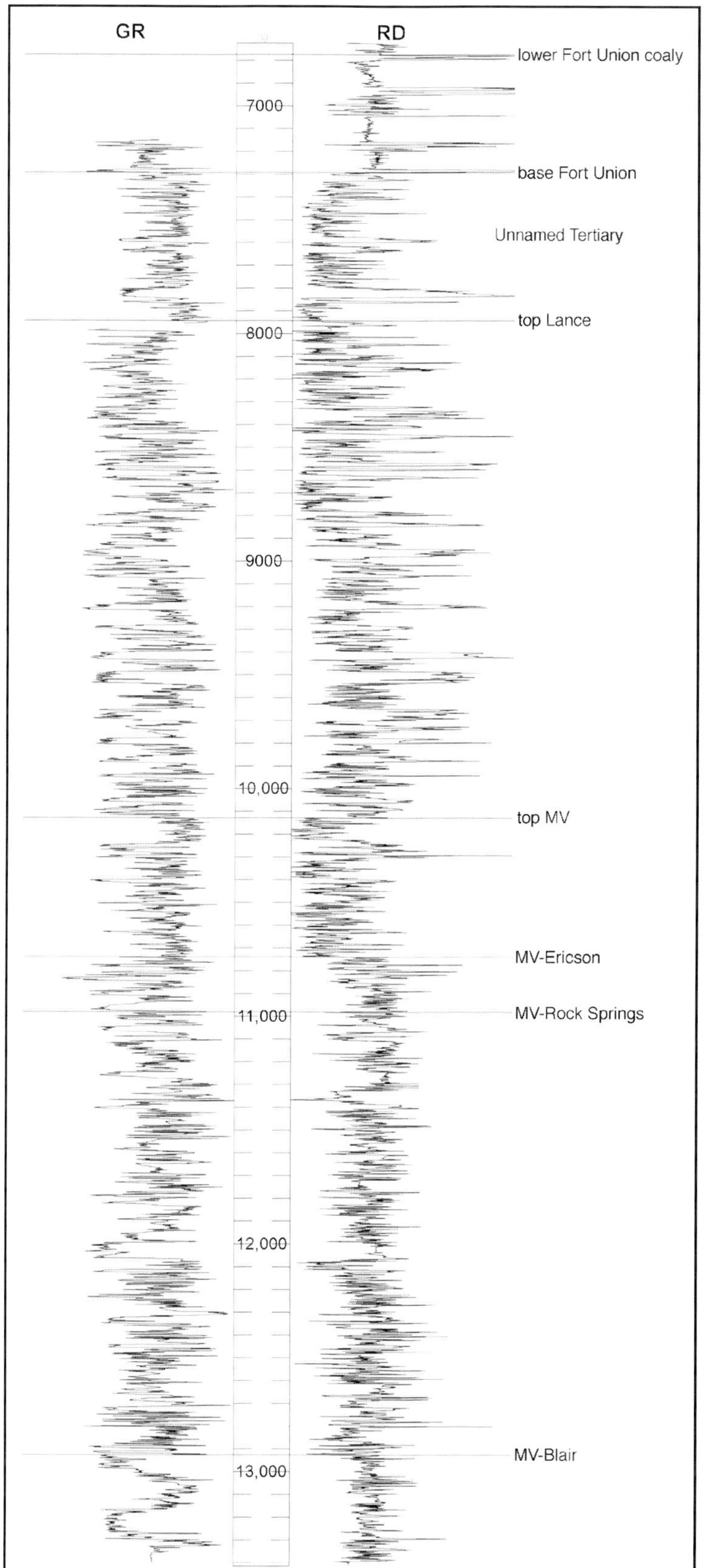

Figure 4. Type log from Jonah field. Well shown is the Yellow Point 11-13 (NE SW Sec. 13, T28N, R109W). Log display shows gamma-ray curve on the left side and deep resistivity on the right. Minor ticks in depth track indicate 100-ft (30-m) intervals. Interval depths of 1000 ft (305 m) are labeled. MV = Mesaverde. The base of the Fort Union equals the top of the Unnamed Tertiary unit.

Table 1

Isopach Values, in Feet, for Wells Shown in Figure 3.

	A		A′	B		B′
Isopach Interval	**Well 7-24**	**Well 15-19**	**Well 13-20**	**Well 4-8**	**Well 3-8X**	**Well 5-5**
Top of Fort Union coaly to base of Fort Union	615	646	650	500	567	586
Base of Fort Union to top of Lance	710	760	768	683	682	701

The Rock Springs Formation is composed of fluvial sandstones, siltstones, and mudstones about 1950 ft (590 m) thick. The sandstones consist of very well-sorted, subangular to subrounded, lower to upper fine-grained litharenite. Cements are quartz, mixed-layer illite-smectite, illite, and sparse ferroan dolomite (Webb, 1999, personal communication).

The Blair sandstone is the deepest stratigraphic unit penetrated within the confines of Jonah field. It consists of moderately well- to well-sorted, subangular to subrounded, upper very fine- to upper fine-grained marine sandstones that overlie the Hilliard (Baxter) Shale. The sandstone is cemented by quartz, mixed-layer illite-smectite, illite, and sparse ferroan dolomite (Webb, 1999, personal communication). In the Yellow Point 11-13, the Blair is about 400 ft (120 m) thick.

Lance Formation

The Lance Formation is an entirely nonmarine sequence comprised of fine-grained sandstones intercalated with siltstones and silty mudstones. No true shales are found in the Lance at Jonah (Brown, 2000, personal communication). During the Late Cretaceous, the Wyoming thrust belt and Wind River Mountains were highlands shedding sediment that was deposited on a broad alluvial plain. The principal direction of sediment transport was from north-northwest to south-southeast, parallel to the paleoaxis of the basin. The source of sediment was Mesozoic and Paleozoic rocks that were exposed in highland areas to the east and west. As a result of this provenance, Lance sandstones are composed primarily of quartz and rock fragments, have only minor amounts of feldspar, and are classified as litharenites (Warner, 2000).

Commonly used subdivisions of the Lance are upper Lance, middle Lance, Jonah, Yellow Point, and Wardell, from top to bottom (Robinson et al., 1996). The complexity of the internal stratigraphy makes the fieldwide use of these subunits exceedingly difficult. Therefore, no attempt was made to characterize the Lance with respect to these subunits.

Unnamed Tertiary

The Unnamed Tertiary (Law and Johnson, 1989) is a transitional unit that overlies the Lance Formation and underlies the Tertiary Fort Union Formation. It ranges in thickness from about 600 ft (180 m) at the updip end of the field to about 1000 ft (300 m) at the downdip margin. It is composed of isolated fluvial sand bodies intercalated with thick sections of nonmarine mudstone. In the upper portion of this unit, the sandstones are commonly arkosic, indicating a change in provenance from Lance time. This change is the result of erosion and unroofing of the Precambrian core of the Wind River uplift (Prensky, 1989).

The Unnamed Tertiary was considered to be the top seal for Jonah field during the early years of development. Gas flares and sustained gas background are earmarks of overpressure and commonly occurred near the transition between the Unnamed Tertiary and the underlying Lance Formation. This chapter presents a different view; the Unnamed Tertiary behaves as a pressure continuum with the underlying Lance Formation.

Fort Union

The thickness of the Fort Union Formation is not defined here because of lack of an obvious and continuous boundary with the overlying Wasatch Formation. In the updip part of the field, the lower portion of the Fort Union is composed of arkosic sandstones intercalated with mudstone and several laterally persistent coal beds. The base of the lowest coal is considered to mark the boundary

Table 2

Compositional Analysis of Gas Produced from the Jonah Federal 1-5 Well (W/2 Section 5, T28N, R108W) (June 25, 1993).

	mol%
Methane	91.557
Ethane	4.956
Propane	1.544
Iso-butane	0.337
n-Butane	0.333
Iso-pentane	0.121
n-Pentane	0.097
Hexanes+	0.079
Nitrogen	0.204
Carbon dioxide	0.772
Total	100
Sample pressure	575 psig
Sample temperature	75°F (23.8°C)
Z factor	0.9976

between the Fort Union and the underlying Unnamed Tertiary. The coal beds pinch out on the downdip side of the field. The trend of the pinch-out extends from Sec. 19, T29N, R108W to Sec. 4, T28N, R108W. Northeast of this pinch-out, the equivalent section of the Fort Union is composed of arkosic sandstones and mudstones and grades into thinly and chaotically bedded muddy sandstones, which is interpreted to have been deposited in an alluvial-fan setting.

HYDROCARBON SOURCE, MIGRATION, AND RESERVOIR CONDITIONS

Hydrocarbon Liquid Yield versus Depth

Hydrocarbons produced from Jonah are rich gas and condensate. Table 2 shows an analysis of the produced gas from the Jonah 1-5 well (W/2 Sec. 5, T28N, R108W). The lower portion of the Lance Formation produces about four times more hydrocarbon liquid relative to the volume of gas than do the upper portions of the formation (Figure 5). Drill-cuttings analysis confirms the increase in hydrocarbon staining and greater amounts of hydrocarbon residue in the lower Lance. These observations suggest that vertical fractionation of liquid-rich hydrocarbons has occurred in the Lance.

Proprietary geochemical analysis performed for BP America indicates that the gas in the Lance is cracked oil (W. Hanson, 2001, personal communication; Hanson et al., 2004). The most likely horizon containing liquid-prone source rocks at sufficient temperature to be undergoing catagenesis is the Hilliard Shale, which underlies the Mesaverde Group. Another possible source rock interval is the Lower Cretaceous Mowry Shale. Average depth to the Frontier Formation (base of the Hilliard) in Jonah field is about 21,000 ft (6400 m). Based on extrapolation of temperature-gradient data from the Lance Formation, it is estimated that the temperature in the Hilliard ranges from about 260 to more than 360°F (112 to more than 168°C) (Figure 6).

Subsurface Reservoir Pressure

Pressure data from wells in the field (Figure 7A) exhibit a good relationship between pressure and depth

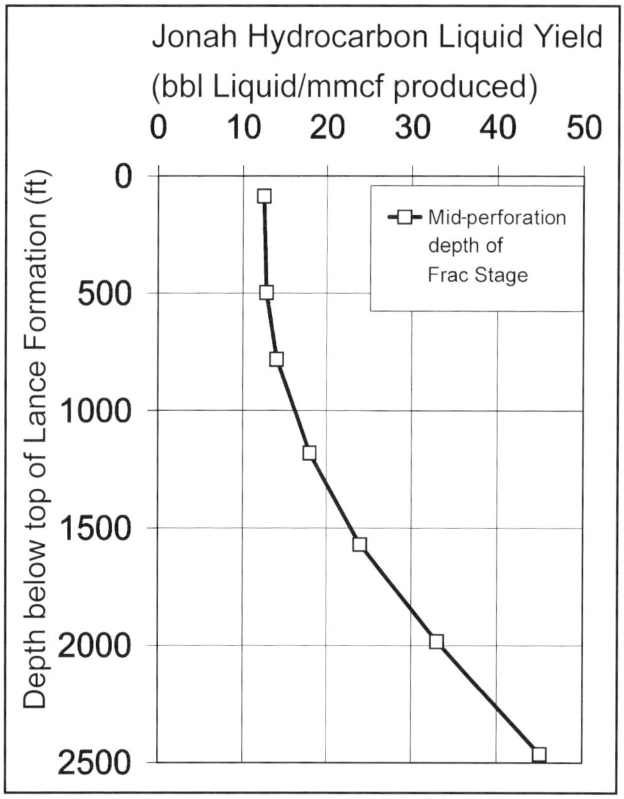

Figure 5. Generalized relationship of liquid hydrocarbon yield versus depth below the top of the Lance Formation for Jonah field. This relationship is observed when each completion interval is isolated from other intervals and is produced for at least several days to clean up the produced fluids following fracture stimulation.

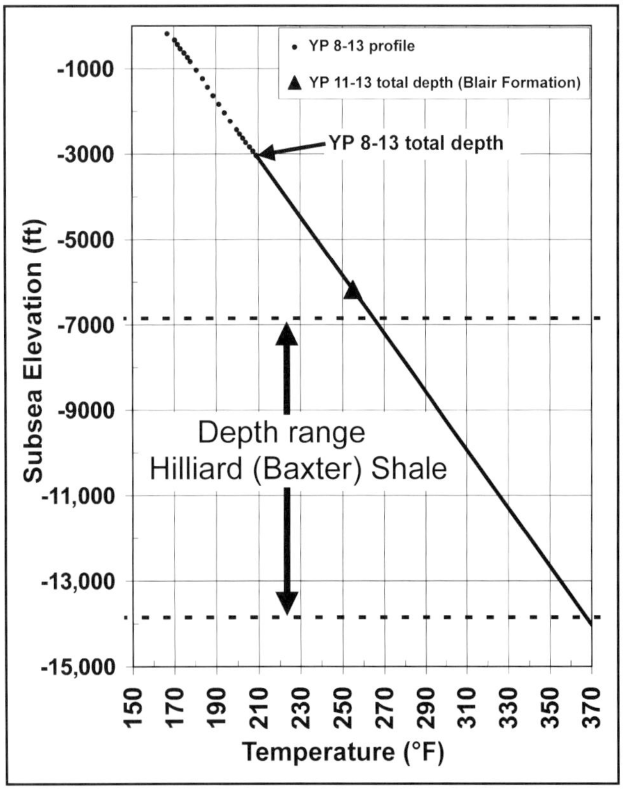

Figure 6. Temperature profile from Jonah field. The upper portion is data acquired inside casing from the Yellow Point 8-13 well (SE NE Sec. 13, T28N, R109W). The data were acquired approximately 20 days after the casing was cemented in place. No circulation of fluids occurred inside the casing between finishing the cementing operation and running the temperature profile. The temperature profile is augmented by a bottomhole data point from the cement bond log run in the Yellow Point 11-13 (NE SW Sec. 13, T28N, R109W), which was run 25 days after cementing operations. The temperature profile is extrapolated below the Yellow Point 11-13 data point. A range of subsea elevations for the Hilliard Shale, a potential source rock, is shown.

(Figure 7B). On the average, the pressure in the Lance Formation increases by about 1.0 psi/ft. This rate of pressure increase is more than two times the normal hydrostatic gradient of 0.433 psi/ft. The high pressure gradient (rate of pressure change with change in depth) in the Lance probably originates from an imbalance between hydrocarbon influx from source rocks below the field and hydrocarbon leak-off above the field. Spencer (1987) indicated that present-day or recently active hydrocarbon generation and migration cause most overpressure conditions in the Rocky Mountain region. The authors consider that present-day migration of hydrocarbons at a rate that exceeds leak-off is the most likely case for Jonah field. Within the Lance, the low permeability in the mudstone portion of the formation controls the overall gradient of more than 1 psi/ft. By contrast, any set of pressure data from a contiguous and interconnected body of sandstone should plot on a gas gradient (about 0.1 psi/ft at reservoir conditions).

Individual pressure data points are commonly expressed as a gradient, where pressure is divided by the depth from surface to the test interval, yielding a pounds-per-square-inch-per-foot value for that point. It would be more accurate to characterize this style of data expression as a pressure-depth ratio, reserving the term "gradient" for a slope determined from multiple, acceptably colinear data points. Figure 7C illustrates that when pressure–depth ratio is plotted against depth below the base of the Fort Union, the data converges on a value of 0.433 at the base of the Fort Union. A pressure–depth ratio of 0.433 is expected when hydrostatic (normal pressure) conditions exist and the potentiometric surface is coincident with the topographic surface. This demonstrates that the top of overpressure is stratigraphically controlled in Jonah field and occurs near the base of Fort Union. This convergence toward hydrostatic conditions, without any abrupt change in pressure, also demonstrates that there is no top seal in the traditional sense of the word. Indeed, the top seal to the hydrocarbon system in Jonah field is the combination of all of the mudstones throughout the upper Mesaverde, Lance, and Unnamed Tertiary formations. The reservoir sandstones at Jonah are intercalated with the top seal. The base of the Fort Union is a key horizon for the pressure structure of Jonah field because the Fort Union is dominated by sandstone, either well-sorted, clean fluvial sandstones deposited in an alluvial-plain setting or poorly sorted, chaotically deposited alluvial-fan deposits. Because of this dominance of sandstone in the Fort Union, there is no laterally persistent capability to restrain vertical migration of hydrocarbons and no capacity to hold overpressure.

Figure 7D shows pressure data that are plotted on the *Y*-axis as depth below base of Fort Union and plotted on the *X*-axis as the difference between the observed pore pressure and calculated hydrostatic pressure at the base of the Fort Union. These data have been separated into several groups with different pressure gradients. Accounting for the structural position of each well in the data set results in a higher pressure gradient and a significantly higher correlation coefficient than was estimated from the data set shown in Figure 7B. The lowest trend (the lowest pressure extrapolation for a given depth) is the central group. This group exhibits a pressure gradient of 1.04 psi/ft. The next higher trend is called the near SBF and exhibits a pressure gradient of 1.2 psi/ft. The highest trend is a data

(A)

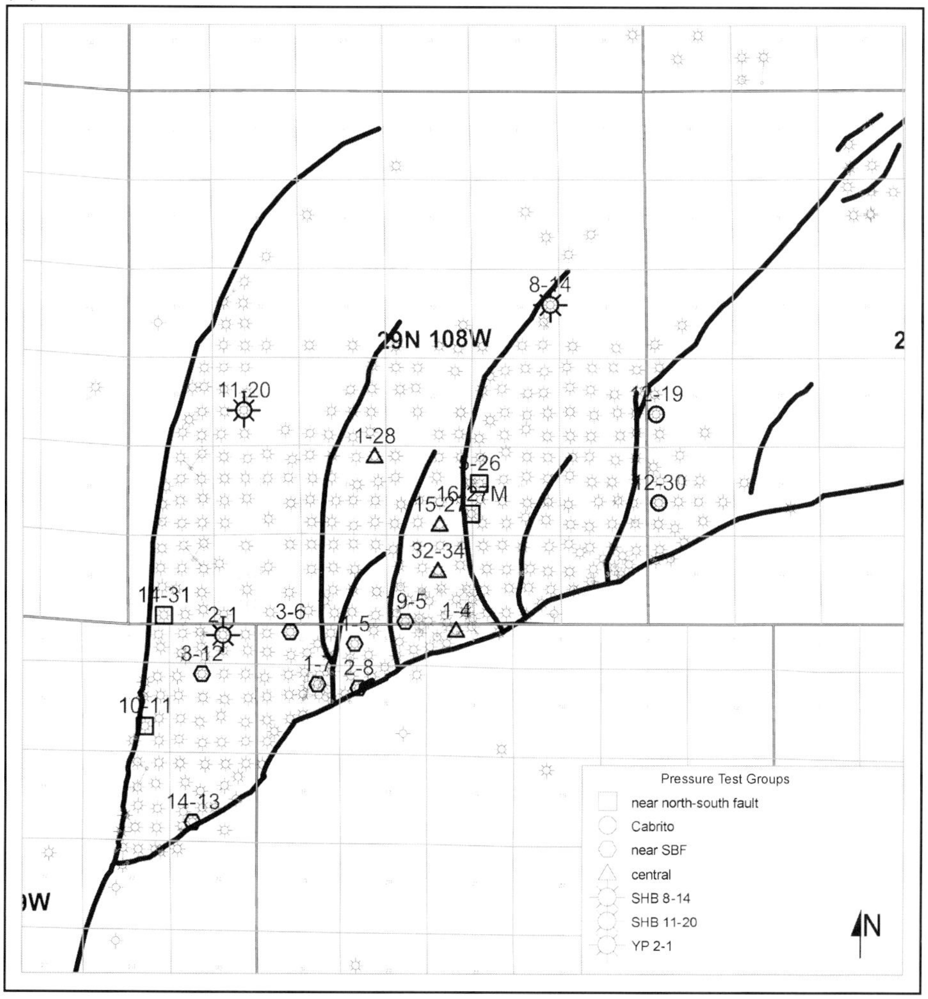

Figure 7. (A) Location map of wells with pressure data. Wells 1-28, 15-27, 32-34, and 1-4 comprise the central group shown in (D). Wells 1-5, 2-8, 1-7, 3-6, and 14-13 comprise the near SBF (south-bounding fault) group shown in (D). Wells 10-11, 14-31, 5-26, and 16-27M comprise the near north–south fault group. Multiple data points from wells 2-1, 11-20, and 8-14 are shown as groups in (D) (cont.).

set from wells located close to the east side of a north–south arcuate fault in the field. The gradient exhibited by data from this group is 1.35 psi/ft. Data from other wells and groups of wells are presented on Figure 7D but are not part of the groups for which correlations have been shown. Comparison of these correlated trends indicates that regions near the north–south-bounding and internal faults, particularly those closer to the intersection with the SBF, exhibit higher pressures for a given stratigraphic depth and higher pressure gradients than regions interior to the field. One possible explanation for these differences is that present-day hydrocarbon migration into the Lance Formation is accentuated in the regions adjacent to, and east of, the bounding faults and interior compartment faults.

Typically, strike-slip fault zones initiate with a series of synthetic and antithetic shears that are ultimately bypassed by the throughgoing master fault (Harding, 1973). This process leaves abandoned ends of the initial shears on both sides of the master fault. These abandoned shears may be localized avenues for vertical hydrocarbon migration into the Lance Formation. The abandoned shears would affect a small region immediately adjacent to the master fault, whether it is one of the external bounding faults or an internal fault.

Permeability versus Depth

Productivity per net foot of reservoir tends to be lower from sandstones in the lower portion of the Lance Formation. This is caused by a reduction in permeability with depth (Figure 8) and the negative effect that greater liquid content has on relative permeability. Lance sandstones at greater depths exhibit greater deformation of the rock

(B)

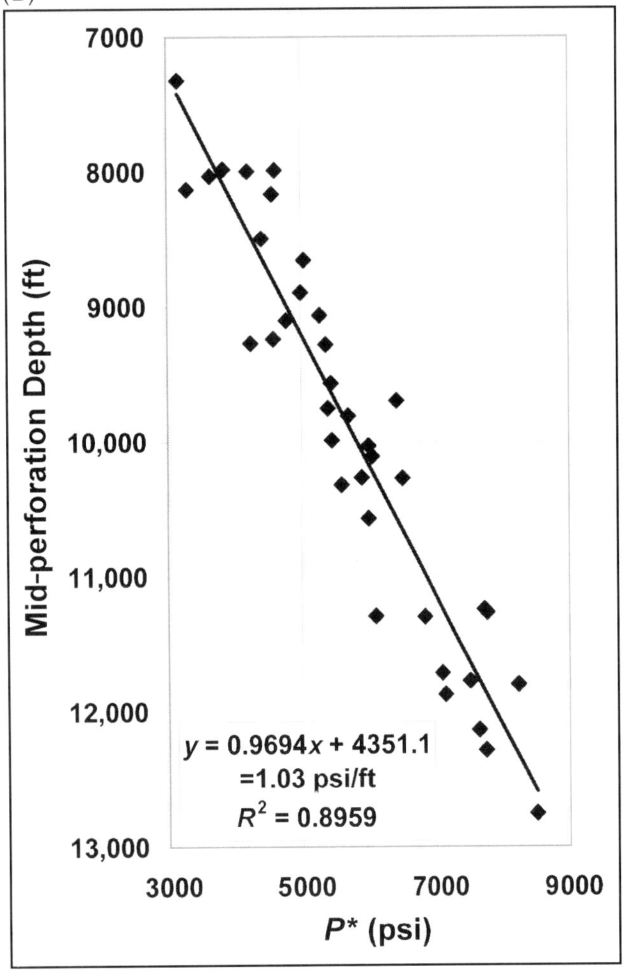

Figure 7. (cont.) (B) Graph of pore pressure (P^*) versus depth for selected wells in Jonah field. The slope of the best-fit line yields an average pressure gradient of 1.03 psi/ft. P^* is the extrapolated stabilized pressure estimated to be the ultimate result of the buildup or fall-off measurements.

fragments that make up a significant fraction of total rock volume (TerraTek, Inc., 1994, personal communication). The reduction in permeability, combined with a lower amount of net pay in the lower Lance, results in lower production rates.

Lateral Variation in Well Performance

Estimated ultimate recovery (EUR) values of individual wells in the field show a consistent pattern. Figure 9 is an EUR contour map with data from wells drilled prior to year 2001. Although there have been many more wells drilled, the pre-2001 group represents a relatively consistent set of well completions. The map pattern demonstrates how internal faults subdivide Jonah field into smaller compartments. Each compartment exhibits increasing productivity in the updip direction, reaching maximum values near the fault that forms the updip limit of that compartment.

Given consistent completion techniques, variation in EUR is controlled by two factors: (1) variation in reservoir volume contacted by a well bore and (2) the quality (permeability) of the reservoir rock. Figure 10A is a net sandstone isopach map (gamma ray < 75 API), and Figure 10B is net pay isopach. Net pay is a subset of net sandstone and is determined by petrophysical calculation, the details of which are not discussed here. Petrophysical cutoffs used to make these net pay summations are as follows: effective porosity greater than 4%, water saturation less than 65%, and bulk volume water (water saturation times porosity) less than 0.04. Figure 11 is an isopach map that shows lateral variation in the average hydrocarbon fraction ($S_g\Phi$) of the net pay. The values are determined by calculating the average gas saturation (S_g) for all the net pay volume and multiplying that number by the average porosity (Φ) determined from the same volume. The resulting value is the fraction of the total net pay volume that contains hydrocarbons. It should be noted that many of the wells in the field do not penetrate the entire Lance section, but the net pay potential of the portion not penetrated is considered small. Figures 10–12 represent data from the top of Lance to the total depth penetrated, or to the base of the Lance, whichever is shallower.

$S_g\Phi$ is a good measure of relative reservoir quality. Reservoir rocks with intergranular porosity typically exhibit a trend of increasing permeability with increasing porosity (Figure 8). This relationship in Jonah has been verified by Yin et al. (2001). Evans (1997, personal communication) found that resistivity could be correlated to permeability in Lance sandstone core samples. High values of resistivity correlated with high permeability. Resistivity is a principal component in the petrophysical calculation of S_g; therefore, higher resistivity values equate to higher S_g. Increases in S_g and/or Φ should reflect rock with higher permeability.

$S_g\Phi$ for Jonah wells is higher when the percentage of clay cement is lower and when grain size and resultant pore throats are larger. Figure 11 shows that there is more that a 50% change in $S_g\Phi$ across the field. The pattern shown in Figure 11 is similar to the EUR and well pattern shown in Figure 9. The reservoir quality, as indicated by $S_g\Phi$, appears to play a significant role in the overall well performance. This relationship is consistent with the lack of fracturing observed among the variety of subsurface data acquired at Jonah. No significant fracturing has been

(C)

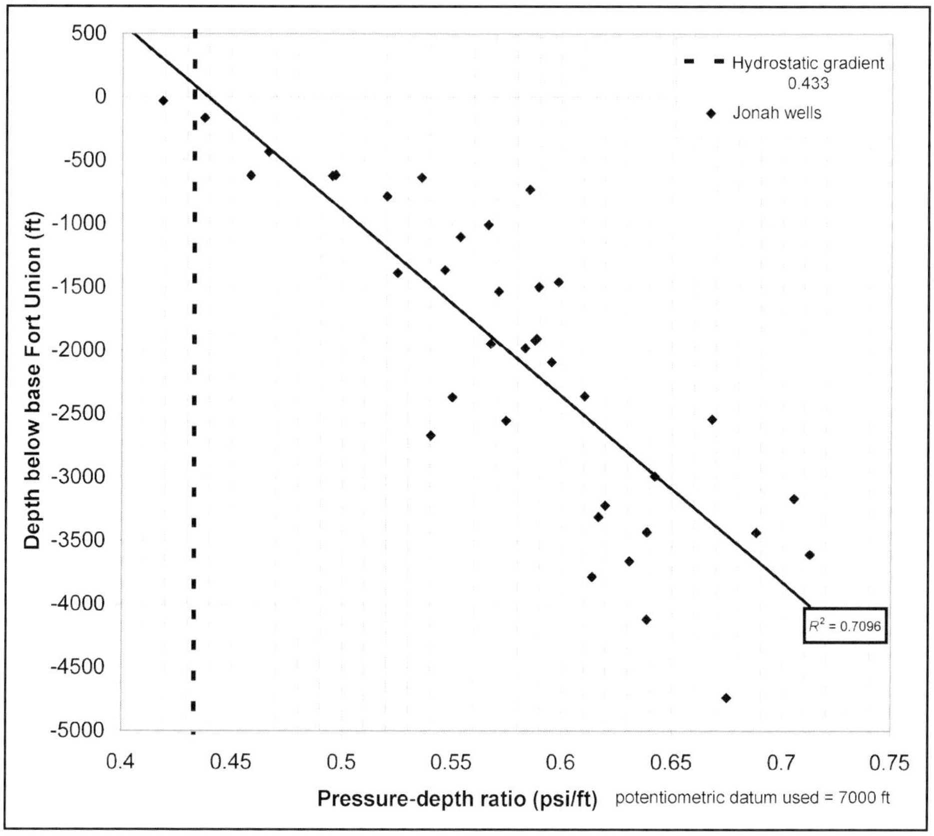

(D)

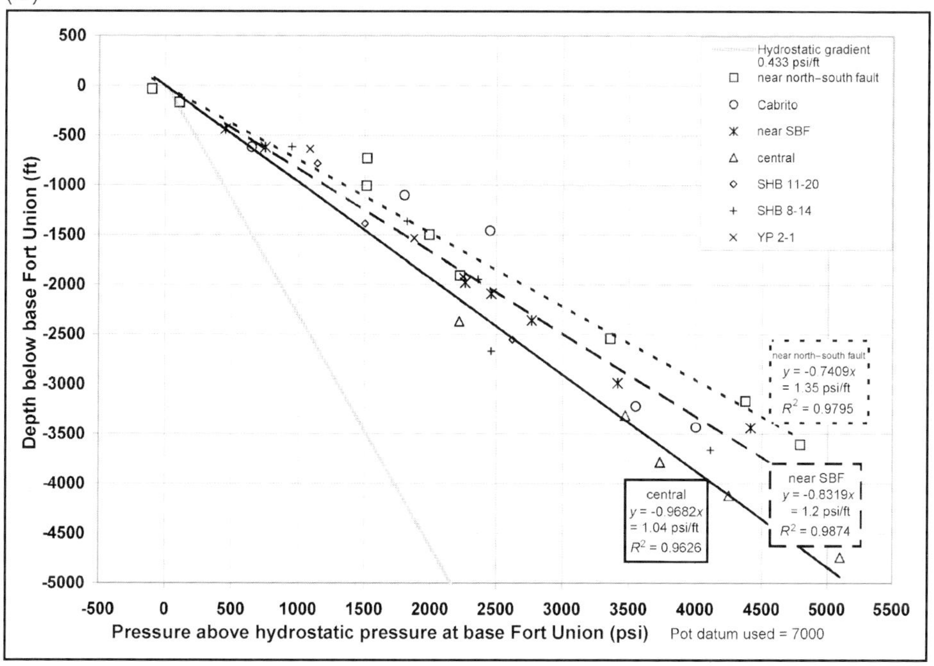

Figure 7. (cont.) (C) Graph of pressure/depth ratio. Pressure/depth ratio (psi/ft) is plotted versus depth below the base of the Fort Union. This graph indicates that pressure-depth ratios for depths below a datum of 7000 ft (2140 m) converge toward hydrostatic conditions as the depth below the base of the Fort Union is reduced. The base of the Fort Union is the depth below which pressures deviate from normal hydrostatic pressure and begin to exhibit overpressured conditions. (D) Graph of pressure data wherein each pressure point is expressed as the difference between the observed pressure and the calculated hydrostatic pressure at the base of the Fort Union (using a potentiometric datum of 7000-ft (2140-m) elevation). These data are graphed versus the depth below the base of Fort Union. The data are separated into groups that are shown in (A). Three of these groups have linear correlations, and the resultant pressure gradient is shown. A standard hydrostatic gradient (0.433 psi/ft) is also shown.

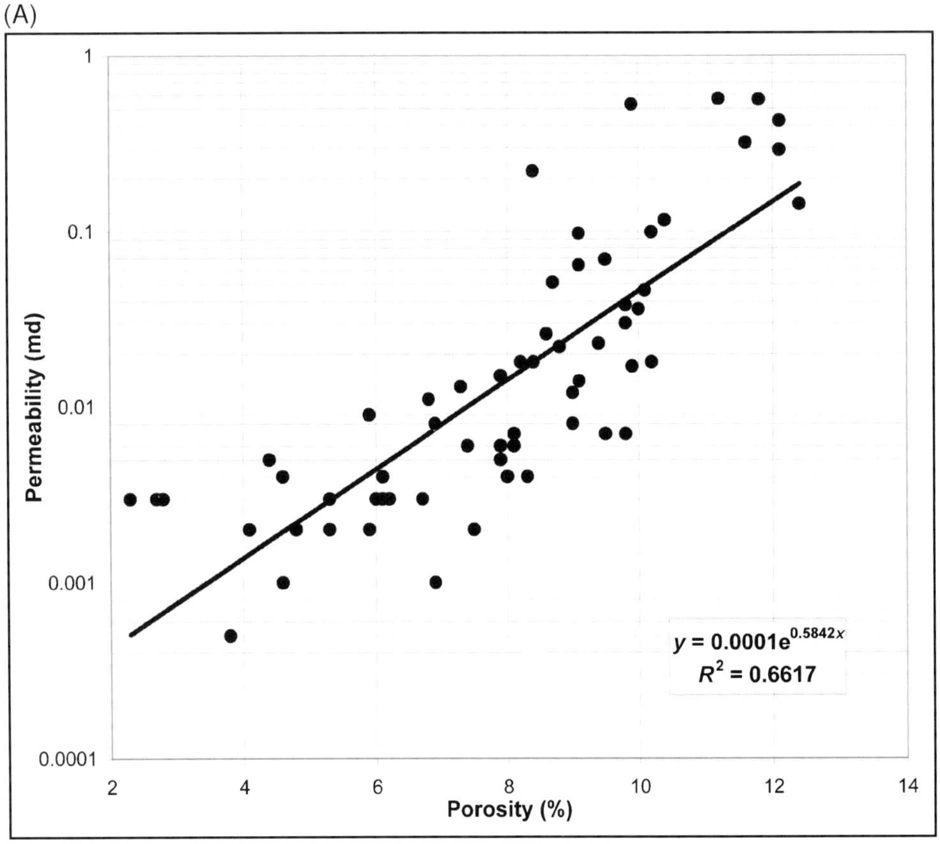

Figure 8. Core plug data from two wells, Jonah Federal 2-5 (W/2 Sec. 5, T28N, R108W) and Stud Horse Butte 11-27 (NE SW Sec. 27, T29N, R108W). Data were acquired at 4000-psi confining pressure. (A) Graph of porosity versus permeability. (B) Graph of permeability versus depth. The data form four depth clusters. A line representing maximum permeability versus depth based on maximum values from four data clusters is shown. Average permeability is plotted at a midpoint depth for each of the four depth clusters. These points are connected by a near-linear trend. (C) Graph of porosity versus depth.

observed from cores in the Lance at Jonah. Image logs do not indicate the presence of fracturing, nor do pressure responses during pump-in tests acquired prior to fracture stimulation. Additionally, production data do not support fractures being a significant part of the gas-recovery process. Matrix permeability dominates the production of gas from the Lance Formation at Jonah field.

The combination of the net pay thickness and hydrocarbon fraction results in a value that reflects the total hydrocarbon volume ($S_g\Phi H$, also known as the hydrocarbon pore volume, HPV). Figure 12 is a map showing the lateral variation of $S_g\Phi H$ in the field. Comparison of the patterns from Figures 9, 11, and 12 shows that EUR is not simply a function of total hydrocarbon volume ($S_g\Phi H$) but is strongly influenced by the reservoir quality (Figure 11).

Gas in Place and Recoverable Reserves

The $S_g\Phi H$ map (Figure 12) represents a total Lance Formation reservoir volume of approximately 700,000 ac ft (9300 km^2 m) or about 30 bcf of reservoir volume. Applying uniform reservoir conditions of 5000 psi, 185°F (85°C), and a Z factor of 1.0 to this volume yields an estimate of gas in place that exceeds 8.3 tcf. A summation of EUR values for 364 wells completed in the field prior to July 2002 yields a recoverable gas reserve estimate of 2.31 tcf. The average gas recovery per well is about 6.35 bcf.

LANCE FORMATION DEPOSITIONAL MODELS

The areal extent of a reservoir sandstone body that is penetrated in a well exerts a significant control on the maximum volume of hydrocarbons the well may recover from that sandstone body. Controls on depositional patterns and resulting sand body geometry in the Lance Formation include depositional slope, subsidence rate, sediment load, sediment-transport rate (controlled by runoff and climate), and provenance. The authors have studied the Lance sandstone geometry in selected areas of Jonah

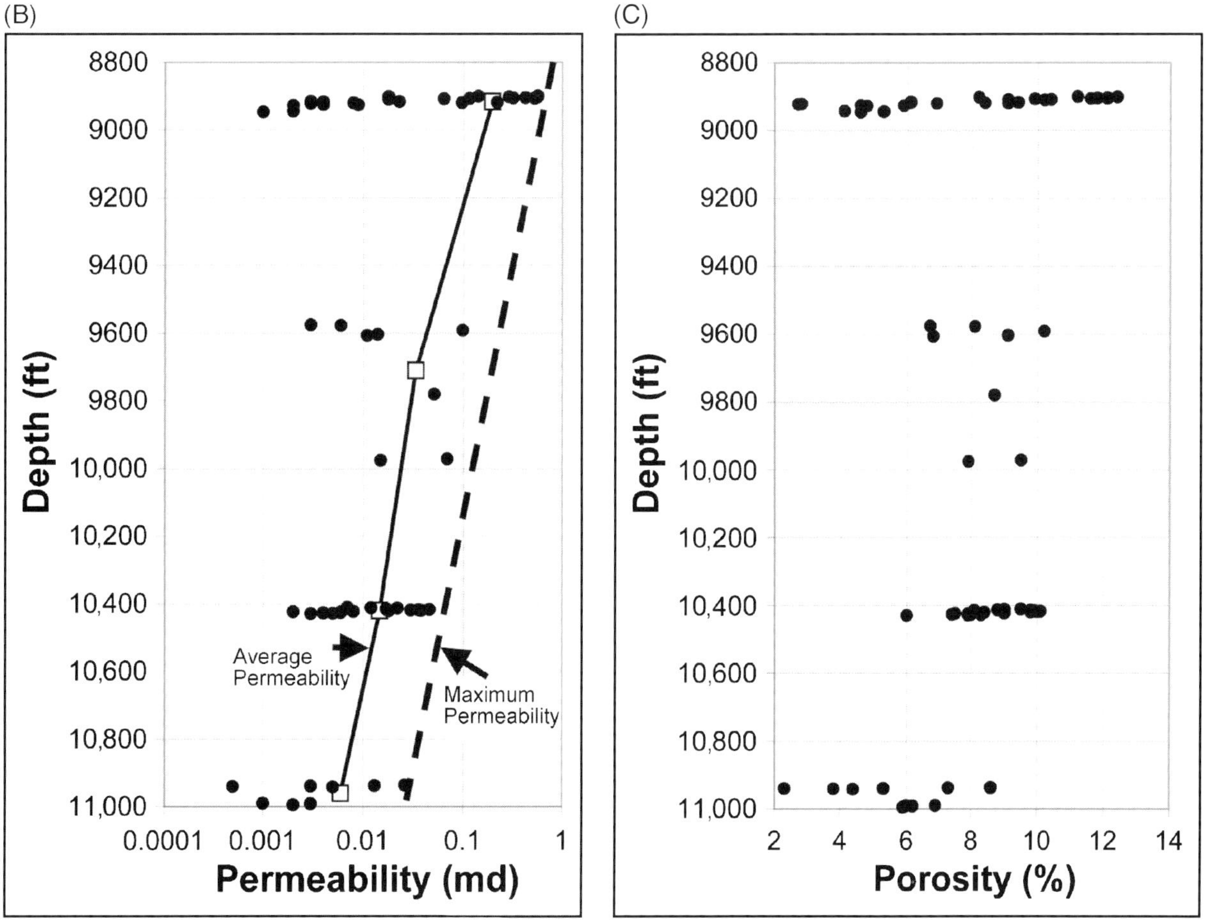

Figure 8. (cont.).

field and have contrasted a model by Robinson and McCabe (1997), with another by Lorenz et al. (1985). The Lorenz model is based on subsurface and outcrop studies of Upper Cretaceous Mesaverde Group sandstones interpreted to be deposited in a meandering fluvial environment. The Mesaverde Group in their study contains about 40% net sandstone of the gross interval thickness. The model by Robinson and McCabe involved sandstone bodies deposited by low-sinuosity, sandy, braided streams in the Salt Wash Member of the Jurassic Morrison Formation. The Salt Wash Member in their study area exhibits a net-to-gross sandstone ratio of about 75%. The median net sandstone content of the Lance Formation at Jonah is 46% (Figure 13).

The Lance Formation at Jonah field has been described by Finch et al. (1997) as having been deposited in a combination of meandering and braided fluvial settings. In their description, isolated fining-upward packages as much as 20 ft (6 m) thick represent deposition in a meandering fluvial environment. These packages can amalgamate into units as much as 50 ft (15 m) thick. Stacked sandstone units that attain a thickness of 100 ft (30 m) or more and consist of coarser grain sizes represent deposition in a braided fluvial environment. Based on these observations, there is significant overlap in the field between meandering and braided stream environments. Grain size alone cannot be used to indicate the depositional environment of the Lance Formation because of a provenance that consisted entirely of sedimentary rocks already restricted in grain size. Grain sizes larger than medium are sparsely seen in cuttings during drilling of the Lance Formation (Brown, 2000, personal communication; Shanley, 2001). This observation suggests that a majority of the sand bodies were deposited by meandering

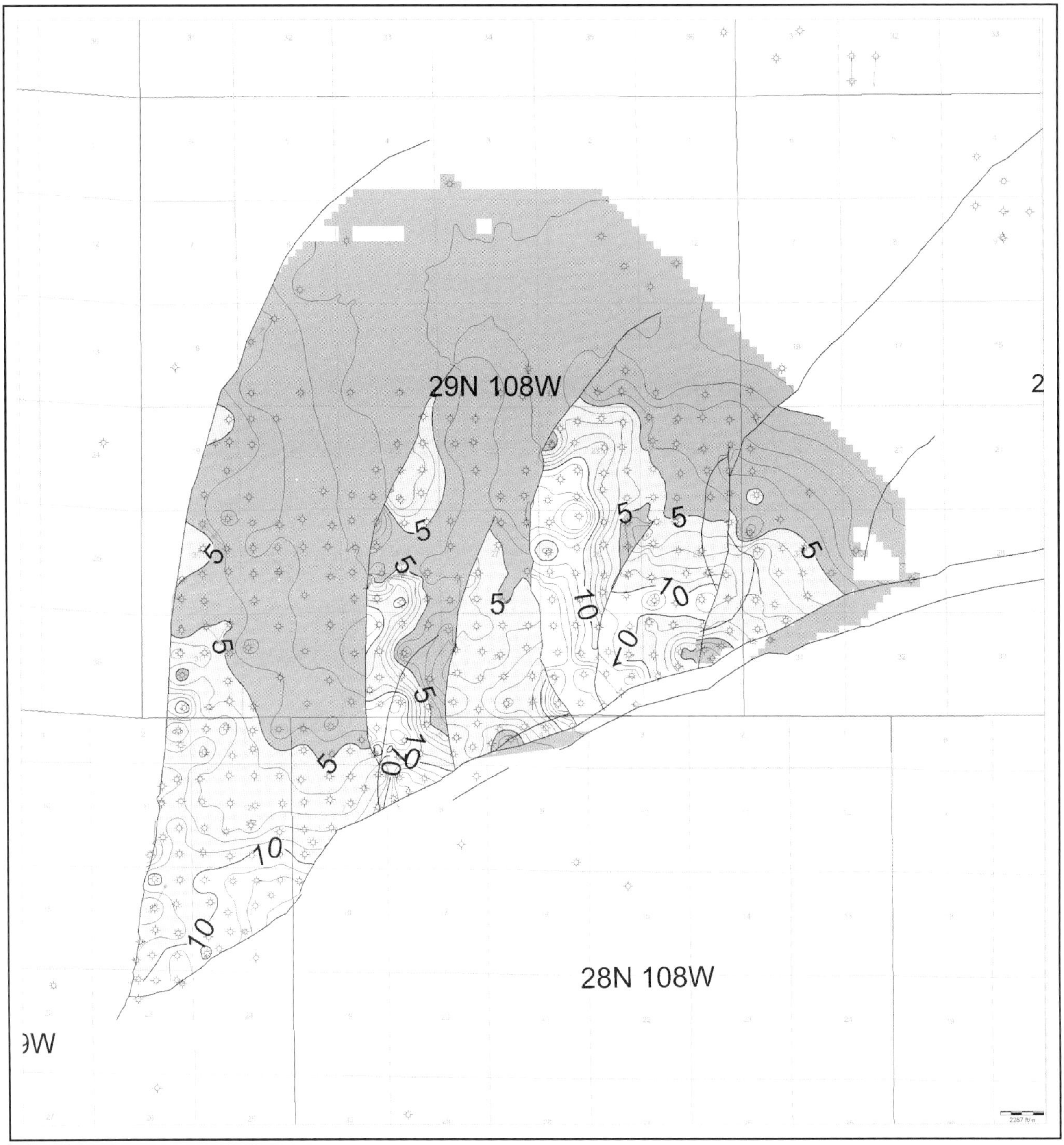

Figure 9. Map showing iso-EUR (estimated ultimate recovery) for wells drilled prior to 2001 in Jonah field. Contour interval is 1 bcf/well. The downdip limit of the field is located at the synclinal axis that separates Jonah field from the Pinedale anticline (Figure 2A).

streams. In core, the largest clasts are mudstone fragments that represent either rip-ups from the channel floor or material that has slumped from the channel banks. Comparison of several numeric models describing sand body geometry as it relates to the Lance Formation has been presented by Shanley (2001; 2004).

(A)

Figure 10. (A) Isopach map of net sandstone (gamma ray <75 API) for the Lance Formation at Jonah field. Contour interval is 100 ft (30 m). (B) Isopach map of net pay for the Lance Formation at Jonah field. Net pay is a subset of net sandstone determined by petrophysical calculation. Contour interval is 50 ft (15 m).

Sandstone bodies deposited in meandering channels will have a higher degree of lateral discontinuity than those deposited in a braided fluvial system. As a result, a single well will contact less contiguous reservoir volume than a similar well in braided fluvial deposit. Reservoir simulations from a meandering-channel model based on

(B)

Figure 10. (cont.).

parameters described by Lorenz et al. (1985) and honoring Jonah well control generated reasonable production-history matches. These simulations, as well as a comparison of the gas in place to recoverable reserve estimates near the end of the 40-ac (16-ha) development, point to the need for development on spacing significantly less than 40 ac (16 ha) to effectively capture the resource.

CONCLUSIONS

Jonah field is a significant hydrocarbon accumulation in the Rocky Mountain region. It is a structural-stratigraphic trap that is a very distinct sweet spot in a much larger basin-center system of pervasive gas saturation in the Lance Formation. Pressure data suggest that

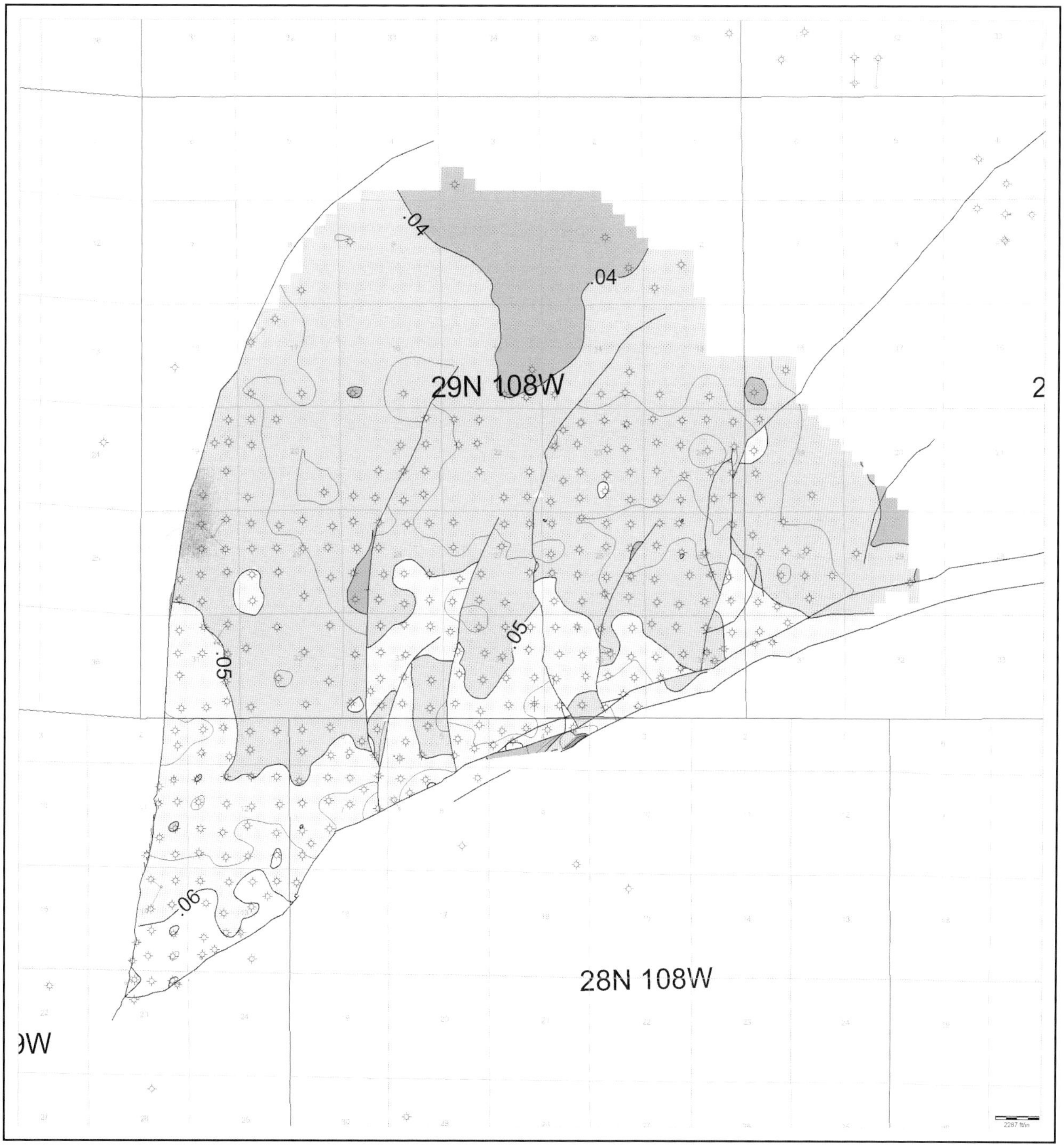

Figure 11. Isopach map of average $S_g\Phi$ for the net pay in the Lance Formation at Jonah field. Contour interval is 0.005 (0.5%).

the process of migrating hydrocarbons into the Jonah compartment is proceeding today. Infill drilling at spacing less than 40 ac (16 ha) will be necessary to adequately produce the gas trapped in the Lance Formation at Jonah field.

The key elements that make Jonah field a significant resource area are the following:

- Overpressure: Increases storage capacity and may help preserve porosity and permeability.

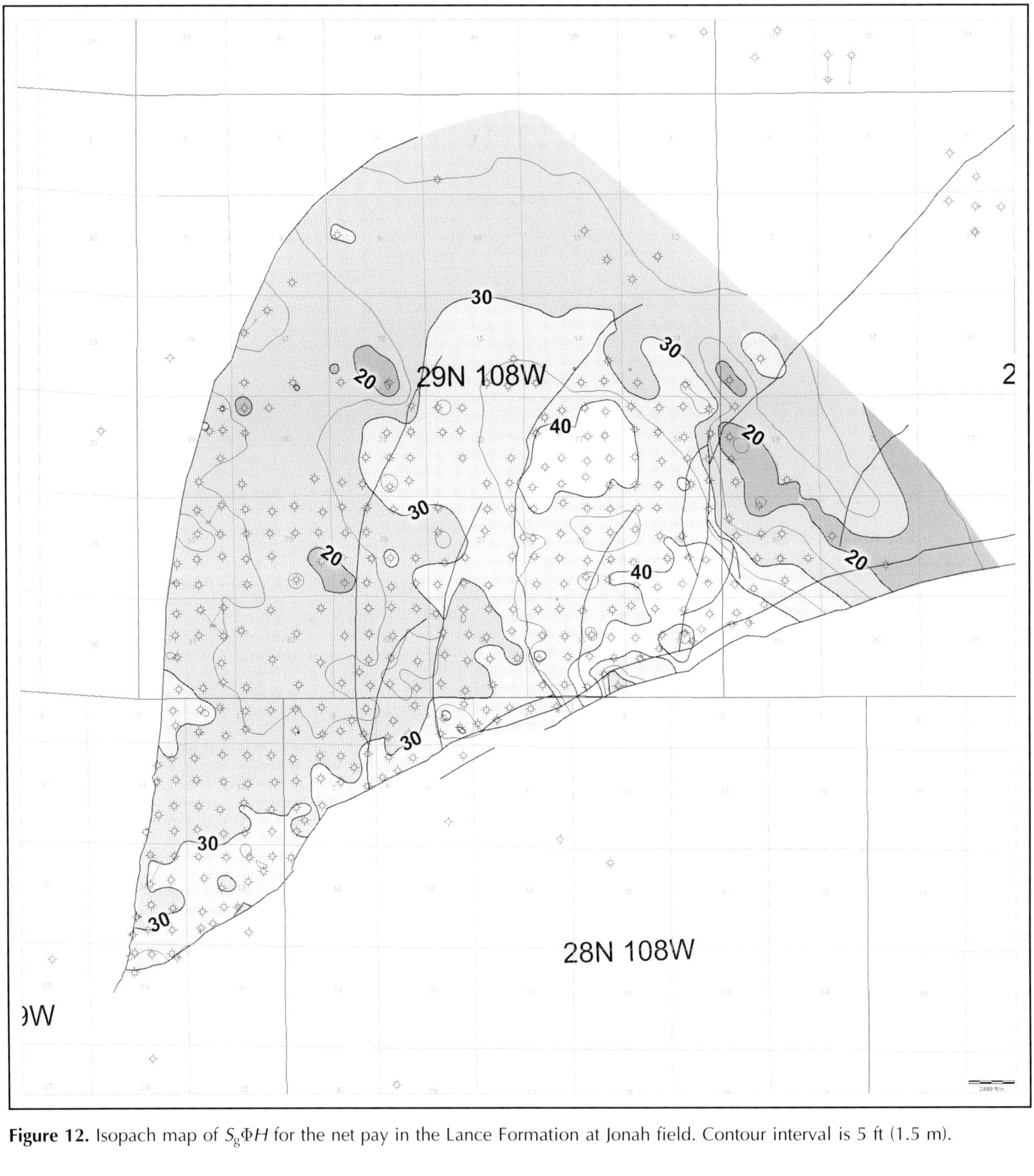

Figure 12. Isopach map of $S_g\Phi H$ for the net pay in the Lance Formation at Jonah field. Contour interval is 5 ft (1.5 m).

- Large porosity and thickness: The Lance Formation near the depositional axis of the Hoback basin is thick (as much as 3000 ft [915 m]) and contains large amounts of porous sandstone. The permeability of these sandstones is very low; single sand body pay zones would not be commercially attractive.

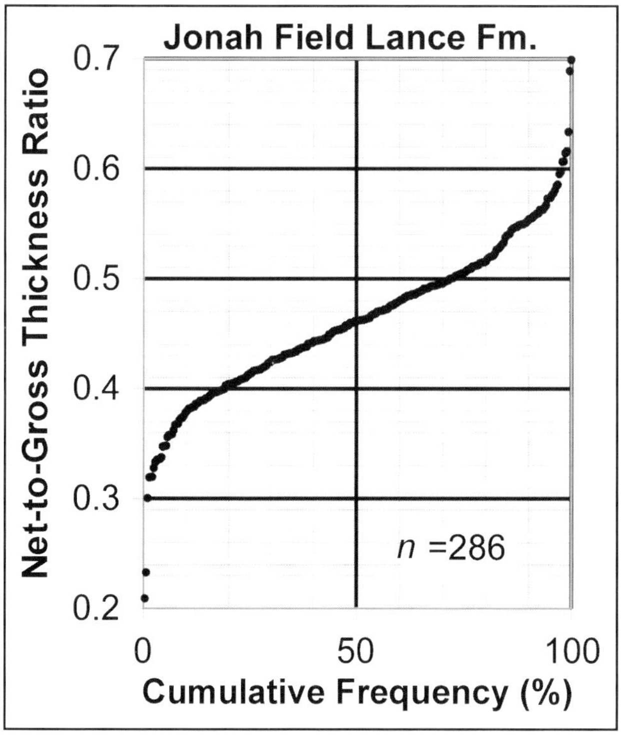

Figure 13. Cumulative frequency plot of net-to-gross thickness ratio for an interval from the top of the Lance to the top of the Mesaverde, or total depth, whichever is less. Net sandstone is defined by gamma ray < 75 API.

- Stimulation technology: Without modern completion techniques, the field was overlooked, but recent completion technology has transformed the field into one of the gems of the Rocky Mountain region.

REFERENCES CITED

Dorr, J. A. Jr., D. R. Spearing, and J. R. Steidtmann, 1977, Tectonic and synorogenic depositional history of the Hoback basin and adjacent areas, *in* E. L. Heisey, D. E. Lawson, E. R. Norwood, P. H. Wach, and L. A. Hale, eds., Rocky Mountain thrust belt geology and resources: Wyoming Geological Association Guidebook, p. 549–562.

Finch, R. W., W. W. Aud, and J. W. Robinson, 1997, Evolution of completion and fracture-stimulation practices in Jonah field, Sublette County, Wyoming, *in* E. B. Coalson, J. C. Osmond, and E. T. Williams, eds., Innovative applications of petroleum technology in the Rocky Mountain area: Rocky Mountain Association of Geologists Guidebook, p. 13–24.

Hanson, W. B., V. Vega, and D. Cox, 2004, Structural geology, seismic imaging, and genesis of Jonah field: A giant tight-gas accumulation, Wyoming, U.S.A., *in* J. W. Robinson and K. W. Shanley, eds., Jonah field: Case study of a tight-gas fluvial reservoir: AAPG Studies in Geology 52, p. 61–92.

Harding, T. P., 1973, Newport-Inglewood trend— An example of wrenching style of deformation: AAPG Bulletin, v. 57, p. 97–116.

Law, B. E, and R. C. Johnson, 1989, Structural and stratigraphic framework of the Pinedale anticline, Wyoming, and the multiwell experiment site, Colorado, *in* Geology of tight gas reservoirs in the Pinedale anticline area, Wyoming, and at the multiwell experiment site, Colorado: U.S. Geological Survey Bulletin, v. 1886, p. B1–B11.

Lorenz, J. C., D. M. Heinze, J. A. Clark, and C. A. Searls, 1985, Determination of widths of meander-belt sandstone reservoirs from vertical downhole data, Mesaverde Group, Piceance Creek basin, Colorado: AAPG Bulletin, v. 69, p. 710–721.

Prensky, S. E., 1989, Gamma-ray well-log anomaly in the northern Green River basin of Wyoming, *in* Geology of tight gas reservoirs in the Pinedale anticline area, Wyoming, and at the multiwell experiment site, Colorado: U.S. Geological Survey Bulletin, v. 1886, p. H1–H21.

Robinson, J. W., and P. J. McCabe, 1997, Sandstone-body and shale-body dimensions in a braided fluvial system: Salt Wash Sandstone Member (Morrison Formation), Garfield County, Utah: AAPG Bulletin, v. 81, p. 1267–1291.

Robinson, J. W., D. Delozier, and R. Finch, 1996, Integrated reservoir description and analysis of the Lance Formation, Jonah field, Sublette County, Wyoming: AAPG Bulletin, v. 80, p. 980.

Shanley, K. W., 2001, Fluvial reservoir description for a tight-gas giant: Jonah field, Green River basin, Wyoming, *in* J. W. Robinson and K. W. Shanley, eds., Tight gas fluvial reservoirs: A case history from the Lance Formation, Green River basin, Wyoming: AAPG Short Course 2, AAPG Annual Meeting, Denver, Colorado, unpaginated.

Shanley, K. W., 2004, Fluvial reservoir description for a giant, low-permeability gas field: Jonah field, Green River basin, Wyoming, U.S.A., *in* J. W. Robinson and K. W. Shanley, eds., Jonah field: Case study of a tight-gas fluvial reservoir: AAPG Studies in Geology 52, p. 159–182.

Spencer, C. W., 1987, Hydrocarbon generation as a mechanism for over pressuring in the Rocky Mountain region: AAPG Bulletin, v. 71, p. 368–388.

Warner, E. M., 2000, Structural geology and pressure compartmentalization of Jonah field based on 3-D seismic data and subsurface geology, Sublette County, Wyoming: The Mountain Geologist, v. 37, p. 15–30.

Yin, P., S. M. Stookey, R. C. Surdam, and T. Dunn, 2001, Tight gas sandstone of the Lance Formation in Jonah field— a core from the Jonah 2-5 well, *in* J. W. Robinson and K. W. Shanley, eds., Tight gas fluvial reservoirs: A case history from the Lance Formation, Green River basin, Wyoming: AAPG Short Course 2, AAPG Annual Meeting, Denver, Colorado, unpaginated.

CHAPTER 6

Structural Geology, Seismic Imaging, and Genesis of the Giant Jonah Gas Field, Wyoming, U.S.A.

WILLIAM B. HANSON
Consultant and Independent Geologist, Vernonia, Oregon, U.S.A.

VICTOR VEGA
BP America Production Company, Houston, Texas, U.S.A.

DENNIS COX
BP America Production Company, Houston, Texas, U.S.A.

ABSTRACT

Jonah field, Wyoming's second largest gas producer, is a structurally controlled trap located in the northwestern part of the Green River basin. Gas and condensate are produced from innumerable latest Cretaceous and early Tertiary overpressured tight-gas sandstones at depths of 7300–12,800 ft (2200–3900 m). Jonah field is remarkable for many reasons, including the large per-well reserves (relative to other tight-gas reservoirs), hundreds of feet of net pay, and a gross producing interval as great as 4000 ft (1220 m) thick. These superlative production characteristics exist, despite the fact that the structural trap is subtle and locally cryptic.

Advanced seismic techniques define the Jonah trap boundaries and add value when they are used to position wells in proximity to the faults and on subtle structures. One of the main seismic techniques is a three-dimensional, broadband, amplitude-based coherency algorithm that has edge-detection capabilities. This algorithm analyzes the reflector amplitude gradient and records the lateral change in amplitude with azimuthal angle, which allows the interpreter to illuminate a particular feature from the optimal angle to reveal the maximum detail in the data.

Gas entrapment at Jonah field is enabled by two bounding faults. Fault throw is variable but commonly less than 200 ft (60 m), and the major faults are nearly vertical. Two field-bounding fault zones, the west fault and south Jonah fault, intersect updip toward the southwest to create the overall wedge-shaped trap. The updip edge of a tilted fault block underlies the prolific Stud Horse Butte anticline and the Cabrito nose trends. The principal in-field faults terminate at the south Jonah fault to form four compartments, each comprised of a northeast-plunging, faulted nose or homocline bounded on the west and south by faults. The Jonah faults juxtapose high- and low-reserve wells; high-reserve wells are concentrated on the east side of the north- and northeast-trending faults, regardless of their sense of displacement.

The south Jonah fault is probably a left-lateral, wrench-fault zone. The south Jonah fault was active concurrently with Lance Formation deposition, resulting in thicker Lance north of the fault in the center of the field and thin along the updip edge of the tilted block. Subsequent post-Paleocene motion on the fault caused formation of the Stud Horse Butte anticline, which is evident in the basal Tertiary strata.

Jonah is a multipay field with numerous productive lenticular, fluvial sandstones. Whereas local structural features control the Jonah trap, regional structural elements shaped the fluvial system that deposited the reservoir sandstones. Regional structural features also affected the burial history that resulted in petroleum generation. Gas isotope composition and the thermal maturity of the producing strata indicate that the field produces hydrocarbons formed in deeper strata. The gas field shows evidence of paleostructural growth, which,

Jonah Field: Case Study of a Giant Tight-Gas Fluvial Reservoir, J. W. Robinson and K. W. Shanley, eds.: AAPG Studies in Geology 52 and Rocky Mountain Association of Geologists 2004 Guidebook

in combination with the production pattern along the faults and the "bottoms-up" origin of the gas, may explain the enigmatic charging of these low-permeability sandstone bodies according to structural position, i.e., segregation by buoyancy. In this scenario, gas emplacement occurred before the reduction of sandstone permeability to the present condition.

INTRODUCTION

Jonah field is located in one of the deepest parts of the Green River basin, approximately 60 mi (97 km) north-northwest of Rock Springs, Wyoming, in Sublette County. The Pinedale anticline and the Wind River uplift are to the east, and the La Barge platform and Sevier fold and thrust belt are located to the west (Figure 1). Jonah field has rapidly developed into one of the largest natural gas fields in Wyoming since its rediscovery by McMurry Oil Company in 1993 (Robinson, 2000). This field now has more than 350 producing gas wells. Well density is 40 ac (0.16 km^2) (two wells allowed per 80-ac [0.32-km^2] spacing unit), and 20-ac (0.08-km^2) well density is being tested. Jonah field produces gas and condensate at depths of 7300–12,800 ft (2200–3900 m) from a gross productive interval ranging from approximately 1700 to 4000 ft (520 to 1220 m). More than 90% of the production is from multiple, lenticular, overpressured fluvial sandstones in the latest Cretaceous Lance Formation (Figure 2). The Lance and Mesaverde pay sandstones are tight-gas reservoirs based on an average in-situ permeability of less than 0.1 md, as determined by core analysis and well tests (Montgomery and Robinson, 1997; C. Christensen, 1999, personal communication). During 2001, basal Tertiary low-permeability sandstones became significant gas producers and recompletion targets. These basal Tertiary reservoirs are correlated to the Paleocene-age Unnamed Tertiary unit of Law (1979) and Law and Johnson (1989). A few wells are also completed in sandstones in the upper part of the Mesaverde Group. The base of commercial gas production is not clearly established at Jonah field, although the frequency of occurrence of highly productive Upper Cretaceous sandstones decreases below approximately 11,000 ft (3400 m). Commercially significant gas production now occurs as deep as 12,800 ft (3900 m) on the northeast edge of the field, and strong gas shows have been encountered in the deepest wells drilled in the field. Several completions in the Mesaverde Group, including a few in the Ericson Sandstone, have produced gas at noncommercial rates. Only one well has fully penetrated the Mesaverde Group in the Jonah field (McMurry Oil Company, Yellow Point 11-13, Sec. 13, T28N, R109W, 13,408 ft [4087 m] total depth).

Jonah field was first thought to be a stratigraphic trap. It has also been described as a sweet spot in a basin-centered gas accumulation (Bowker and Robinson, 1996) and a structural trap bounded by wrench faults (Warner, 1997, 1998). Warner (1998, p. 45) described the west, updip boundary of the field as "a scissors fault (or possibly a wrench fault with little lateral displacement) bearing southwest to northeast . . . within one curving fault plane," and described the south boundary of Jonah field to be "a wrench fault that is near-vertical and extends from the basement to the surface. Segments appear to be imbricated and sinusoidal rather than merely vertical."

This chapter discusses the structural geology of Jonah field and the role of structural elements in its genesis. We employ an integrated interpretation of well data, production data, and three-dimensional (3-D) seismic data in the context of the regional geology to describe aspects of the field and to present ideas concerning the origin of this important gas accumulation. We use a proprietary 3-D seismic survey acquired during 1996, 1997, and 1999; digital well-log data from more than 300 wells; 450 ft (137 m) of full diameter core from five wells; and sidewall cores from one additional well. These data have been integrated in state-of-the-art computer workstation tools employed during our development of Jonah field at 80- and 40-ac (0.32- and 0.16-km^2) well density during the past 3.5 yr (1999–2002).

REGIONAL SETTING

Rapid subsidence of the Green River basin in front of the Wind River uplift, in combination with slower subsidence and episodic uplift of the La Barge platform, created an asymmetric downwarp during the latest Cretaceous and early Tertiary. Regional northeast dip at Jonah field is primarily a product of basin-margin loading by the Wind River uplift (Shuster and Steidtmann, 1988), a crustal-scale element carrying Precambrian basement and Phanerozoic sedimentary cover (Stone, 1993). Figure 3 is a generalized structural cross section across Jonah field and the Wind River thrust. The Wind River uplift moved southwestward a distance of at least 14 mi (23 km) along the Wind River thrust during the Laramide orogeny (Blackstone, 1993), thereby driving asymmetric subsidence that accommodated northeastward-thickening wedges of coeval strata. Figure 4, a detailed structure map of Jonah field, shows the regional tilt effected by emplacement of

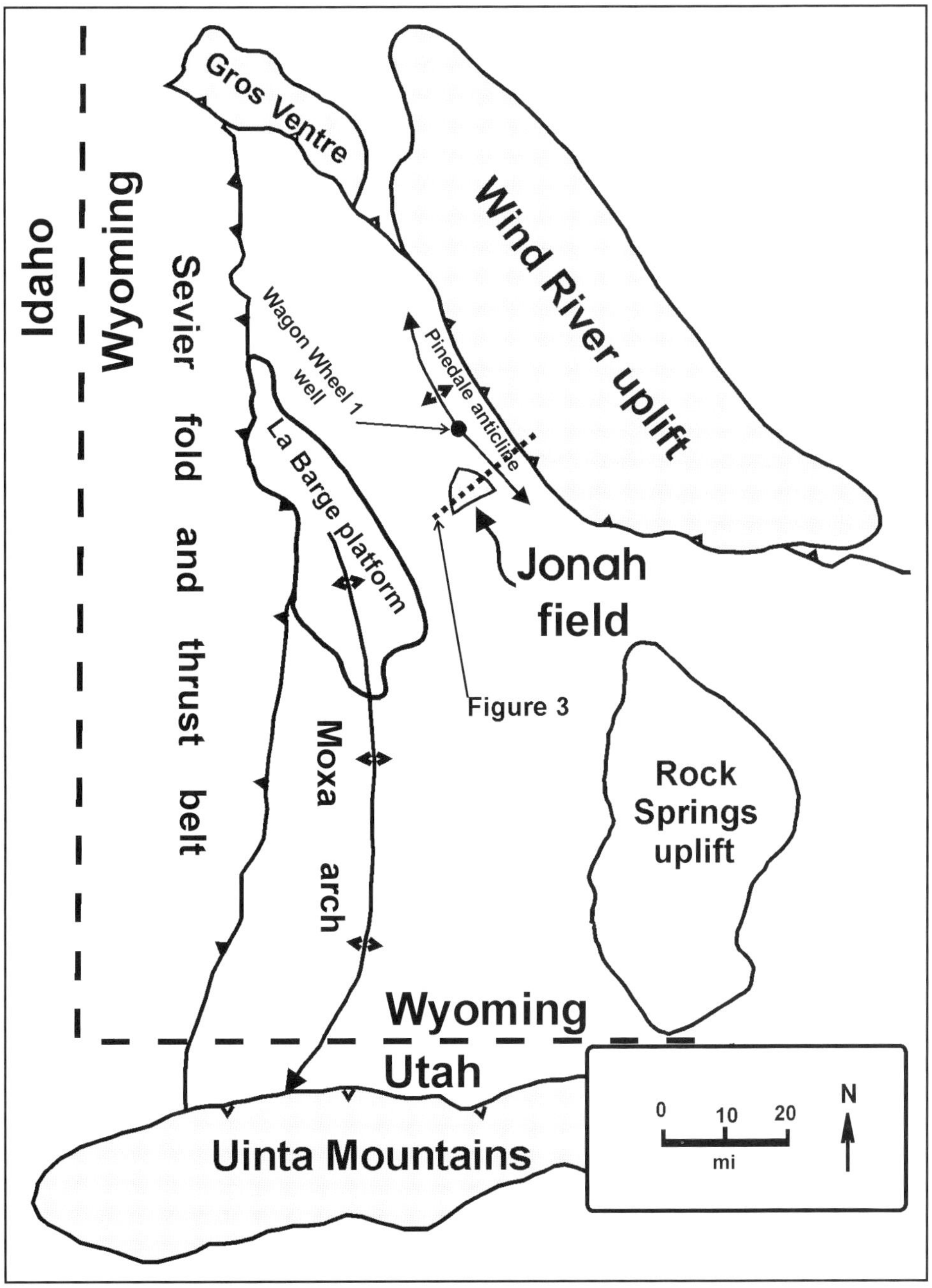

Figure 1. Location map for Jonah field.

the Wind River uplift. Figure 5 demonstrates that the latest Cretaceous Lance Formation thickens significantly downdip across Jonah field and toward the Wind River uplift. Movement of the Darby thrust, one of the Sevier overthrust sheets west of the Green River basin, was, in part, coeval with the deposition of Lance sediments (Royse, 1993).

Dip variations at Jonah field result from local structural details superimposed on the regional tilt. Figure 4 shows the structure at the top of the gas-producing interval. The dip is 300–400 ft/mi (60–80 m/km) in the eastern part of the field (T29N, R108W), at a position approximately 15 mi (24 km) southwest of the surface trace of the Wind River thrust. In the southwestern part of Jonah field, dip rate (in the basal Tertiary section) is approximately 100 ft/mi (20 m/km), where local folding reduces regional dip (northeast part of T28N, R109W, labeled "Yellow Point nose" in Figure 5).

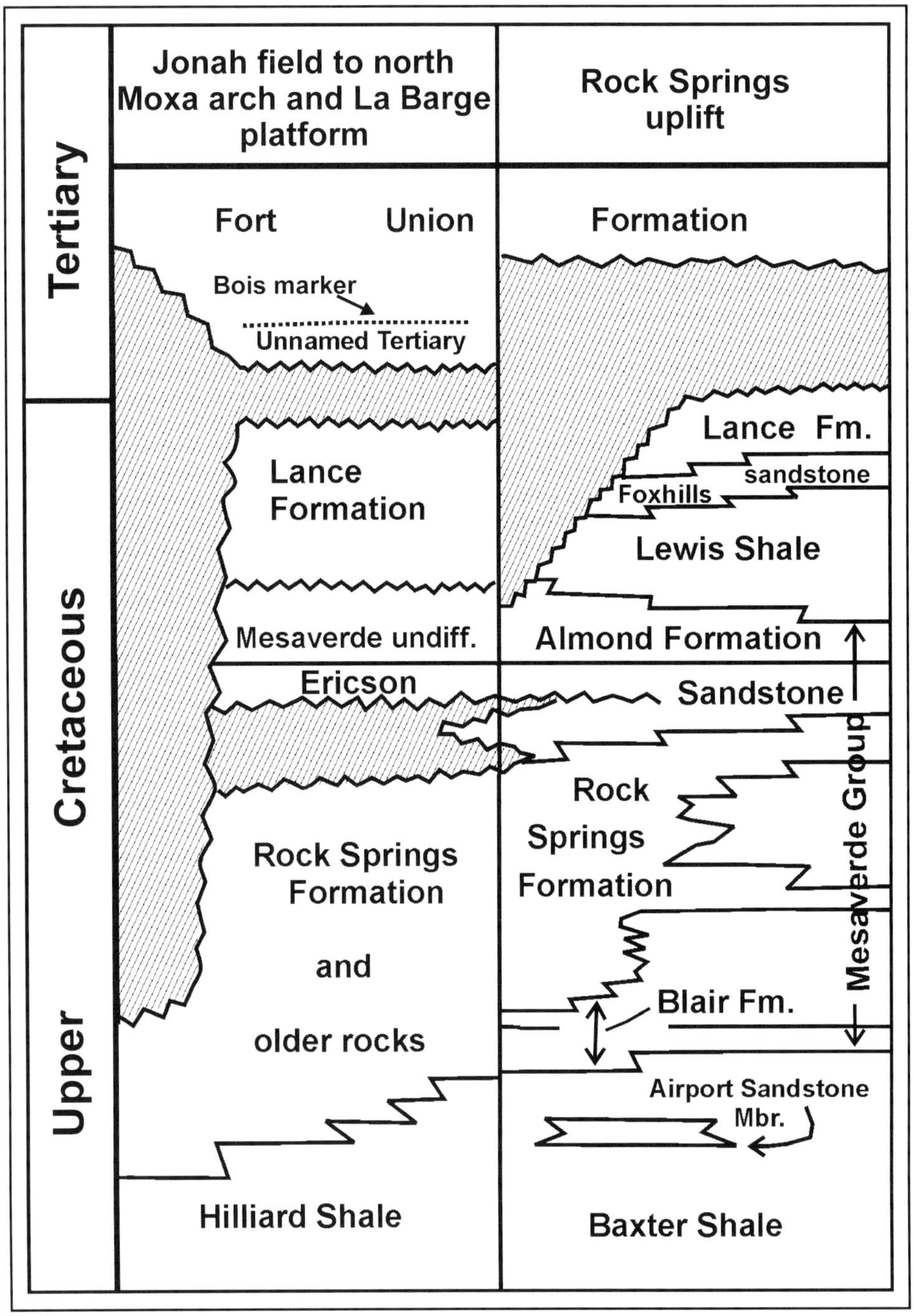

Figure 2. Stratigraphic columns for Jonah field area and the Rock Springs uplift. Modified from Law (1984) and Law and Johnson (1989).

TECTONIC CONTROL OF LANCE PALEOGEOGRAPHY

Asymmetric Subsidence

The pronounced asymmetric subsidence that was created by early movement of the Wind River uplift exerted a first-order influence on the Lance Formation fluvial system. Table 1 compares the thickness of selected stratigraphic units across Jonah field. Note that the Lance Formation thickens significantly from 2600 ft (790 m) on the west to 3740 ft (1140 m) in the east. Thickness values are derived from an integrated interpretation of well and seismic data. The base of the Lance Formation is at the top of a sequence of distinctive, high-amplitude reflections. This

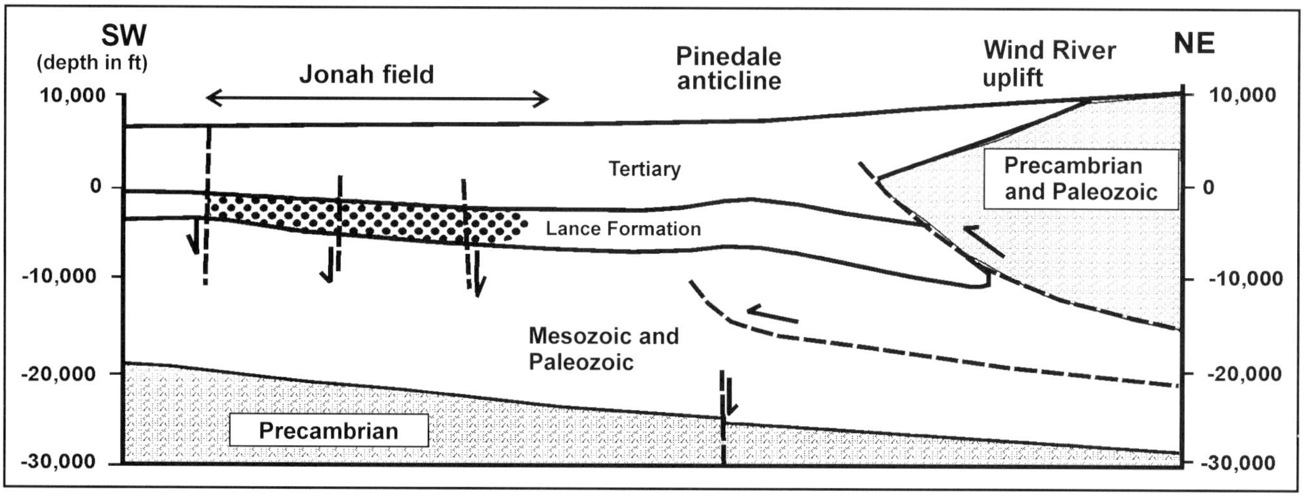

Figure 3. Diagrammatic true-scale cross section from Jonah field to the Wind River uplift based on well data, 3-D seismic data at Jonah field and Pinedale anticline, selected 2-D seismic lines, and Gries (1983, figure 26). Location of cross section is shown in Figure 1.

stratigraphic position fits regional well-log correlations and is above the Ericson Sandstone, as seen in key wells at Jonah field and on the Pinedale anticline. Approximately 3 mi (9 km) farther east on the Pinedale anticline, the Lance is 4500 ft (1370 m) thick (Hanson, 2001), whereas the Lance is absent on the La Barge platform (Figure 1) west of Jonah field (McDonald, 1973). Figure 5 also illustrates the persistent northeastward thickening of the combined Late Cretaceous and basal early Tertiary strata. The Lance Formation may exceed 6000 ft (1830 m) farther east in the footwall of the Wind River thrust. Hanson, et al. (1994) identified a similar northeastward-thickening wedge of latest Cretaceous strata (Lewis, Fox Hills, and Lance) that developed east of the Rock Springs uplift. The thickest part of this wedge is in the northeast corner of the Red Desert basin against the Rawlins uplift, adjacent to an area known to have been tectonically active during latest Cretaceous (Lance and Fox Hills) deposition (Reynolds, 1976). The northeast-thickening stratigraphic units at Jonah field indicate that the Wind River uplift is one of the older foreland uplifts, and that it had incipient development during the Late Cretaceous.

Paleogeography and Sediment Dispersal

During the time of the Lance Formation (late Maastrichtian), the Rocky Mountain region was undergoing a change in overall paleogeography from a broad, north-south, epicontinental seaway and foreland basin (connecting the polar sea with the proto-Gulf of Mexico) to a region of uplifts and intermontane basins. The long-lived Sevier fold and thrust belt supplied sediment to a broad piedmont alluvial plain and coastal plain (Roberts and Kirschbaum, 1995). The Wind River and Granite Mountains were being thrust southwestward and southward over the flanking basins. The great Cretaceous seaway was forced out of the region as the crust was broken into numerous uplifts and basins. Depositional space was filled by syntectonic detritus derived from adjacent and distant uplifts. Based on stratigraphic and sedimentologic measurements, Connor (1992) showed an emergent Granite Mountains uplift, shedding coarse-grained Lance sediment north and northeast into the eastern Wind River basin. Therefore, it appears that the Greater Green River basin and the Wind River basin were starting to be structurally and topographically separated during the Maastrichtian, as the incipient Laramide uplifts formed. Sediment was rapidly shed from the rising uplifts and generally filled the basins to above sea level. One exception is the Lewis Shale seaway, located east-southeast of Jonah field, where tectonic subsidence outpaced sediment supply, resulting in a deep seaway, complete with steep-faced deltas that built southward from the Granite Mountains and debouched prodeltaic sediment gravity flows into relatively deep water to form sandstones of the Lewis Shale. In fact, this persistent Late Cretaceous tectonic feature created a deeply embayed shoreline for the early Maastrichtian *Baculites clinolobatus* zone (Cobban et al., 1994). Lance deposition at Jonah field was located northwest of this shoreline in alluvial-plain and upper coastal-plain environments, where fluvial channels, flood-basin splays, flood-plain mudstones, and paleosols accumulated. The nonmarine facies and reservoir characteristics

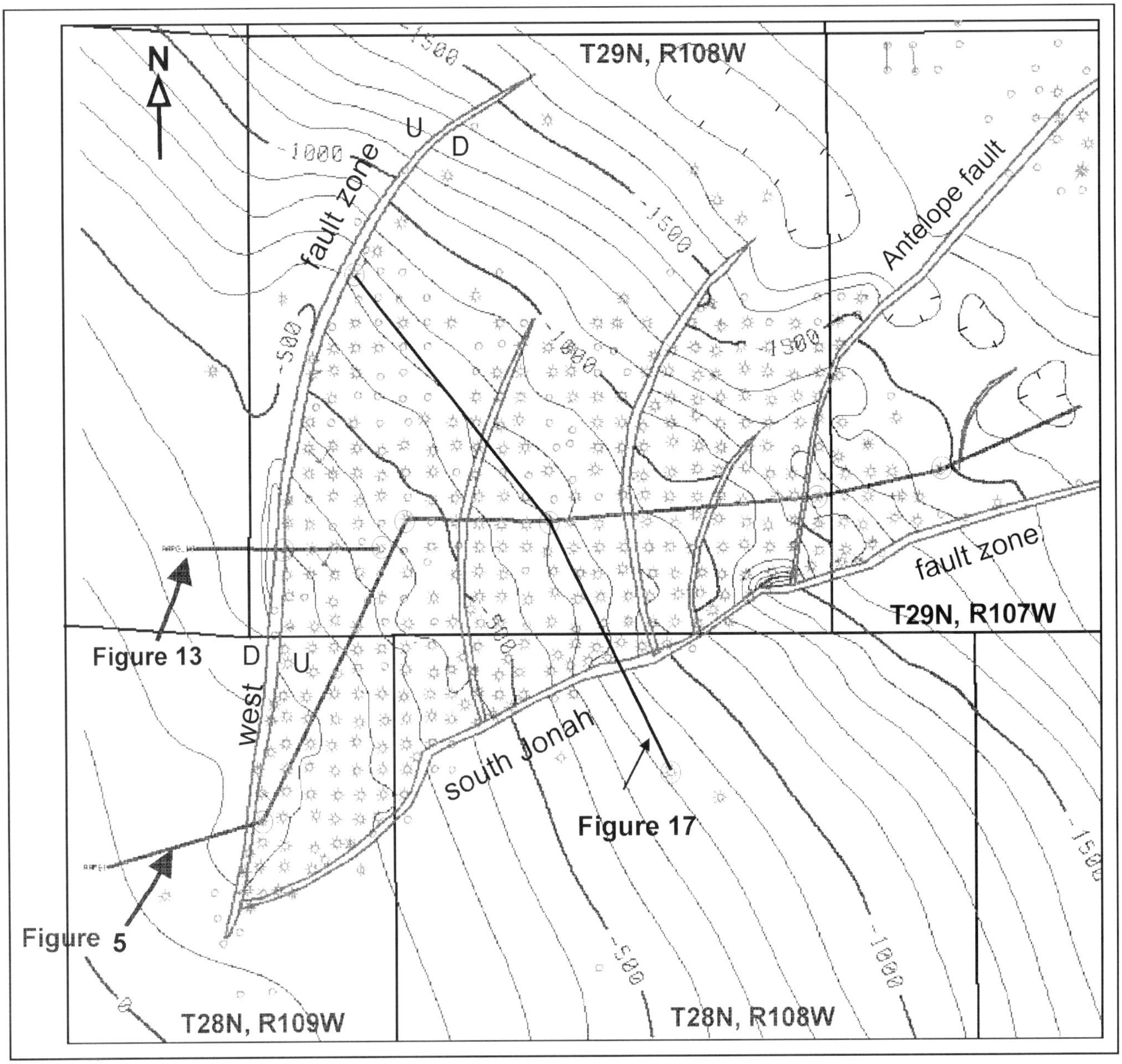

Figure 4. Structure contour map of Jonah field drawn on a marker bed in the Unnamed Tertiary unit (Bois marker) near the top of the Lance Formation. Contour interval is 100 ft (30 m). The major faults for Jonah field are shown. The field is developed with 40-ac (0.16-km^2) well density.

at Jonah field are discussed in several recent publications, including Bowker and Robinson (1996), Robinson et al. (1996), Evans and Foulk (1997), Montgomery and Robinson (1997), Robinson and Evans (1997), Shanley (2001), and other chapters in this volume. Reservoir rock at Jonah field is almost totally fluvial channel-fill facies, although Shanley (2004) has identified a minor proportion of probable crevasse splay sandstones. Based on a 75° API unit gamma-ray cutoff for corrected and normalized well-log curves, net sandstone thickness in the Lance Formation at Jonah field is 600–1150 ft (180–350 m), comprising 20–40% of the formation thickness. Bowker and Robinson (1996) reported that sediment transport was generally to the southeast and east. Montgomery and Robinson (1997)

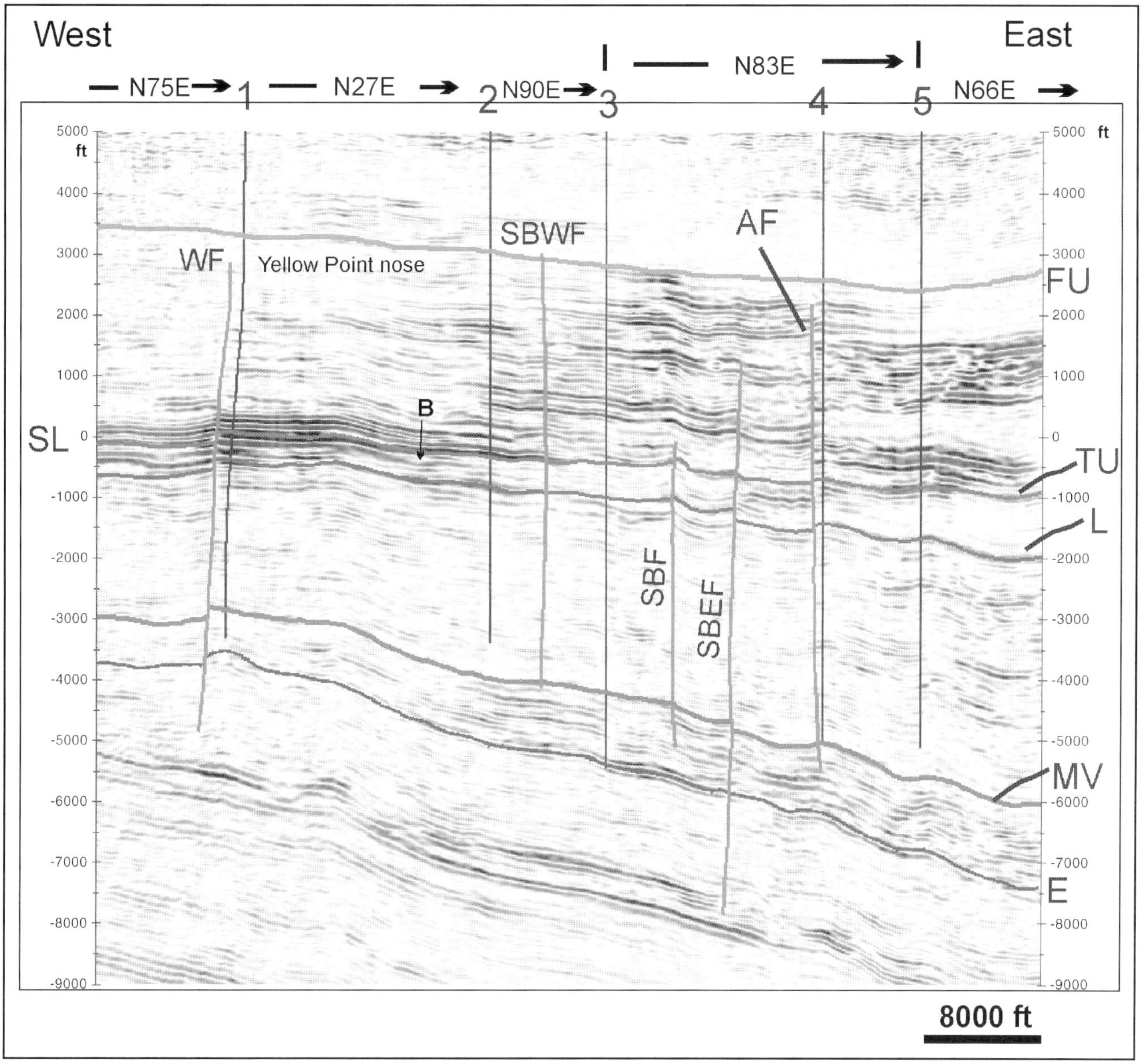

Figure 5. Depth-converted seismic profile through Jonah field selected from the 3-D seismic volume. Location is shown in Figure 4. Seismic profile is 12.1 mi (20 km) long with a vertical exaggeration of 4:1. Wells on this profile are numbered 1 to 5. Elevation along the vertical axis in 1000-ft (300-m) increments with sea level marked SL. Interpreted horizons are FU = Fort Union Formation; TU = Unnamed Tertiary unit; L = Lance Formation; MV = top of Mesaverde Group; E = Ericson Sandstone. Yellow Point nose is between wells 1 and 2, where B indicates the Bois marker horizon in the basal Tertiary interval. Five of the six major Jonah field faults are shown. WF = west fault; SBWF = Stud Horse Butte west fault; SBF = Stud Horse Butte fault; SBEF = Stud Horse Butte east fault; AF = Antelope fault.

reported southeastward-directed cross-bedding in an oriented core from the McMurry Oil Jonah 2-5 well.

The fluvial system responsible for the Lance Formation reservoir sandstones was a persistent system with medium-sized river channels (Shanley, 2001) that flowed generally southeastward along the eastern edge of the rapidly subsiding basin, toward the Lewis embayment. This model proposes that the Lance fluvial system emanated from a major structural reentrant in the Sevier fold and thrust belt approximately 60 mi (96 km) northwest of Jonah

Table 1

Comparison of Thickness of Selected Lower Tertiary and Upper Cretaceous Stratigraphic Units at Three Locations along a Dip Traverse across Jonah Field. Well Locations are YP 11-13 in Sec. 13, T28N, R109W; SHB 13-27 in Sec. 27, T29N, R108W; Cab 15-19 in Sec. 19, T29N, R107W. Values are True Vertical Thickness in Feet (0.3 m) as Determined from Integrated Interpretation of Well Logs and 3-D Seismic Data.

	Southwest YP 11-13	**Middle SHB 13-27**	**East CAB 15-19**
Unnamed Tertiary	360	530	1150
Lance	2570	3220	3740
Mesaverde Group	3100	3600	3970

field, in the Hoback Canyon area. At Hoback Canyon, paleocurrent orientations in older Upper Cretaceous fluvial channel-fill sandstones show southeastward sediment transport, as summarized at locality 2 in Figure 6 (Shuster and Steidtmann, 1987). Lance-age sediments are largely eroded beneath Paleocene conglomerate in the Hoback Canyon area outcrops, but the Harebell and Pinyon conglomerates, which are trapped in the hinterland and foreland of the Sevier fold and thrust belt, record a Late Cretaceous fluvial megafan that entered the Green River basin through the structural reentrant (Decelles and Cavazza, 1999). Figure 6 summarizes our paleogeographic model. We cite the following observations and concepts:

(1) The northwest- to southeast-oriented region of asymmetric subsidence in front of the incipient Wind River uplift extended into the deep-water Lewis embayment. This is the typical position for a persistent fluvial system, the axis of a basin, or subsiding foredeep.

(2) Southeastward-directed paleocurrents are recorded in Campanian and Santonian fluvial channels at Hoback Canyon northwest of Jonah field (Shuster and Steidtmann, 1987). These measurements are at locality 2 in Figure 6 and are interpreted to record persistent fluvial systems that emanated from a structural reentrant during the Late Cretaceous. Strike valleys and "piggyback" basins of the Sevier fold-and-thrust belt provided the drainage basin to support the persistent Lance fluvial system.

(3) A net sandstone isolith map (not provided in this chapter), constructed using normalized gamma-ray well-log curves for the Lance Formation, shows a strikingly elongate "thick" comprised of 900–1500 ft (270–460 m) of net sandstone oriented parallel to the Wind River Mountains. This net sandstone thick extends from the northern part of the Pinedale anticline (T33N, R109W) southeastward through the eastern half of Jonah field (T29N, R108W). This feature continues farther south at least to T28N, R108W and T28N, R107W.

(4) Interpretation of the Lance Formation cross-bed orientations using high-resolution borehole image logs (Formation MicroImager, FMI) from several wells at Jonah field and on the adjacent south part of the Pinedale anticline shows primarily bimodal (east and southwest) dips. These dips are interpreted to record the sloping margins of longitudinal sand bars

►

Figure 6. Selected Late Cretaceous paleocurrent orientations in southwestern Wyoming. The sediment dispersal directions for Late Cretaceous nonmarine rocks at localities 1–6 are discussed in the text. The heavy dashed line, which arcs around the Rock Springs uplift, is the position of a marine shoreline at the early Maastrichtian *Baculites clinolobatus* zone (Cobban et al., 1994). This shoreline defines the embayment of the Lewis seaway on the south side of the Granite Mountains uplift. Localities are 1 = Hams Fork conglomerate (DeCelles and Cavazza, 1999); 2 = Hoback Canyon (Schuster and Steidtmann, 1987); 3 = Jonah field Formation MicroImager (FMI) well logs; 4 = Rock Springs uplift (Hodges and Knutson, 1981) and the senior author's observations; 5 = Red Desert basin Maastrichtian strata (Van Horn and Shannon, 1989); 6 = Wind River basin (Gillespie and Fox, 1991).

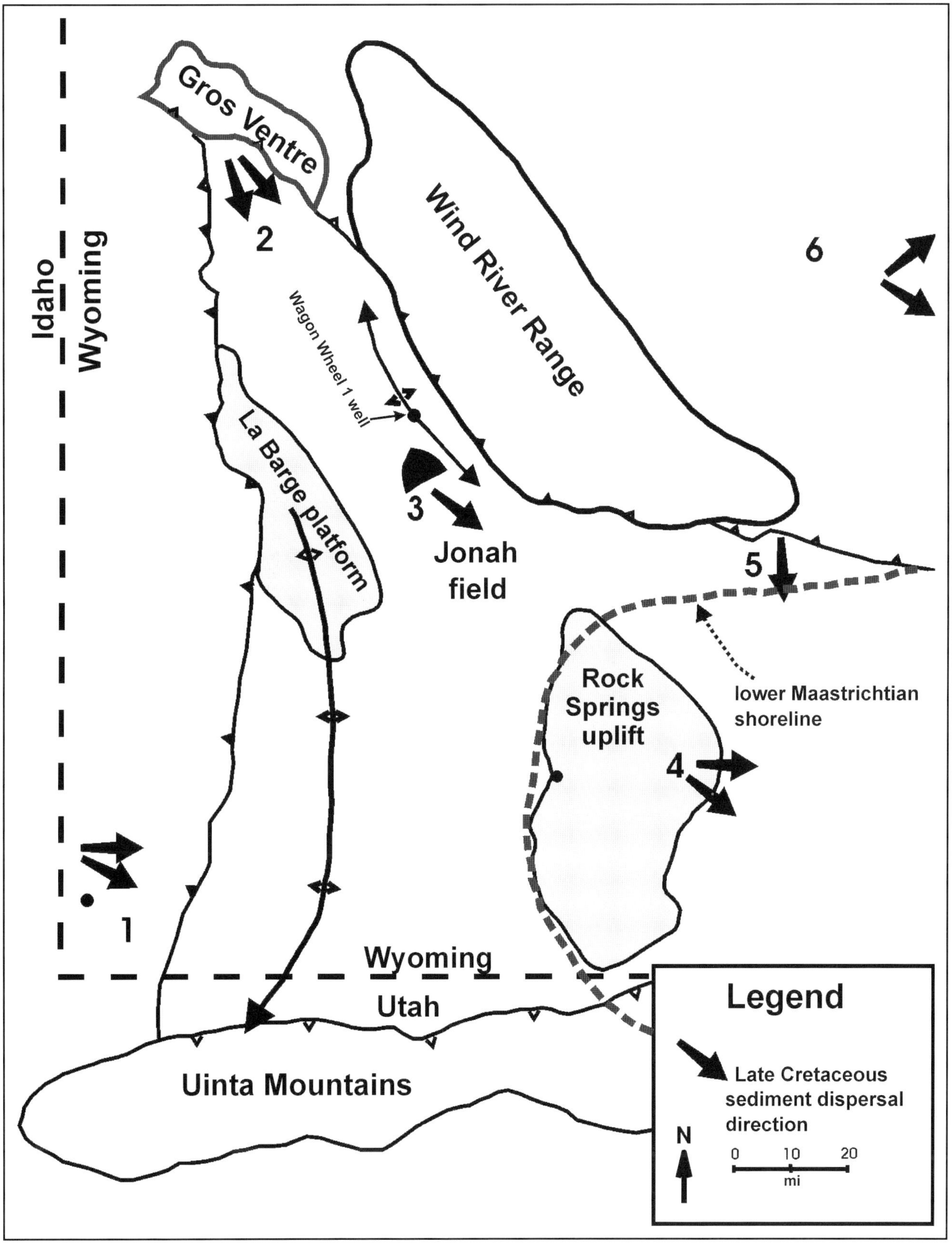
Gros Ventre
Wind River Range
Idaho
Wyoming
Wagon Wheel 1 well
La Barge platform
Jonah field
Rock Springs uplift
lower Maastrichtian shoreline
Wyoming
Utah
Uinta Mountains
1
2
3
4
5
6
Legend
Late Cretaceous sediment dispersal direction
N
0 10 20
mi

and lateral accretion surfaces in south–southeast-flowing, low-sinuosity fluvial channels (S. Hansen, 2000, personal communication). For example, Figure 7 shows a histogram of 682 cross-bed dip orientations recorded by an FMI survey in a well located in the southwestern part of Jonah field (Amoco Yellow Point 8-14, Sec. 14, T28N, R109W). Cross-bed inclinations were measured over the entire Lance interval (2650 ft; 810 m) and have a mean dip direction to the east–southeast (110°) but are bimodal to the east–northeast and southwest, reflecting the opposite flanks of longitudinal bars and accretion surfaces in a southeastward-flowing fluvial system.

(5) Sandstone bodies in the Lance Formation on the Rock Springs uplift record sediment dispersal to the east (Hodges and Knutson, 1981), toward the shoreline of the epicontinental seaway and the Lewis Shale embayment. This is indicated at locality 4 in Figure 6. Locality 5 in Figure 6 indicates the generally southward or southwestward progradation of deltaic and prodeltaic turbidite and sediment gravity flow sandstones into the deep-water Lewis seaway (Van Horn and Shannon, 1989). Sediment dispersal was away from the emerging Granite Mountains uplift and northern part of the Rawlins uplift.

(6) Southeastward-trending elongated stratigraphic bodies in the upper Lance Formation at Jonah field have been detected with 3-D seismic data using a maximum coherency technology (Vega and Hanson, 2001; Vega, et al., 2001).

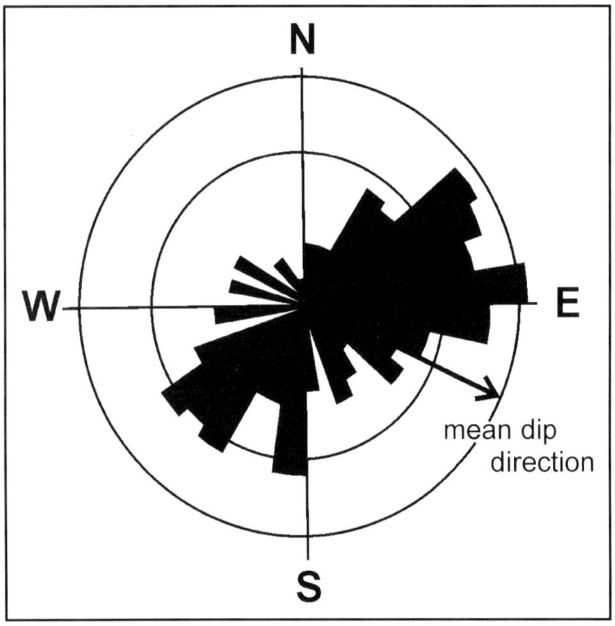

Figure 7. Rose diagram for 682 cross-bed dips in the Lance Formation at 7500–10,150 ft (2290–3100 m) in the Corona Unit 8-14 well (Sec. 14, T28N, R109W) in the southwestern part of the field. Cross-bed dips are from the FMI log. Mean dip direction is to the east–southeast. The bimodal dip pattern is interpreted to represent slopes on the sides of longitudinal bars of a low-sinuosity fluvial system, which flowed toward the southeast, parallel to the front of the Wind River uplift.

Northeastward-directed paleocurrents reported by Montgomery and Robinson (1997) may indicate sediment delivery by fluvial systems flowing more directly eastward from the Sevier fold-and-thrust belt, redirection of fluvial channels by localized syndepositional faulting along the south Jonah fault, or dips on the margins of longitudinal sand bars instead of cross-bed foresets.

STRUCTURE

Overall Structural Form

Jonah field is a subtle but complicated structural feature. Two near-vertical fault zones converge and intersect updip, toward the southwest, forming a pie-shaped, fault-bounded homocline (Figure 4). Figure 4 is a structure contour map of the field drawn on a marker bed in the Unnamed Tertiary unit (Bois marker) near the top of the Lance Formation and was constructed by integrating more than 300 digital well logs with depth-converted 3-D seismic data. It reveals a broad, low-relief, northeast-plunging structural nose in the area of the field. The Sand Draw syncline, at the downdip northeast edge of the field, separates it from the adjacent Pinedale anticline. We identify six important faults or fault zones and six folds or noses, which are shown in Figure 8. The Yellow Point nose is present in Jonah field, as well as outside of the field in the hanging wall of the west fault. The Sand Draw nose and the Stud Horse Butte syncline are en echelon folds oriented at acute angles to the south Jonah fault.

Field Boundaries

The updip and lateral fault-bounded edges of Jonah field are sharply defined and near vertical, although the south Jonah fault is locally complex and characterized by a zone of branching and anastomosing splays. Drill-hole data reveal the remarkably abrupt and updip edges of Jonah field where prolific gas wells exist along both the west fault and south Jonah fault in close proximity to uneconomic gas wells (Figure 4) (Warner, 1998).

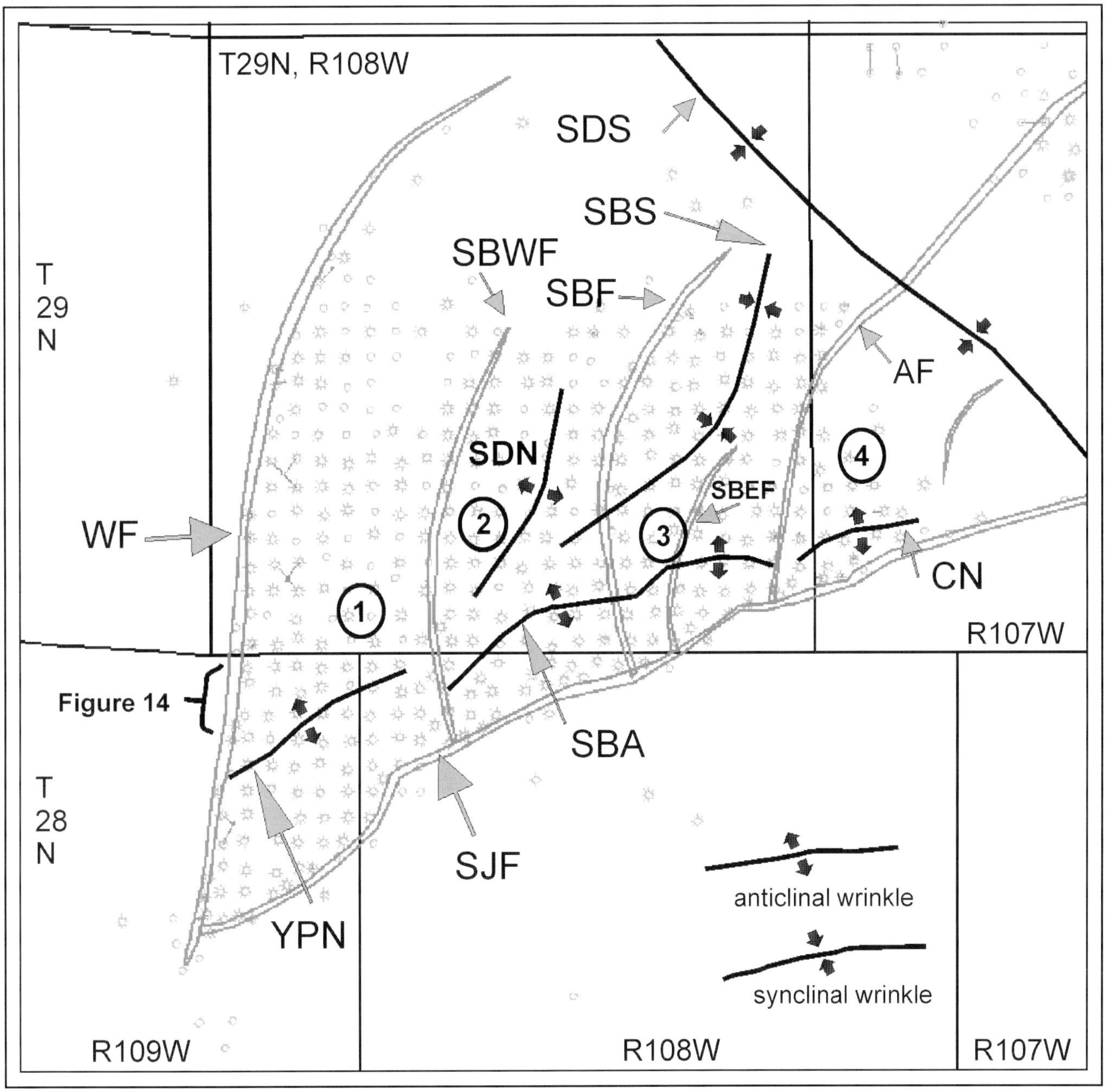

Figure 8. Map of Jonah field showing the principal faults and the small anticlinal and synclinal folds discussed in the text. WF = west fault; SBWF = Stud Horse Butte west fault; SBF = Stud Horse Butte fault; SBEF = Stud Horse Butte east fault; AF = Antelope fault; SJF = south Jonah fault; SDS = Sand Draw syncline; SBS = Stud Horse Butte syncline; SDN = Sand Draw nose; CN = Cabrito nose; SBA = Stud Horse Butte anticline; YPN = Yellow Point nose.

In the southwest part of the field (Figure 4; T28N, R109W), the field-bounding south Jonah fault has approximately 30 ft (9 m) of throw but is the ultimate trap for a 2000-ft (610-m)-thick highly productive section of Lance sandstones capable of 5–10 mmcf/day per well. These prolific gas wells are located adjacent to wells south of this fault, which yield noncommercial gas shows. This cryptic fault segment, a single planar feature, is detected with advanced seismic techniques; the throw is confirmed by maps on the basal Tertiary Bois marker (Figure 4). In contrast, the central section of the south Jonah fault is a broad zone of deformation 2000–4000 ft (610–1220 m) wide. In Figure 4, the position of the south Jonah fault marks the edge of the gas field at the stratigraphic level of the Bois marker. In deeper strata, the broad band of deformation comprising the south Jonah fault zone is displaced slightly south of the fault location shown in Figure 4. The west fault zone is less complex than the south Jonah fault and is generally a simple offset in Tertiary and Upper Cretaceous strata. This fault reaches the surface and influences landforms. A narrow anticline or counterregional monocline is associated with this fault where fault throw is small.

The downdip edge of Jonah field has a gradual decrease in productivity compared to the abrupt fault-bounded margins. Four wells drilled on the west limb of the Sand Draw syncline in the northeast corner of T29N, R108W (initial potential approximately 0.6–1 mmcf/day) are interpreted to be within the Jonah trap. These wells are less productive than wells drilled 1 mi (1.6 km) to the southwest, where equivalent strata are 400 ft (120 m) structurally higher.

Regional Strike Section (2-D Seismic Line)

Seismic line 803b, shown in Figures 9 and 10, is oriented approximately along regional structural strike. Jonah field is located between shotpoints 920 and 1020 (Figure 10). The Bois marker horizon at approximately 1500 ms is a mapping horizon above the top of the Lance Formation in the basal part of the Tertiary (Figure 2). Lance Formation gas reservoirs are located between the Bois marker and the Ericson Sandstone horizon. Note that the Lance interval has relatively poor reflector continuity compared to overlying and underlying strata. The west fault is located at approximately shotpoint 920, and a shallow horst is associated with this fault zone between 800 and 1000 ms. Displacement on both faults is subtle. Reflectors offset by the south Jonah fault are seen at least as deep as 2950 ms in this figure. Overall, the structural form is that of a broad, low-relief, slightly asymmetric nose.

Dip Cross Section (3-D Arbitrary Line)

Figure 5 is a dip-oriented seismic profile, with a vertical exaggeration of 4:1 and a total length of 12.1 mi (19.5 km), which traverses Jonah field from southwest to east. A depth-converted arbitrary seismic line from the 3-D seismic depth volume forms the backdrop, which is tied to five wells numbered 1–5. Gas production from the Lance Formation occurs in the interval between the top of the Lance (L) and the top of the Mesaverde Group (MV). The Lance Formation and the Unnamed Tertiary unit thicken eastward as previously discussed. B indicates the Bois marker-mapping horizon. The deepest stratigraphic penetration in Jonah field is a single well drilled to the top of the Hilliard Shale (−6000 ft; −1830 m) near well 1 on this figure. Well 3 penetrates the Ericson Sandstone near the center of the field. The syncline separating Jonah field from the Pinedale anticline is on the east side of well 5, which is at the eastern downdip limit of commercial production as presently known.

This seismic profile shows five of the six major faults. Note that these faults are nearly vertical but have little or no apparent displacement of the high-amplitude, basal Mesaverde reflections that are 1500–2000 ft (457–610 m) below the Ericson Sandstone (E). The west fault forms part of the trap for the gas field and dips slightly to the west at this location. Both the Stud Horse Butte fault and the Stud Horse Butte east fault locally have "tent-shaped" or apical anticlinal wrinkles. The small tent-shaped anticlines are found associated with other small faults in Jonah field and suggest local transpression of strike-slip or oblique-slip faults. Note that the seismic signature of the Fort Union Formation changes across the Antelope fault.

Identifying Faults with Coherency Technology

Faults form the critical trap for Jonah field and control significant production variations. A 3-D, broadband, amplitude-based, coherency algorithm with edge-detection capabilities (Marfurt and Kirlin, 2000) is one of the main tools we have used to interpret the faults. This algorithm examines the reflector amplitude gradient and captures the lateral change in amplitude at any azimuthal angle, i.e., at any horizontal orientation. This technique, which uses a time slice of seismic data or a mapped horizon, allows the interpreter to analyze or illuminate a particular feature from the optimal angle to see the maximum detail in the data. For example, Figure 11 shows views of the same area and seismic volume illuminated at 45 and 135°. The area illuminated at an azimuth of 135° clearly shows the Stud Horse Butte fault trending approximately normal to the direction of illumination, whereas the same area illuminated at an azimuth of 45° shows a fluvial channel but not the fault.

This coherency technology has enabled us to successfully execute precise drilling targets, with target radius as small as 40 ft (12 m), along the field boundaries and on the more prolific side of in-field faults. Wells A, B, and D in Figure 11 are on the east side of the Stud Horse Butte fault, and wells C and E are nearby on the west side. The fault dips approximately 85° to the east at this location. Production histories given in Figure 12 show that comparable wells on the eastern side of the Stud Horse Butte fault

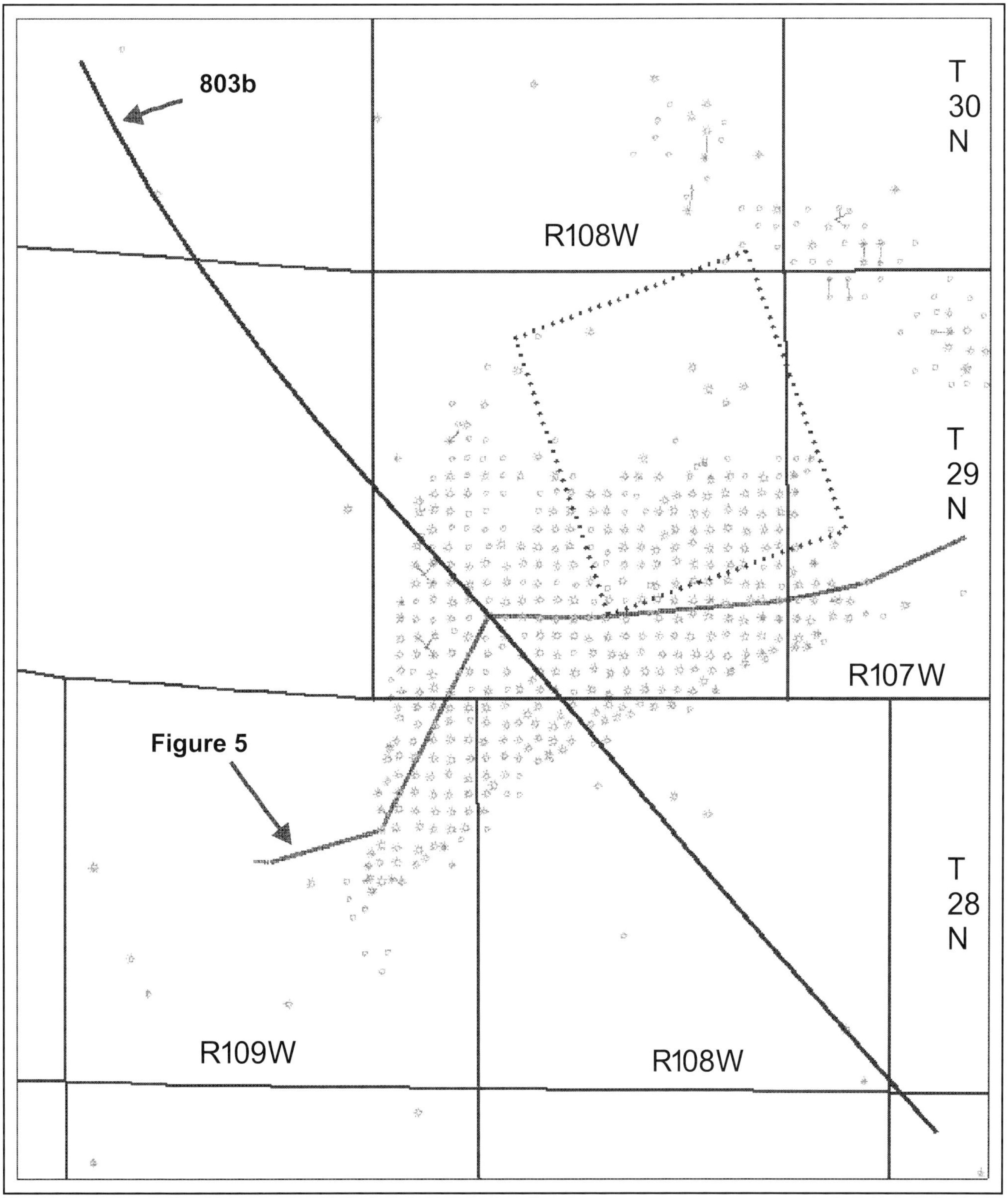

Figure 9. Location map showing the regional 2-D seismic line 803b and the dip-oriented seismic profile of Figure 5. The dashed rectangular outline is the area of the coherency maps shown in Figure 11.

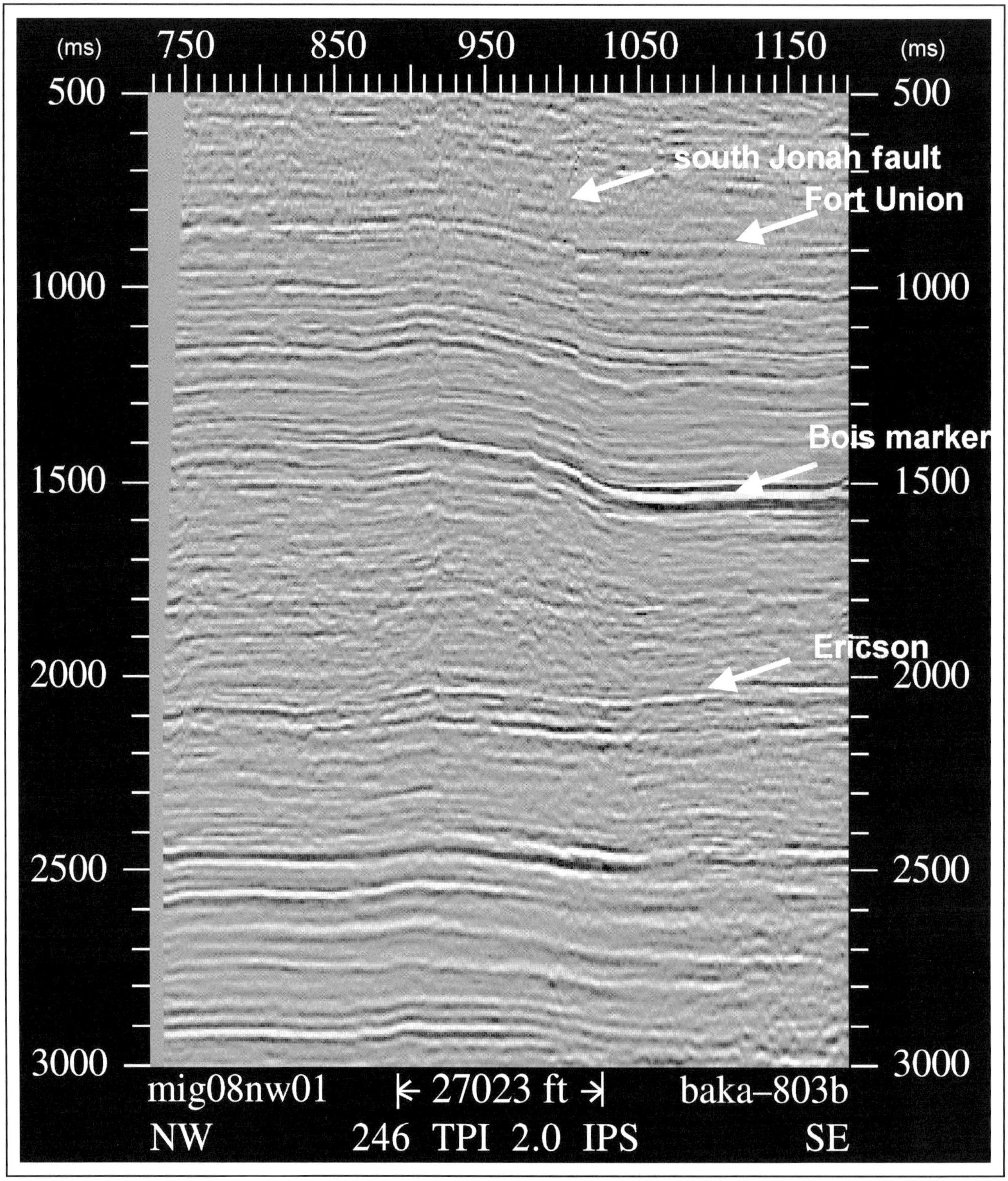

Figure 10. Regional 2-D seismic line 803b. Location of the line is shown in Figure 9. Northwest is on the left, and southeast is on the right. This line is 19.6 mi (32 km) long, time migrated, and the vertical scale is two-way traveltime in milliseconds (ms). Vertical exaggeration is approximately 7.4:1 above the Bois marker. The position of the south Jonah fault and selected horizons are identified. Jonah field is located between shotpoints 920 (west fault) and 1020 (south Jonah fault), as given along the top of the figure. Lance Formation reservoir sandstones are located between 1500 and 1900 ms.

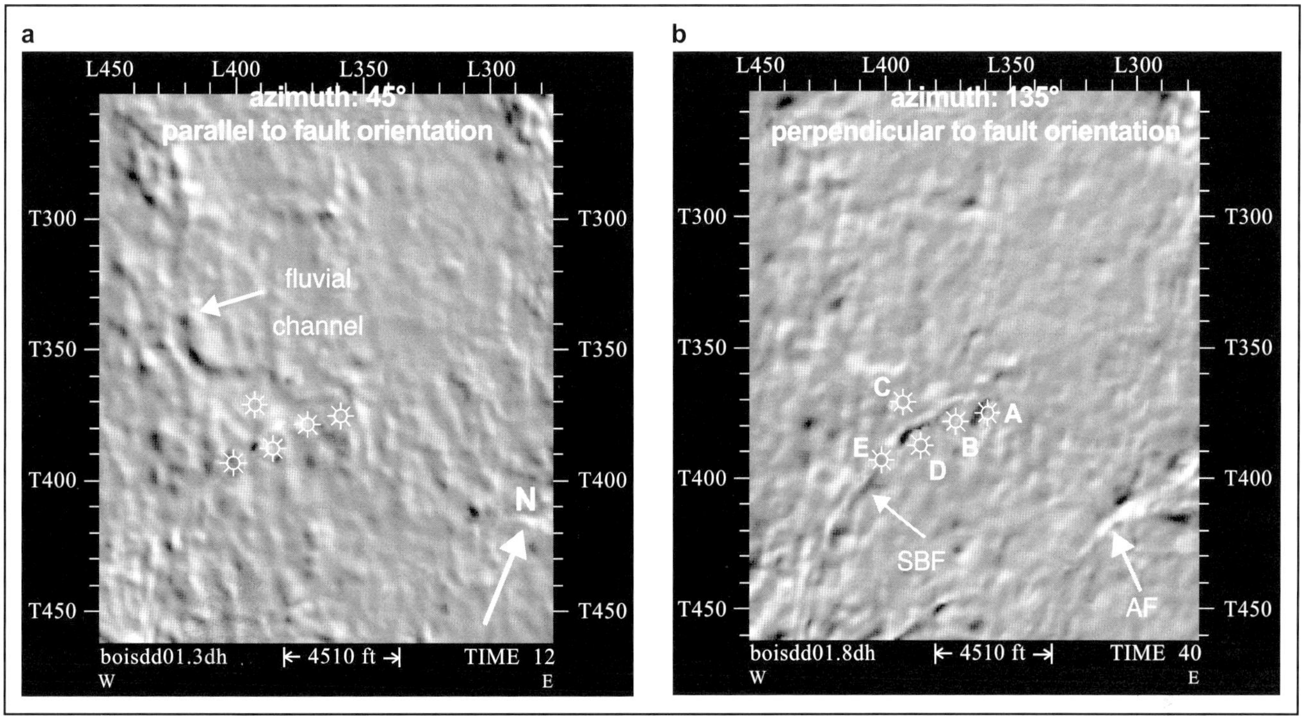

Figure 11. Two maps of azimuthal coherency computed from the 3-D seismic volume for an area in northeastern Jonah field shown by the dotted outline in Figure 9. These images show amplitude coherency computed at azimuths of (a) 45 and (b) 135°. Coherency is computed for a narrow time slice in the basal Tertiary section near the Bois marker horizon. Production for wells A, B, C, D, and E is discussed in the text and shown in Figure 12. Note that linear features oriented approximately normal to the direction for which amplitudes are analyzed ("illumination" direction) are preferentially resolved. SBF and AF identify the Stud Horse Butte fault and the Antelope fault, respectively. The north end of the Stud Horse Butte east fault is also visible but is not labeled.

have greater initial rates and estimated ultimate recoveries (EUR) than those located nearby on the west side. Superior production is on the east side of this fault despite the fact that this is the downdropped block. This pattern of superior production continues along the entire eastern side of the Stud Horse Butte fault for 3.25 mi (5 km) to the intersection with the south Jonah fault. The contrast in EUR across this fault increases toward the south, where wells on the east have EUR of 10–15 bcf, and adjacent wells on the west recover 1–5 bcf.

The Stud Horse Butte west fault and the west fault have superior gas production on the eastern side, even along the northern part of the west fault, where the eastern side is down. At least four of the principal faults or fault zones show preferential entrapment of gas on their east or north side, regardless of the amount or direction of fault throw. This pattern results from hydrocarbons being trapped against the faults in the general direction of westward hydrocarbon migration.

The amplitude-based coherency technique can also reveal details of stratigraphic features such as the low-sinuosity fluvial channel shown in Figure 11. However, the heterogeneity and scale of stratification in the Lance Formation makes this approach difficult to apply throughout the field.

West Fault

The west fault is a critical part of the updip trap for Jonah field. It is actually comprised of two faults and a minor fold that act as a transfer zone. It is not a scissors fault. The west fault juxtaposes normally pressured Lance Formation on the west with overpressured and commercially productive sandstones on the east. In the northern downdip region, commercial production has not been found on the east side of the fault below approximately −900 ft (−270 m) subsea, based on the three wells drilled in the northwestern part of T29N, R108W (Figure 4). The west fault is nearly vertical and extends from the surface, where it influences geomorphology, to at least the depth of the Hilliard Shale. Above the Lance Formation, displacement on the fault is normal, with downward displacement

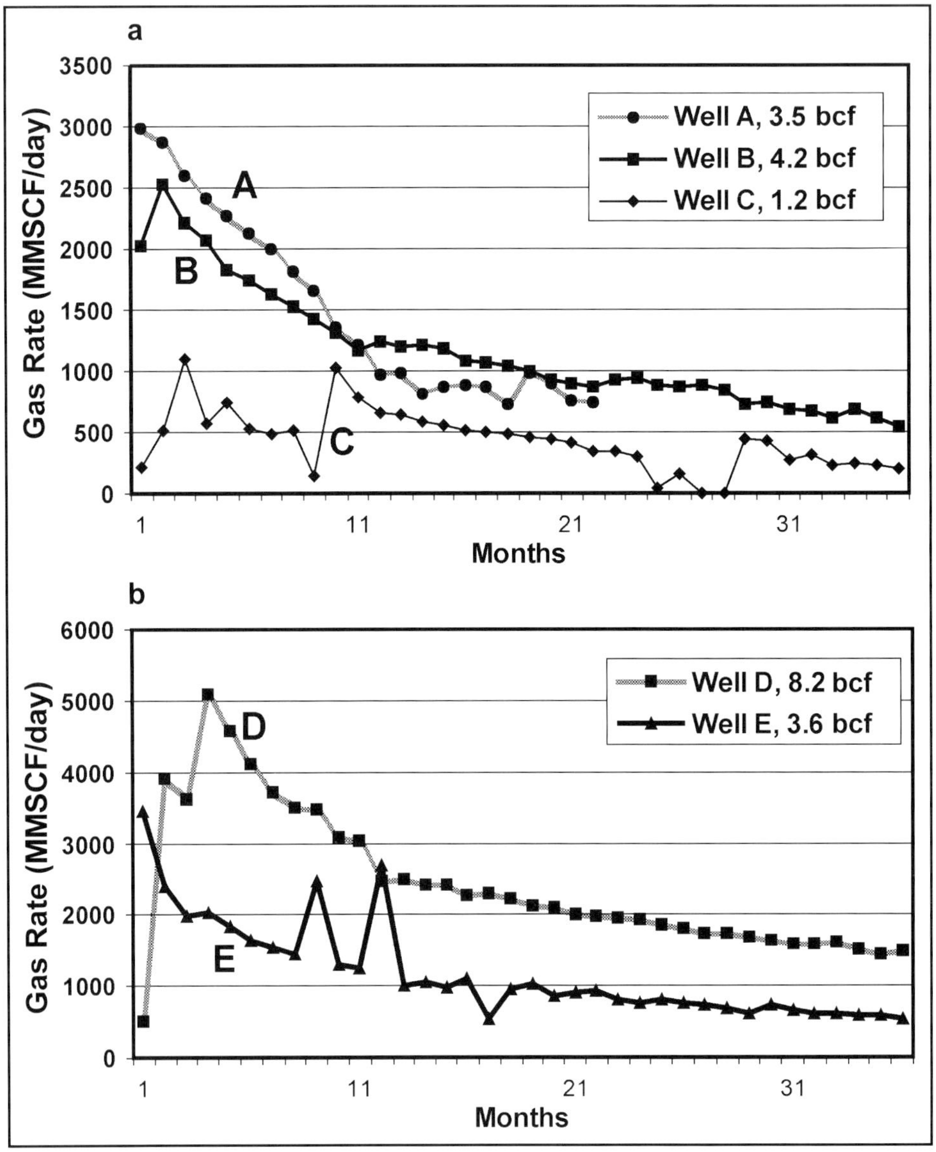

Figure 12. Comparison of production among wells on both sides of the Stud Horse Butte fault. Locations of these wells are in Figure 11. Graph a compares three downdip wells, where well C is on the west side of the fault and wells A and B are on the downthrown east side of the fault. Graph b compares two wells located a short distance to the southwest, where wells E and D are on the west and east sides of the fault, respectively.

of the footwall block. The shallow, normal displacement may result from stretching above a deep-seated strike-slip or oblique-slip fault, but the horizontal slip on such an associated wrench fault must be relatively small. Within the Lance Formation, seismic data show subtle indications of arching of strata and possibly structural imbrication by small thrust faults, suggestive of an incipient convergent (transpressional) strike-slip fault. Figure 13 is a cross section showing the west fault at a location approximately 3.8 mi (6.1 km) from the intersection with the south Jonah fault. Note the arching of upper Lance and basal Tertiary beds and possible thrust fault imbrication of beds between −2000 and −6000 ft (−610 and −1830 m) subsea. In addition, note that vertical displacement is small or nil in the basal strata of the Rock Springs Formation at −6000 to −7000 ft (−1830 to −2130 m) subsea, although coherency technology shows a discontinuity indicative of the fault trace. Figure 10 crosses the northern part of the west fault at shotpoint 920 where throw is, in total, down to the east. A small horst is visible at the top of the Fort Union Formation (850 ms), and a narrow, pointed anticlinal wrinkle occurs along the fault in the lower Fort Union through upper Lance strata at 1150–1600 ms.

Some Jonah field operators have been highly successful by designing wells to intersect productive Lance strata within 300 ft (90 m) or less on the eastern side of west fault. The thick pay section, available well control, 3-D seismic data, and drilling technology allow this precise

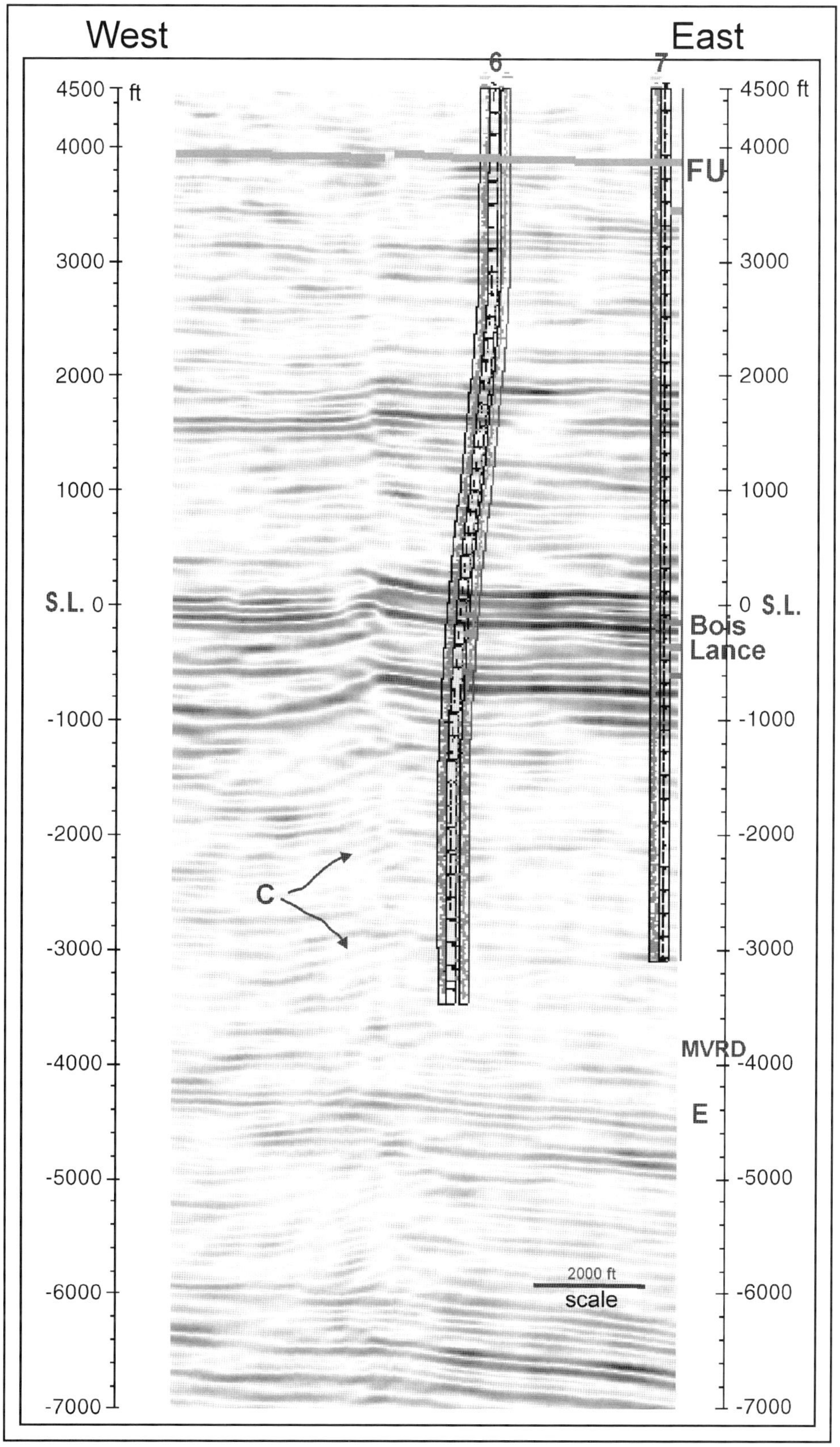

Figure 13. East–west cross section across the west fault of Jonah field showing well logs with a depth-converted seismic backdrop extracted from the 3-D seismic volume. Location of section is shown in Figure 4. Cross section is 1.9 mi (3 km) long, with vertical exaggeration 2:1. Elevation is labeled along the sides of the section in 1000-ft (300-m) increments. Horizons are marked as follows: FU = top of Fort Union; Bois = Bois marker in the basal part of the Tertiary section; Lance = top of Lance Formation; MVRD = top of Mesaverde Group; E = Ericson Sandstone. Basal marine sandstones of the Rock Springs Formation are at −6000 to −7000 ft (−1830 to −2140 m). Interpreted compressional structures are shown at the arrows labeled C.

development and field extension work. The west fault was intersected by a few wellbores in the process of drilling directional wells to small bottomhole targets. Distinctive stratigraphy in the lower part of the Fort Union Formation, including an interval with multiple coal beds, allows fault throw to be accurately determined with well logs and seismic data. Five wells drilled in Sec. 2 and 11, T28N, R109W are correlated in Figure 14. The west fault intersects wells C and D at the positions indicated by the arrows. Note the cutout of coal beds between 6500 and 7500 ft (1980 and 2290 m) in well C. The throw on this part of the fault is down to the west, with normal displacement of approximately 200 ft (60 m), which agrees with the throw estimated from depth-converted 3-D seismic data. The amount of throw intersected by well D is indicated by downward displacement of the sandstone at the top of the lower Fort Union Formation at approximately 5700 ft (1740 m).

Figure 15 shows the amount and direction of fault throw along the length of the west fault. Fault throw is interpreted with closely spaced arbitrary seismic lines taken from the depth-converted 3-D volume, and displacement is recorded at or near the Bois marker in an interval of high-amplitude, laterally persistent reflections. Within the Lance Formation, displacement can be difficult to determine because of discontinuous seismic reflections and lack of marker beds. Vertical displacement of the strata underlying the Lance Formation is apparently small. Figure 15 shows that displacement varies from south to north. The fault has a maximum throw of 300 ft (91 m) near the top of the Lance. A detailed study of several east–west arbitrary seismic lines reveals that the west fault is a composite of two separate faults. Reversal of fault throw occurs at approximately 26,000 ft (7900 m) from the south end of the west fault, where displacement is transferred, to both local folding and to the en echelon northern fault segment, which has normal displacement down to the east. Remarkably, this feature is an effective trap regardless of the direction of throw and even where the throw is nil (Figure 15). The amount of strike-slip displacement, if any, is unknown. The ratio of maximum throw to total fault length is approximately 0.01, which suggests that minimal fault-parallel stretch is required to accommodate the observed throw. Stewart (2001) has observed that this is a favorable setting for entrapment of hydrocarbons.

South Jonah Fault

General Characteristics

The south Jonah fault is complex and has had a protracted, episodic, and complicated displacement history. This feature is a fault zone and was interpreted to be a left-lateral wrench fault by Warner (1998), who cites evidence for 0.5–1.0 mi (0.8–1.6 km) of horizontal displacement. The south Jonah fault is more than 14 mi (20 km) long and extends east of Jonah field across the south plunge of the Pinedale anticline (T29N, R107W), which it appears to predate. The western terminus of the fault is in T28N, R109W, as revealed by 3-D seismic data and shown in Figure 4. Fault motion during the Lance, Fort Union, and post-Fort Union times is indicated by thickness and facies changes and by displaced strata. The central part of the fault is a broad zone of deformation locally 2000–4500 ft (610–1370 m) wide such as in Sec. 4, T28N, R108W. In this area, the south Jonah fault contains multiple strands of branching or anastomosing breaks that bifurcate vertically and laterally, typical of a strike-slip fault. The fault zone is marked by slightly offset, or terminated, short reflectors and includes small-scale reverse and normal faults in pre-Tertiary strata and mostly extensional faulting of the Tertiary section. The south Jonah fault can be seen as deep as 4200 ms on 3-D seismic survey in the Cabrito unit area (southeast part of T29N, R108W, Figure 4).

The western part of this fault zone is a left-stepping segment apparently containing a single fault plane, which is nearly subseismic but is, nonetheless, a remarkable and abrupt boundary for the field. Vertical displacement across the south Jonah fault ranges from less than 50 ft (15 m) in the west (T28N, R109W) to a few hundred feet in the east (northwest part of T28N, R107W).

Warner (1998) reported intense silica cementation near the south Jonah fault. A well drilled in the fault zone (Sec. 36, T29N, R108W) also encountered intensely silicified Lance sandstones with virtually destroyed reservoir properties. However, other wells drilled in similar positions have encountered excellent reservoirs and production.

Evidence for Late Cretaceous Displacement

Evidence for paleostructural development at Jonah field is present, where at least a portion of the south Jonah fault was active during the latest Cretaceous. Figure 16 shows an isopach of a unit comprised of the Lance Formation plus a basal Tertiary unit (interval from Bois marker to the top of Lance). The map illustrates thickening north of the south Jonah fault in the center of the field and a narrow thinning along the southeast edge. Figure 17 is a strike-oriented cross section across the field, located by Figures 4 and 16. The Mesaverde (MV), Ericson (E), and the basal Mesaverde Group reflections (−7000 ft; −2100 m) are offset by the south Jonah fault. The direction of throw in these deeper strata is down to the north, opposite the displacement in the overlying Tertiary beds (+3500 to −1000 ft; +1100 to −300 m). Figures 4 and 17 show that the overall structure in the basal Tertiary at Jonah field is

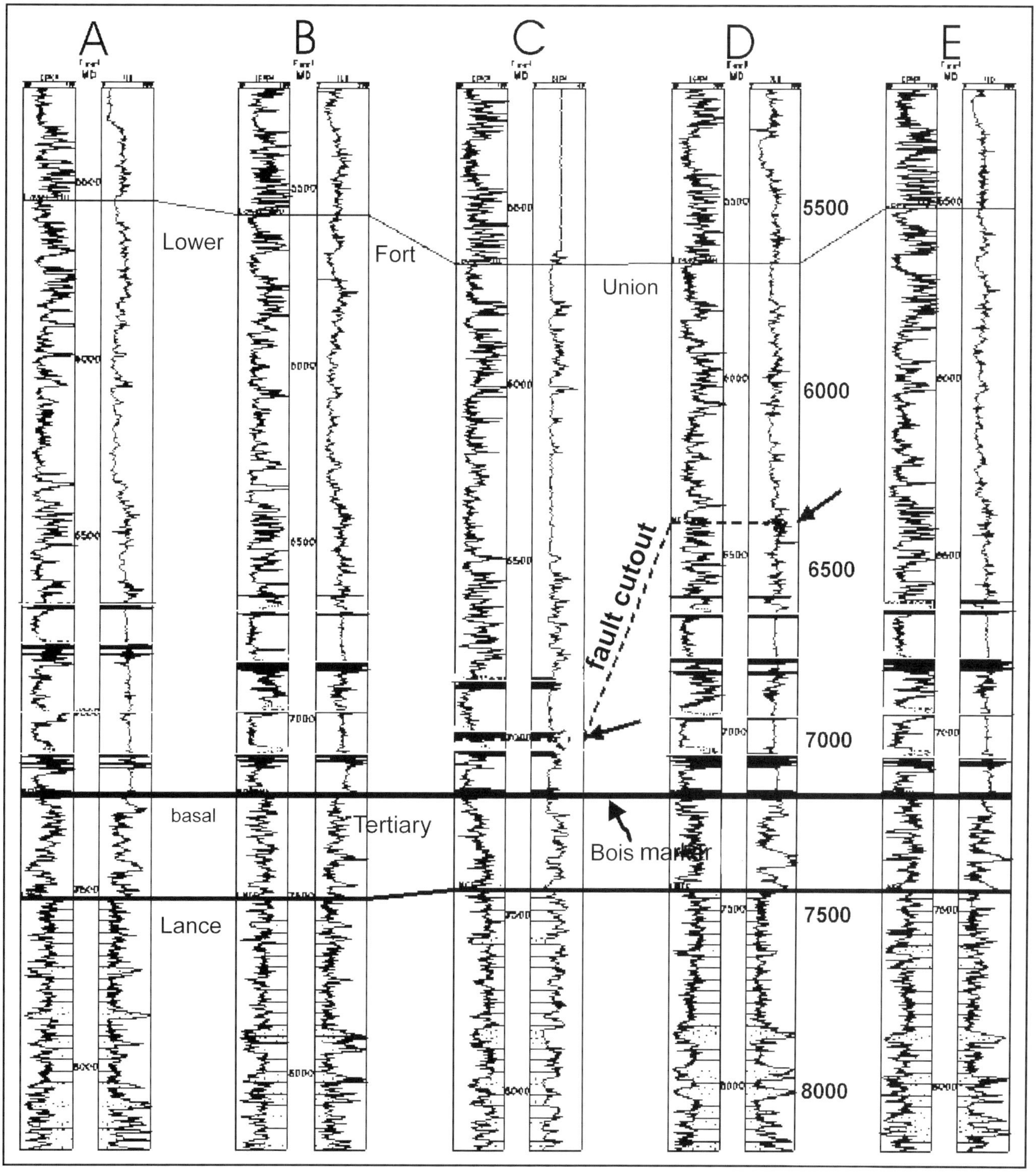

Figure 14. Correlation of five well logs from an area along the west fault in Sec. 2 and 11, T28N, R109W, in the southwestern part of Jonah field. General location of wells is shown in Figure 8. Section is flattened on the Bois marker horizon. Wells C and D drilled through the west fault in the basal part of the Tertiary section are at the positions shown by the two arrows. Note that the top of the lower Fort Union is dropped down in these two wells, and that coal beds have been cut out in well C. The west fault is a normal fault at this level, and the throw is about 200 ft (60 m), down to the west.

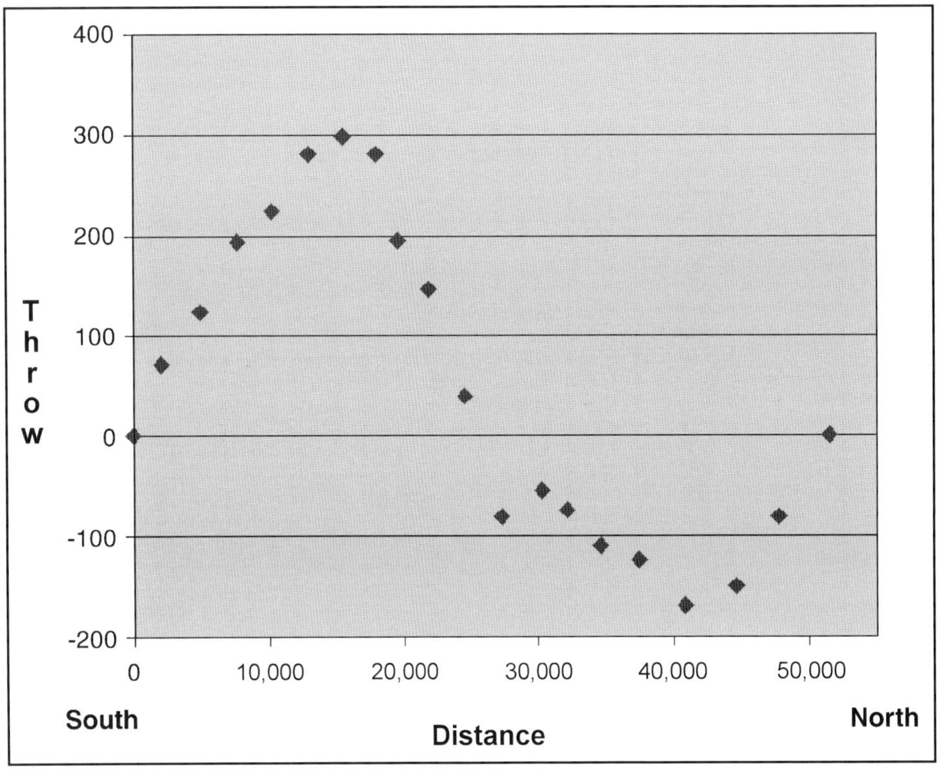

Figure 15. Graph of the amount and direction of fault displacement for the west fault zone. Positive displacement values indicate relative upward displacement of the east side of the fault. A reversal of fault throw is shown at a location approximately 26,000 ft (7930 m) from the south end and records the transfer of displacement to a second fault, which is down on the east.

nearly horizontal, except for the dip into the north side of the south Jonah fault. Most of the underlying Lance interval dips strongly northward away from the fault, and thinning in the upper part of the Lance is evident along the updip edge of the tilted block (northwest side of the south Jonah fault). Early Tertiary inversion accentuated the Lance thickening to form the Sand Draw nose. Northward tilt of the block and south dip into the south Jonah fault formed the Stud Horse Butte anticline, which is a prolific gas-producing trend.

The south Jonah fault has a protracted and complicated displacement history, with block tilting and warping during the Maastrichtian (down on north), influence of Fort Union facies, and the inversion of the Lance thickening during the Paleogene. Regional asymmetric northeastward thickening of the Lance, which was previously discussed, plus this local upper Lance thickening show that some movement on the south Jonah fault occurred during Cretaceous emplacement of the Wind River uplift.

Given that the south Jonah fault is a wrench fault formed during Laramide contractional tectonics, it may be assumed logically to be a left-lateral, convergent wrench driven by application of northeast–southwest-oriented, regional compressive stresses. However, the south Jonah fault and associated structures shown in Figures 4 and 8 are an unusual (nonideal) wrench-fault geometry (e.g., compare to Stone, 1969; Wilcox, et al., 1973). This could be a result of complications derived from preexisting structural fabric in the basement and sedimentary section, or it might be caused by counterclockwise rotation of regional compressive stress during the Laramide (Gries, 1983). It is possible that the south Jonah fault was subject to right-lateral and left-lateral displacement at different times if horizontal contractional stresses rotated from east–west to northeast–southwest or north–south. Two structural noses in the south part of Jonah field (Stud Horse Butte anticline and Cabrito nose) may result from transpressional stresses locally modified (counterclockwise rotation) as a result of extension imposed by bending over the edge of the inverted Lance thickening. This could explain the creation of these fold axes nearly parallel to, instead of being at acute angles to, the south Jonah fault.

Local Pull-apart Structure and Complex Displacement History

Figure 18 is a detailed structure contour map drawn on the Bois marker and shows the south Jonah fault and three secondary faults in addition to the Stud Horse Butte anticline, Cabrito nose, and south end of the Stud Horse Butte syncline. Figure 19, a cross section with a vertical

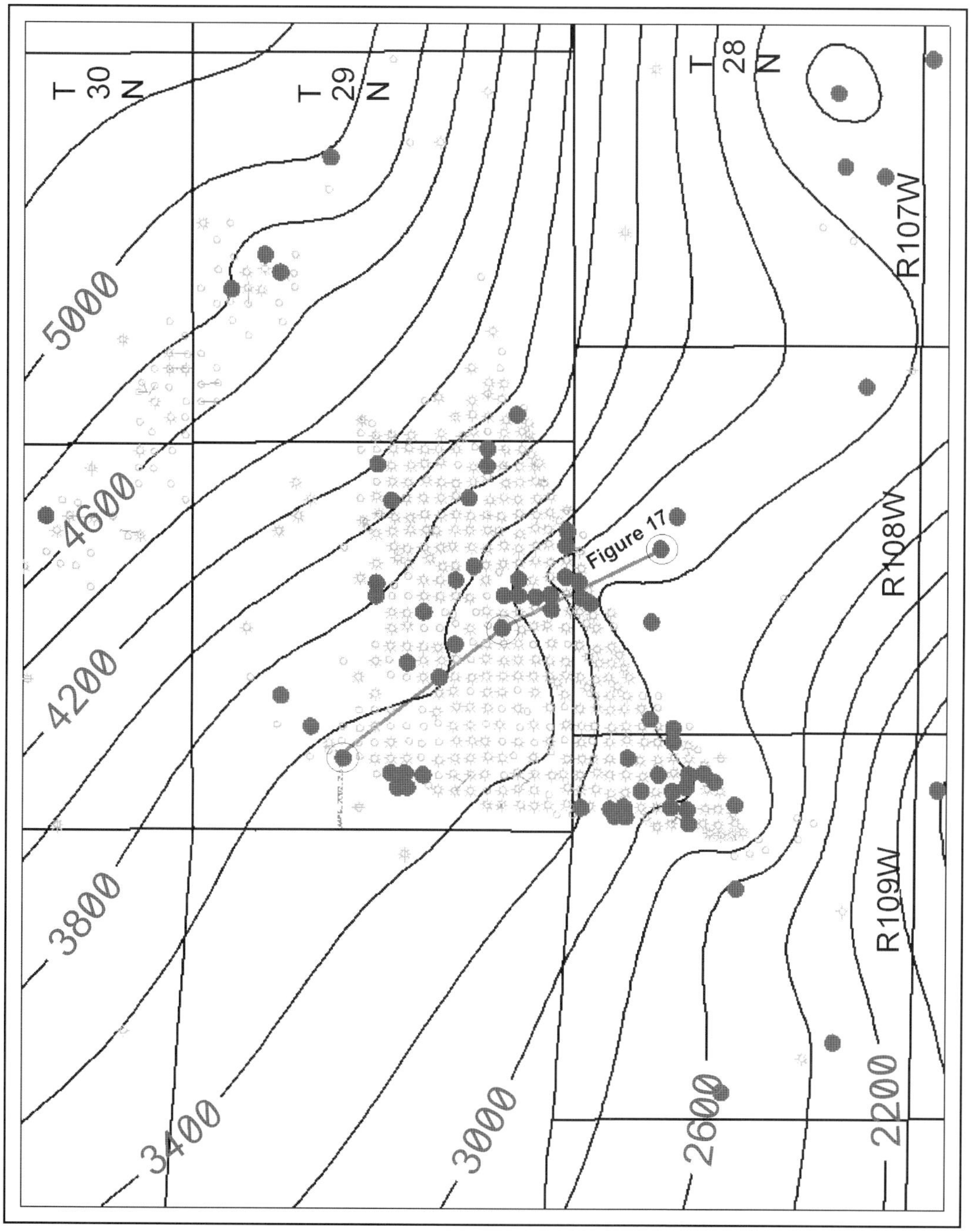

Figure 16. Isopach map of the Lance Formation and basal Tertiary unit. The top of isopach interval is the Bois marker horizon (see Figure 13) and the base is the top of the Mesaverde (base of Lance). Contour interval is 200 ft (60 m). Black dots are locations of control wells that penetrated the base of Lance. Note the narrow zone of thinning along the southeast edge of the field. Location of Figure 17 is shown.

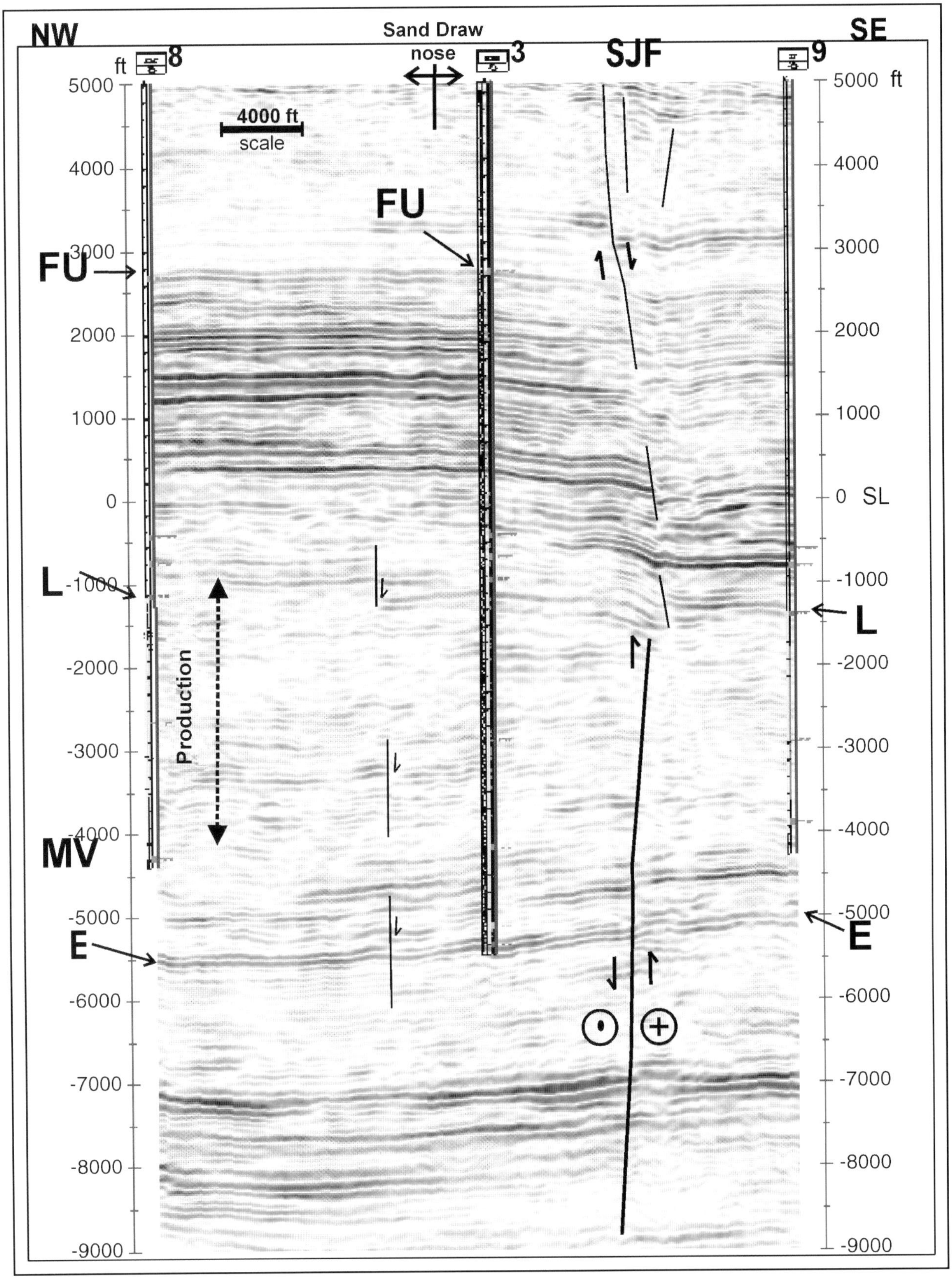
NW
Sand Draw
nose
SE
SJF
8
3
9
4000 ft
scale
FU
FU
L
L
MV
E
E
Production
5000 ft
0 SL
-9000

exaggeration of 2:1, shows the south Jonah fault, where there is approximately 200 ft (60 m) of throw down to the south. High-amplitude reflections of the Mesaverde Group are offset at this location, creating a structurally high fault block edge beneath the south margin of Jonah field.

The bold arrow in Figure 18 indicates the position of a small, anomalous structural low associated with a bend in the south Jonah fault. This is the lowest structural feature in this part of the gas field and is interpreted to be a pull-apart graben or sag (Aydin and Nur, 1982). A small segment of the fault has a north-facing convex shape, which acted as a releasing bend (Biddle and Christie-Blick, 1985). Figure 20 is a cross section at the western end of this graben. The graben is visible as a minor syncline at the Bois marker horizon, and a pronounced graben exists in the lower part of the Fort Union. The fault comprising the north side of this graben is antithetic to and merges with the south Jonah fault at a level above the top of the Lance. These relationships support the interpretation of the south Jonah fault as a left-lateral wrench.

Fault zone complexities are common to strike-slip faults. Variable displacement, vertically and horizontally along the fault, results from strain partitioning at fault bends, causing local areas of compression and extension. In Figure 20, note that the Stud Horse Butte east fault is a simple feature having approximately 100 ft (30 m) of throw. This simple fault does not extend as far downdip or vertically as some of the other faults, and it is the only one of the principal in-field faults that does not juxtapose areas of high- and low-EUR wells. It is surrounded by excellent gas wells.

Possible Displacement Transfer

Three-dimensional seismic data show that the south Jonah fault terminates in the northeastern part of T28N, R109W, where it meets the west fault (Figure 4). Horizontal displacement on the south Jonah fault is probably small, perhaps a few hundred feet. Throw decreases westward from a broad band of deformation in the central part of the south Jonah fault (T28N, R108W) toward the southwest, where it steps to an en echelon, nearly subseismic segment. This part of the south Jonah fault zone has less than 50 ft (15 m) of throw and is located in the northeast part of T28N, R109W (Figure 4). A small amount of displacement is supported by lack of offset of the west fault in Sec. 23, T28N, R109W. The cryptic southwestern part of the fault clearly exists, as demonstrated by 3-D seismic and precise well-log correlation in the thick Lance pay section. Advanced seismic techniques show that this western segment is probably a single, simple fault plane.

Strain along south Jonah fault may be distributed westward by gentle folding in the field. Opposite the transition from the central to the western segments of the south Jonah fault, thrust faulting exists deep below the Lance Formation. In addition, there is broad folding of Lance and post-Lance strata in the southwestern apex of the field. The folding and thrusting may record the transfer of displacement from strike-slip to other compressional structures. In Sec. 6 and 7, T28N, R108W, there is an abrupt northeastward increase of dip located above primarily west-verging thrust faults. The thrust faults are evident in the Hilliard Shale and in the Frontier Formation, and the overlying strata form two northwest-plunging noses arched above the duplicated beds. At a depth of approximately 5000 ft (1500 m) in NE/4 Sec. 1, T28N, R109W, the Fort Union Formation is broadly folded along this same northwest-trending feature. Onlap of this fold by fluvial beds in the upper Fort Union Formation shows that deformation was synchronous with deposition of the Paleocene strata. Total shortening among these contractional features appears to be no more than 1000 ft (305 m). Orientations of these fold axes fit a left-lateral wrench system, wherein the principal horizontal stress is perpendicular to the primary folds and thrust faults during the early stage of wrenching. More work is needed to fully test this concept.

◀

Figure 17. Northwest–southeast section across Jonah field and the south Jonah fault (SJF) (Location shown in Figures 4, 16). The section is 6.0 mi (9.6 km) long, and vertical exaggeration is 4:1. Three wells are shown on a backdrop of depth-converted seismic data from the 3-D volume. Well 3 is the same as well 3 in Figure 5. Lance production interval is shown by the vertical dashed line. Horizons shown are FU = Fort Union; L = Lance; MV = Mesaverde Group; and E = Ericson Sandstone. Basal marine sandstones of the Rock Springs Formation are at −7000 to −7500 ft (−2140 to −2290 m). The Stud Horse Butte west fault is barely visible between wells 8 and 3 (intermittent line and arrows), where it has less than 100 ft (30 m) of throw, down to the west. Within Jonah field, strata in the lower Fort Union are nearly horizontal except for southeast dip near the SJF. Most of the Lance section dips continuously to the northeast, creating a structurally high fault block edge along the SJF. Note the thinning and truncation in the upper Lance interval on the northwest side of SJF. Although throw is barely evident at −7000 to −9000 ft (−2140 to −2740 m) on this section, coherency techniques reveal a prominent discontinuity through these strata that extends much deeper into the sedimentary section.

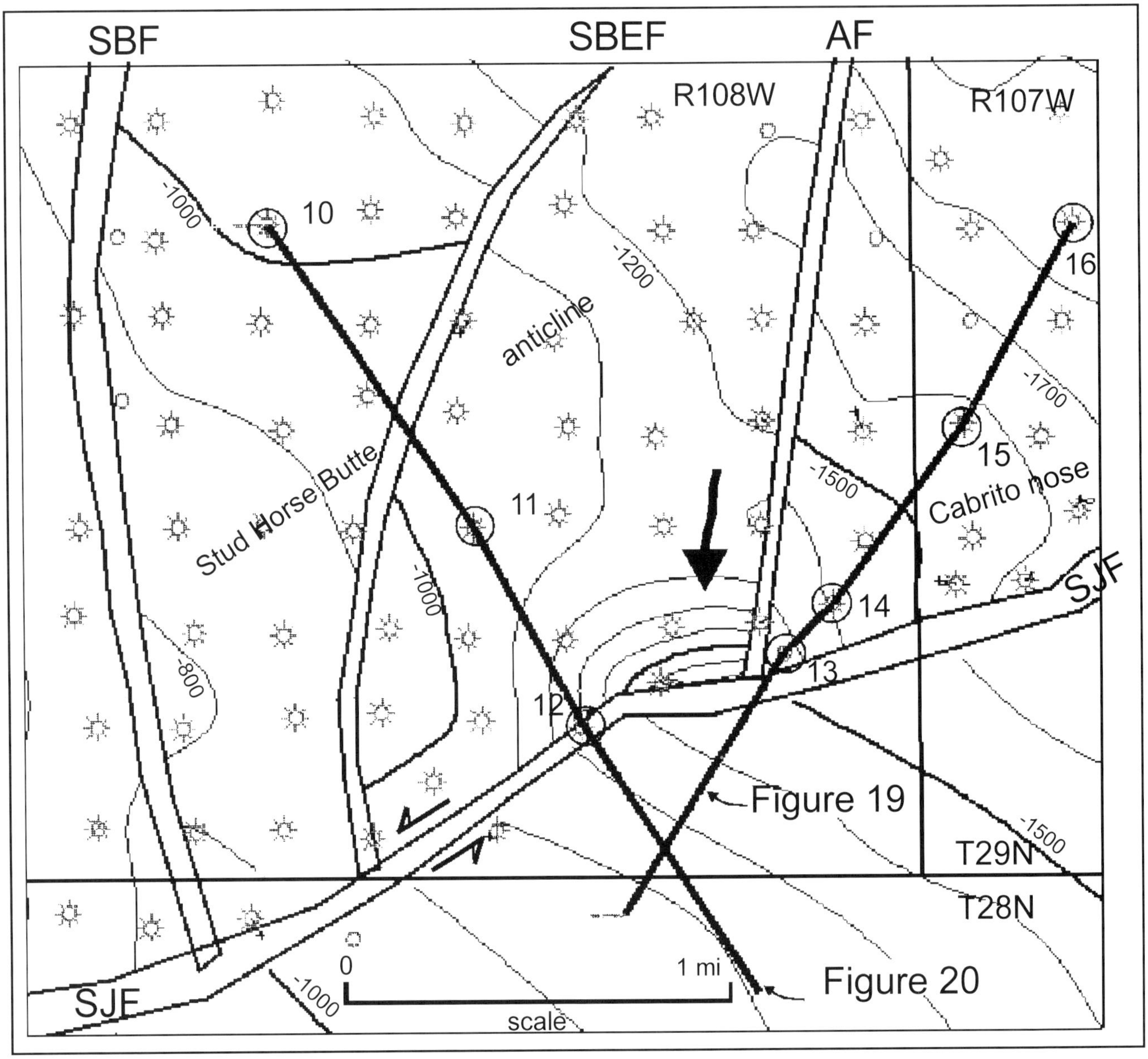

Figure 18. Detailed structure map on the Bois marker of an area along the south Jonah fault (SJF) in the eastern part of Jonah field. Contour interval is 100 ft (30 m). The bold arrow indicates an anomalous structural low along the south Jonah fault in section 36. Faults are labeled SJF = south Jonah fault; AF = Antelope fault; SBEF = Stud Horse Butte east fault; SBF = Stud Horse Butte fault. SJF is the southern boundary for Jonah field.

Antelope Fault, Stud Horse Butte Fault, and Stud Horse Butte West Fault

The arcuate nature of faults (Figure 4) in Jonah field is especially interesting because these faults are nearly vertical. The in-field faults may be curved for any of several reasons, including (1) lateral change in the stress field approaching the south Jonah fault and (2) interaction of two or more faults to create late-stage composite features. At a distance from the south Jonah fault, the orientation of the secondary faults is compatible with their origin as antithetic Reidel shears (Mandl, 1988). Approaching the south Jonah fault, these secondary faults bend into the south Jonah fault and terminate against it (Figures 4, 8). This indicates that the south Jonah fault is an older feature, and that it altered the local stress field at shallow depth, where overburden stress was relatively low.

Small, tent-shaped folds locally exist astride in-field faults, including the Antelope fault, Stud Horse Butte fault, and Stud Horse Butte west fault. For example, Figure 5 shows a tent-shaped fold on the Stud Horse Butte fault. These features are prevalent along the Antelope fault and may result from transpression. Displacement on the Antelope fault varies in magnitude along its length and appears to diminish with depth and with distance from the south Jonah fault. Maximum displacement of a few hundred feet exists at the south end of this fault, as shown in Figure 18. With coherency technology, the Antelope fault is seen as deep as 3400 ms near the south Jonah fault. The maximum depth at which coherency reveals the fault trace and at which fault throw is visible generally decreases with distance from the south Jonah fault. Displacement changes from west-side down in the south to east-side down in the north. Examination of seismic coherency maps at various levels in the 3-D seismic volume shows that the Antelope fault may have evolved to its present arcuate shape from multiple, linear trends. Below the Lance, the Antelope fault is linear, trends S45°W, and approaches the south Jonah fault at an angle of approximately 20°. This linear fault segment shows a sense of displacement down to the east along its length. At shallower levels, the Antelope fault is arcuate, convex to the west (Figure 8), and appears to be the result of two faults that have merged in the eastern part of T29N, R108W. A curious small syncline in the northeast portion of Section 25 is present on the upthrown west side of this fault but actually marks the downthrown east side of the earlier linear fault. Therefore, two originally separate faults have merged. This explains the apparent reversal of throw along the feature. The Antelope fault appears to have responded, over time, to a changing stress field in proximity to the older south Jonah fault.

JONAH PETROLEUM SYSTEM

Thermal Maturity and Origin of the Hydrocarbons

Jonah field produces rich gas with approximately 10 bbl condensate/mmcf gas. Average gas composition is approximately 91% C_1, 8.5% C_{2+}, and 0.5% CO_2. The methane isotope composition is indicative of thermogenic gas ($\delta^{13}C$ ratio = −36‰ [PDB]). Advanced carbon isotopic analysis of the larger gas molecules indicates that 90% of the gas is derived from the thermal cracking of oil generated by vitrinite group organic matter. The thermal cracking occurred at high temperatures (L. Dzou, 2001, personal communication). Bowker and Robinson (1996) and Warner (1998) observed that much of the Lance Formation is less thermally mature than the probable source rocks for the produced hydrocarbons. Warner (1998) (Figure 6) shows that vitrinite reflectance in rocks shallower than 11,000 ft (3400 m) is indicative of early maturity for gas that was created by catagenesis of humic organic matter. Bowker and Robinson (1996) reported that the carbon isotopic composition of the Jonah gas does not become heavier with depth. They proposed that the Jonah hydrocarbons were derived from more mature rocks in the lower Lance and upper Mesaverde Group. The Mesaverde Group contains significant coal beds and carbonaceous shale, as seen on wire-line logs from deep wells in the Jonah area, such as McMurry Oil Company Yellow Point 11-13 (Sec. 13, T28N, R108W). Table 2 shows selected samples taken from full diameter core material. These samples represent all of the lithologies that have the potential to contain a significant amount of carbonaceous or sapropelic organic matter. Note that Table 2 includes some upper Mesaverde coaly material.

Our limited investigations of Lance Formation thermal maturity, in combination with analyses of the isotopic composition of produced gases, indicate that Lance reservoirs contain gas and condensate generated predominantly from deeper and hotter, more thermally mature rocks. The source materials were comprised of humic organic matter. Jonah hydrocarbons migrated upsection and possibly laterally from the east into the present reservoirs. This "bottoms-up" sourcing fits a model in which reservoir sandstones were charged with gas prior to complete compaction, diagenesis, and reduction of reservoir characteristics (porosity, permeability, and pore-throat size) to the present condition.

Trap, Seal, and Hydrocarbon Emplacement

Warner (1998, 2000) identifies a shale in the upper Lance as the top seal for the Jonah trap based on the need to increase drilling mud density upon penetrating overpressured gas-bearing sandstone bodies in the top part of the Lance Formation. The interpreted top seal is nonmarine mudstone with sandstone beds, as seen in Figure 14 at 7500–7800 ft (2300–2380 m). However, this mudstone-dominated unit is not present throughout the field and is lithologically indistinguishable from other parts of the formation. During the past 2 yr, many wells have been completed in lenticular, fluvial-sandstone bodies in the basal Tertiary overlying the Lance shale (Figure 14). The fluid-pressure gradient in the upper part of the Lance is commonly approximately 0.56 psi/ft in areas of typical gas production and reservoir quality. The fluid-pressure gradient is at least 0.70 psi/ft in the lower Lance at 11,800 ft (3600 m) in the McMurry Oil Company Stud Horse Butte 5-26 (Sec. 26, T29N, R108W), and it is 0.68 psi/ft in the

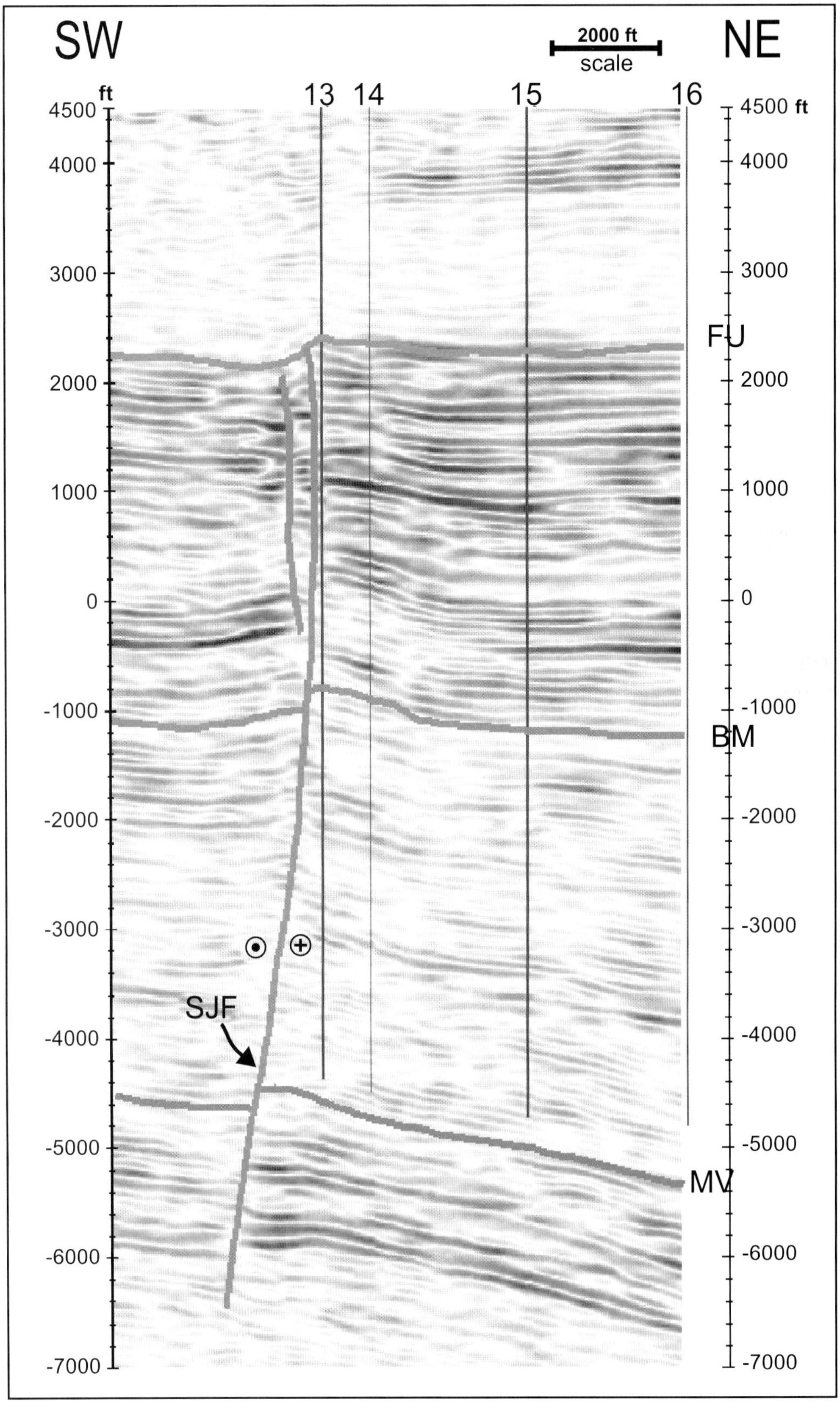
SW
2000 ft
scale
NE
ft
4500
4000
3000
2000
1000
0
-1000
-2000
-3000
-4000
-5000
-6000
-7000
13
14
15
16
4500 ft
4000
3000
2000
1000
0
-1000
-2000
-3000
-4000
-5000
-6000
-7000
FU
BM
MV
SJF

upper Mesaverde Group at 12,165 ft (3710 m) in the Amoco Stud Horse Butte 5-9 (Sec. 9, T29N, R108W). The fluid-pressure gradient increases with depth in the Lance Formation at Jonah field, as shown by drilling mud weights and formation pressure measurements (D. DuBois, 2003, personal communication). We propose that the Lance section, in total, is an effective seal, which also contains intercalated, lenticular, low-permeability sandstones. In this model, the Lance and basal Tertiary are both seal and reservoir, and the overlying Fort Union Formation is a leaky conduit. The large, thick, and laterally continuous coalesced sheets of meander belt sandstone bodies in the Fort Union Formation are remarkably ineffective traps, as indicated by the lack of gas production and the minor hydrocarbon shows found by wells drilled within and around Jonah field and elsewhere in the Green River basin. The conspicuous lack of gas saturation and lack of fluid overpressure in the Fort Union Formation may result from the efficiency of these laterally continuous and interconnected sandstone bodies as conduits for fluid migration. Vega et al. (2001) showed a laterally extensive, thick sheet of upper Fort Union sandstone extending from Jonah field to the Pinedale anticline. Fluid hydrocarbons are able to move buoyantly updip within these beds because of the high ratio of sandstone to mudstone, relatively high permeability, likely survival of sandstone-to-sandstone contacts across small-scale faults, and lateral migration pathways among large, interconnected sandstone bodies. Commercial production and drilling experience in the region confirms that the Fort Union Formation generally requires a structural closure to entrap gas around Jonah field. Whereas Lance and basal Tertiary mudstones provide the ultimate top seal for Jonah, the Fort Union is a conduit that enables buoyant fluids to escape updip once they are in this pipeline. Terminal traps for the Fort Union Formation are found on the La Barge platform (McDonald, 1973).

The Lance Formation is composed of 25–40% sandstone interbedded with mudstone and siltstone. These sandstones are highly lenticular and are mostly unconnected rock bodies (Shanley, 2001). Despite the occurrence of gas pay throughout a gross interval of 2000–4000 ft (610–1220 m), Jonah field does not have a single hydrocarbon column. There would be a greater degree of overpressure in the upper Lance if the field contained a single hydrocarbon column. Figure 8 displays four fault-bounded compartments, labeled 1 to 4. Note that the total structural relief across the gas field is approximately 1700 ft (520 m), and the downdip limit of commercial production, as shown by the limits of 40-ac (0.16-km^2) development wells, varies among the four compartments. Interestingly, in each compartment, the structural relief (near top of the Lance Formation) is less than 750 ft (230 m) from updip sealing fault to the downdip limit of commercial production. This may reflect the limiting seal capacity of the faults in combination with internal formation heterogeneity. A layered reservoir volume, having a dominant proportion of sealing layers, has more potential for hydrocarbon entrapment by faulting than does a connected reservoir volume 2000–4000 ft (610–1220 m) thick, for which a single fault seal imperfection will result in trap failure. Therefore, the stratigraphic heterogeneity of the Lance fluvial facies, including the 60–75% mudstone content, is critical for entrapment of the gross reservoir thickness found at Jonah field.

Fault block 3 (Figure 8), bounded by the Stud Horse Butte fault and Antelope fault, is one of the most prolific areas in the field because of sandstone quality and high net sandstone thickness (as much as 1150 ft [350 m] of sandstone). Fault block 3 has about 750 ft (230 m) of structural relief on the Bois marker horizon (Figure 4) between the updip south Jonah fault and the downdip limit of high-quality (>3 bcf/well) production. The change from high-quality production to marginal or subeconomic production is caused by a degradation in rock quality (low permeability to gas) and is abrupt, generally occurring across a single 40-ac (0.16-km^2) tract at about −1550 ft (−470 m) structural elevation (on Bois marker). The structurally high corner of fault block 3 is more than 700 ft (210 m) low to the highest part of the field, and it is approximately the same elevation as the downdip limit of commercial production in the western part of the field within fault block 1, i.e., between the west fault and Stud Horse Butte west fault.

A gas column of approximately 2000 ft (610 m) is required to create the buoyancy necessary to saturate the Jonah tight-gas reservoirs, given the existing pore-size distribution. A gas column of this magnitude would challenge the sealing capacity of the trap and possibly breach it. We propose that hydrocarbon was emplaced by buoyant

◀

Figure 19. Southwest–northeast depth-converted seismic profile across the south Jonah fault (SJF) in the eastern part of the field. The location of the section is shown in Figure 18. Seismic profile is from the 3-D seismic volume. The locations of four wells on the seismic line are shown with thin vertical lines. Vertical exaggeration is 2:1. Interpreted horizons are FU = top of Fort Union; BM = Bois marker in the basal Tertiary section; and MV = top of Mesaverde Group.

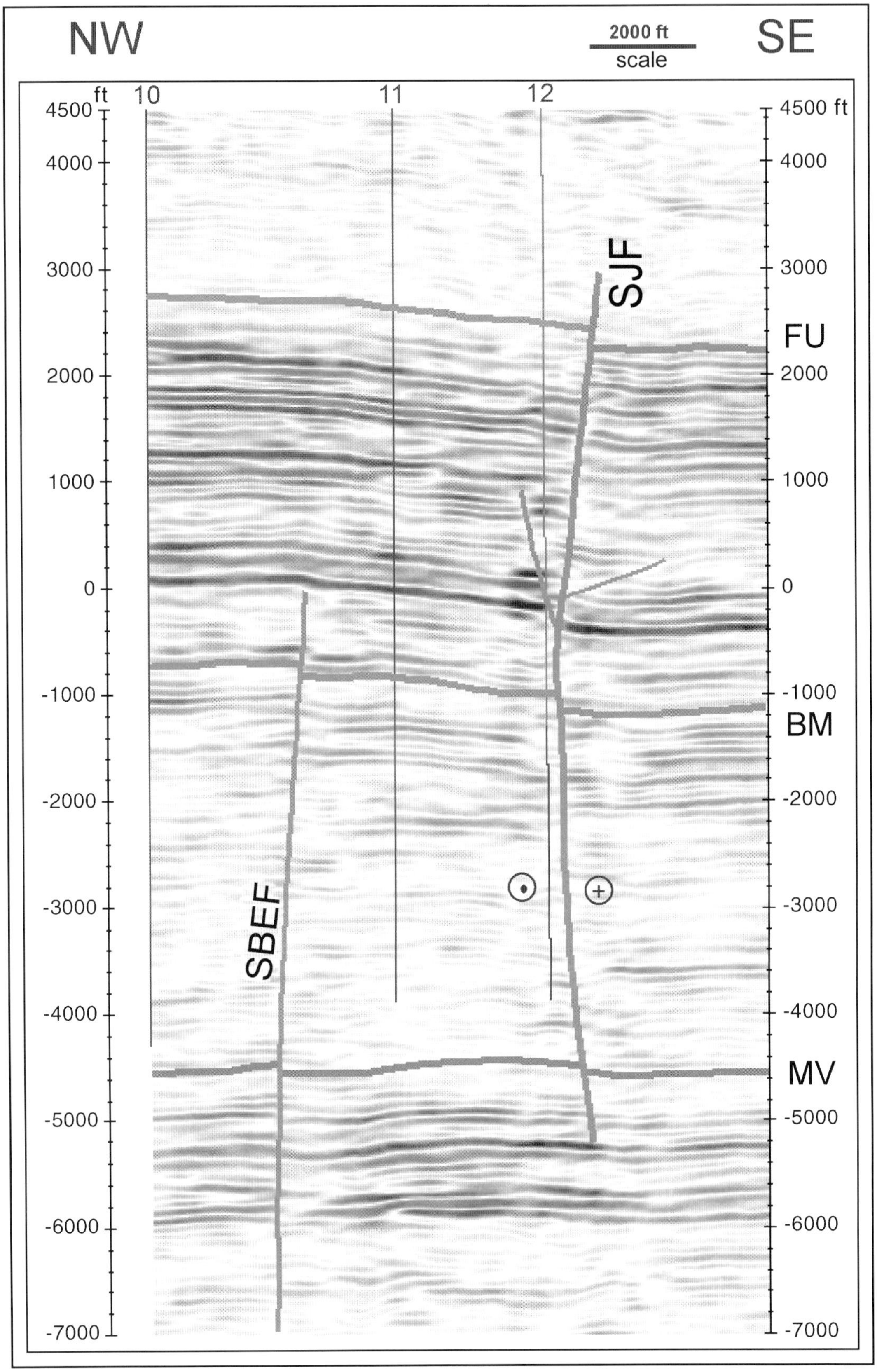
NW
2000 ft
scale
SE
ft
10
11
12
4500
4000
3000
2000
1000
0
-1000
-2000
-3000
-4000
-5000
-6000
-7000
4500 ft
SJF
FU
BM
SBEF
MV

Table 2

Vitrinite Reflectance and Total Organic Carbon Measurements for Selected Core Samples in Jonah Field.

Well Name	Location	Depth (ft)	Formation and Lithology	R_o (%)	Total Organic Carbon (wt.%)
Cabrito 15-25	Sec. 25, T29N, 108W	10,590–10,591	Lance, medium gray siltstone to silty mudstone	0.69–0.97 (few measurements)	1.04
Corona 1-11	Sec. 11, T28N, R109W	8883–8884	Lance, medium dark gray silty mudstone	0.89	0.48
Corona 7-14	Sec. 14, T28N, R109W	10,507–10,507.6	Mesaverde, dark gray silty carbonaceous shale	0.83	1.46
Corona 7-14	Sec. 14, T28N, R109W	10,521–10,521.3	Mesaverde, rooted silty claystone	0.82	1.44
Corona 7-14	Sec. 14, T28N, R109W	10,523–10,524.6	Mesaverde, carbonaceous shale and coal	0.85	24.64
Stud Horse Butte 11-27	Sec. 11, T29N, 108W	10,943.5–10,944.1	Lance, dark gray claystone	0.85	0.72
Stud Horse Butte 11-27	Sec. 11, T29N, 108W	10,944.5–10,945	Lance, dark gray claystone	0.78	1.70
Stud Horse Butte 11-27	Sec. 11, T29N, 108W	10,965.2–10,966.2	Lance, dark gray silty carbonaceous mudstone	0.71	2.48

migration and gravity segregation of fluids in the Jonah traps. Updip buoyant movement of hydrocarbons is also indicated by the concentration of superior gas production on the eastern and northern sides of Jonah faults, regardless of the direction or amount of fault throw. This process required early gas migration and saturation of the Lance sandstones before degradation of reservoir properties to the present condition. Paleostructural features prepared the Jonah area for early hydrocarbon entrapment. Regional dip, early development of the high block edge along the south Jonah fault, and gentle warping of the basal Tertiary strata directed the charge of gas from deeper source beds.

CONCLUSIONS

Jonah is a large and locally subtle, structural trap, which produces from a great thickness of low-permeability, lenticular, fluvial sandstones near the deepest part of the Green River basin. The gross-producing interval ranges from 1700 to 4000 ft (520 to 1220 m) thick and is found at depths ranging from 7300 to 12,800 ft (2200 to 3900 m). Structure provides critical trap elements and also shaped the fluvial system (reservoirs) of the Lance Formation. The combination of faulting and lenticular reservoir sandstone bodies enables entrapment of gas. Fault throw is commonly less than 200 ft (60 m). Production is correlated

◄

Figure 20. Northeast–southwest depth-converted seismic profile across the south Jonah fault (SJF) and Stud Horse Butte East Fault (SBEF) at the location shown in Figure 18. The locations of three wells on the seismic line are shown with thin vertical lines. Vertical exaggeration is 2:1, and interpreted horizons are the same as Figure 19. Note that the SBEF is up on the northwest, with approximately 100 ft (30 m) of throw. Despite the lack of visible offset of the upper Mesaverde reflections at −5200 to −6000 ft (−1590 to −1830 m) by the SJF, seismic coherency shows a fault break through these horizons.

to structural position, and high-quality gas pay is less common below 11,000 ft (3400 m), but the base of commercial production is not established. Seismic coherency technologies have been effective tools for mapping the faults and have added value when used to locate precise wellbore targets during field development and extension. The major faults are almost vertical, and most extend deeply into the sedimentary section, as shown by our coherency work.

Pronounced, asymmetric subsidence induced by the emerging Wind River uplift exerted a first-order control on the Lance Formation fluvial system. The Lance Formation thickens systematically northeastward toward the Wind River uplift. The Lance fluvial system flowed southeastward in front of the Wind River uplift, along the eastern side of the asymmetrically subsiding basin, toward the Lewis Shale seaway. This process resulted in an elongate net sandstone thickness (600–1150 ft [180–350 m] of net sandstone), which passes through Jonah field.

The overall shape of the Jonah structure, which is mapped at the top of the producing interval, is a broad, low-relief, northeast-plunging nose or homocline bounded by faults. Two nearly vertical fault zones intersect updip toward the southwest to form the wedge-shaped trap. These faults are probably associated with a wrench zone, as proposed by Warner (1998). The west fault zone is the west boundary of the field and is comprised of two faults with opposing throw and a minor fold at the transfer zone between the faults. The west fault, as with most of the Jonah faults, is an effective trap even where throw is downward toward the gas field or very small (<100 ft; <30 m). The secondary faults at Jonah field bend into the south Jonah fault zone, the southern boundary of the field, and they terminate against it, indicating that the south Jonah fault is an older feature, and that it altered the local stress field during formation of the in-field faults.

Isotopic analyses of produced gases, in combination with the thermal maturity of the producing strata, indicate that the field produces hydrocarbons originating in deeper and hotter, more thermally mature source rocks. Superior gas production is concentrated along the eastern and northern edges of the Jonah faults, regardless of the direction or amount of fault throw. The production patterns, composition of the produced gases, thermal maturity of the reservoir strata, and the downdip limits of production in separate fault blocks indicate that hydrocarbons migrated toward the southwest.

Structural relief across the Jonah gas accumulation is approximately 1700 ft (520 m) (at top of production interval), but the field is comprised of four fault-bounded compartments, each with no more than approximately 750 ft (230 m) of structural relief across the gas production. Elevation at the downdip limit of quality production varies among these four fault-bounded compartments. A hydrocarbon column of 2000 ft (610 m) would be required to saturate the Jonah tight-gas sandstones. These gas production patterns suggest that buoyant movement and gravity segregation of fluids in the Lance Formation occurred before maximum burial and degradation of reservoir properties. We envision that hydrocarbon emplacement occurred from the bottom upward and from northeast to southwest. Early Tertiary paleostructural development explains the enigmatic filling of these tight-gas sandstones according to structural position. In this scenario, the Lance reservoirs were filled before final reduction of sandstone permeability. The early faulting and gentle warping were factors in concentrating the hydrocarbon charge. Additional integrated studies employing 3-D seismic data may reveal other examples of subtle structural control of natural gas and oil resources in Rocky Mountain foreland petroleum systems.

ACKNOWLEDGMENTS

The work discussed in this chapter was completed while the senior author was employed by BP America Production Company in the capacity of senior geological associate. We thank BP America for permission to publish this work and for the privilege to work the Jonah asset for the past 3.5 yr (1999–2002). Ira Pasternack, geologist, friend, and colleague, introduced the senior author to the Bois marker and Lance stratigraphy at Jonah field and supported a solid "handoff" of the Jonah asset after the merger of Amoco and BP in early 1999. His dedication to the project is greatly appreciated. Paul Boissevain, after whom the Bois marker is named, is acknowledged for the early 3-D seismic work at Jonah and for preparing data showing that the Jonah faults may originate from the basement. Greg Partyka applied immensely skillful use of geophysical technology to the Jonah asset during 1999–2000. We thank Chris Christiansen for prolific petrophysical work, keen intellect and computing skills, and for many fervent conversations. Terry Young, BP geologist, is thanked for reading the manuscript draft and for being a talented, cordial, and dedicated colleague on the Jonah–South Pinedale anticline team. Appreciation is extended to Mike Mellor and Mai Nguyen of BP, who worked continuously and skillfully to organize records and enter data into the workstation environment during the fast pace of Jonah development. S. M. Hansen of Schlumberger is thanked for interpreting sedimentary structures with the FMI logs. We thank Dean DuBois and John Robinson for helpful and cordial conversations over the past few years. Keith Shanley is thanked for his astute interpretation of Jonah cores, probing conversations, and for reviewing this chapter. This manuscript benefited from a rigorous and challenging review by Donald S. Stone, and we appreciate his considerable effort to improve it.

Finally, McMurry Oil Company of Casper, Wyoming, is acknowledged for the skill to recognize, and actions to "rediscover," the existence of Jonah field during the early 1990s.

REFERENCES CITED

Aydin, A., and A. Nur, 1982, Evolution of pull-apart basins and their scale independence: Tectonics, v. 1, p. 91–105.

Biddle, K. T., and N. Christie-Blick, 1985, Glossary— Strike-slip deformation, basin formation, and sedimentation, *in* K. T. Biddle and N. Christie-Blick, eds., Strike-slip deformation, basin formation and sedimentation: SEPM Special Publication 37, p. 375–386.

Blackstone, D. L., 1993, The Wind River Range, Wyoming: An overview, *in* B. Stroock and S. Andrew, eds., 50th Field Conference Guidebook, Wyoming Geological Association, p. 121–140.

Bowker, K. A., and J. W. Robinson, 1996, Jonah field: A shallow sweetspot in the basin-centered gas accumulation of the northern Green River basin, Wyoming: AAPG Bulletin, v. 81, p. 1219, and Expanded Abstracts Volume, p. 25–30.

Cobban, W. A, E. A. Merewether, T. D. Fouch, and J. D. Obradovich, 1994, Some Cretaceous shorelines in the western interior of the United States, *in* M. V. Caputo, J. A. Peterson, and K. J. Franczyk, eds., Mesozoic systems of the Rocky Mountain region, U.S.A.: Rocky Mountain Section SEPM, p. 393–413.

Connor, C. W., 1992, The Lance Formation— petrography and stratigraphy, Powder River basin and nearby basins, Wyoming and Montana: U.S. Geological Survey Bulletin, v. 1917-I, p. 1–17.

DeCelles, P. G., and W. Cavazza, 1999, A comparison of fluvial megafans in the Cordilleran (Upper Cretaceous) and modern Himalayan foreland basin systems: Geological Society of America Bulletin, v. 111, p. 1315–1334.

Evans, L. W., and L. S. Foulk, 1997, Electrical resistivity reservoir description of the Lance Formation, Jonah field, Sublette County, Wyoming: AAPG Bulletin., v. 81, p. 1222.

Gillespie, J. M., and J. E. Fox, 1991, Tectonically influenced sedimentation in the Lance Formation, eastern Wind River basin, Wyoming: The Mountain Geologist, v. 28, p. 53–66.

Gries, R. R., 1983, North-south compression of Rocky Mountain foreland structures, *in* J. D. Lowell, ed., Rocky Mountain foreland basins and uplifts: Rocky Mountain Association of Geologists, p. 9–32.

Hanson, W. B., 2001, Log cross section, exhibit 10 and Lance isopach, exhibit 11, *in* Amoco Production Company Exhibits for Wyoming Oil and Gas Conservation Commission Docket 288-2001, Application to create spacing, Jonah/Antelope area, August 7, 2001. Documents are in files of the Wyoming Oil and Gas Conservation Commission, Casper, Wyoming.

Hanson, W., B. V. Carrillo, and S. Ritger, 1994, Evolution of the Wamsutter arch, Wyoming foreland, U.S.A.: Influence on formation of stratigraphic traps in Upper Cretaceous Almond shoreline sandstones: Annual Meeting of American Association of Petroleum Geologists-Society of Sedimentary Geology-Energy Minerals Division-Division of Professional Affairs-Division of Environmental Geosciences (AAPG-SEPM-EMD-DPA-DEG) Abstracts with Programs, p. 163.

Hodges, L. T., and C. F. Knutson, 1981, Tight-gas sandstone channel continuity and directivity, Upper Cretaceous and Paleocene, Greater Green River basin, Wyoming, *in* Proceedings of the Society of Petroleum Engineers/Department of Energy Low Permeability Gas Reservoirs Symposium: SPE paper 9844, p. 165–175.

Law, B. E., 1979, Section B–B', subsurface and surface correlations of some Upper Cretaceous and Tertiary rocks, northern Green River basin, Wyoming: U. S. Geological Survey Open-File Report 79-1689, 2 sheets.

Law, B. E., 1984, Relationships of source-rock, thermal maturity, and overpressuring to gas generation and occurrence in low-permeability Upper Cretaceous and lower Tertiary rocks, Greater Green River basin, Wyoming, Colorado, and Utah, *in* J. Woodward, F. F. Meissner, and J. L. Clayton, eds., Hydrocarbon source rocks of the greater Rocky Mountain region: Rocky Mountain Association of Geologists, p. 469–490.

Law, B. E., and R. C. Johnson, 1989, Structural and stratigraphic framework of the Pinedale anticline, Wyoming, and the multiwell experiment site, Colorado, *in* B. E. Law and C. W. Spencer, eds., Geology of tight gas reservoirs in the Pinedale anticline area, Wyoming, and at the multiwell experiment site, Colorado: U. S. Geological Survey Bulletin, v. 1886, p. B1–B11.

Mandl, G, 1988, Mechanics of tectonic faulting: New York, Elsevier, 407 p.

Marfurt, K., and L. Kirlin, 2000, 3-D broad-band estimates of reflector dip and amplitude: Geophysics, v. 65, p. 305–320.

McDonald, R. E., 1973, Big-Piney La Barge producing complex, Sublette and Lincoln Counties, Wyoming, *in* E. M. Schell, ed., The geology and mineral resources of the Green River basin: 25th Field Conference Guidebook, Wyoming Geological Association, p. 57–77.

Montgomery, S. L., and J. W. Robinson, 1997, Jonah field, Sublette County, Wyoming: Gas production from overpressured Upper Cretaceous Lance sandstones of the Green River basin: AAPG Bulletin, v. 81, p. 1049–1062.

Reynolds, M. W., 1976, Influence of recurrent Laramide structural growth on sedimentation and petroleum accumulation, Lost Soldier area, Wyoming: AAPG Bulletin, v. 60, p. 12–33.

Roberts, L. N., and M. A. Kirschbaum, 1995, Paleogeography of the Late Cretaceous of the western interior of middle North America; coal distribution and sediment accumulation: U. S. Geological Survey Professional Paper 1561, 115 p.

Robinson, J. W., 2000, Discovery of Jonah field, Sublette County, Wyoming: The Mountain Geologist, v. 37, p. 135–143.

Robinson, J. W., and L. W. Evans, 1997, Reservoir lithofacies in the Lance Formation, Jonah field, Sublette Co., Wyoming: AAPG Bulletin, v. 81, p. 1232.

Robinson, J. W., D. L. Delozier, and R. Finch, 1996, Integrated reservoir description and analysis of the Lance Formation at Jonah field, Sublette County, Wyoming: AAPG Bulletin, v. 80, p. 980.

Royse, F. Jr., 1993, An overview of the geologic structure of the thrust belt in Wyoming, northern Utah, and eastern Idaho, *in* A. W. Snoke, J. R. Steidtmann, and S. M. Roberts, eds., Geology of Wyoming: Geological Survey of Wyoming Memoir 5, p. 272–311.

Shanley, K. W., 2001, Fluvial reservoir description for a tight-gas giant: Jonah field, Green River basin, Wyoming, *in* J. W. Robinson and K. W. Shanley, eds., Tight gas fluvial reservoirs: A case history from the Lance Formation, Green River basin, Wyoming: AAPG Annual Meeting Short Course 2, Rocky Mountain Association of Geologists, p. 1–16.

Shanley, K. W., 2004, Fluvial reservoir description for a giant, low-permeability gas field: Jonah field, Green River basin, Wyoming, U.S.A., *in* J. W. Robinson and K. W. Shanley, eds., Jonah field: Case study of a tight-gas fluvial reservoir: AAPG Studies in Geology 52, p. 159–182.

Shuster, M. W., and J. R. Steidtmann, 1987, Fluvial-sandstone architecture and thrust-induced subsidence, northern Green River basin, Wyoming, *in* F. G. Etheridge, R. M. Flores, M. D. Harvey, and J. N. Weaver, eds., Recent and ancient nonmarine depositional environments: Models for exploration: SEPM Special Publication 39, p. 279–285.

Shuster, M. W., and J. R. Steidtmann, 1988, Tectonic and sedimentary evolution of the northern Green River basin, western Wyoming: Geological Society of America Memoir 171, p. 515–529.

Stewart, S. A., 2001, Displacement distributions on extensional faults: Implications for fault stretch, linkage, and seal: AAPG Bulletin, v. 85, p. 587–599.

Stone, D. S., 1969, Wrench faulting and Rocky Mountain tectonics: The Mountain Geologist, v. 6, p. 67–79.

Stone, D. S., 1993, Basement-involved thrust-generated folds as seismically imaged in the subsurface of the central Rocky Mountain foreland, *in* C. J. Schmidt, R. B. Chase, and E. A. Erslev, eds., Laramide basement deformation in the Rocky Mountain foreland of the western United States: Geological Society of America Special Paper 280, p. 271–318.

Van Horn, M. D., and L. T. Shannon, 1989, Hay Reservoir field: A submarine fan gas reservoir within the Lewis Shale, Sweetwater County, Wyoming, *in* J. L. Eisert, ed., Gas resources of Wyoming: 40th Field Conference Guidebook, Wyoming Geological Association, p. 155–180.

Vega, V. H., and W. B. Hanson, 2001, Jonah field area: Geophysical imaging of trap boundaries and reservoir heterogeneity, *in* J. W. Robinson and K. W. Shanley, eds., Tight gas fluvial reservoirs: A case history from the Lance Formation, Green River basin, Wyoming: AAPG Annual Meeting Short Course 2, Rocky Mountain Association of Geologists, section 3, 6 p.

Vega, V. H., W. B. Hanson, G. Partyka, C. J. Christensen, and D. L. Cox, 2001, Jonah field area: Seismic coherency cube technology used to define trap boundaries and stratigraphic features: Annual Meeting AAPG-SEPM-EMD-DPA-DEG Abstracts with Programs, p. A207.

Warner, E. M, 1997, Geology of Jonah field, a major gas accumulation in the Upper Cretaceous Lance Formation, Sublette County, Wyoming, *in* E. B. Coalson, J. C. Osmond, and E. T. Williams, eds., Innovative applications of petroleum technology in the Rocky Mountain area: Rocky Mountain Association of Geologists, p. 1–12.

Warner, E. M., 1998, Structural geology and pressure compartmentalization of Jonah field, Sublette County, Wyoming, *in* R. M. Slatt, ed., Compartmentalized reservoirs in Rocky Mountain basins: Rocky Mountain Association of Geologists, p. 29–46.

Warner, E. M., 2000, Structural geology and pressure compartmentalization of Jonah field based on 3-D seismic data and subsurface geology, Sublette County, Wyoming: The Mountain Geologist, v. 37, p. 15–30.

Wilcox, R. E., T. P. Harding, and D. R. Seely, 1973, Basic wrench tectonics: AAPG Bulletin., v. 57, p. 74–96.

CHAPTER 7

Burial-history Modeling of the Jonah Field Area: An Unusual and Possibly Unique Gas Accumulation in the Green River Basin, Wyoming

ROBERT J. COSKEY
Platte River Associates. Inc., Boulder, Colorado, U.S.A.

ABSTRACT

The hydrocarbon accumulation at Jonah field is the result of a complex series of temporal and spatial geologic events. This chapter is a preliminary investigation using burial-history modeling and petroleum-systems analysis to evaluate the generative potential of source rocks in the Jonah and Pinedale anticline areas and determine the timing of generation and expulsion. To test the ability of the Lance to self-generate significant hydrocarbons through thermal maturation, burial-history models were constructed with source intervals containing type III kerogen and average initial total organic carbon values ranging from 1.00 to 1.25%. Models built using optimistic charge parameters show that in the Jonah area, the Lance is capable of generating 1.79 tcf gas and, depending on the saturation threshold applied, can only expel from 0 to 0.64 tcf gas. Source rock pyrolysis data indicate that the Lance does not contain sufficient organic material, and burial-history models calibrated with vitrinite reflectance data suggest that potential Lance source material was not exposed to the thermal conditions necessary to generate and expel the quantities of hydrocarbons estimated to be present in the Jonah trap. Generation, expulsion, and migration (vertical and/or horizontal) from the Mesaverde Group and Rock Springs Formation or deeper hydrocarbon sources such as the Mowry Shale are necessary to account for the in-place hydrocarbons. The source potential of the coal-bearing lower Mesaverde is significantly greater that that of the Lance, and modeling suggests that in the Jonah area, it is capable of generating 5.48 tcf and expelling 4.06–5.12 tcf gas.

Modeling suggests that within the Jonah field area, the source rocks included in the Lance, upper Mesaverde, and lower Mesaverde formations can only provide the Jonah trap with between 2.61 and 3.98 tcf gas in place. This gas volume is significantly less than current in-place estimates of 8.3 tcf and suggests that gas must be generated and migrated from either deeper sources such as the Hilliard Shale or Mowry Shale and/or have migrated to the Jonah trap from a larger fetch area.

INTRODUCTION

Jonah field, which is located southwest of the axis of the Pinedale anticline in the northwestern Green River basin (Figure 1), has produced 719 bcf gas and 7.48 MMBO (Wyoming Oil and Gas Conservation Commission [WOGCC], 2003) and is estimated by DuBois et al. (2004) to contain 8.3 tcf gas in place. The main producing reservoir in the field is the Upper Cretaceous (Maastrichtian) Lance Formation, which is anomalously overpressured. A secondary producing reservoir is the Mesaverde Formation.

Previous work concerning petroleum systems in the northern Green River basin includes Law (1984) and Spencer (1987). This work supports the presence of a regional, basin-centered hydrocarbon system that has been generating gas from the early Tertiary to the present. Source rock characteristics have been quantified by Law (1984); burial history has been investigated from a structural perspective by Lickus et al. (1984), Naeser (1984),

Jonah Field: Case Study of a Giant Tight-Gas Fluvial Reservoir, J. W. Robinson and K. W. Shanley, eds.: AAPG Studies in Geology 52 and Rocky Mountain Association of Geologists 2004 Guidebook

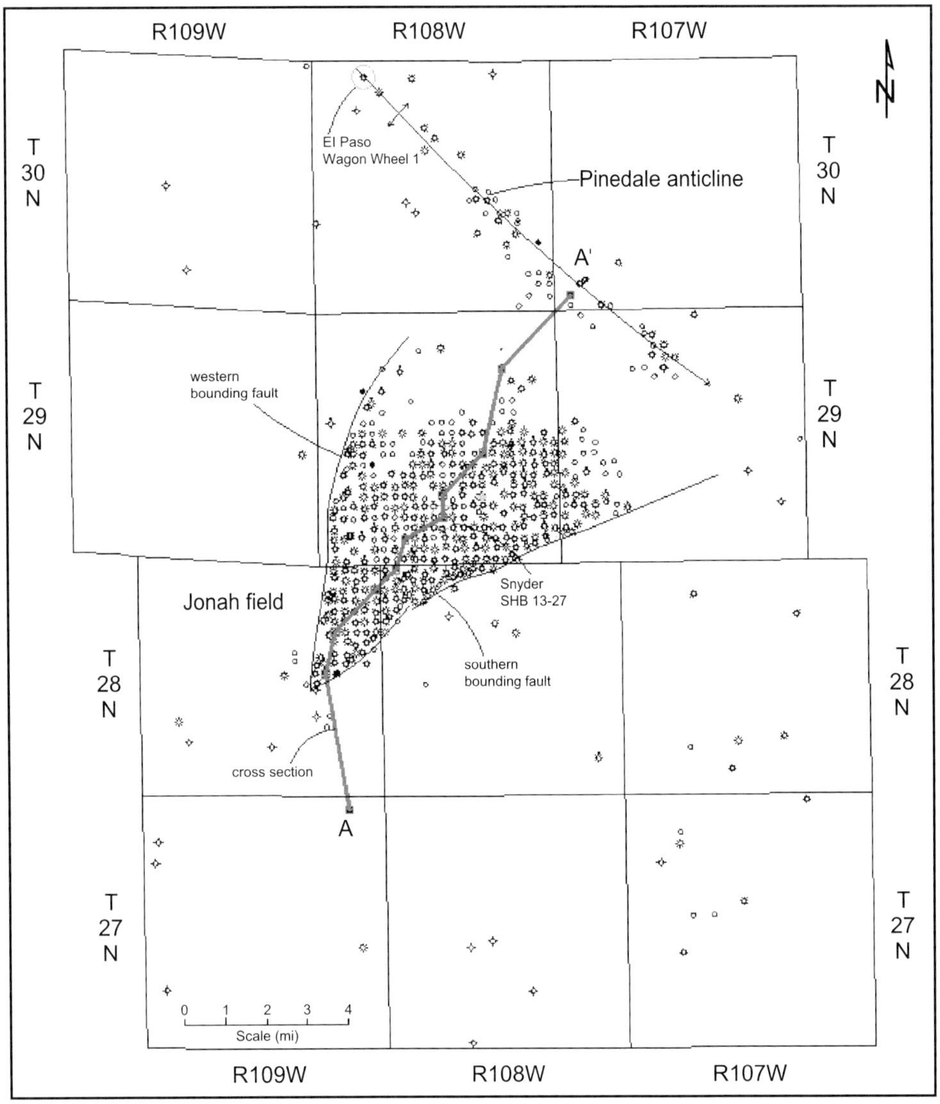

Figure 1. Index map, Jonah field and Pinedale anticline area, Sublette County, Wyoming. AA′ is the location of cross section (Figure 3).

and Dickinson (1989), and from a diagenetic perspective by Pollastro and Barker (1984). Additionally, work by Bowker and Robinson (1996) and Montgomery and Robinson (1997) provided the background data necessary to construct the burial-history models that form the basis of this chapter. The goals of this investigation are to (1) aggregate available modeling calibration data, (2) identify potential source rocks, (3) model the burial history of the Jonah and Pinedale areas, (4) estimate the timing of generation for the multiple source rocks, (5) evaluate the generative capacity of source rock packages, (6) test via modeling whether the Lance Formation can act as a source, (7) determine what other local source rocks have the capacity to generate the amount of gas estimated to be in place at Jonah, and (8) review the local geology and identify those geologic elements that appear instrumental to the trap. This modeling study quantifies the timing of generation and expected volumes of hydrocarbons from each potential source rock, providing insights into the petroleum systems in the northern Green River basin.

LOCAL STRATIGRAPHY

The stratigraphy and depositional environments of Tertiary and Cretaceous strata in the northern Green River basin have been previously discussed by others, (i.e., Curry,

Formation Name		Geologic Age	Beginning Age	End Age	General Lithology
Eroded post-Pliocene		post-Pliocene	5(?)	0	fluvial- and alluvial-fan intebedded sandstone, siltstone, and shale deposits?
Hiatus			17	5(?)	
Miocene rocks		Miocene	24	17	fluvial- and alluvial-fan deposits
Hiatus			48	24	
Bridger, Green River, and Wasatch		upper – lower Eocene	50	48	fluvial- and alluvial-fan sandstone, minor fresh water lacustrine limestone, marl, and shale
Wasatch Formation unconformity		lower Eocene	54	50	current surface fluvial- and alluvial-fan sandstone, minor freshwater lacustrine limestone, marl, and shale
Fort Union or undiferentiated Paleocene		middle Paleocene	65	54	discontinuous fluvial conglomerates, sandstone, and mudstones, minor coals
Erosion Paleocene			68	65	
Lance Formation unconformity		Upper Cretaceous – Maastrichtian	73	68	Discontinuous fluvial-, alluvial-plain sandstone and mudstones
Mesaverde Group	Ericson Sandstone upper Mesaverde	Upper Cretaceous – Campanian	76	73	Amalgamated fluvial-braided stream sandstones
	Rock Springs Formation lower Mesaverde	Upper Cretaceous – Campanian	86	76	marginal-marine sandstone, shale, and coals
Hilliard Shale		Upper Cretaceous – mostly Santonian	90	86	homogeneous marine shales
Frontier Formation		Upper Cretaceous	95	90	deltaic sandstones and shales
Mowry		Lower Cretaceous	99	95	marine siliceous shale, marginal-marine sandstone.

Figure 2. Local stratigraphy used in modeling. Absolute ages from Dickinson (1989).

1973; Dickinson, 1989; Law and Johnson, 1989; Prensky, 1989). Information about potential source rocks, reservoirs, and seals was derived from Montgomery and Robinson (1997) and is readdressed in this chapter only in the context of burial-history modeling. The lithology and rock property data for the underlying rock units were aggregated to assign geologic age and thermal properties of the stratigraphic section and to provide additional constraints on burial-history models. Figure 2 shows the generalized stratigraphic column and geologic ages used in burial-history modeling.

GEOLOGIC COMPONENTS CRITICAL FOR ACCUMULATION

This study uses burial-history modeling to evaluate source efficiency and hydrocarbon charge in the Jonah field–Pinedale anticline area. The focus was to determine those geologic factors that exert a primary control on the occurrence of the gas accumulation. Charge volume and timing elements were addressed through a rigorous modeling effort constraining each model with calibration data, including stratigraphic thickness, lithologic composition, total organic carbon (TOC), vitrinite reflectance, pyrolysis, and temperature data. Potential trapping elements were addressed through an analysis of regional stratigraphy and local structural features.

Structural and Stratigraphic Elements

Jonah is bounded by two faults (Warner, 1998) that provide lateral seals to the accumulation (Figure 1). The structural timing and persistence of the sealing capacity of these faults is critically important but cannot solely explain the presence of the accumulation, i.e., a vertical "seal" is also required to form the trap. Regional cross section AA′ (Figure 3) illustrates the stratigraphic geometries and unconformities in the Jonah area. These correlations suggest that the top of the Lance Formation is truncated by a regional unconformity. The interpreted geometry indicates that the upper Lance shaly zone has been truncated updip of Jonah field. The presence of this

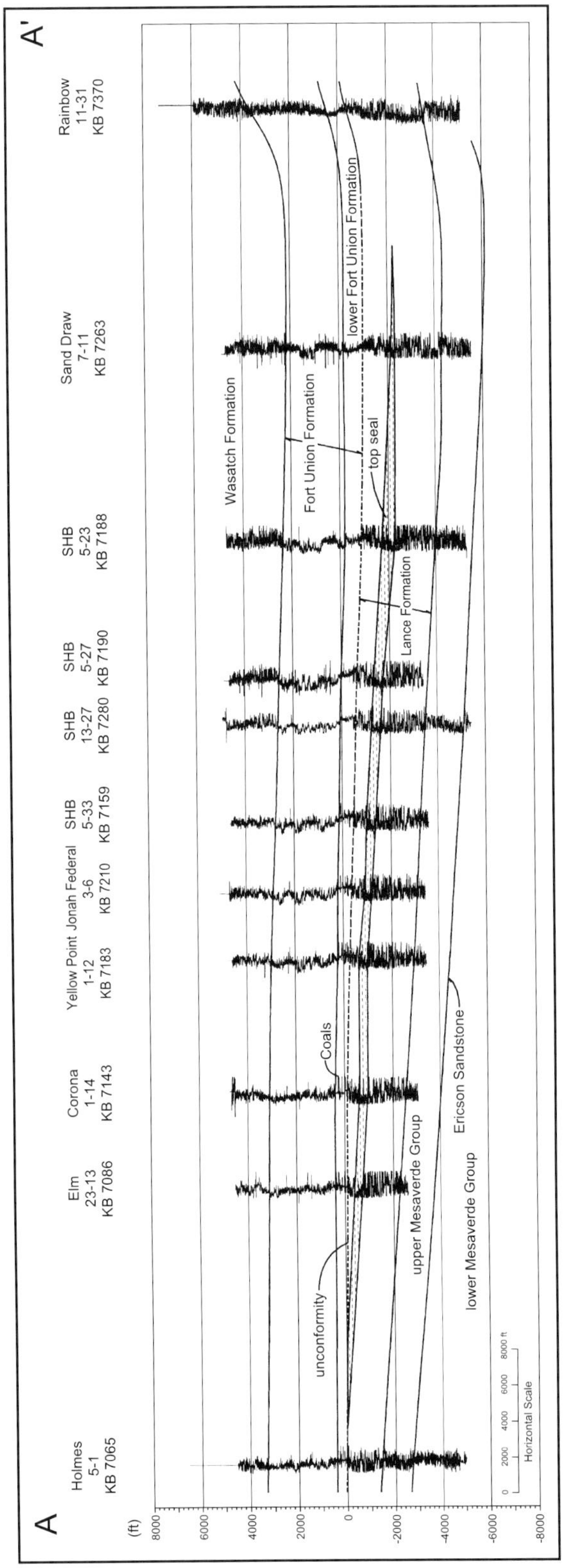

Figure 3. Cross section AA'. Cross section is oriented southwest-northeast, running downdip from the apex of the Jonah trap showing the interpreted termination by unconformity of the upper Lance top seal. Dashed pattern is the shaly unit in the upper Lance. See Figure 1 for the location of the cross section.

shale unit is a critical factor because it serves a top seal to the gas accumulation. At Jonah, preservation of Lance overpressure has been maintained by a greater gas supply (generation, expulsion, and migration) than loss through the top seal. South of Jonah, where the shaly zone has been completely removed and the lateral bounding faults are absent, the Lance Formation is normally pressured. Although the shaly zone is widespread at Jonah, log data suggest that the upper Lance contains numerous fine-grained sandstones that, given time and tectonics, will allow even this seal to leak.

Source Rock(s)

Fort Union Formation

The Paleocene Fort Union Formation in the northern Green River basin has been described by Law (1984) and Curry (1973) as containing a thick (more than 3000 ft [900 m]) of nonmarine strata. Previous workers have not described the Fort Union as containing rich source material. In the southern part of the Jonah area, however, the lower Fort Union contains numerous coaly beds. Two wells in the Jonah field area have data that specifically address the source rock potential of the Fort Union Formation. In the Jonah Federal 1-4 (Lot 3, Sec. 4, T28N, R108W), vitrinite reflectance data from coals at 5055 and 7635 ft (1541 and 2327 m) have measured R_o that range from 0.56 to 0.72 (D. Battin, 2002, personal communication) (Table 1). In the Stud Horse Butte (SHB) 1-28 (NE NE Sec. 28, T29N, R108W), similar data show R_o of 0.54 at 8000 ft (2438 m) and 0.58 at 8130 ft (2478 m) (Table 1). These data suggest that the Fort Union coals never reached sufficient thermal maturity for significant thermogenic gas generation. For this reason, the Fort Union was considered as an overburden and is not regarded as a viable, local source rock.

Lance Formation

The Cretaceous Lance Formation was deposited in a nonmarine, oxygenated, alluvial-plain environment (Montgomery and Robinson, 1997). A review of available mud logs indicates that although carbonaceous partings are described, there are no thick coals or carbonaceous shales in the Lance. Mud logs describe the intercalated shales as shale or shaly siltstone: light–medium gray to off white, clay rich, splintery, sub-blocky, occasionally calcareous, with traces of carbonaceous material. In the middle portion of the Lance, shales are described as shale: medium–dark gray, light green-gray, dark brown, sub-platy, tabular, soft, earthy, carbonaceous/coaly (in part), silty, slightly calcareous. The A core cut in the Jonah 2-5 (NW SE Sec. 5,

T28N, R108W) penetrated several root-mottled, dark gray to black, shaly zones, interpreted as flood-plain deposits (Yin, et al, 2001). Rock-Eval TOC data from the SHB 5-26 (SW NW Sec. 26, T29N, R108W) and SHB 1-28 (NE NE Sec. 28, T29N, R108W) (Table 2) show that present-day TOC ranges from 2.37% (upper shaly Lance) to 0.24% by weight (average, 0.57%; n = 20). Hydrogen index (HI) values for the same samples are low (113–24; average, 55.9; n = 20). T_{max} values for these same samples are suspect because of low TOC, rendering HI values (HI = $S_2 \times 100/\text{TOC}$) suspect. Based on observations made on overpressured rocks in the northern Green River basin, Spencer (1987) stated that source rocks must (1) generate gas faster than it is lost from the system and (2) have an average initial TOC of 2.0% or greater in order for generative overpressure to occur. Based on these criteria, available Rock-Eval data, and mud-log descriptions, the Lance Formation does not appear to be an effective source rock, and it is unlikely that it could act alone to source the Jonah accumulation. In addition, vitrinite reflectance data (Figure 4; Table 1) and burial-history modeling show that in the central portion of Jonah field, the Lance scarcely attained a thermal maturity sufficient to generate and, more importantly, expel gaseous hydrocarbons.

Upper Mesaverde Formation

Undifferentiated upper Mesaverde shales (this includes those organic shales that are positioned beneath the Lance and above the Ericson Sandstone [see Figure 4, DuBois et al., 2004], possibly the lateral equivalent of the Lewis Shale) may contain sufficient organic material to generate and expel hydrocarbons. Vitrinite reflectance data (Figure 4; Table 1) indicate that the upper Mesaverde Formation (above the Ericson Sandstone) attained a maturity of 0.73–1.0% R_o. Mud logs describe the potential source rocks of the upper Mesaverde as shale and silty shale, light-medium gray, black, dark gray, gray-black, sub-platy, carbonaceous, coaly inclusions, occasional scattered 'free' coals. Given these observations, the upper Mesaverde must be considered a potential source rock.

Lower Mesaverde and Rock Springs Formation

Law (1984) and Hanson et al. (2004) suggest that the lower Mesaverde section contains abundant humic coals. Pyrolysis data published by Law (1984) show that the Ericson and Mesaverde formations contain coals and carbonaceous shales with present-day bulk TOC values from about 1 to greater than 8%. Mud logs from the Yellow Point Federal 11-13 (NE SW Sec. 13, T28N, R109W) document the presence of coals (Figure 5) in the lower Mesaverde section, with weak fluorescence and light brown oil and condensate stain observed in sandstones adjacent to these coals (Figure 5). These data suggest that the coals and carbonaceous shales may contain sufficient exinite or liptinite to facilitate the generation of some liquid hydrocarbons. Total organic carbon data from the El Paso Wagon Wheel 1 (SE NW Sec. 5, T30N, R108W) (Law, 1984) and maturity data (Table 1) indicate that the lower Mesaverde shaly unit is one of the major hydrocarbon-generating source rocks in the Jonah area.

Hilliard and Baxter Formations

In the northwestern Green River basin, this thick (>3700 ft [>1100 m] in the Jonah area), dominantly marine unit contains dispersed organic matter with present-day TOC ranging from 0.25 to 1.39% (Law, 1984). Although these TOC values indicate that the rock is capable of generating hydrocarbons, expulsion and migration from thick, relatively homogeneous shales, such as the Hilliard and Baxter, are difficult and are an issue for additional study. An observation for consideration is the evidence from seismic data that some of the faults in the Jonah field area penetrate the Hilliard and Baxter section and deeper areas. These faults and associated fracture systems may be the conduits allowing vertical migration of gas from the massive Hilliard Shale. Hanson et al. (2004) suggest that the gas produced from the Lance is, at least in part, the product of secondary cracking from liquid hydrocarbons. Modeling shows that, even with an estimated average TOC of 1%, the great thickness of Hilliard is capable of generating significant volumes of gas (see Probabilistic Modeling). However, without additional gas fingerprinting data, the assumption that gas derived from the Hilliard made its way into the Jonah trap is, at this time, speculation.

Mowry Shale

In the northwest Green River basin, penetrations of the Mowry are extremely sparse. Burtner and Warner (1984) provide a single point of geochemical data from the International Nuclear 1 Government Swanson well (Center NE Sec. 22, T25N, R110W) where they report a present-day TOC of 1.4% for the Mowry. This relatively low TOC for the Mowry may be caused by (1) its elevated thermal maturity, as indicated by a T_{max} of 441°C, and/or (2) dilution of organic matter from westward-derived clastics, and/or (3) the broad sample interval of 10,820–11,170 ft (3298–3405 m). Given the thickness, potential richness, and burial history, the Mowry is capable of generating hydrocarbons. However, there is a high degree of uncertainty of whether conduits exist or existed to facilitate vertical migration of hydrocarbons to charge younger reservoirs such as the Lance.

Table 1

Vitrinite Reflectance Data (R_o) from the Jonah Field and Pinedale Anticline Areas, Sublette County, Wyoming. SHB 13-27 Data Provided by McMurry Energy, 2002, personal communication; SHB 1-28 Data Provided by EnCana, U.S.A., 2003, personal communication; Jonah Federal 1-4 Data Provided by D. Battin, 2002, personal communication.

Well Name	Well Abbreviation	Point	Measured Depth (ft)	R_o	Formation	Comments
Stud Horse Butte (SHB)	13-27	1	7074	0.64	Fort Union	McMurry (2002)
	13-27	2	7507	0.74	Fort Union	McMurry (2002)
	13-27	3	7796	0.64	Fort Union	McMurry (2002)
	13-27	4	8293	0.70	Lance	McMurry (2002)
	13-27	5	8324	0.63	Lance	McMurry (2002)
	13-27	6	8926	0.64	Lance	McMurry (2002)
	13-27	7	9239	0.67	Lance	McMurry (2002)
	13-27	8	9907	0.75	Lance	McMurry (2002)
	13-27	9	9907	0.79	Lance	McMurry (2002)
	13-27	10	9935	0.84	Lance	McMurry (2002)
	13-27	11	10,771	0.81	Lance	McMurry (2002)
	13-27	12	11,121	0.79	Lance	McMurry (2002)
	13-27	13	11,496	0.81	Lance	McMurry (2002)
	13-27	14	11,655	0.83	upper Mesaverde	McMurry (2002)
	13-27	15	12,110	0.95	Ericson	McMurry (2002)
	13-27	16	12,242	0.97	lower Mesaverde	McMurry (2002)
	13-27	17	12,357	0.91	lower Mesaverde	McMurry (2002)
	13-27	18	12,578	0.97	lower Mesaverde	McMurry (2002)
	13-27	19	12,781	1.03	lower Mesaverde	McMurry (2002)
Stud Horse Butte (SHB)	1-28	1	8000	0.54	Fort Union	EnCana (2003)
	1-28	2	8130	0.58	Fort Union	EnCana (2003)
	1-28	3	8300	0.63	Lance(?)	EnCana (2003)
	1-28	4	8500	0.62	Lance(?)	EnCana (2003)
	1-28	5	8700	0.69	Lance	EnCana (2003)
	1-28	6	8900	0.65	Lance	EnCana (2003)
	1-28	7	9150	0.68	Lance	EnCana (2003)
	1-28	8	9450	0.73	Lance	EnCana (2003)
	1-28	9	9700	0.74	Lance	EnCana (2003)
	1-28	10	9900	0.76	Lance	EnCana (2003)
	1-28	11	10,180	0.80	Lance	EnCana (2003)
	1-28	12	10,500	0.76	Lance	EnCana (2003)
	1-28	13	10,900	0.78	Lance	EnCana (2003)

Table 1

Vitrinite Reflectance Data (R_o) from the Jonah Field and Pinedale Anticline Areas, Sublette County, Wyoming. SHB 13-27 Data Provided by McMurry Energy, 2002, personal communication; SHB 1-28 Data Provided by EnCana, U.S.A., 2003, personal communication; Jonah Federal 1-4 Data Provided by D. Battin, 2002, personal communication (cont.).

Well Name	Well Abbreviation	Point	Measured Depth (ft)	R_o	Formation	Comments
	1-28	14	11,200	0.82	Lance	EnCana (2003)
	1-28	15	11,470	0.88	upper Mesaverde	EnCana (2003)
	1-28	16	11,730	0.88	upper Mesaverde	EnCana (2003)
	1-28	17	12,100	0.92	upper Mesaverde	EnCana (2003)
	1-28	18	12,668	0.96	Ericson	EnCana (2003)
Ross	33-7	1	2700	0.44	Eocene(?)	McMurry (2002)
	33-7	2	3448	0.49	Eocene(?)	McMurry (2002)
	33-7	3	5257	0.53	Wasatch	McMurry (2002)
	33-7	4	7315	0.58	Lance	McMurry (2002)
	33-7	5	8530	0.72	Lance	McMurry (2002)
	33-7	6	9510	0.68	Lance	McMurry (2002)
	33-7	7	10,895	0.73	upper Mesaverde through lower Mesaverde	McMurry (2002)
	33-7	8	11,393	0.82	upper Mesaverde through lower Mesaverde	McMurry (2002)
	33-7	11	11,817	1.24	Carlile(?)	McMurry (2002), from southwest core
	33-7	9	11,905	0.89	Carlile(?)	McMurry (2002)
	33-7	10	12,039	1.15	Frontier	McMurry (2002), from southwest core
Jonah Federal	1-4	1	2565	0.46	Eocene(?)	D. Battin (2002)
	1-4	2	3015	0.53	Eocene(?)	D. Battin (2002)
	1-4	3	3525	0.57	Eocene(?)	D. Battin (2002)
	1-4	4	4185	0.57	Wasatch(?)	D. Battin (2002)
	1-4	5	4600	0.51	Wasatch(?)	D. Battin (2002)
	1-4	6	5055	0.56	Fort Union(?)	D. Battin (2002)
	1-4	7	5565	0.61	Fort Union(?)	D. Battin (2002)
	1-4	8	6075	0.59	Fort Union	D. Battin (2002)
	1-4	9	6495	0.59	Fort Union	D. Battin (2002)
	1-4	10	7035	0.62	Fort Union	D. Battin (2002)

Table 1

Vitrinite Reflectance Data (R_o) from the Jonah Field and Pinedale Anticline Areas, Sublette County, Wyoming. SHB 13-27 Data Provided by McMurry Energy, 2002, personal communication; SHB 1-28 Data Provided by EnCana, U.S.A., 2003, personal communication; Jonah Federal 1-4 Data Provided by D. Battin, 2002, personal communication (cont.).

Well Name	Well Abbreviation	Point	Measured Depth (ft)	R_o	Formation	Comments
	1-4	11	7335	0.72	Fort Union	D. Battin (2002)
	1-4	12	7635	0.67	Fort Union	D. Battin (2002)
	1-4	13	8115	0.78	Lance	D. Battin (2002)
	1-4	14	8115	0.76	Lance	D. Battin (2002)
	1-4	15	8325	0.69	Lance	D. Battin (2002)
	1-4	16	8775	0.73	Lance	D. Battin (2002)
	1-4	17	9435	0.64	Lance	D. Battin (2002)
	1-4	18	9585	0.73	Lance	D. Battin (2002)
	1-4	19	9825	0.67	Lance	D. Battin (2002)
	1-4	20	10,105	0.67	Lance	D. Battin (2002)
	1-4	21	10,555	0.69	Lance	D. Battin (2002)
	1-4	22	10,790	0.79	Lance	D. Battin (2002)
	1-4	23	10,845	0.69	Lance	D. Battin (2002)
	1-4	24	11,005	0.86	Lance	D. Battin (2002)
	1-4	25	11,105	0.74	Lance	D. Battin (2002)
	1-4	26	11,305	0.89	Lance	D. Battin (2002)
	1-4	27	11,455	0.79	Lance	D. Battin (2002)
	1-4	28	11,735	0.98	upper Mesaverde	D. Battin (2002)
	1-4	29	11,925	0.99	Ericson	D. Battin (2002)
	1-4	30	11,995	0.95	Ericson	D. Battin (2002)
	1-4	31	12,195	0.94	lower Mesaverde	D. Battin (2002)
	1-4	32	12,435	1.09	lower Mesaverde	D. Battin (2002)
Wagon Wheel	WW 1	1	5028	0.51	Fort Union	Dickinson (1989)
	WW 1	2	5072	0.46	Fort Union	Dickinson (1989)
	WW 1	3	7048	0.68	Unnamed Tertiary	Dickinson (1989)
	WW 1	4	7130	0.74	(Law, 1984)	Dickinson (1989)
	WW 1	5	7137	0.70		Dickinson (1989)
	WW 1	6	7388	0.72		Dickinson (1989)
	WW 1	7	7395	0.76		Dickinson (1989)
	WW 1	8	8073	0.78	Lance through Ericson	Dickinson (1989)
	WW 1	9	8135	0.77		Dickinson (1989)
	WW 1	10	8984	0.85		Dickinson (1989)

Table 1

Vitrinite Reflectance Data (R_o) from the Jonah Field and Pinedale Anticline Areas, Sublette County, Wyoming. SHB 13-27 Data Provided by McMurry Energy, 2002, personal communication; SHB 1-28 Data Provided by EnCana, U.S.A., 2003, personal communication; Jonah Federal 1-4 Data Provided by D. Battin, 2002, personal communication (cont.).

Well Name	Well Abbreviation	Point	Measured Depth (ft)	R_o	Formation	Comments
	WW 1	11	10,145	0.95		Dickinson (1989)
	WW 1	12	10,173	0.91		Dickinson (1989)
	WW 1	13	11,024	0.96		Dickinson (1989)
	WW 1	14	11,032	0.98		Dickinson (1989)
	WW 1	15	11,051	1.03		Dickinson (1989)
	WW 1	16	13,097	1.22		Dickinson (1989)
	WW 1	17	13,141	1.15		Dickinson (1989)
	WW 1	18	13,198	1.13		Dickinson (1989)
	WW 1	19	14,902	1.47	lower Mesaverde	Dickinson (1989)
	WW 1	20	14,914	1.48	lower Mesaverde	Dickinson (1989)
	WW 1	21	14,918	1.53	lower Mesaverde	Dickinson (1989)
	WW 1	22	14,923	1.48	lower Mesaverde	Dickinson (1989)
	WW 1	23	14,930	1.55	lower Mesaverde	Dickinson (1989)
	WW 1	24	14,942	1.49	lower Mesaverde	Dickinson (1989)
	WW 1	25	16,068	1.61	lower Mesaverde	Dickinson (1989)
	WW 1	26	16,078	1.61	lower Mesaverde	Dickinson (1989)
	WW 1	27	16,101	1.65	lower Mesaverde	Dickinson (1989)
	WW 1	28	17,173	1.77	lower Mesaverde	Dickinson (1989)
	WW 1	29	17,960	1.98	Hilliard Shale	Dickinson (1989)
	WW 1	30	17,965	1.85	Hilliard Shale	Dickinson (1989)

Table 2

Rock-Eval Data from Snyder Stud Horse Butte 5-26 (SW NW Sec. 26, T29N, R108W) and McMurry Stud Horse Butte 1-28 (NE NE Sec. 28, T29N, R108W) Wells. Stud Horse Butte 5-26 Data Provided by McMurry Energy, 2002, personal communication; Stud Horse Butte 1-28 Data Provided by EnCana, U.S.A., 2003, personal communication.

Well Abbreviation	Measured Depth (ft)	S_1 (mg/g)	S_2 (mg/g)	S_3 (mg/g)	T_{max} (°C)	TOC (wt.%)	Hydrogen Index	Oxygen Index	Productivity Index	Formation
5-26	7940	0.07	0.31	0.33	435	0.29	107	114	0.18	Fort Union
5-26	8185	0.34	0.83	1.41	433	0.70	119	201	0.29	Fort Union
5-26	8515	0.20	0.66	0.67	433	0.59	112	114	0.23	?
5-26	8715	0.14	2.68	0.54	438	2.37	113	23	0.05	shaly upper Lance
5-26	8935	0.03	0.29	0.96	441	0.51	57	188	0.09	Lance
5-26	9255	0.03	0.26	0.73	436*	0.40	65	182	0.10	Lance
5-26	9825	0.06	0.19	0.40	443*	0.26	73	154	0.24	Lance
5-26	10,220	0.06	0.18	0.46	435*	0.34	53	135	0.25	Lance
5-26	10,415	0.05	0.24	0.50	440*	0.38	63	132	0.17	Lance
5-26	10,985	0.06	0.22	0.86	444*	0.39	56	221	0.21	Lance
5-26	11,715	0.23	1.22	0.62	448	1.40	87	44	0.16	Lance
1-28	8000	0.05	0.30	0.12	436*	0.41	73	29	0.14	Fort Union
1-28	8130	0.07	1.15	0.12	438	1.23	93	10	0.06	Fort Union
1-28	8300	0.06	0.98	0.17	441	1.28	77	13	0.06	shaly upper Lance
1-28	8500	0.02	0.27	0.15	439	0.57	47	26	0.07	shaly upper Lance
1-28	8700	0.01	0.20	0.14	442*	0.35	57	40	0.05	upper Lance
1-28	8900	0.01	0.07	0.17	443*	0.25	28	68	0.12	upper Lance
1-28	9150	0.02	0.07	0.15	424*	0.25	28	60	0.22	upper Lance
1-28	9450	0.01	0.07	0.15	446*	0.27	26	56	0.12	upper Lance
1-28	9700	0.02	0.06	0.16	446*	0.25	24	64	0.25	upper Lance
1-28	9900	0.07	0.19	0.14	405*	0.24	79	58	0.27	upper Lance
1-28	10,180	0.01	0.08	0.10	443*	0.25	32	40	0.11	middle Lance
1-28	10,500	0.02	0.13	0.10	446*	0.29	45	34	0.13	Lance and Yellow Point
1-28	10,900	0.03	0.14	0.10	449*	0.29	48	34	0.18	Lance and Wardell
1-28	11,200	0.08	0.63	0.16	450	1.03	61	16	0.11	Lance and Wardell
1-28	11,470	0.05	0.21	0.15	457*	0.42	50	36	0.19	upper Mesaverde
1-28	11,730	0.06	0.30	0.12	451	0.56	54	21	0.17	upper Mesaverde
1-28	12,100	0.06	0.29	0.15	457*	0.47	62	32	0.17	upper Mesaverde
1-28	12,668	0.07	0.31	0.20	459	0.51	61	39	0.18	Ericson

* T_{max} unreliable because of low TOC.

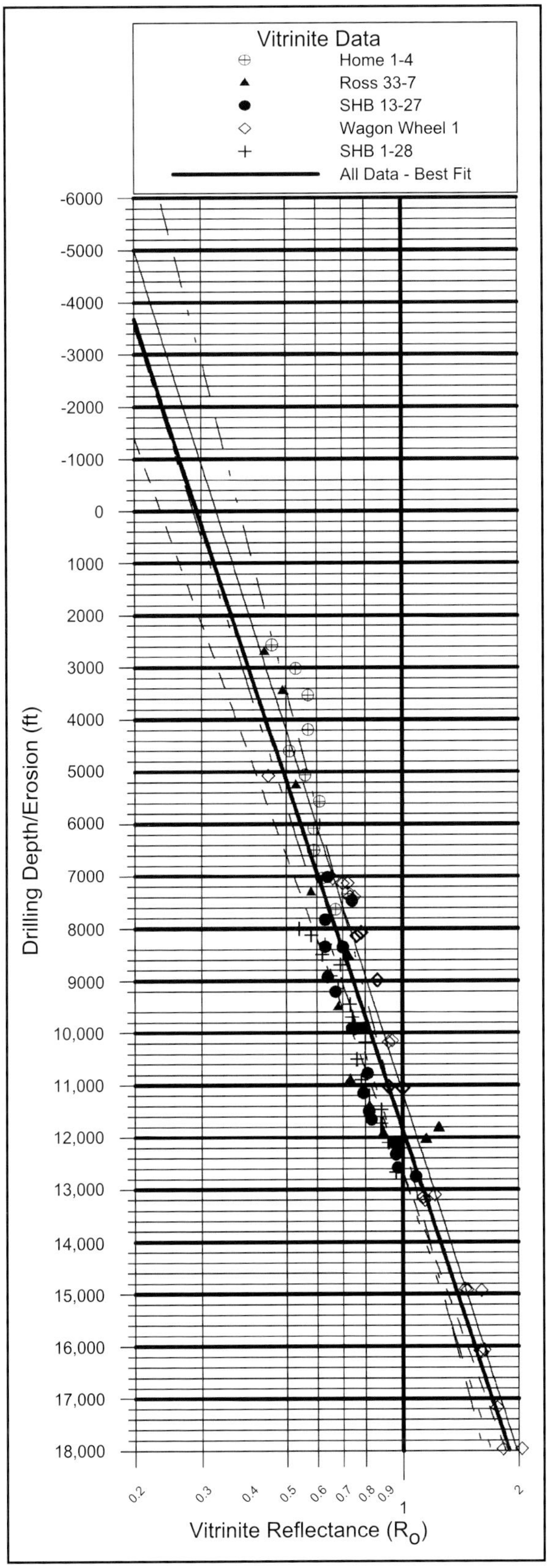

Presence of Sufficient Reservoir Pore Volume Concurrent with Delivery of Hydrocarbon Charge

The majority of the gas produced at Jonah field is from the numerous sand bodies that occur at varying vertical frequency throughout the Lance section. Present-day reservoir porosity ranges from 5 to 12%, and unstimulated, ambient permeability ranges from 3 to 20 μd (Bowker and Robinson, 1996). A compaction history profile suggests that the porosity and permeability of the Lance sandstones were slightly better concurrent with expulsion of gas from the downdip generative kitchen in the Pinedale area than they were when source rocks beneath the Lance began generating in the Jonah area.

Generation, Migration, and Trap Timing

Based on vitrinite reflectance data, Bowker and Robinson (1996) estimated that hydrocarbon generation from the lowermost Lance and upper Mesaverde began in the early Tertiary and ceased during the Pliocene, concurrent with uplift and exhumation of the area. A one-dimensional (1-D) burial-history model of the SHB 13-27 well in Jonah field suggests that the earliest gas generation from the lower Mesaverde source interval began at approximately 62 Ma. Maximum gas generation was nearly time coincident with expulsion at approximately 47 Ma. It should also be noted that at the SHB 13-27 location, the lower Mesaverde source never enters the main gas-generation window, i.e., R_o is greater than 1.3%.

BURIAL-HISTORY MODELING

Burial-history modeling was performed using Platte River Associates, Inc, BasinMod© BMOD (release dated August 2002). Two BMOD 1-D models were constructed: one on the Snyder Oil SHB 13-27 (SW SW Sec. 27, T29N, R108W) and another on the El Paso Natural Gas Wagon Wheel 1 (SE NW Sec. 5, T30N, R110W) to examine the timing of hydrocarbon generation and expulsion in the Jonah and Pinedale areas. These wells were selected for modeling because of the amount of calibration data

◄

Figure 4. Composite vitrinite reflectance (R_o) plot of data available in the Jonah field and Pinedale anticline areas. Projection of best fit R_o gradients above the current surface indicates that 1400 to greater than 6000 ft (430 to >1830 m) of post-Wasatch–age rocks have been eroded.

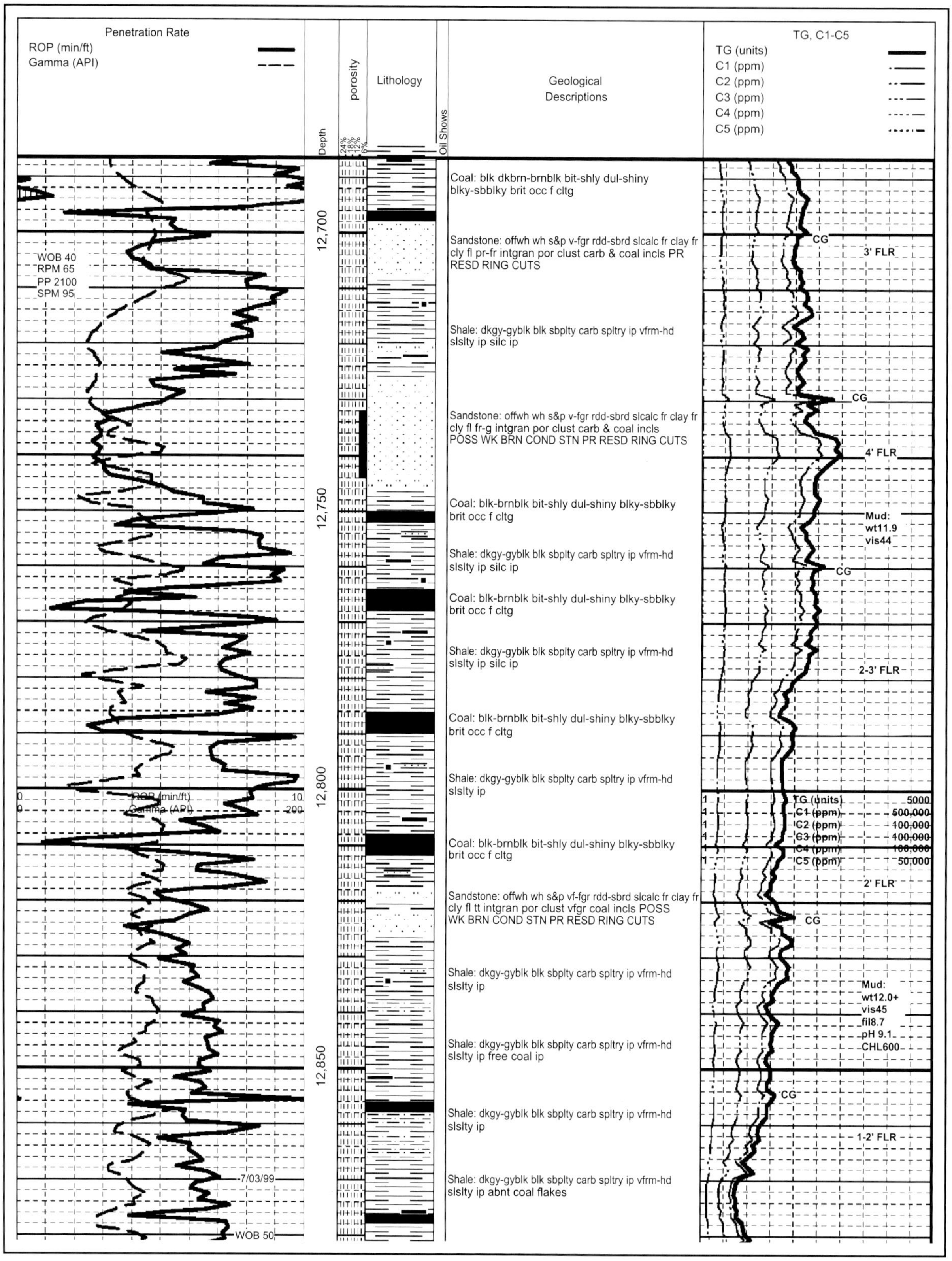

Figure 5. Section of mud log from Yellow Point Federal 11-13 (NE SE Sec. 13, T28N, R109W), illustrating the presence of coals and organic-rich shales in the lower portion of the Rock Springs Formation. Note indications of brown condensate stain in sandy units.

Table 3

Greater Jonah Area; Model Calibration Data Availability, Data Type, and Well Locations.

Operator (Original)	Well Name	American Petroleum Institute (API) Number	Location	Well Abbreviation	Data Type(s)
Chevron-Snyder	Stud Horse Butte 13-27	4903521359	SW SW Sec. 27, T29N, R108W	13-27	R_o
El Paso	Wagon Wheel 1	4903520124	SE NW Sec. 5, T30N, R108W	WW 1	R_o and temperature
Home Petroleum	1-4 Jonah Federal	4903520926	NE SW Sec. 4, T28N, R108W (irregular section)	1-4	R_o
Chevron	33-7 Ross	4903521336	NW SE Sec. 7, T30N, R110W	33-7	R_o and temperature
Snyder	Stud Horse Butte 5-26	4903521374	SW NW Sec. 26, T29N, R108W	5-26	Rock-Eval
McMurry Oil	Stud Horse Butte 1-28	4903521890	NE NE Sec. 28, T29N, R108W	1-28	Rock-Eval and R_o
Hay Leonard	Vible	4903520606	SW NW Sec. 11, T31N, R109W	V 1	temperature
McMurry Oil	Jonah 2-8	4903521214	Center NW Sec. 8, T28N, R108W	2-8	temperature
Alpine Gas	Billy 33-16	4903521342	NW SE Sec. 16, T31N, R111W	33-16	temperature
McMurry Oil	Yellow Point Federal 11-13	4903521887	NE SE Sec. 13, T28N, R109W	11-13	mud log

(bottomhole temperature [BHT], vitrinite reflectance, and Rock-Eval pyrolysis; Table 3) available either in the modeled wellbore or from nearby wells. Tables 4 and 5 summarize the modeling assumptions, stratigraphy, lithologies, and geologic ages used to construct the models.

Calibration Data

Bottomhole temperature data were measured during long-term pressure buildup tests from several wells in the Jonah area (Figure 6). These data show a fairly consistent geothermal gradient between wells within and outside of Jonah field. The Vible 1 well in the Pinedale area (SE NE Sec. 11, T31N, R109W) appears to have a slightly higher temperature gradient.

Vitrinite reflectance data were available for five wells in the Jonah field and Pinedale anticline areas (Figure 4; Table 1). Extrapolation of best fits through these data projected to R_o of 0.2 at depths of 1400 to greater than 6000 ft (430 to >1830 m) provides a minimum estimate of the amount of section removed by erosion (Dow, 1977). The best fit line through all R_o data indicates that approximately 3600 ft (1100 m) of section has been removed.

Within Jonah field, pyrolysis data (Table 2) were only available on well cuttings from the Snyder SHB 5-26 (SW NW Sec. 26, T29N, R108W) and McMurry SHB 1-28 (NE NE Sec. 28, T29N, R108W) wells. The SHB 5-26 well is located only 5700 ft (1700 m) northeast and approximately 100 ft (30 m) downdip of the SHB 13-27 well. Because of the proximity of these two wells, pyrolysis data from the SHB 5-26 well were incorporated into the SHB 13-27 1-D BMOD model. Of the 11 pyrolysis data points, six T_{max} values are suspect because of low TOC values.

Constructing Northwest Green River Basin Burial-history Models

Burial-history or maturity modeling is particularly sensitive to several parameters, including burial and uplift history, rock thermal conductivity, and the thermal history or heat flow. Cluff (2002) used sonic- and resistivity-log data to conclude that 2000 to more than 3000 ft (610 to more than 910 m) of erosion has occurred. Naeser (1984), using fission-track data, suggests that during the last 2–4 m.y., rocks in the Pinedale area have cooled by at least 20°C (68°F). This temperature change may be the result of the removal of approximately 5000 ft (1500 m) of sediment based on the present-day geothermal gradient. Dickinson (1989) postulated the removal of more or less 500 ft (150 m) of Wasatch and younger section based on higher paleoheat flow. However, Dickinson also suggested

Table 4

Basic Modeling Assumptions Used in SHB 13-27 and Wagon Wheel 1 1-D Burial History Models.

Modeling Parameter	SHB 13-27	Wagon Wheel 1
Heat flow	46 mW/m^2 constant through the model's time	46 mW/m^2 constant through the model's time
Geothermal calculation	transient heat flow	transient heat flow
Present-day surface temperature	60°F	60°F
Compaction method	mechanical	mechanical
Porosity reduction method	exponential	exponential
Permeability calculation	modified Kozeny-Carman	modified Kozeny-Carman
Kerogen expulsion	saturation	saturation
Calculation depth intervals	200 ft	200 ft
Calculation time intervals	2 m.y.	2 m.y.
Pliocene and post-Pliocene uplift and erosion	3540 ft	5050 ft

that without an increased paleoheat flow, significantly more burial and subsequent erosion would have been necessary to account for the observed vitrinite reflectance readings in the Wagon Wheel well. Our own work with vitrinite reflectance data versus depth (Figure 4) indicates that there has been from 1400 to more than 6000 ft (430 to more than 1830 m) of erosion in the Jonah and Pinedale areas. Although Dickinson (1989) suggests that the observed, relatively high R_o data at the Wagon Wheel well can be explained using a higher paleoheat flow, several time-variant heat-flow scenarios were tested. The best model calibration, i.e., modeled R_o maturity and temperature as compared to the measured vitrinite reflectance and subsurface temperature data, was achieved using a 46-mW/m^2 (4.27-mW/ft^2) heat flow throughout the time extent of the models (Table 4). In the Jonah area, the SHB 13-27 burial-history model was constructed with a scenario that removed approximately 3500 ft (1100 m) of section, using vitrinite reflectance data from the modeled well to support this construct. In the Wagon Wheel model, 5050 ft (1540 m) of post-Wasatch was eroded. Table 5 summarizes the stratigraphy and thicknesses used in the burial-history models.

Assumptions made about the thermal conductivity of the sedimentary section and the rate of transfer of heat through the section also affects the outcome of the model with R_o dependent on heat flow. Alternatively, if the thermal conductivity is varied (by modifying the lithology of stratigraphic units), heat flow may have to be adjusted to match the modeled maturity to the available calibration data. The stratigraphic section in the Jonah area is highly heterogeneous both laterally and vertically. However, within the model, large, vertical packages of rock are modeled with average rock properties. For example, when the Lance is considered as a homogeneous rock unit, the average lithology assigned is that of siltstone. The thermal conductivity of siltstone, however, is substantially higher than a mix of sandstone, siltstone, and shale. Therefore, the Lance Formation reservoirs were modeled as stratigraphic units containing 34% sandstone, 33% siltstone, and 33% shale, and potential source rock shales were modeled as 65% shale and 35% siltstone. Lithologies of other formations in the overburden were modeled in a similar manner (Table 5).

In the Jonah area, the variable with the greatest uncertainty is heat flow, both Paleocene and present day. Only one present-day heat-flow point (Pollack, et al, 1991) was located near the area, in Sec. 2, T32N, R107W (Table 6). Those data were collected from a borehole in the Precambrian basement rocks, along the southwest flank of the upthrust Wind River Mountains, and are probably not representative of present-day heat flow in the adjoining Green River basin. These data were only used as constraint of the maximum present-day heat flow. A constant heat flow of 46 mW/m^2 (4.27 mW/ft^2), derived from the present-day geothermal gradient, was applied throughout the model's time. Several permutations of time-variant heat flow were tested, with heat flow increasing or decreasing through time; all of these variations in heat flow produced calculated maturities with a high degree of divergence to the available calibration data.

Table 5

Burial-history Stratigraphy (Ages, Thickness, and Lithologies) Used in 1-D Models (SHB 13-27 and El Paso Wagon Wheel 1). The Frontier Formation Top Was Used as Model Bottom in Each 1-D Model, Except Those Models Used to Test the Generative Capacity and Timing of Generation in the Mowry Shale.

	El Paso Wagon Wheel 1 (SE NW Sec. 5, T31N, R111W) KB 7089			Stud Horse Butte 13-27 (SW SW Sec. 27, T29N, R108W) KB 7280			
Formation or Event	**Depth (ft)**	**Thickness (ft)**	**Beginning Age (m.y.)**	**Depth (ft)**	**Thickness (ft)**	**Beginning Age (m.y.)**	**Modeled Lithology**
Late erosion		−5050	5		−3540	5	34% sandstone, 33% siltstone, 33% shale
Hiatus		0	17		0	17	
Miocene		550	24		540	24	34% sandstone, 33% siltstone, 33% shale
Hiatus		0	48		0	48	
Wasatch deposit		4500	50		3000	50	75% sandstone, 25% shale
Wasatch	0	4495	54	0	4530	54	75% sandstone, 25% shale
Fort Union	4495(?)	3025	65	4530	3760	65	50% sandstone, 25% siltstone, 25% shale
Hiatus		0	67.9			67.9	
Lance	7520	600	68	8290	500	68	34% sandstone, 33% siltstone, 33% shale
Lance Source 1	8120	300	68.625	8790	300	68.625	65% shale, 35% siltstone
Lance Reservoir 1	8420	600	69.25	9090	500	69.25	34% sandstone, 33% siltstone, 33% shale
Lance Source 2	9020	300	69.875	9590	300	69.875	65% shale, 35% siltstone
Lance Reservoir 2	9320	600	70.5	9890	500	70.5	34% sandstone, 33% siltstone, 33% shale
Lance Source 3	9920	300	71.125	10,390	300	71.125	65% shale, 35% siltstone
Lance Reservoir 3	10,220	600	71.75	10,690	500	71.75	34% sandstone, 33% siltstone, 33% shale
Lance Source 4	10,820	300	72.375	11,190	300	72.375	65% shale, 35% siltstone
Lance Reservoir 4	11,120	600	73	11,490	119	73	34% sandstone, 33% siltstone, 33% shale
Up Cretaceous Mesaverde Formation NSR	11,720	600	73.25	11,609	225	73.25	34% sandstone, 33% siltstone, 33% shale
Upper Cretaceous Mesaverde Formation Source	12,320	345	73.5	11,834	345	73.5	65% shale, 35% siltstone
Upper Cretaceous Mesaverde Formation NSR	12,665	500	74	12,179	230	74	34% sandstone, 33% siltstone, 33% shale
Ericson	13,165	300	75	12,409	300	75	100% sandstone
Upper Rock Springs	13,465	2000	79	12,709*	980	79	50% sandstone, 25% siltstone, 25% shale
Rock Springs SR	15,465	580	83	13,689*	580	83	65% shale, 35% siltstone
Lower Rock Springs	16,045	2225	86	14,269*	600	86	75% sandstone, 25% shale
Hilliard Shale	18,270	3730	90	14,869*	3700	90	85% shale, 15% siltstone
Frontier Sandstone	22,000	1	90.01	18,569*	850	95	50% sandstone, 25% siltstone, 25% shale
Mowry Shale				19,419*	350	99	85% shale, 15% siltstone
Muddy Sandstone				19,769*	1	100	100% sandstone

* Projected tops in SHB 13-27 well.

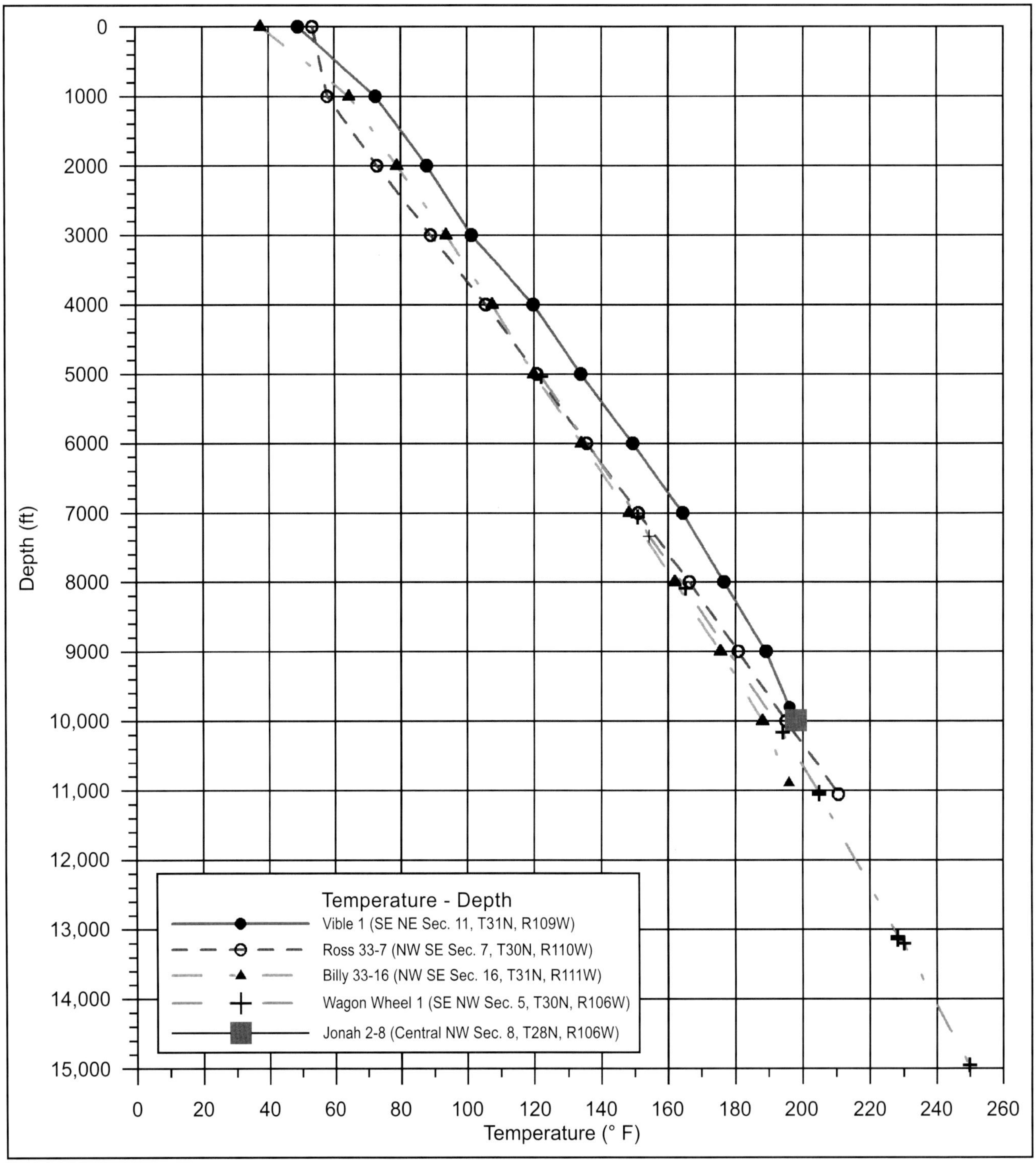

Figure 6. Composite temperature-depth plot for wells in the Jonah area. Most of the data were collected on long-term pressure buildup tests and closely represent stabilized bottomhole temperatures.

Table 6

Heat-flow Data (Pollack, et al., 1991). This Location is Considerably North and East of the Jonah Field Area and is on the Precambrian Core of the Wind River Mountains. Because of the Location, it Was Not Used for Model Calibration Purposes in the Jonah or Pinedale Areas. The Data Were Only Used as a Constraint of the Maximum Present-day Heat Flow.

Latitude	Longitude	Heat Flow (mW/m^2)	Temperature Gradient (mK/m)	Data Point ID	Elevation (ft)	Minimum Depth (ft)	Maximum Depth (ft)	Number of Readings	Conductivity
42.7667	−109.567	51		US3110	7277				
42.7667	−109.567	57	19	A	7277	1000	4449	27	3.01
42.7667	−109.567	47	16	B	7277	6850	9829	29	2.8
42.7667	−109.567	50	18	C	7277	1000	9829	56	2.93

1-D Modeling Results

Jonah Federal Stud Horse Butte 13-27

The SHB 13-27 well was modeled using the lithologies shown in Table 5. Iterative burial-history modeling, matching calculated R_o to measured R_o, suggested that approximately 3500 ft (1100 m) of Wasatch and younger section was removed in the last 5 m.y. (Figure 7). Calibration to vitrinite reflectance, T_{max}, and present-day BHT data was accomplished using a constant heat flow of 46 mW/m^2 (4.27 mW/ft^2) throughout the modeled period (Figure 8). It was unnecessary to alter heat flow during the last 5 m.y. of uplift to match the calibration data. The timing generation and expulsion for all the source rocks included in the SHB 13-27 model are summarized in Table 7. Using a saturation threshold of 0.20 (the percentage of source rock pore space that must be saturated with hydrocarbon before expulsion can occur), expulsion only occurs after achieving maximum generation. After the main generation and expulsion event (50–40 Ma), there is a 30-m.y. period of low volume expulsion, followed by a drop in rate concurrent with Pliocene uplift (Figure 9). This period of continuous, low-volume expulsion may supply enough gas to maintain a pressured accumulation with concurrent bleed-off through the top seal.

The SHB 13-27 model transformation ratios show that the majority of generation and expulsion occurs after the maximum burial and compaction of the Lance reservoir rocks. All of the source rocks modeled, with the possible exception of the lower Mesaverde, are still capable of generating additional hydrocarbons. Within the Lance, however, expulsion only occurs from lowermost source units. During the period of rapid burial, the maturity of the lowermost Lance source unit (Figure 10, dashed lines) only reaches an R_o of 0.75% at 44 Ma and gradually increases to R_o of 0.92 at present day. The maturity at the top of the Lance only reaches R_o of more or less 0.72 at present day. The modeled R_o values are slightly high relative to the measured vitrinite data (R_o), suggesting that the assumed 3500 ft (1100 m) of uplift and erosion may be a slight overestimate. The minimal amount of dispersed organic material and the poor quality (low TOC) of that organic matter, coupled with the immature nature of the Lance, suggest that it is highly unlikely that the Lance generated enough hydrocarbons to fill available pore space and account for the volumes of gas found at either Jonah or Pinedale.

Pinedale Anticline: Wagon Wheel 1

The Wagon Wheel 1 model was constructed with the identical lithologic properties and formation ages to those

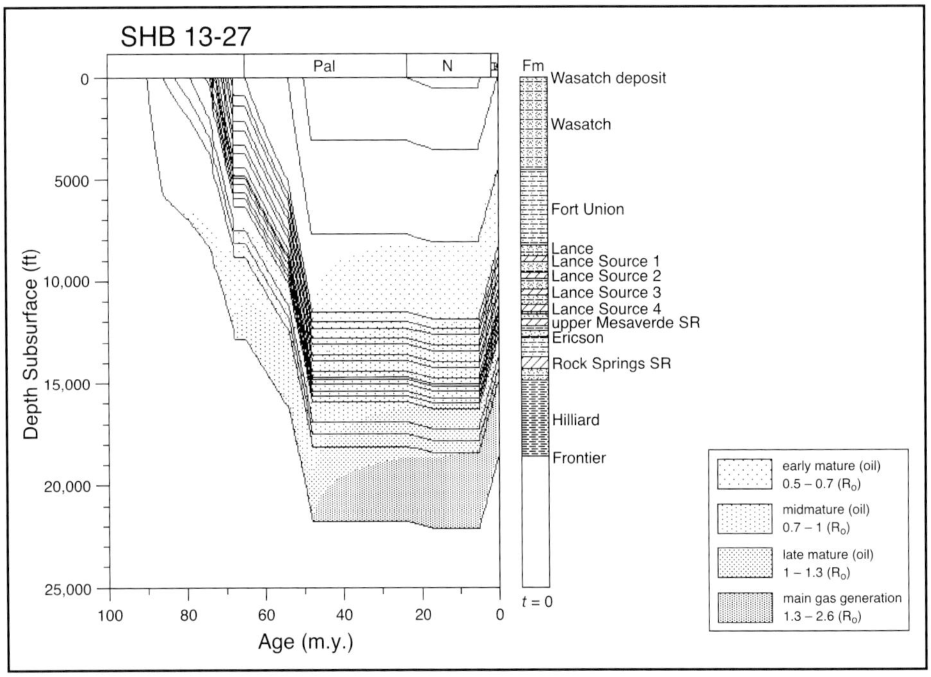

Figure 7. Burial-history plot SHB 13-27 (SW SW Sec. 27, T29N, R108W). This model incorporates the uplift and erosion of approximately 3500 ft (1100 m) of Wasatch and younger rocks during the last 5 m.y. Plot also shows hydrocarbon generation windows for the Hilliard Shale through the Fort Union Formation.

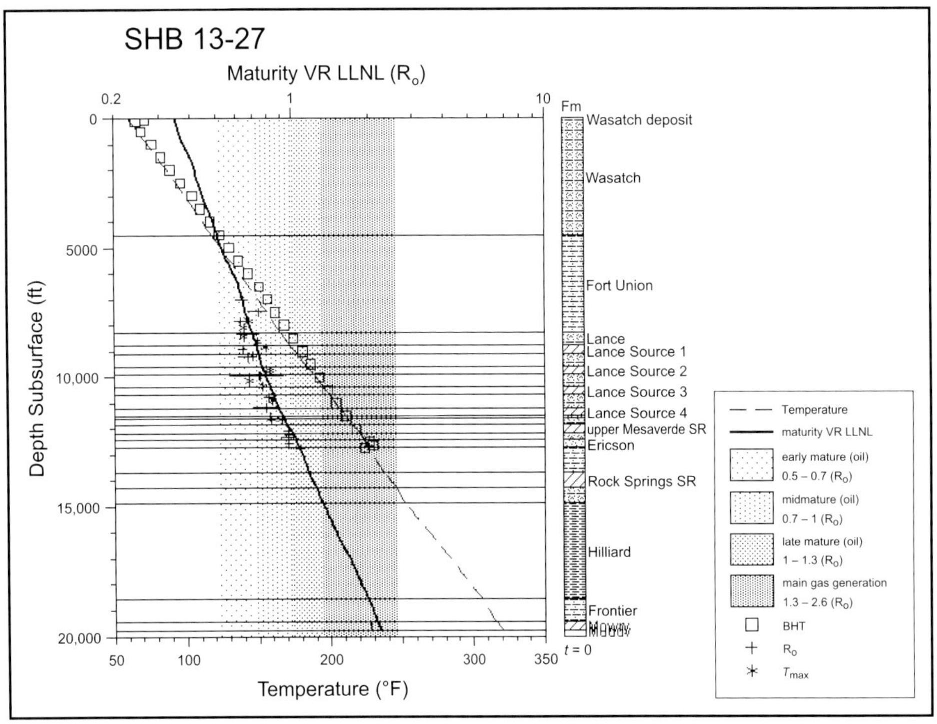

Figure 8. Calibration plot of the SHB 13-27 well (SW SW Sec. 27, T29N, R108W) comparing modeled maturity (R_o) and temperature with the measured data. T_{max} calibration data is from SHB 5-26. Note that most of the probable source rock intervals, with the exception of the Hilliard, never entered the main gas-generation window.

Table 7

SHB 13-27 Model Results. Timing of Gas Generation and Expulsion for Modeled Source Rocks (SR) with Variation in Saturation Threshold from 0.20 (Conservative) to 0.05 (Optimistic). Early Generation Time Was Picked Where Calculated Generation Curve Begins to Rapidly Increase from Very Low Rates of Generation. Multiple Maximum Values Occur Where More than One Generation and Expulsion Peak Can Be Observed Through Time. Shaded Areas Represent Combinations of Source and Saturation Thresholds Where No Expulsion Occurs. With Decreasing Saturation Threshold, Deeper Source Rocks Expel Earlier, and Relatively Immature Shallower Sources Begin to Expel Hydrocarbons.

			Time of Maximum Gas Expulsion Rate (Ma)		
Source Rock	**Time to Earliest Gas Generation (Ma)**	**Time of Maximum Gas Generation Rate (Ma)**	**Saturation Threshold 0.20**	**Saturation Threshold 0.10**	**Saturation Threshold 0.05**
Lance SR 1	51.6	43.5 and 14.0	–	–	–
Lance SR 2	52.5	45.5 and 14.5	–	–	–
Lance SR 3	53.5	45.9 and 14.9	–	–	19.0
Lance SR 4	55.9	47.8 and 14.5	–	7.0	39.0 and 19.0
Upper Cretaceous Mesaverde Formation	58.9	45.9 and 14.9	39.0 and 19.0	45.0 and 19.0	47.0 and 19.0
Lower Cretaceous Mesaverde Formation (LRS)	67.8	47.5 and 14.5	47.0 and 19.0	49.0	49.0
Hilliard	74.5	66.4, 57.8, and 49.3	51.0 and 21.0	53.0 and 19.0	59.0 and 49.0

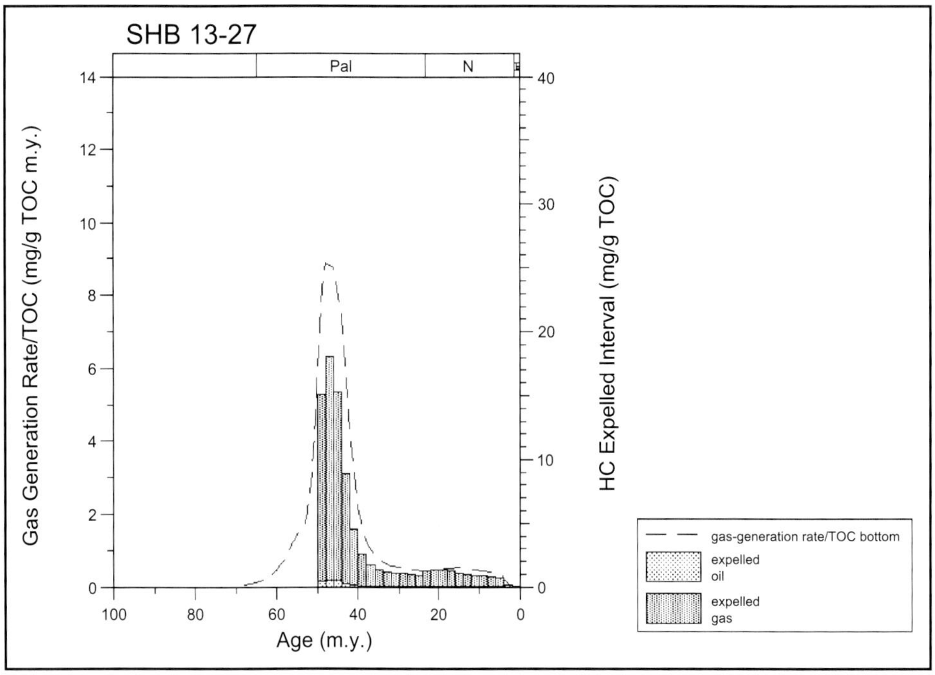

Figure 9. Gas-generation rate and time interval of expelled hydrocarbon plot for the SHB 13-27 well (SW SW Sec. 27, T29N, R108W). The lower Mesaverde (lower Rock Springs) gas generation and expulsion is shown per 2-m.y. time interval. Generation begins as early as 70 Ma, peaking at 47 Ma concurrent with peak expulsion rate. This model was run with a saturation threshold of 0.20. The time of generation and expulsion in this model is partially controlled by the modeled position of the source in the lower Mesaverde (approximately midway between the base of the Ericson Sandstone and the top of the Hilliard Shale).

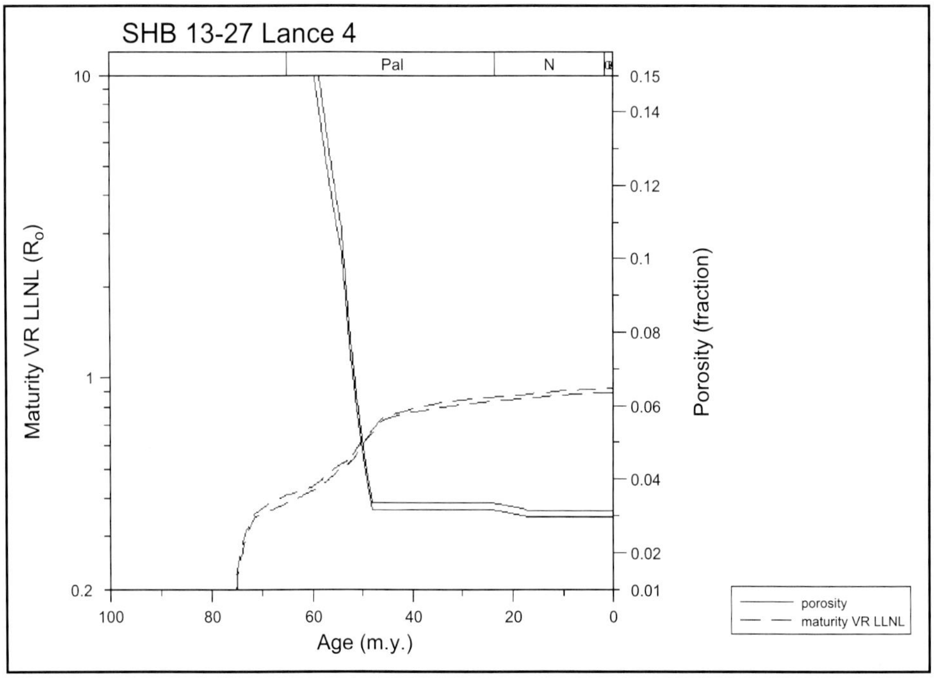

Figure 10. Calculated maturity and porosity plot for the SHB 13-27 (SW SW Sec. 27, T29N, R108W). Plot shows that by the time this lower Lance source rock reaches a maturity of R_o 0.75, it is also nearly fully compacted.

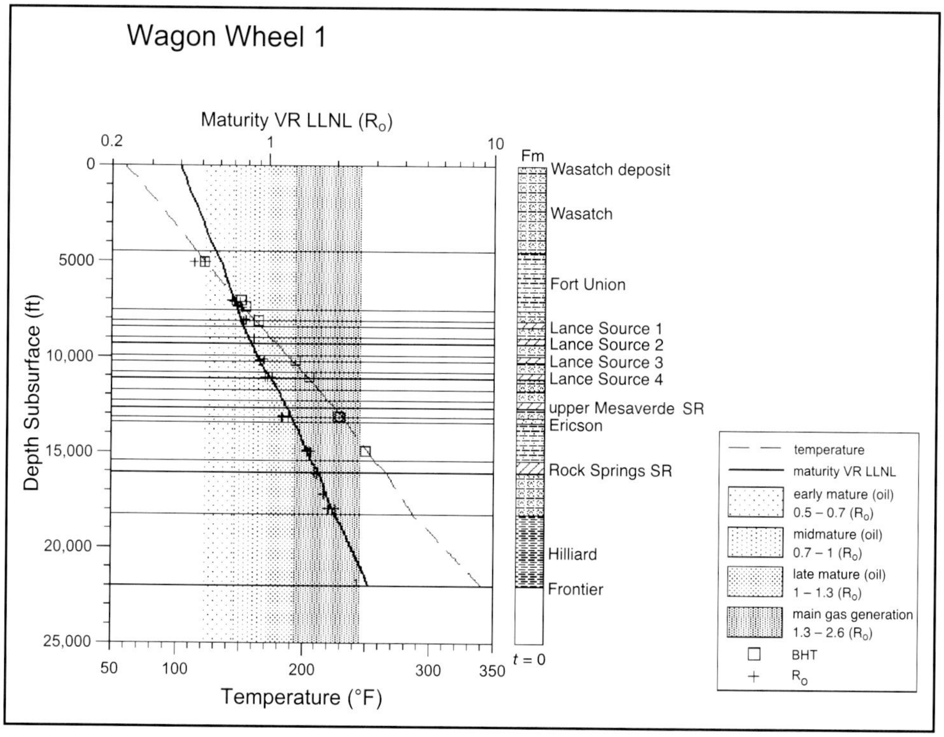

Figure 11. Calibration plot of the Wagon Wheel 1 well (SE NW Sec. 5, T30N, R108W) comparing modeled R_o maturity and temperature to measured vitrinite reflectance (R_o) and corrected bottomhole temperature data.

used in the SHB 13-27 model, with formation tops and thicknesses taken from Law and Johnson (1989). In this model, 5050 ft (1540 m) of Pliocene uplift and erosion were included in the burial history based on (1) the observed local uplift of the Pinedale anticline and (2) the observed vitrinite reflectance profile (Figure 4). Figure 11 shows the resulting calibration plot. Because of the increased burial depth relative to the SHB 13-27 well (Figure 12), the lower Mesaverde source begins generating gas at about 70 Ma and begins expelling as early as 54 Ma (Figure 13). Unlike the SHB 13-27 model, the Wagon Wheel 1 model shows that the lower Mesaverde source rock has passed through the main gas-generation thermal window (Figure 12). Table 8 summarizes the timing of early and maximum gas generation and expulsion for all the source rocks included in the model.

The models suggest that because of a greater depth of burial, generation and expulsion from the lower Mesaverde occurred approximately more than 4 m.y. earlier in the Pinedale area than in the Jonah area. A comparison of Figure 14a and b shows that although both models contain identical source rocks in the lower Mesaverde, the amount of gas expelled in the Pinedale area is significantly greater than that expelled in the Jonah area. Calculated transformation ratios indicate that in the Pinedale area, the lower Mesaverde is nearing generative completion while the same interval is still capable of generation in the Jonah area. It should also be noted that in both areas, because of uplift, erosion, and cooling, generation of gas has decreased to a very low rate or may have ceased.

PROBABILISTIC MODELING

To test the multiple source rock generative potential in the Jonah and Pinedale areas, probabilistic models (simulations) were constructed using Platte River Associates, Inc., BasinMod Risk© (released August 2002). In these simulations, hydrocarbon charge elements of TOC, source area, and migration efficiency, as well as the trap elements of reservoir area, reservoir gross thickness, net-to-gross ratio, porosity, pore pressure, reservoir temperature, and seal efficiency, are input as distributions instead of discrete values (Tables 9, 10). The calculated output values (gas generated, gas expelled, gas in place, etc.) are also continuous distributions and are summarized in Tables 11 and 12. In the following discussion, all estimated values of generated gas (in trillion cubic feet) and expelled gas (in trillion cubic feet) are P_{50} values taken from simulations. Figure 15, showing expelled gas volume for the lower Rock Springs, is representative of

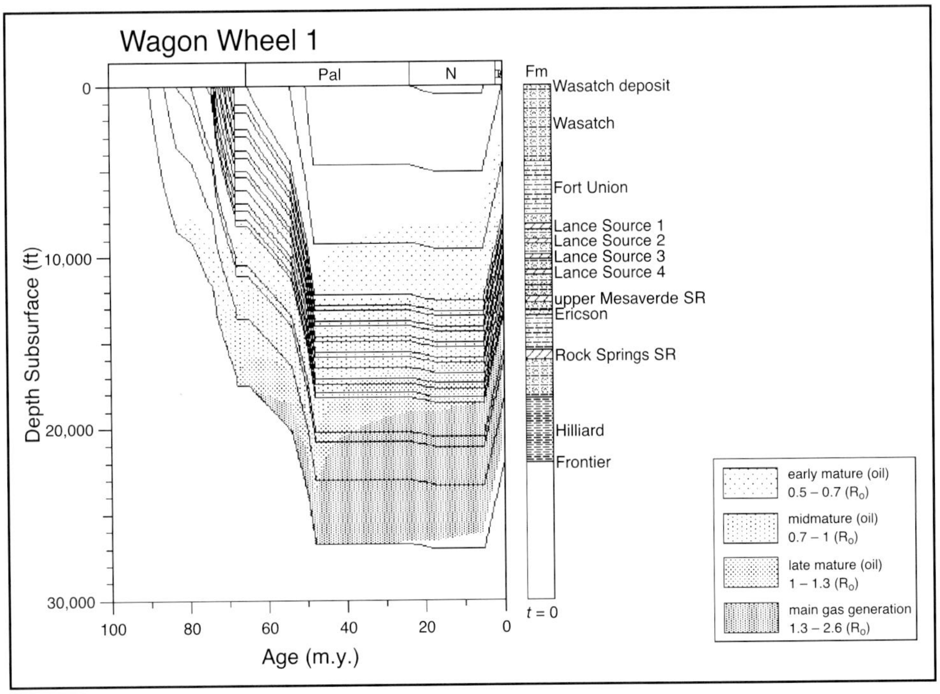

Figure 12. Burial-history plot of the Wagon Wheel 1 well (SE NW Sec. 5, T30N, R108W). Burial history includes the uplift and erosion of approximately 5000 ft (1500 m) of Wasatch and younger rocks during the last 5 m.y. Plot also shows hydrocarbon-generation windows for the Hilliard Shale through the Fort Union Formation.

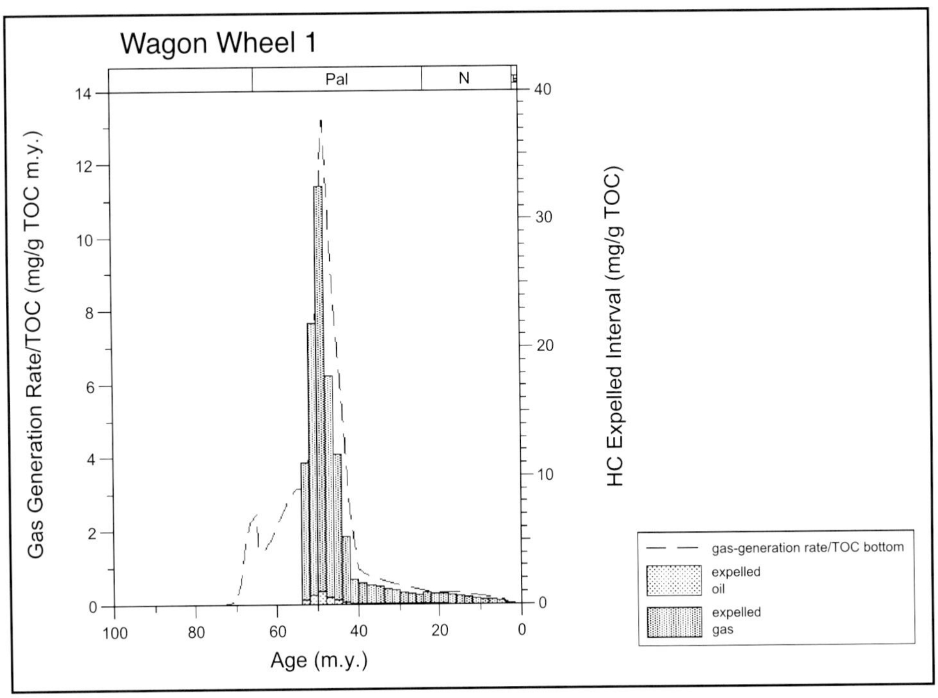

Figure 13. Gas-generation rate and time interval of expelled hydrocarbon plot for the Wagon Wheel 1 well. Lower Mesaverde (lower Rock Springs) source rock gas generation and expulsion is shown per 2-m.y. time interval. Note early generation (approximately 75–70 Ma) and expulsion beginning at about 56 Ma. This model was run with a saturation threshold of 0.20. The time of generation and expulsion in this model is partially controlled by the modeled position of the source in the lower Mesaverde (approximately midway between the base of the Ericson Sandstone and the top of the Hilliard Shale).

Table 8

Wagon Wheel 1 Model Results. Timing of Generation and Expulsion for Modeled Source Rocks (SR) with Variation in Saturation Threshold from 0.20 (Conservative) to 0.05 (Optimistic). Early Generation Time was Picked Where Calculated Generation Curve Begins to Rapidly Increase from Very Low Rates of Generation. Multiple Maximum Values Occur Where More than One Generation and Expulsion Peak Can be Observed Through Time. Shaded Areas Represent Combinations of Source and Saturation Threshold Where no Expulsion Occurs. With Decreasing Saturation Threshold, Deeper Source Rocks Expel Earlier and Relatively Immature Shallower Sources Begin to Expel Hydrocarbons.

			Time of Maximum Gas Expulsion Rate (Ma)		
Source Rock	**Time to Earliest Gas Generation (Ma)**	**Time of Maximum Gas Generation Rate (Ma)**	**Saturation Threshold 0.20**	**Saturation Threshold 0.10**	**Saturation Threshold 0.05**
Lance SR 1	50.7	44.2 and 14.1	–	–	–
Lance SR 2	51.1	45.5 and 14.5	–	–	17.0
Lance SR 3	52.1	46.4 and 14.2	–	9.0	37.0 and 19.0
Lance SR 4	53.2	46.9 and 14.9	–	35.0 and 19.0	43.0 and 19.0
Upper Cretaceous Mesaverde Formation	59.5	48.7 and 15.0	47.0 and 19.0	47.0	47.0
Lower Cretaceous Mesaverde Formation (LRS)	69.7	65.0 and 48.3	49.0	49.0	49.0
Hilliard	80.2	69.2 and 43.1	53.0 and 45.0	67.0, 53.0, and 43.0	69.0, 53.0, and 43.0

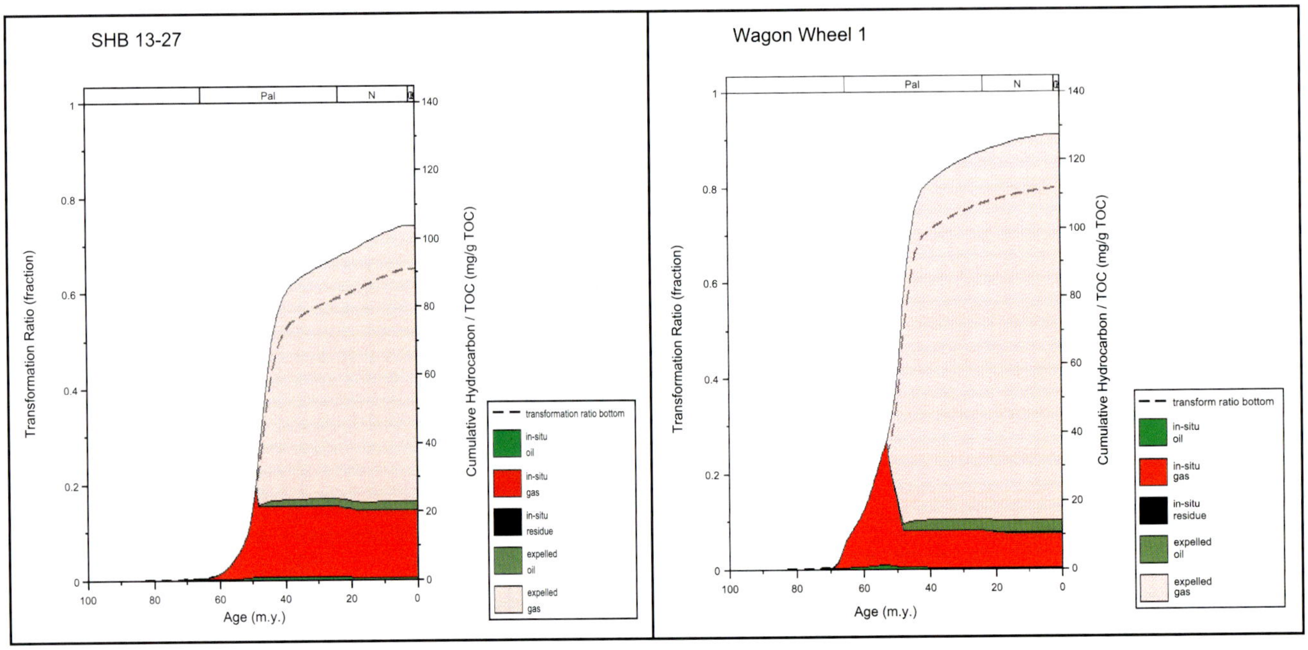

Figure 14. (a, b) Lower Mesaverde source transformation ratios and cumulative hydrocarbon plots. (a) SHB 13-27 well (SW SW Sec. 27, T29N, R108W) and (b) Wagon Wheel 1 well (SE NW Sec. 5, T30N, R108W). Because of the differences in burial histories, gas generation at the Wagon Wheel 1 area begins approximately 4 m.y. earlier than in the area around the SHB 13-27. Both the amount of gas expelled and the degree of kerogen transformation to hydrocarbon are greater in the Wagon Wheel 1 area.

the iterative output of RISK modeling. In the Jonah field Risk model, source parameters of area, thickness, and richness and reservoir parameters of area, thickness, and porosity were assigned to the stratigraphy in the triangular structural block confined by the Jonah bounding faults (Figure 1). The Jonah model was designed to test (1) the generative capacity of each source contained in modeled area and (2) given distributions of expulsion and migration efficiencies as to what each source could contribute to the in-place gas volume. In the Pinedale

Table 9

SHB 13-27 and Wagon Wheel 1 RISK Model-simulation Charge-elements Input Data. Charge Elements That Were Held Constant: Effective Source Ratio = 1.0, Source Lithologies (as Specified in 1-D Models), Kerogen Types (as Specified in 1-D Models). For the Wagon Wheel Model, Source Rock Area Was Held Constant at 23,040 ac (93 km^2) (One Township).

Source Name	TOC (%)	Area (ac)
Lance Source 1	0.75–**1.00**–1.25	20,000–**24,000**–26,000
Lance Source 2	0.75–**1.00**–1.25	20,000–**24,000**–26,000
Lance Source 3	1.00–**1.25**–1.50	20,000–**24,000**–26,000
Lance Source 4	1.00–**1.25**–1.50	20,000–**24,000**–26,000
Upper Mesaverde Source	2.00–**2.25**–3.00	20,000–**24,000**–26,000
Lower Mesaverde Source	2.00–**2.25**–3.00	20,000–**24,000**–26,000
Hilliard Shale	0.75–**1.00**–1.25	20,000–**24,000**–26,000

Overall migration efficiency applied to all models = 0.50–**0.60**–0.70. Mode values are shown in bold.

Table 10

RISK Model-simulation Trap Element Distributions Sampled for the SHB 13-27 Model. Probability Distributions Were Sampled, through Latin Hypercube Methods, Reducing the Number of Iterations Necessary per Model to 1000 Versus 5000+ for a Monte Carlo Approach.

Trap Element	Minimum	Maximum	Mode	Constant
Hydrocarbon saturation				0.80
Pore pressure (psi)	3700	7000	**5500**	
Reservoir temperature (°F)	200	230	**220**	
Gas recovery efficiency	0.25	0.50	**0.33**	
Reservoir porosity	0.05	0.12	**0.09**	
Reservoir net to gross	0.25	0.33	**0.40**	
Reservoir area (ac)	20,000	26,000	**24,000**	
Reservoir thickness (ft)	2400	3000	**2600**	
Gas seal efficiency	0.80	0.90	**0.85**	

area, the intent was to simulate source rock generation and expulsion capacity per 36 mi^2 (93 km^2) without consideration of the reservoir properties.

DISCUSSION

Can the Lance Act as a Self-source?

DuBois et al. (2004) estimate that the Lance reservoir at Jonah field contains at least 8.3 tcf gas in place, with an ultimate recovery of 2.31 tcf gas. One-dimensional burial-history models and probabilistic simulations were constructed to test if the Lance generative capacity and thermal history can produce the gas charge observed at Jonah. In the initial 1-D burial-history models, the entire Lance section was assigned 0.75–1.50% TOC and subjected to burial. In those models, the mix of sandstone, siltstone, shale, and kerogen produced an average lithology that, with burial, generated, but never expelled, hydrocarbons. The compaction of the average lithology never reduced the porosity to the point where there was enough hydrocarbons generated to saturate source rock pore space to the levels (5–20%) necessary to expel. In the final 1-D models, the Lance was subdivided into four source and four reservoir units, with the lithologic characteristics shown in Table 5. This allowed the shaly source rocks to compact and expel hydrocarbons. In those simulations, the Lance source units were modeled containing type III, gas-prone kerogen, having an initial TOC distribution from 0.75 to 1.50%. The probabilistic model was then subjected to the same burial and thermal conditions as all other SHB 13-27 1-D models. Table 11 summarizes the results of probabilistic modeling. Daly and Edman (1987) suggest that the maximum TOC loss caused by maturation for a type III kerogen is 20%. Based on their method and an average measured TOC of 0.57%, the modeled 0.75–1.50% initial TOC is probably an overestimation, and the resulting gas generated from the Lance should be considered as the maximum volume possible.

Between simulation runs, all variables were held constant, with the exception of saturation threshold. Simulation 2002/11/13 20:43:01 (Table 11) assumed a saturation threshold of 0.20; in this model, gas was generated in the Lance source intervals but never reached a saturation level sufficient to expel into adjacent reservoirs. Only by lowering the saturation threshold to 0.05 (simulation 2002/11/13 21:41:31) did the Lance expel gas. In this simulation, the Lance generated a total of 1.79 tcf and expelled 0.64 tcf gas (Table 11), significantly less than what is observed at Jonah. In addition, migration efficiency and seal leakage would act to reduce the total trapped gas in place to a much smaller amount.

The low initial TOC values, the oxidizing depositional environment, and the low thermal maturity of the Lance (as supported by vitrinite reflectance) all suggest that although the shaly zones of the Lance may contribute a minimal volume to local charge, they cannot provide the in-place gas reserves as estimated by DuBois et al. (2004). In addition, the low generative capacity of the Lance suggests that hydrocarbon generation from in-situ sources cannot be the sole origin of overpressuring at

Table 11

Simulation Results for the SHB 13-27 Model (SW SW Sec. 27, T29N, R108W). Volumetric Values are P_{50} Volumes for the Entire Jonah Trap Area. Trap Area Was One of the Uncertainties Modeled and Varied from 20,000 to 26,000 ac (81 to 105 km^2). Only in the Most Optimistic Expulsion Scenario (Saturation Threshold = 0.05, and Contribution from All Potential Source Intervals) Does the Gas in Place Even Begin to Approach the 8.3-tcf Estimates Provided by DuBois et al. (2004). An Additional Simulation (2002/11/14 19:10:32) was Run to Test the Expulsion Capacity of the Hilliard at a Higher Saturation Threshold. At a Saturation Threshold of 0.50, the Hilliard Shale Generated 17.59 tcf (Standard temperature and pressure [STP]) and Expelled 4.17 tcf (STP).

	Model Charge Elements					Model Trap Elements			
Run	**Source Unit**	**TOC Distribution**	**Saturation Threshold**	**Generated Gas (tcf)**	**Expelled Gas (tcf)**	**Total Migrated Gas Charge**	**Gas in Place (tcf)**	**Oil in Place (MMBO)**	**Recovered Gas (tcf)**
1	Lance Source Rock 1	0.75–1.0–1.25	0.05	0.17	0				
1	Lance Source Rock 2	0.75–1.0–1.25	0.05	0.29	0				
1	Lance Source Rock 3	1.0–1.25–1.5	0.05	0.55	0.18				
1	Lance Source Rock 4	1.0–1.25–1.5	0.05	0.78	0.46				
1	Upper Cretaceous Mesaverde Formation	2.0–2.5–3.0	0.05	2.36	2.04				
1	Lower Rock Springs Formation	2.0–2.5–3.0	0.05	5.48	5.12				
Simulation 2002/11/13 21:41:31 Subtotals				**9.63**	**7.80**	**4.68**	**3.98**	**21.35**	**1.36**
1	Hilliard	0.75–1.0–1.25	0.50	17.38	15.99				
Simulation 2002/11/13 21:41:31 Totals				**27.01**	**23.79**				
2	Lance Source Rock 1	0.75–1.0–1.25	0.10	0.17	0				
2	Lance Source Rock 2	0.75–1.0–1.25	0.10	0.29	0				
2	Lance Source Rock 3	1.0–1.25–1.5	0.10	0.55	0				
2	Lance Source Rock 4	1.0–1.25–1.5	0.10	0.78	0.15				
2	Upper Cretaceous Mesaverde Formation	2.0–2.5–3.0	0.10	2.34	1.72				
2	Lower Rock Springs Formation	2.0–2.5–3.0	0.10	5.49	4.77				
Simulation 2002/11/13 21:11:07 Subtotals				**9.62**	**6.64**	**3.99**	**3.38**	**18.17**	**1.16**
2	Hilliard	0.75–1.0–1.25	0.10	17.50	14.71				
Simulation 2002/11/13 21:11:07 Totals				**27.12**	**21.35**				
3	Lance Source Rock 1	0.75–1.0–1.25	0.20	0.17	0				
3	Lance Source Rock 2	0.75–1.0–1.25	0.20	0.29	0				
3	Lance Source Rock 3	1.0–1.25–1.5	0.20	0.55	0				
3	Lance Source Rock 4	1.0–1.25–1.5	0.20	0.78	0				
3	Upper Cretaceous Mesaverde Formation	2.0–2.5–3.0	0.20	2.36	1.09				
3	Lower Rock Springs Formation	2.0–2.5–3.0	0.20	5.48	4.06				
Simulation 2002/11/13 20:43:01 Subtotals				**9.63**	**5.15**	**3.07**	**2.61**	**13.97**	**0.90**
3	Hilliard	0.75–1.0–1.25	0.20	17.38	11.84				
Simulation 2002/11/13 20:43:01 Totals				**27.01**	**16.99**				

Table 12

Simulation Results for the Wagon Wheel 1 Model (SE NW Sec. 5, T30N, R108W). These Simulations were Run to Test the Generative Capacity of the Source Intervals in the Pinedale Area. The Volumetric Values are P_{50} Volumes and are in Units of Trillion Cubic Feet per 23,040 ac (93 km^2) (One Township). Only the Charge Aspects of the Simulation were Considered. Using an Overall Migration Efficiency Distribution of 0.5–0.7 (Mode = 0.6), the Simulations Predict that for the Source Rocks in the Pinedale Area, 5.03–6.46 tcf (Per Source Township) of Charge is Available to the Trap. Subtotal Values are Summations of all Modeled Source Rocks, with the Exception of the Hilliard Shale. Values in the "Totals" Line Include the Possible Contribution from the Hilliard Shale. An Additional Simulation (2002/11/14 19:50:22) was Run to Test the Expulsion Capacity of the Hilliard at a Higher Saturation Threshold. At a Saturation Threshold of 0.50, the Hilliard Generated 19.62 tcf (STP) and Expelled 14.50 tcf (STP)/23,040 ac.

	Model Charge Elements						Model Trap Elements		
Run	Source Unit	TOC Distribution	Saturation Threshold	Generated Gas (tcf)	Expelled Gas (tcf)	Total Migrated Gas Charge	Gas in Place (tcf)	Oil in Place (MMBO)	Recovered Gas (tcf)
1	Lance Source Rock 1	0.75–1.0–1.25	0.05	0.24	0				
1	Lance Source Rock 2	0.75–1.0–1.25	0.05	0.40	0.05				
1	Lance Source Rock 3	1.0–1.25–1.5	0.05	0.73	0.44				
1	Lance Source Rock 4	1.0–1.25–1.5	0.05	1.04	0.80				
1	Upper Cretaceous Mesaverde Formation	2.0–2.5–3.0	0.05	3.10	2.90				
1	Lower Rock Springs Formation	2.0–2.5–3.0	0.05	6.79	6.61				
Simulation 2002/11/10 16:17:45 Subtotals				**12.30**	**10.80**	**6.46**			
1	Hilliard	0.75–1.0–1.25	0.05	19.44	18.91				
Simulation 2002/11/10 16:17:45 Totals				**31.93**	**29.71**				
2	Lance Source Rock 1	0.75–1.0–1.25	0.10	0.24	0				
2	Lance Source Rock 2	0.75–1.0–1.25	0.10	0.40	0				
2	Lance Source Rock 3	1.0–1.25–1.5	0.10	0.73	0.14				
2	Lance Source Rock 4	1.0–1.25–1.5	0.10	1.04	0.55				
2	Upper Cretaceous Mesaverde Formation	2.0–2.5–3.0	0.10	3.10	2.70				
2	Lower Rock Springs Formation	2.0–2.5–3.0	0.10	6.81	6.43				
Simulation 2002/11/14 12:00:28 Subtotals				**12.32**	**9.82**	**5.88**			
2	Hilliard	0.75–1.0–1.25	0.10	19.47	18.44				
Simulation 2002/11/14 12:00:28 Totals				**31.93**	**25.30**				
3	Lance Source Rock 1	0.75–1.0–1.25	0.20	0.24	0				
3	Lance Source Rock 2	0.75–1.0–1.25	0.20	0.40	0				
3	Lance Source Rock 3	1.0–1.25–1.5	0.20	0.73	0				
3	Lance Source Rock 4	1.0–1.25–1.5	0.20	1.04	0.07				
3	Upper Cretaceous Mesaverde Formation	2.0–2.5–3.0	0.20	3.10	2.28				
3	Lower Rock Springs Formation	2.0–2.5–3.0	0.20	6.81	6.07				
Simulation 2002/11/14 14:30:11 Subtotals				**12.32**	**8.42**	5.03			
3	Hilliard	0.75–1.0–1.25	0.20	19.63	14.53				
Simulation 2002/11/14 14:30:11 Totals				**31.95**	**23.93**				

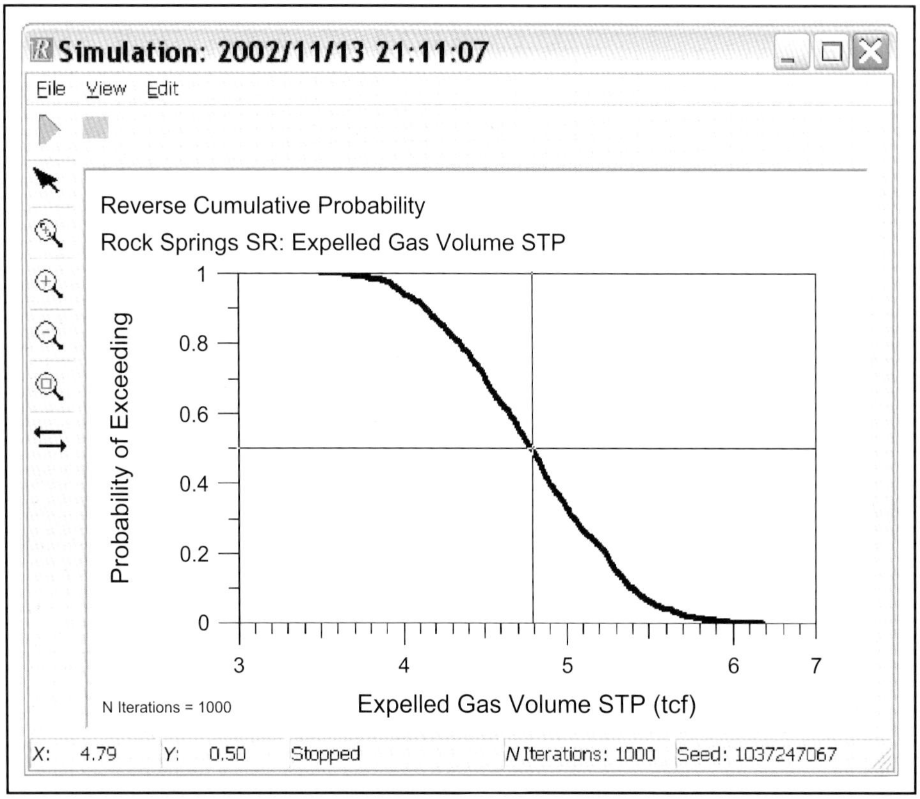

Figure 15. Expelled gas STP (tcf) from simulation run 2002/11/06 18:18:29 on SHB 13-27. P_{50} = 4.79 (tcf at STP) of gas expelled from lower Mesaverde (lower Rock Springs) source. Saturation threshold was fixed at 0.10 for this simulation run.

Jonah field. The values of source rock thickness, richness, and saturation threshold that were incorporated in the models were very optimistic, and yet insufficient charge was liberated from the Lance shales. These results suggest that the Lance cannot be adequately self-sourced, and the field must have received significant charge from other sources. This, in turn, implies migration from deeper sources along a combination of vertical and/or lateral migration pathways. Hanson et al. (2004) also suggest that some of the Lance gas at Jonah was the result of secondary cracking of liquid hydrocarbon and vertical migration from deeper source rocks and reservoirs.

Other Potential Source Rocks

Within the upper Mesaverde Formation, coals and shaly intervals that appeared to have source rock potential (based on gamma-ray logs and mud-log description) were isolated into a single unit and placed in the middle of the upper Mesaverde section. Individual coals were not modeled as discrete beds, but instead, their TOC contribution was averaged, along with other lower TOC shale and silty shale. In the SHB 13-27 model, out of a total 800 ft (243.8 m) of upper Mesaverde, 345 ft (105 m) was assigned as source rock with initial TOC from 2 to 3%, having an average of 2.5%. Because of the organic richness and thermal maturity of this unit, the resulting model generated 2.36 tcf (STP) or more of gas and expelled between 1.09 and 2.04 tcf (P_{50} values), depending on saturation threshold (Table 11).

A similar approach of combining potential source rocks was applied to the lower Mesaverde Formation (Rock Springs Formation). In the SHB 13-27, out of a total 1220 ft (370 m) of lower Mesaverde, 580 ft (180 m) was assigned as source rock with initial TOC from 2 to 3%, having an average of 2.5%. The model with this source rock suggests that this unit is capable of generating 5.48 tcf or more of gas and expelling between 4.06 and 5.12 tcf (P_{50} values), depending on saturation threshold (Table 11). Additional models were built to test the feasibility of generating enough hydrocarbons from source rocks in the Jonah structural block as confined by the southeast- and western-bounding faults and the plunge of the Pinedale anticline. Table 11 illustrates that only in the most optimistic cases can the indigenous source rocks from Lance through lower Mesaverde generate, expel, and effectively migrate enough gas to approach the 8.3-tcf gas in place suggested by DuBois et al. (2004). The three SHB 13-27 simulations estimate total migrated gas charge of 3.07–4.68 tcf, gas in place of 2.61–3.98 tcf, and recoverable reserves of 0.90–1.36 tcf.

The generative potential of the Hilliard Shale was modeled at Jonah using thicknesses taken from the Wagon Wheel (Pinedale) well. The Hilliard was assumed to contain an average of 1% TOC over the entire thickness of the unit, and the models were subjected to the same burial history and thermal conditions as all other models (Table 4). In the Wagon Wheel area, the model predicts that the Hilliard Shale could generate 19.5 tcf + 89 MMBO per township (36 mi^2), and in the Jonah area, because of less burial, the Hilliard is predicted to generate 17.4 tcf + 30 MMBO per township.

Additional Considerations Regarding Source and Migration

Modeling of the Wagon Wheel 1 well shows a higher degree of thermal maturity and greater amount of expulsion in the upper and lower Mesaverde sources than for the Jonah area (Table 12). The Wagon Wheel simulation was run with identical source rock characteristics as used in the SHB 13-27 model; however, the source rock area was normalized at 23,040 ac (93 km^2) (one township). The three Wagon Wheel simulations estimate total migrated gas charge of 5.03–6.46 tcf per township from all source intervals, excluding the Hilliard Shale. Regional geology and basin paleogeometry suggest that during generation, a large fetch area was available, and the amount of expelled gas across a multitownship area is much more than that necessary to charge the Jonah trap. Migration of hydrocarbons in the Jonah area is difficult to quantify; however, burial history suggests that lower lithostatic pressures would have existed upward and southwest of the present-day Pinedale anticline area. This regional pressure profile may facilitate migration from the generative kitchen toward updip traps. In addition to those conduits provided by faults and fractures, the Upper Cretaceous Ericson Sandstone may have acted as a carrier bed to facilitate more efficient migration from the generative kitchen to the trap. DuBois et al. (2004) suggest that the faults bounding the Jonah trap were active during the Late Cretaceous–early Tertiary and may have created a structural condition resulting in slightly improved reservoir characteristics in the Jonah fault block. These faults predate expulsion from Upper Cretaceous sources and could act as conduits for the vertical migration from carrier beds and/or deeper, actively generating sources into the Jonah trap.

Most of the structural development of the Pinedale anticline occurred after maximum gas generation had occurred in the Pinedale area. Therefore, migrating hydrocarbons would have direct, updip access to the Jonah block and adjacent areas. The gas expulsion wave emanating from the greater Pinedale area may have charged a much larger area than that currently observed. During and after migration, the upper Lance top seal acted to impede vertical migration or leakage; however, this unit is absent by erosion southwest of Jonah. If migrating hydrocarbons reached this escape point, gas would leak out of the Lance. During the time of maximum burial, areas to the northwest and southeast of Jonah may have received and maintained a charge if (1) the upper Lance top seal was effective and (2) lateral seals confined the migrating gas. The discontinuous sands in the Lance could be charged while at or near maximum burial depth; however, upon uplift, without the benefit of top and lateral seals, the gas would gradually bleed off.

The Wagon Wheel model predicts early expulsion and subsequent migration beginning approximately 54 Ma in the Pinedale area (and possibly earlier to the northeast of the Pinedale anticline axis). At this time, the Lance reservoirs in the Jonah area had only reached a burial depth of approximately 8000 ft (2400 m) (middle Lance) and would not have been fully compacted. Modeled Lance reservoir porosities range from 23% (shallowest) to 16.5% (deepest) in the SHB 13-27 well at 56 Ma. Between 56 and 48 Ma, the middle Lance underwent rapid burial of an additional 5500 ft (1700 m) at the SHB 13-27 well. The differences in burial history between the Pinedale and Jonah areas allowed lithostatic pressure, hydrocarbon generation (concentration), and expulsion rates to be greater in the Pinedale area than in the Jonah area. Expelled hydrocarbons may have migrated updip in the direction of lower pressure and, in the case of the Lance, toward higher reservoir porosity charging Lance reservoirs during burial. As burial continued, Lance porosity was further reduced through compaction, and diagenetic processes, potentially resulting in higher pore pressure accumulation, were preserved. Any additional hydrocarbon charge during the period of burial would have occurred at increasingly higher pressures. Critical to the preservation of a gas accumulation is the quality of the top and lateral seals. Imperfections in the seal, in the form of fractures, or the incorporation of permeable lithologies will diminish the effectiveness of the seal. Downdip of Jonah, where the top-of-flare surface is dropping faster than structural dip, the upper Lance top seal appears to contain less shale. If the paucity of shale in the top seal equates with lower sealing efficiency, then gradual leakage may explain why, in the northeast portion of Jonah, the top of flare is converging on the regional overpressure depth.

Additional models were run to test the generative capacity of the Mowry Shale. Assumptions were that the Mowry was 350 ft (107 m) thick, having 2.5% initial TOC and a mixed kerogen consisting of 30% type II and 70%

type III. Modeling suggests that the onset of oil generation occurred at 80 Ma, with the main pulse of generation and maximum expulsion at 74–61 Ma; by 44 Ma, the majority of oil generation is complete. These models were also normalized to determine generation per unit area and suggest that approximately 615 million bbl (P_{50}) of oil were generated per township. One-dimensional models suggest that at maximum burial (5 Ma), the base of the Mowry reached a temperature of 370°F (188°C), decreasing to about 320°F (160°C) at present, corresponding to a maximum R_o of 2.22. The R_o history suggests that any oil retained in the Mowry or trapped in adjacent reservoirs (Dakota, Muddy, or Frontier) would have been cracked to secondary thermogenic gas by approximately 44 Ma. Biomarker data may help determine whether Mowry-derived gas found its way into the Jonah trap.

MODELING ASSUMPTIONS AND PITFALLS

The shortfall in modeled gas-in-place volumes versus that determined as suggested by drilling-based volumetrics (DuBois et al., 2004) may be caused by the assumptions inherent to the modeling process. The volume of generated hydrocarbons is heavily dependent on the initial TOC of the source rock intervals. Very little TOC data is available in the study area, and existing data are all from mature source rock intervals. This makes extrapolation of TOC to initial conditions difficult. Based on the probable depositional environments, the values used in our models were considered realistic to optimistic.

In these models, a range of saturation thresholds from 0.05 to 0.20 were tested. Very little data exists to rigorously quantify this aspect of modeling in highly heterogeneous source rocks; however, saturation threshold must be greater than zero, or source rocks would be devoid of hydrocarbons in pore space, and hydrocarbon shows would be sparse. The assumptions made concerning saturation threshold significantly affect the volumes of hydrocarbons available to migrate (Tables 11, 12).

The models in the Jonah and Pinedale areas are complex models with multiple source rocks and nondiscrete carrier beds. Given the model's stratigraphic range, it is difficult to assign a single migration efficiency value; however, the current modeling algorithms only allow for a single migration efficiency distribution per model. Instances where the source rocks are proximal to the reservoir(s) may necessitate the use of higher migration efficiencies, whereas migrating moving gas from deeper source rocks through undersaturated reservoirs may be far less efficient. The Risk© models used a 0.5–0.7 (0.6 mode) distribution as a compromise solution.

Local Geologic Conditions

Generation, expulsion, and vertical migration of gas from the Hilliard and Baxter shales may have contributed a significant amount of gas to the trap. If these shales can effectively expel gas, a much larger volume of gas could be available for vertical migration, eliminating the need for lateral migration. Biomarker data may help to determine if the Hilliard and Baxter are contributing to the Jonah charge volume.

The presence of lower Fort Union coals in the southern part of the field area may affect the local geothermal gradient. Continuous temperature log data shows a gradient dogleg adjacent to the coaly section with a temperature increase of 9 to 10°F (−12.8 to −12.2°C) below the coals. This effect was also observed by Nuccio and Johnson (1989) at the multiwell experiment (MWX) site, northwest Colorado, where the Cameo-Fruitland coals caused a more or less 10°F (−12.2°C) dogleg shift in the temperature profile. The low thermal conductivity of the coals produces a thermal blanket that causes a local, positive warping of the isotherms. Source rocks that might otherwise be marginally mature may begin to generate gas because of this thermal insulation effect, and the local charge may be enhanced.

OBSERVATIONS AND CONCLUSIONS

Geologic evidence and the burial-history modeling suggest the following for the Jonah accumulation:

(1) The low thermal maturity and low TOC in the Lance suggests that it is not self-sourced and contributes only a minor amount of gas to the overall hydrocarbon charge.
(2) Mud-log data document the presence of coals and carbonaceous shales in the upper and lower Mesaverde that could be source rocks for gas. Vitrinite reflectance data and modeled maturity indicate that these coals and shales are sufficiently mature to have generated and expelled gas.
(3) The estimated gas in place exceeds the charge capacity of local Upper Cretaceous source rocks. This indicates that a fetch area larger than the current field boundary was available, and/or multiple deeper sources (Hilliard and/or Mowry) combined to charge the trap.
(4) Modeling indicates that for Upper Cretaceous source rocks, maximum hydrocarbon generation occurred earlier in the Pinedale area (47 Ma) than in the Jonah area (39 Ma), assuming a saturation

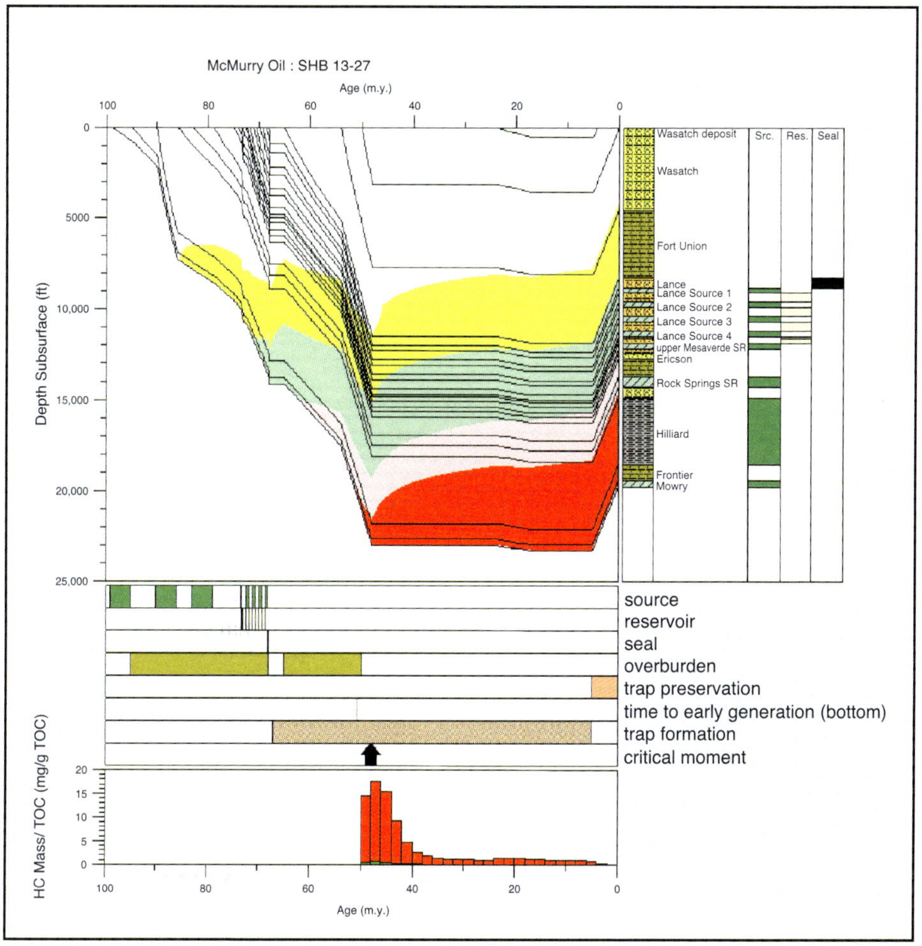

Figure 16. Petroleum systems diagram, SHB 13-27 (SW SW Sec. 27, T29N, R108W), illustrating the timing of events in the Jonah field area. Motion on the bounding faults may have affected the sealing capacity of these faults controlling the effectiveness of the trap through time; hence, there is a high degree of uncertainty on the timing of trap formation. Critical moment is that time when maximum generation or expulsion and trap are coincident. Modeled expulsion (lower panel on diagram) is for the Rock Springs source rock (lower Mesaverde).

threshold of 0.20. Expulsion of hydrocarbons from the lower Mesaverde source in the Pinedale area occurred prior to and during maximum burial, prior to reservoir degradation of the Lance in the Jonah area. Charging of the Lance with gas occurred during a period of rapid burial and increased pressure.

(5) Regional pressure and structural gradients at the time of expulsion were favorable for direct gas migration toward the Jonah area.

(6) Deposition, preservation, and effectiveness of the upper Lance top seal (shaly zone) appears to be essential to the accumulation but only when coupled with the timing, presence, and sealing efficiency of the lateral bounding faults.

(7) The Ericson Sandstone member of the Mesaverde may have acted as a carrier bed in the migration process.

(8) The tectonics involved in the formation of the Pinedale anticline (Late Cretaceous to Eocene; Law and Johnson, 1989) may have generated a stress field, creating and/or reactivating the Jonah bounding fault systems, enhancing migration routes enabling hydrocarbon recharge.

(9) Pliocene uplift placed the gas-charged Lance reservoir in a position of disequilibrium relative to regional pressure gradients. The effectiveness of the top and lateral seals preserved an accumulation that would have otherwise bled off or possibly migrated into another trap during late-stage uplift and erosion.

(10) In the Jonah area, the upper and lower Mesaverde source units are actively generating gas, and oil trapped in deeper source rocks or traps may be cracking to gas supplying a continuous charge to the overlying Jonah trap.

The petroleum systems diagram (Figure 16) illustrates the timing of events critical to the Jonah accumulation. Of all the components, trap formation and preservation have the highest degree of temporal uncertainty. The timing of the stratigraphic components of the trap is known within reasonable limits; however, the timing of other trap components, such as diagenetic porosity reduction or enhancement and the development of effective lateral seals, is much more uncertain. Modeling shows that the bulk of hydrocarbon generation and expulsion from the lower Mesaverde occurred between 50 and 40 Ma. For the Jonah accumulation to form, the receiving trap needed to be effective during and after this time, or alternatively, the expelled gas was somehow sequestered, only to be released at a later time to spill into the Jonah trap. To date, the Jonah accumulation is anomalous within the Rockies. However, the individual petroleum system components, i.e., presence of source(s), hydrocarbon generation and expulsion, presence and position of seals, migration routes, structural trapping elements, etc., are all present to one degree or another in the Greater Green River basin and other Rocky Mountain basins. A thorough understanding of the critical components and how those components combine in both space and time will help in future exploration efforts.

ACKNOWLEDGMENTS

I thank Douglas Battin, Barry McBride, Fred Meissner, Steve Kneller, John Robinson, Keith Shanley, and Gary Stewart for their time, geological insight, and frequent input throughout the course of writing this chapter. Their local knowledge was instrumental in defining the geologic constraints that were applied to the burial-history models. Platte River Associates, Inc., provided the necessary time, critical review, modeling software, and technical support for model development and refinement. EnCana, U.S.A. provided additional pyrolysis data that aided in model calibration.

REFERENCES CITED

Bowker, K. A., and J. W. Robinson, 1996, Jonah field: A basin-centered sweetspot in the northern Green River basin, Wyoming, *in* S. Longacre, B. Katz, R. Slatt, and M. Bowman, eds., Compartmentalized reservoirs: Their detection, characterization and management: 1997 AAPG Rocky Mountain Section Meeting Official Program Book and Extended Abstracts Volume, p.25–30.

Burtner, R. L., and M. A. Warner, 1984, Hydrocarbon generation in Lower Cretaceous Mowry and Skull Creek shales of the northern Rocky Mountain area, *in* J. Woodward, F. F. Meissner, and J. L. Clayton, eds., Hydrocarbon source rocks of the greater Rocky Mountain region: Rocky Mountain Association of Geologists Symposium, p. 449–467.

Cluff, R. M., 2002, The origin of Jonah field: Evidence for pressure leakage from sonic and resistivity log anomalies in the northern Green River basin, Wyoming (abs.): Rocky Mountain Association of Geologists Luncheon Presentation, http://www.rmag.org (accessed June 1, 2002).

Curry, W. H. III, 1973, Late Cretaceous and early Tertiary rocks southwestern Wyoming: Wyoming Geological Association Guidebook, p. 79–86.

Daly, A. R., and J. D. Edman, 1987, Loss of total organic carbon from source socks during thermal maturation (abs.): AAPG Bulletin, v. 71, no. 5, p. 546.

Dickinson, W. W., 1989, Analysis of vitrinite maturation and Tertiary burial history, northern Green River basin, Wyoming, *in* Geology of tight gas reservoirs in the Pinedale anticline area, Wyoming, and at the multiwell experiment site, Colorado: U.S. Geological Survey Bulletin, v. 1886, p. F1–F13.

Dow, W. G., 1977, Kerogen studies and geological interpretations: Journal of Geochemical Exploration, v. 7, p. 79–99.

DuBois, D. P., P. J. Wynne, T. M. Smagala, J. L. Johnson, K. D. Engler, and B. C. McBride, 2004, Geology of Jonah field, Sublette County, Wyoming, *in* J. W. Robinson and K. W. Shanley, eds., Jonah field: Case study of a tight-gas fluvial reservoir: AAPG Studies in Geology 52, p. 37–60.

Hanson, W. B., V. Vega, and D. Cox, 2004, Structural geology, seismic imaging, and genesis of the giant Jonah gas field, Wyoming, U.S.A., *in* J. W. Robinson and K. W. Shanley, eds., Jonah field: Case study of a tight-gas fluvial reservoir: AAPG Studies in Geology 52, p. 61–92.

Law, B. E., 1984, Relationships of source-rock, thermal maturity, and overpressuring to gas generation in low-permeability Upper Cretaceous and lower Tertiary rocks, Greater Green River basin, Wyoming, Colorado and Utah, *in* J. Woodward, F. F. Meissner, and J. L. Clayton, eds., Hydrocarbon source rocks of the greater Rocky Mountain region: Rocky Mountain Association of Geologists Symposium, p. 469–490.

Law, B. E., and R. C. Johnson, 1989, Structural and stratigraphic framework of the Pinedale anticline, Wyoming, and the multiwell experiment site, Colorado, *in* Geology of tight gas reservoirs in the Pinedale anticline area, Wyoming, and at the multiwell experiment site, Colorado: U.S. Geological Survey Bulletin, v. 1886, p. B1–B11.

Lickus, M. R., M. J. Pawlewicz, B. E. Law, and W. W. Dickinson, 1984, Thermal maturity map, northern Green River basin, Wyoming, *in* B. E. Law, ed., Geological characteristics of low-permeability Upper Cretaceous and lower Tertiary rocks in the Pinedale anticline area, Sublette County, Wyoming: U.S. Geological Survey Open-file Report 84-753, p. 60–65.

Montgomery, S. L., and J. W. Robinson, 1997, Jonah field, Sublette County, Wyoming: Gas production from overpressured Upper Cretaceous Lance sandstones of the Green River basin: AAPG Bulletin, v. 81, p. 1049–1062.

Naeser, N. D., 1984, Fission-track ages from the Wagon Wheel No. 1 well, northern Green River basin, Wyoming: Evidence for recent cooling, *in* B. E. Law, ed., Geological characteristics of low-permeability Upper Cretaceous and lower Tertiary rocks in the Pinedale anticline area, Sublette County, Wyoming: U.S. Geological Survey Open-file Report 84-753, p. 66–77.

Nuccio, V. F., and R. C. Johnson, 1989, Thermal history of selected coal beds in the Upper Cretaceous Mesaverde Group and Tertiary Wasatch Formation, multiwell experiment site, Colorado, in relation to hydrocarbon generation, *in* Geology of tight gas reservoirs in the Pinedale anticline area, Wyoming, and at the multiwell experiment site, Colorado: U.S. Geological Survey Bulletin, v. 1886, p. L1–L8.

Pollack, H. N., S. J., Hurter, and J. R. Johnson, 1991, A new global heat flow compilation: Department of Geological Sciences, The University of Michigan Ann Arbor, Michigan, ftp://ftp.ngdc.noaa.gov/Solid_Earth/Global_Heatflow (accessed June 22, 2004).

Pollastro, R. M., and C. E. Barker, 1984, Geothermometry from clay minerals, vitrinite reflectance, and fluid inclusion— Applications to the thermal and burial history of rocks cored from the Wagon Wheel No. 1 Well, Green River basin, Wyoming, *in* B. E. Law, ed., Geological characteristics of low-permeability Upper Cretaceous and lower Tertiary rocks in the Pinedale anticline area, Sublette County, Wyoming: U.S. Geological Survey Open-file Report 84-753, p. 78–94.

Prensky, S. E., 1989, Gamma-ray well-log anomaly in the northern Green River basin of Wyoming, *in* Geology of tight gas reservoirs in the Pinedale anticline area, Wyoming, and at the multiwell experiment site, Colorado: U.S. Geological Survey Bulletin, v. 1886, p. H1–H21.

Spencer, C. W., 1987, Hydrocarbon generation as a mechanism for overpressuring in the Rocky Mountain region: AAPG Bulletin, v. 71, p. 368–388.

Warner, E. M., 1998, Structural geology and pressure compartmentalization of Jonah field, Sublette County, Wyoming, *in* R. M. Slatt, ed., Compartmentalized reservoirs in Rocky Mountain basins: Rocky Mountain Association of Geologists, p. 29–46.

Wyoming Oil and Gas Conservation Commission (WOGCC), 2003, Jonah field production query, http://wogcc.state.wy.us/ (accessed February 21, 2003).

Yin, P., S. M. Stookey, R. C. Surdam, and T. Dunn, 2001, Tight gas sands of the Lance Formation in Jonah field— A core from the Jonah 2-5, *in* J. W. Robinson and K. W. Shanley, eds., Tight gas fluvial reservoirs: A case history from the Lance Formation, Green River basin, Wyoming, AAPG Short Course Notes.

CHAPTER 8

The Origin of Jonah Field, Northern Green River Basin, Wyoming

ROBERT M. CLUFF
The Discovery Group Inc., Denver, Colorado, U.S.A.

SUZANNE G. CLUFF
The Discovery Group Inc., Denver, Colorado, U.S.A.

ABSTRACT

Prolific production at Jonah field and many other fields in the Green River basin is dependent on the presence of overpressure. In the Gulf Coast and other areas, plots of shale resistivity and shale sonic transit time versus depth have been used to identify overpressured zones. The same technique has been proposed to map overpressure compartments and their boundaries in the Rocky Mountain region using well logs or, alternatively, interval velocities determined from seismic data. At Jonah field, the top of the overpressure (determined by continuous gas flaring during drilling) correlates within a few hundred feet to a drop in shale resistivity and increase in shale transit time. However, studies of nearby wildcat wells and detailed cross sections through both overpressured and normally pressured wells show that the log anomalies extend significantly beyond the overpressured area. Velocity and resistivity changes in the area around Jonah tend to follow a stratigraphic boundary near the base of the Tertiary Fort Union Formation instead of tracking the top of the overpressured volume.

Early studies of Jonah field considered the hydrocarbons in the field to be derived from vertical migration of gas from regional overpressure conditions 2000–3000 ft (610–915 m) deeper. The upward migration was presumed to be controlled by the presence of extensive microfractures that form a leakage chimney between large sealing faults. This study suggests that the log anomaly both within and surrounding Jonah provides an alternative interpretation. Until the middle Tertiary, overpressure conditions extended up to the base of the Fort Union and resulted in undercompaction of Cretaceous shales, as reflected by resistivity and velocity anomalies. Late Tertiary relative uplift initiated slow leakage of the overpressure conditions wherever the system was not tightly sealed. As a result, the top of the overpressure has been dropping with time over most of the northern Green River basin. The sonic and resistivity anomalies are irreversible, and the logs reveal the signature of the former overpressured and undercompacted conditions.

Based on this new model, Jonah field represents a high remnant of the former regional top of overpressure instead of a leakage chimney from a deeper overpressured generation cell. If this model is correct, exploration methods should focus on the seal conditions that prevent leakage instead of fracture models that promote leakage.

INTRODUCTION

Basin-centered Gas Accumulations and the Sweet-spot Concept

Basin-centered gas accumulations have been a major focus of exploration efforts in the Rocky Mountain region since the mid-1980s. The concept of a basin-centered accumulation dates back to the work in the Elmworth area of Alberta, where Masters (1979, 1984) described what was termed a "deep basin gas field" in the western Canada sedimentary basin. Subsequent work in other Rocky Mountain basins by the U.S. Geological Survey and

Jonah Field: Case Study of a Giant Tight-Gas Fluvial Reservoir, J. W. Robinson and K. W. Shanley, eds.: AAPG Studies in Geology 52 and Rocky Mountain Association of Geologists 2004 Guidebook

others (McPeek, 1981; Spencer, 1983; Law, 1984; Law and Dickinson, 1985) described continuous gas accumulations in the central parts of most basins, several of which are not very deeply buried today. Thus, the concept was extended to include "basin-centered gas" as opposed to just deep basin gas. Basin-centered gas accumulations are considered by many to be a worldwide phenomenon common to sedimentary basins throughout the Phanerozic (Law and Spencer, 1998).

Basin-centered gas has never been tightly defined in the literature. Most papers cite the following common characteristics for a basin-centered gas accumulation (e.g., Law, 2002): (1) a large, regionally pervasive gas-saturated section; (2) the entire interval is abnormally pressured, either overpressured or underpressured; (3) fields commonly lack downdip gas-water contacts; (4) gas is produced with minimal associated water; (5) low permeability of both the enclosing strata and the reservoir rocks; and (6) no obvious updip seals or trapping conditions define the gas accumulation. Because of the lack of obvious traps or downdip free water that are characteristic of conventional structural and stratigraphic hydrocarbon accumulations, basin-centered gas has been grouped with other unconventional or continuous hydrocarbon accumulations, including shale gas, coalbed methane, and gas hydrates for resource-assessment purposes. Despite their seemingly continuous nature over large volumes of rock, commercial production is, with few exceptions, restricted to a small fraction of the overall accumulation.

The observation that economic gas production tends to be highly localized in basin-centered accumulations led to the development of the "sweet-spot" concept for low-permeability reservoirs. A sweet spot is an area of enhanced porosity and permeability that is thought to provide higher deliverability and, therefore, better gas recovery than the surrounding areas (Surdam, 1997; Law, 2002). The origin of the term is obscure. The first published use we know of in this context was by Sauvageau (1984) and subsequently in the geological literature by Billingsley and Reinert (1994) for the area of prolific production from the Almond Formation (Cretaceous, Mesaverde Group) near Wamsutter in the eastern Green River basin.

Sweet spots are thought to be controlled by (1) depositional trends of enhanced porosity and permeability; (2) enhanced fracturing and, consequently, higher bulk permeability; or (3) overpressuring and improved reservoir properties in a pressure compartment (Surdam, 1997). Examples of the first two types of sweet spot have been extensively discussed in the literature on basin-centered gas. The third type, an anomaly in an overpressured compartment, is poorly understood. Because several of the larger basin-centered gas fields in the Rocky Mountain region are significantly overpressured, it has been proposed that overpressuring is somehow causally related to the production of sweet spots, and they might be concentrated near the upper boundary of the overpressured intervals (Surdam et al., 1994, 1997; Jiao and Surdam, 1997; Surdam, 1997; Law, 2002).

The idea that relief on the top of the overpressured surface may indicate the location of sweet spots has led to several proposed techniques to identify and map this surface. These include log-based techniques to identify the top of the overpressure from sonic-log, resistivity-log, or density-log signatures (Jiao and Surdam, 1997; Surdam et al., 1997) and seismic techniques based on inversion of two-dimensional (2-D) or three-dimensional (3-D) velocity data to find reversals in the normal trend of increasing velocity with depth (Surdam et al., 2001). This study was part of a regional project to map the top of the overpressure surface across the northern Green River basin using sonic and resistivity logs. The goal of our work was to determine if the overpressure surface had a significant velocity contrast that might be detectable with seismic data and to investigate if known log anomalies that closely coincide with the known top of overpressure in the Jonah–Pinedale area might show some appreciable relief outside the established fields.

JONAH FIELD: AN OVERPRESSURED SWEET SPOT IN THE GREEN RIVER BASIN?

The discovery and early development history of Jonah field was summarized by Robinson (2000). Jonah field was discovered in 1986–1987 with the drilling of two wells by Home Petroleum Corporation, the Jonah Federal 1-4 and 32-34, both of which encountered continuous gas shows and flared gas from about 8500 ft (2600 m) to total depth. The initial completion results of these wells were disappointing, and the field was considered noncommercial. McMurry Oil Company "rediscovered" Jonah field with the completion of the Jonah Federal 1-5 in 1993, which encountered continuous gas shows from 8429 ft (2570 m) to total depth of the well. Subsequent development drilling in 1994–1995 outlined a large, poorly defined area of gas-saturated Lance Formation (Upper Cretaceous), where the top of continuous gas shows began at or a short distance below the top of the Lance. Sandstones in the overlying Fort Union Formation (Paleocene) had low resistivity, appeared wet on logs, yielded only minor or no gas shows while drilling, and were drilled with normal mud weights. An intervening section of shale and shaly sandstones, tentatively correlated with the "Unnamed unit" of Tertiary age on the Pinedale anticline (Law and Johnson, 1989), yielded erratic shows and appeared to be either normally pressured or perhaps transitional into the overpressured section below (Figure 1). The initial interpretation of Jonah field, based on the degree of overpressuring in these three wells, continuous gas shows over thousands of feet, lack of significant water production, and inability to detect any obvious structural or stratigraphic trapping components, was that Jonah must be a basin-centered gas accumulation (Battin and Clark, 1994).

The best evidence for the top of the overpressured section was the onset of continuous mud-log hydrocarbon shows in both sandstones and shales, which generated a constant gas flare through a gas buster and the need to increase mud weight to control the well. McMurry, Snyder Oil, and other operators in the area called this point the "top of flare," and it was widely assumed to be coincident with the top of significant overpressure. Drilling mud weights above the top of flare vary between operators but are generally less than 9.5 lb/gal (1140 kg/m^3), which is equivalent to a pore-pressure gradient of 0.494 psi/ft (11.2 MPa/km). Below the top of flare, the operators commonly mud up to 9.7–10 lb/gal (1160–1200 kg/m^3) and, in many wells, reach weights greater than 11 lb/gal (1320 kg/m^3) at total depth, equivalent to pore-pressure gradients of 0.572 psi/ft (12.9 MPa/km) (Figure 1). Within the field area, there are relatively few publicly available pressure buildup tests. Warner (1998) cited formation pressures in several tests of the early wells equivalent to pore-pressure gradients of 0.54–0.58 psi/ft (12.2–13.1 MPa/km), whereas the shut-in drillstem test pressures for several dry holes in the area surrounding Jonah indicate that the Lance is normally pressured outside the field (Table 1). Montgomery and Robinson (1997) reported that pore-pressure gradients in the field could reach as high as 0.70 psi/ft (15.8 MPa/km) based on the prefracture-stimulation breakdown tests.

In 1994, the McMurry Jonah Federal 4-8 and the Snyder Oil Jonah Federal 3-15 were drilled immediately south of the Home Petroleum wells and the Jonah Federal 1-5 well (Figure 2). The Jonah Federal 4-8 did not have any significant shows down to its total depth of 11,000 ft (3350 m) and had significantly lower porosity and resistivity than the nearby productive wells. The Jonah Federal 3-15 also did not encounter strong gas shows until 11,240 ft (3430 m), more than 2700 ft (820 m) deeper than the top of gas shows in the wells just 1.5 mi (2.4 km) north.

Subsequent reprocessing of a regional northeast–southeast 2-D seismic line revealed a subtle fault separating the productive wells from the two dry holes to the south. This fault zone became known within Snyder Oil as the "Line of Death" and is now known to form the south boundary of the field (Robinson, 2000). In 1996, Amoco acquired a 51-mi^2 (13,200-ha) 3-D seismic survey over Jonah field. Although many interesting details of the structural configuration of the field emerged from these data (Warner, 1998; Vega and Hanson, 2001), the most significant features imaged were the south boundary fault and a previously unrecognized western bounding fault (Figure 2). Amoco was drilling a well, the Corona 7-24, just west of this fault when the preliminary processing results were delivered. The Corona 7-24 proved to be a dry hole with no significant gas shows down to a total depth of 10,214 ft (3113 m), a mirror image of the conditions on the south side of the southern boundary fault. A test in the Corona 7-24 recorded a downhole pressure gradient of only 0.447 psi/ft (10.0 MPa/km) (Table 1).

Since the recognition of these two major bounding faults, all development drilling operations have been confined to the areas north and east of the faults or, more properly, fault systems, where the top of the pressure surface is approximately 2500 ft (760 m) higher than the surrounding region (Bowker and Robinson, 1997; Montgomery and Robinson, 1997; Warner, 1997a, b, 1998). The operators at Jonah quickly developed a simple model for the distribution of overpressure shown in Figure 3, with a more or less flat-topped overpressure surface near the base of the Fort Union Formation. Warner (1998) concluded that the top of the pressure surface is structurally controlled and rises abruptly between the bounding fault blocks (Figure 3). Our detailed mapping of the top of continuous mud-log shows confirms his interpretation (Figure 4).

To date, attempts to extend production south or west of the intersection of these major fault systems have been unsuccessful. The wildcat tests near Jonah field have not encountered overpressure conditions until they penetrated the basal Lance Formation or upper Mesaverde at depths of 11,000 ft (3350 m) or greater (Table 1). Development drilling has proven that Jonah is a fault-bounded pressure compartment, and it is probably the best example in the Rocky Mountain region of a field where overpressure conditions pop up above what has been interpreted as a regional basin-centered gas accumulation. This observation led to the revised interpretation that Jonah may not be a basin-centered gas field but is instead a gas plume or leakage chimney rooted in a deeper, regional basin-centered gas accumulation. Warner (1997a, p. 1236), for example, stated that "minor high-angle faults may be responsible for upward migration of overpressured gas into the thermally

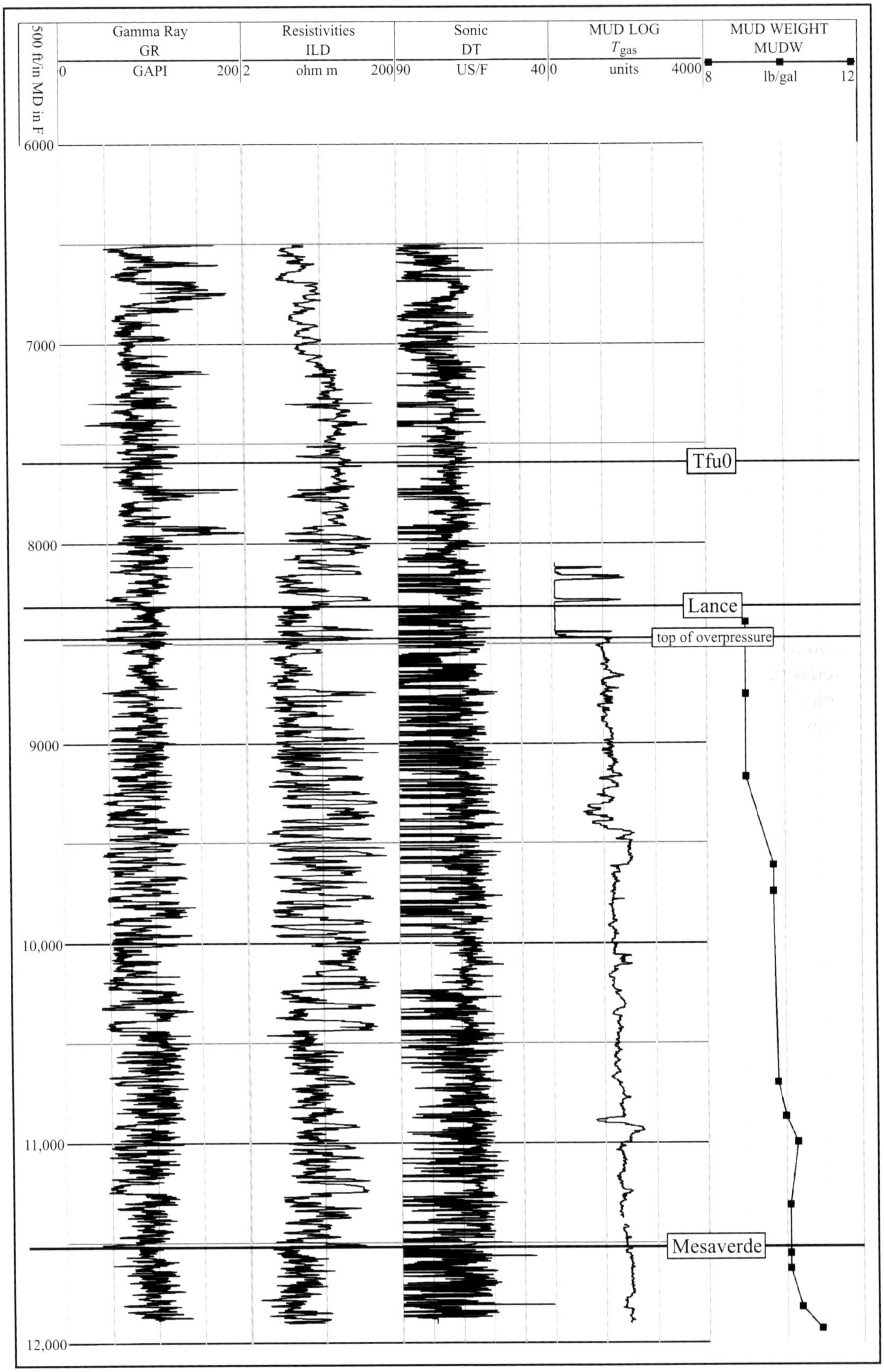
500 ft/in MD in F
Gamma Ray
GR
0 GAPI 200
Resistivities
ILD
2 ohm m 200
Sonic
DT
90 US/F 40
MUD LOG
Tgas
0 units 4000
MUD WEIGHT
MUDW
8 lb/gal 12
6000
7000
8000
9000
10,000
11,000
12,000
Tfu0
Lance
top of overpressure
Mesaverde

immature upper Lance section where it is trapped beneath a shale top seal." Law (2002) also concluded that Jonah is a gas chimney rooted in the regionally pervasive but deeper gas accumulation previously described by Law (1984).

OVERPRESSURE DETECTION FROM SONIC AND RESISTIVITY LOGS

Wire-line logs have been used to detect overpressure, particularly in the Gulf Coast and other Tertiary basins, since at least the mid-1960s (Hottman and Johnson, 1965; MacGregor, 1965; Foster and Whalen, 1966; Ham, 1966). These techniques are based on the response of resistivity, sonic, density, and neutron logs to the porosity of the rocks and their ability to quantify changes in porosity with increasing burial depth and compaction. The loss of porosity in a sedimentary section with increasing burial depth is commonly approximated by an exponential function of the form proposed by Athy (1930):

$$\phi = \phi_0 e^{-bZ} \quad (1)$$

where ϕ is the porosity at depth Z, ϕ_0 is the initial porosity of unconsolidated sediment at the time of deposition (the "mud-line porosity"), and b is a compaction coefficient related to lithology. By taking the logarithm of each side of equation 1,

$$\ln \phi = \ln \phi_0 - bZ \quad (2)$$

a semilog plot of porosity versus depth (depth is linear) is expected to yield an approximately straight line. In the simple case of a continuously subsiding section with no erosional removal of sediments, the slope of the trend is the compaction coefficient, b, and the intercept at depth = 0 is the log of the mud-line porosity (Figure 5A, B). If there has been uplift and erosion, the porosity at the present-day surface will be some value less than the porosity of unconsolidated sediment (Figure 5C). The thickness between the present surface and an extrapolated point where the trend reaches the mud-line porosity, ϕ_0, provides an estimate of the amount of removed section (Magara, 1976).

Several studies have demonstrated a change in the porosity-versus-depth relationship between the first several hundred meters of burial, where mechanical compaction of the sediments is the dominant process, and the deep-burial regime, where chemical compaction is dominant (Burrus, 1998). Although an exponential function can approximate the overall porosity loss with depth, no single set of coefficients fits the entire regime from surface to deep burial. Consequently, in some basins, it may be necessary to fit one equation to the section above 3000 ft (910 m) with a large value for b and a different equation with a low value for b in the deeper rocks. Hunt et al. (1998), on the basis of shale-cuttings data, even suggested that the loss of porosity with depth might be modeled best with a set of piecewise linear instead of exponential equations. Regardless of the subtleties of fitting one or more equations to the overall porosity-versus-depth trend, in the case of the northern Green River basin, several thousand feet of shallow overburden have been erosionally removed by late Tertiary exhumation, and any deviations in the shallow regime have been truncated. This study determined that a single exponential equation can be fit to the entire lower Tertiary section from the base of the surface casing to the top of the Lance Formation. Any error in the model equation at shallow-burial depths will result in increased error in the estimate of removed overburden, but this estimate was not a primary goal of the study.

Using equation 2 to predict porosity is only valid for a single and constant lithology. Different lithologies compact at different rates and have different characteristic values for the constant b. Because of this, basin compaction studies commonly focus on fine-grained rocks and assume all shales in the section compact at similar rates. If there are interbedded shales of distinctly different lithologies, for example, organic-rich shales or hard siliceous shales, they may deviate from the "normal" shale compaction trend. Shale points of similar lithology can be selected using various criteria for shaliness and lithology, such as a gamma-ray (GR) log, spontaneous-potential (SP) log, and neutron-density porosity separation. Several types of geophysical logs are linearly related to porosity and can be substituted into equation 2 as proxies for porosity, including sonic transit time, bulk density, and neutron-porosity logs. Resistivity (or its inverse, conductivity) also varies as a function of porosity, although the relationship is nonlinear, and other effects, such as surface conductivity of the clays, salinity changes of the interstitial waters, and hydrocarbon saturation, all contribute to shale

◀

Figure 1. Type log from the Home Petroleum Jonah Federal 1-4, the first well drilled at Jonah field, showing the major stratigraphic units, key markers, mud-log gas shows (T_{gas}), and mud weight while drilling (MUDW). Tfu0 = base of Tertiary Fort Union Marker; GR = gamma-ray log; ILD = deep induction log; DT = sonic transit time.

Table 1

Formation Pressure Data in the Jonah Field Area.

Well Name	Location (Township, Range, Section)	Test Interval (ft)	Formation	Bottomhole Pressure (psi)	Pressure Gradient (psi/ft)	Estimated Permeability (μd)	Source
Wells inside Jonah							
Jonah Federal 1-4	T28N, R108W, 4	10,204–10,434	Lance	5600	0.543	13	Warner, 1998
Jonah Federal 1-5	T28N, R108W, 5	9695–9924	Lance	5700	0.567	14	Warner, 1998
Jonah Federal 1-6	T28N, R108W, 6	6507–9731	Lance	5450	0.581	7.4	Warner, 1998
Jonah Federal 1-7	T28N, R108W, 7	9556–9580	Lance	5450	0.570	29	Warner, 1998
Jonah Federal 2-5	T28N, R108W, 5	core 9581–9591	Lance			8.2	Warner, 1998
Stud Horse Butte 11-27	T29N, R108W, 27	core 10,411–10,429	Lance			14.3	Warner, 1998
Tests of wells outside Jonah							
Corona Unit 7-24	T29N, 109W, 24	core 9560–9675	Lance			14	Warner, 1998
Corona Unit 7-24	T29N, 109W, 24	9560–9675	Lance	4300	0.447	13	Warner, 1998
Sublette Flat 1	T27N, R107W, 7	6925–7112	Fort Union	2207	0.319		IHS Energy
Sublette Flat 1	T27N, R107W, 7	10,236–10,267	Cretaceous	1985	0.194		IHS Energy
Sublette Flat 1	T27N, R107W, 7	10,369–10,414	Rock Springs	2606	0.251		IHS Energy
Old Road Unit 1	T27N, R108W, 27	8344–8406	Mesaverde	2890	0.346		IHS Energy
Lombard Canyon Unit A-1	T27N, R108W, 33	7944–7981	Ericson	1783	0.224		IHS Energy
Prairie Hen Unit 1-A	T27N, R110W, 30	6202–6264	Mesaverde	2339	0.377		IHS Energy
Wardell Federal 1	T28N, R108W, 9	11,011–11,100	?	5642	0.512		IHS Energy
Golden Gate Unit 1	T28N, R109W, 27	11,598–12,104	Blair	5465	0.471		IHS Energy
Ferry Island Unit 1	T28N, R109W, 29	8581–8614	Lance	3384	0.394		IHS Energy
Pinedale Unit 7	T30N, R108W, 15	8040–8062	Fort Union	3955	0.492		IHS Energy
New Fork Unit 2	T30N, R108W, 2	9463–9516	Fort Union	3325	0.351		IHS Energy
Wagon Wheel 1	T30N, R108W, 5	7340–7400	Fort Union	2297	0.313		IHS Energy
Wagon Wheel 1	T30N, R108W, 5	10,978–11,070	Lance	6748	0.615		IHS Energy

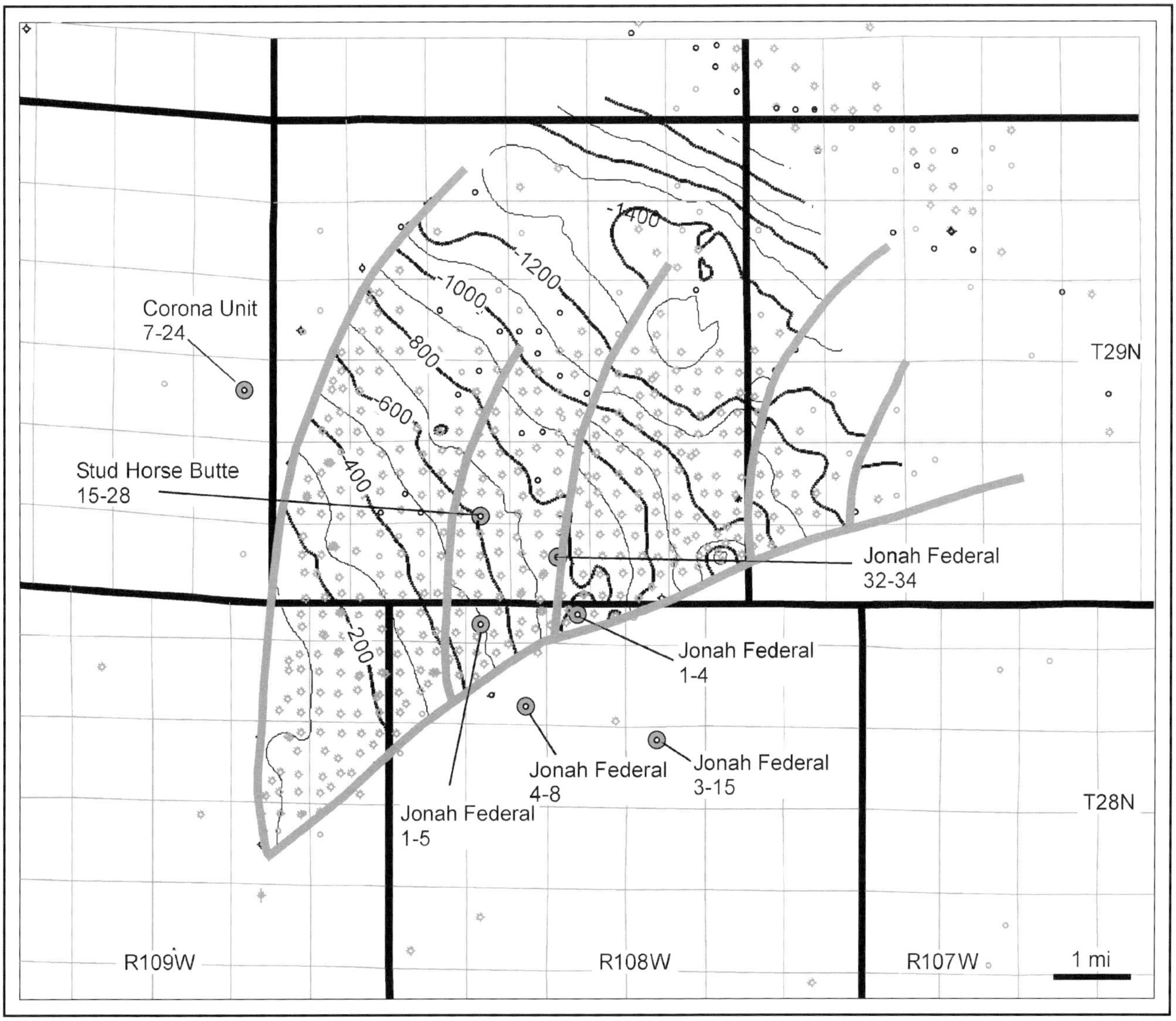

Figure 2. Structure map on the base of the Fort Union Formation (Tfu0 horizon) at Jonah field. Key wells mentioned in the text are highlighted.

conductivity. Both sonic logs and resistivity logs have proven useful for compaction studies and overpressure estimation. Density and neutron logs are considered less reliable because of their sensitivity to borehole conditions and washouts.

For compaction to proceed normally, fluids must be able to freely escape as the section compacts, which implies normal or hydrostatic pore pressures. If a compacting section is unable to continuously expel fluids, usually because of low permeability, rapid deposition, and a lack of continuous carrier beds to conduct the fluids away, part of the overburden will be borne by the interstitial fluids, and overpressure will start to develop. As a result, the rate of compaction is reduced and does not follow an exponential loss of porosity with depth. This deviation from a normal compaction trend caused by vertical loading is known as "disequilibrium compaction." Both qualitative overpressure detection and quantitative estimates of the magnitude of overpressure are based on deviations of the observed compaction-versus-depth trend from an ideal compaction curve fit through the shallow, normally pressured section.

Disequilibrium compaction is thought to be the most common cause of overpressuring in young, rapidly subsiding sedimentary basins, such as the Gulf Coast basin (Osborne and Swarbrick, 1997; Swarbrick and Osborne,

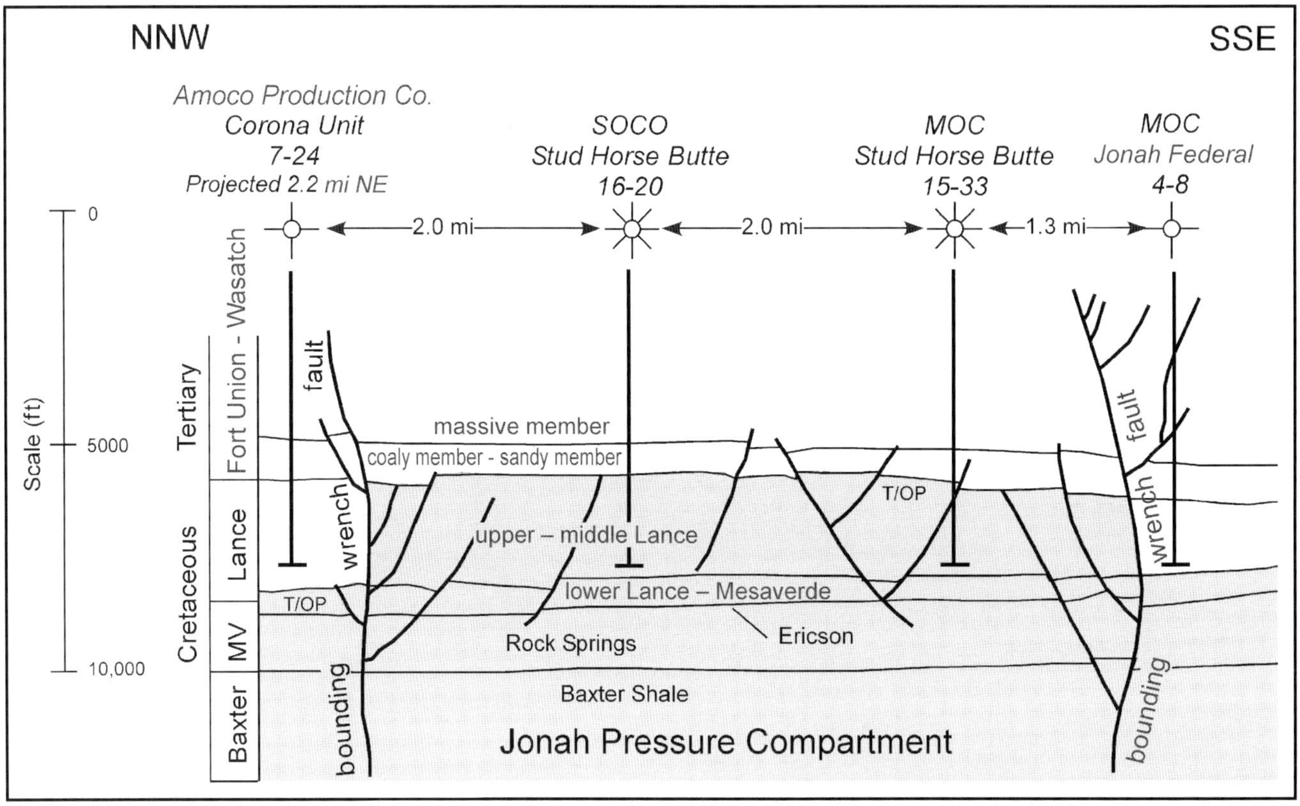

Figure 3. Cross section illustrating the relief on the top of the overpressured section (shaded gray) at Jonah field (modified after Warner, 1998). The top of the overpressure is generally encountered in the shaly Unnamed unit between the massive Fort Union sandstones and the top of the Lance. SOCO = Snyder Oil Company; MOC = McMurry Oil Company; T/OP = top of overpressure.

1998). Other mechanisms proposed as common causes of overpressuring are summarized in Table 2. Numerical models have shown that most of these mechanisms other than disequilibrium compaction are unable to account for the very high overpressures observed in many basins (Swarbrick et al., 2002), including the Greater Green River basin. One notable exception may be fluid expansion resulting from hydrocarbon generation (Spencer, 1987) or thermal cracking of oil to gas in a closed system (Barker, 1990). Gas generation from type I or type II kerogen is associated with a volume expansion of as much as 200% and is capable of generating significant overpressure at least locally within rich source rocks (Burrus, 1998; Meissner, 2000). Law and Dickinson (1985) proposed that basin-centered gas accumulations in the western United States formed by hydrocarbon generation induced overpressure, wherein gas was forced into low-permeability sandstone reservoirs, displacing all movable water and creating regionally extensive overpressure cells. As a note of caution, although the volume expansion associated with kerogen conversion to gas may be large, the gas generated expands into a very much larger volume of reservoir and nonsource rocks and will suffer a substantial pressure drop as it expands.

Sonic-log and Resistivity-log Anomalies in the Jonah Field Area

Figure 6A and B are examples of typical wells in the Jonah area illustrating the sonic-log and resistivity-log anomalies. These logs have been filtered using the GR log to display only points with high shale volume, but no other lithologic criteria were used to discriminate between types of shale. A linear trend of resistivity and sonic transit time is observed from the top of the logged interval to a point near the top of the Lance Formation. At this point, both curves drift off the trend to lower resistivities and higher transit times (lower velocities), and a new, subparallel trend is established within several hundred feet. The top of the overpressure, as determined by continuous mud-log shows and shown by the total gas curve in the far right track, occurs within a few hundred feet

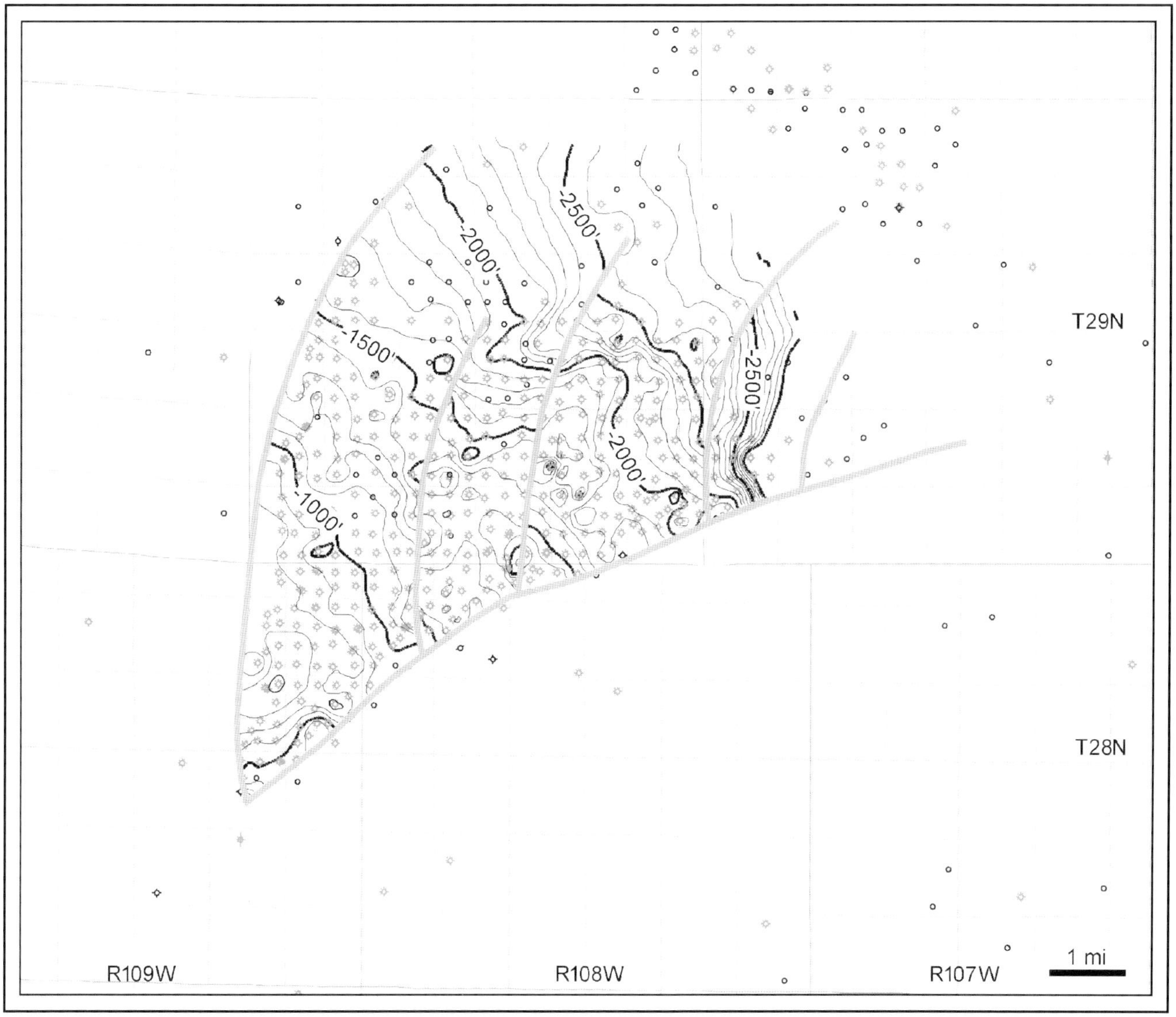

Figure 4. The top of the overpressure determined from continuous mud-log shows at Jonah field.

of the transition to the lower resistivity and velocity trend. Through the transitional zone, the sandstones may yield strong gas shows, but the gas drops back to background levels in the interbedded shales. We interpret the sandstones between the normally pressured Fort Union and the top of the sonic anomaly as slightly overpressured or normally pressured gas sandstones. All of the Jonah field logs we have examined have similar profiles, where the resistivity and the sonic logs break back from the shallow-burial trend at nearly identical depths. This is important because most wells were not logged with a sonic tool, and therefore, resistivity becomes the key measurement for mapping the top of the anomaly.

The sonic trend above the top of the Lance can be projected upward to a theoretical mud-line interval transit time of 189 μs/ft (620 μs/m) (Figure 5C). The difference between the projected sediment thickness to the mud line and the actual thickness to the present-day surface of the Earth provides an estimate of the amount of overburden removed since maximum burial. For the examples shown in Figure 6, more than 5000 ft (1500 m) of section has apparently been removed. One problem with this calculation is the extremely low slope of the trend line. Even small changes in the projected trend to honor values in a particular shale zone can change the removed overburden estimate by hundreds of feet. Furthermore, as discussed previously, any deviation of

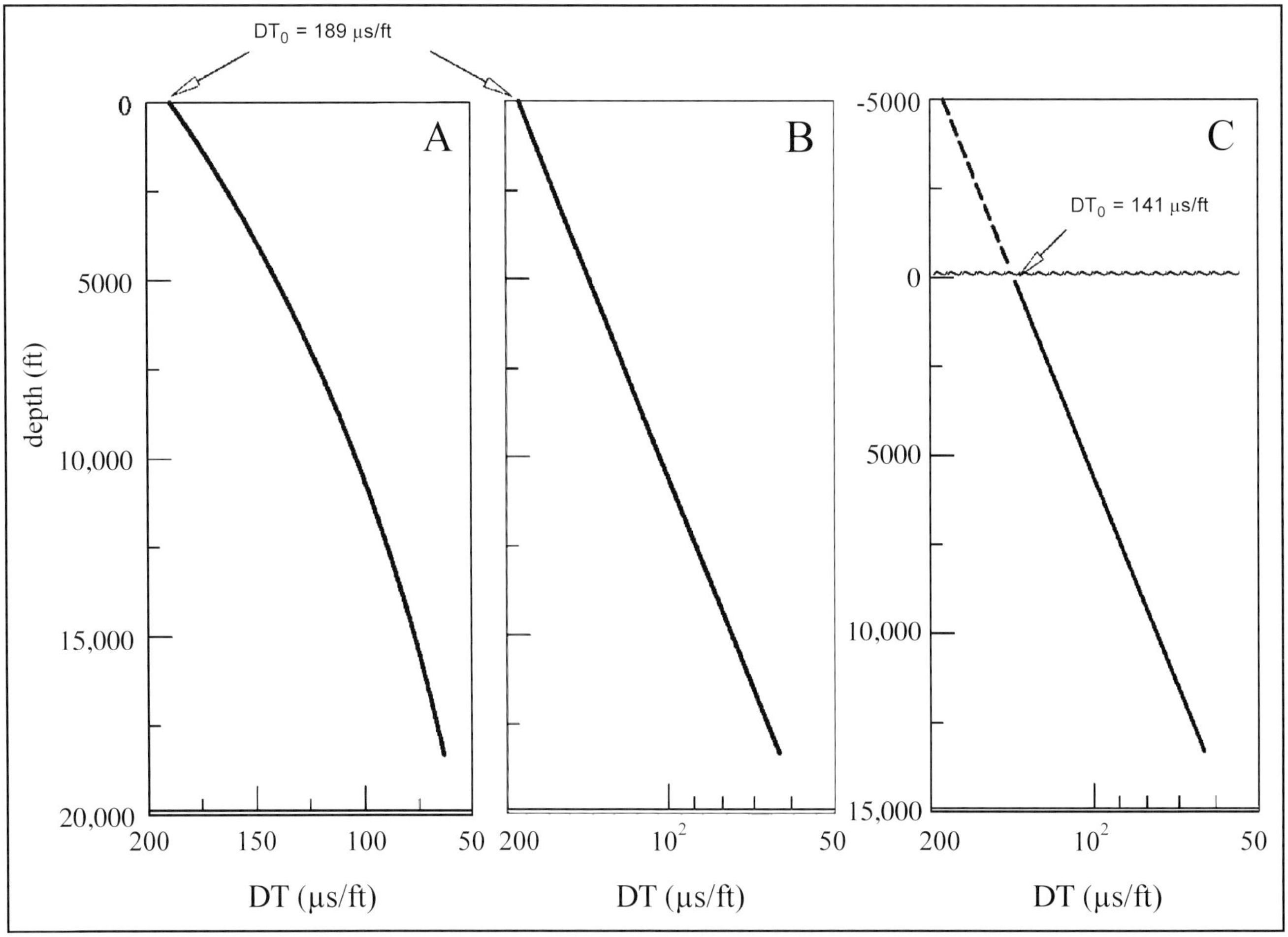

Figure 5. Sonic transit time (DT) vs. depth trends for a compacting section of a uniform shale lithology. (A) DT plotted on a linear scale versus depth displays an exponential curvature; (B) DT on a semilog scale versus depth is a linear trend; (C) trend for the same well if several thousand feet of section was eroded. Projecting the observed DT at the surface (141 μs/ft; 463 μs/m) upward to the original mud-line DT (189 μs/ft; 620 μs/m) yields an estimate of the removed overburden to be 5000 ft (1500 m).

the actual trend from the assumed exponential trend at shallow-burial depths results in additional error in the removed overburden estimates. Dickinson (1989) estimated 1100–1500 m (3600–4900 ft) of erosion at the Wagon Wheel well on the Pinedale anticline. Coskey (2004; Figure 4), using vitrinite reflectance data to estimate the amount of removed overburden in the Jonah–Pinedale area, concluded that from 1400 to more than 6000 ft (430 to >1830 m) of section has been eroded, with a best fit estimate of 3600 ft (1100 m). The timing and precise rates of erosion are unknown, but all the overburden was likely stripped during the post-Miocene regional exhumation of the Rocky Mountain region (Dickinson, 1989).

Because of the coincidence of the log anomaly with the top of the overpressure at Jonah field, it seems reasonable to expect the top of the overpressure to track the log anomalies throughout the northern Green River basin. Figures 7 and 8 are cross sections through Jonah field and adjacent wildcat wells, showing the resistivity-log, sonic-log, and mud-log gas curves. The sonic-log and resistivity-log anomalies do not track the top of the overpressure beyond the limits of the field; instead, they remain at approximately the same stratigraphic position despite the fact that the top of the overpressure deepens by 2000–3000 ft (610–910 m) to the west and south. Mud weights, gas shows, and a few well tests outside the Jonah compartment, but within the anomalous sonic and

Table 2

Common Mechanisms for Abnormal Pressure Development in Sedimentary Basins (after Osborne and Swarbrick, 1997).

Overpressure Mechanisms	Cause	Expected Magnitude (psi)
Stress-related compaction		
Disequilibrium compaction	vertical loading (rapid burial)	thousands, limited by rock failure
Tectonic compression	lateral loading	thousands, limited by rock failure
Fluid-volume increase		
Aquathermal expansion	heating and expansion of water with increasing burial depth and temperature	~100
Mineral transformations	gypsum → anhydrite dehydration	7–100
	smectite → illite transformation	unclear
	smectite dehydration	~100
Hydrocarbon generation	kerogen → gas transformation	70–6000 in source rock itself
	kerogen → oil transformation	70–6000 in source rock itself
	oil → gas (in-situ cracking)	hundreds in reservoir rock
Fluid movement		
Osmotic pressure	salinity contrasts across semipermeable shale membranes	hundreds, but theoretically as much as 3000
Hydraulic or potentiometric head	elevated recharge areas	tens to hundreds, depending on elevation difference
Hydrocarbon buoyancy	hydrocarbon-water density contrasts	1–1000, depending on column height and density difference of phases
Underpressure Mechanisms		
Differential discharge		
Differential gas flow		
Osmosis		
Rock dilatancy	uplift and removal of overburden	
Pore-fluid cooling	cooling and contraction of pore fluids with relative uplift	

resistivity interval, indicate that this section is normally pressured to slightly underpressured (Table 1). The under-compaction signature recorded by the logs and the top of present-day overpressure are obviously decoupled in the area surrounding Jonah field.

DISCUSSION

Sonic and Resistivity Anomalies as Paleopressure Indicators in the Rocky Mountain Region

In many Tertiary basins, including the Gulf Coast region of the United States, burial depths have steadily increased with continuous subsidence. The rate of subsidence and rate of sediment input to these basins have varied over time, but the creation of accommodation space has kept pace. In these basins, which form the overwhelming majority of the case histories for overpressure detection from logs, a close correspondence between the top of the overpressure conditions and the initial deviation from the local compaction trend on resistivity and sonic logs has long been recognized.

In foreland basins, including Laramide basins of the Rocky Mountain region, the geological history is more complex. These basins have typically seen a principal subsidence phase, during which most of the basin-filling sediment was deposited, followed by one or more

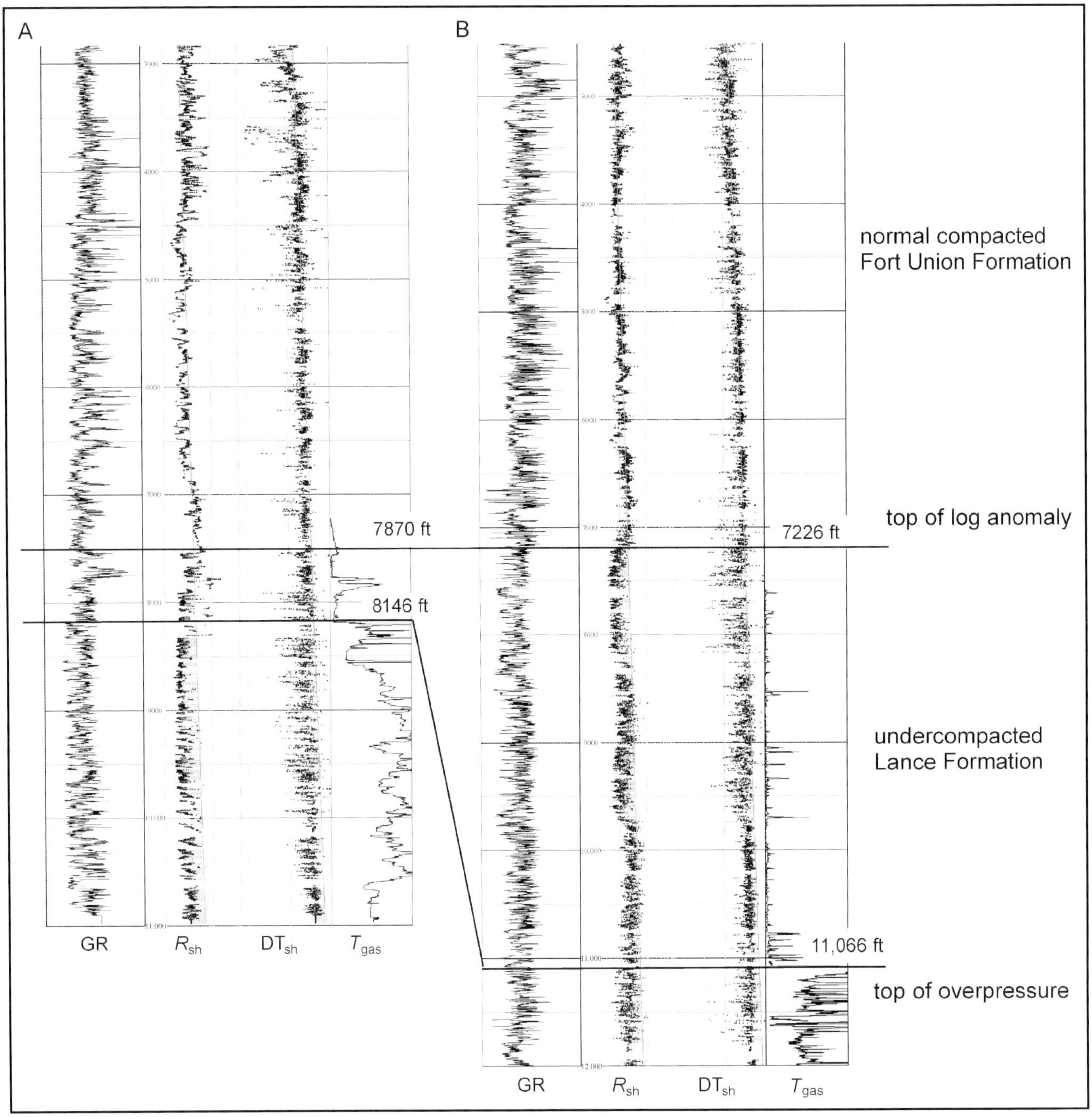

Figure 6. Typical sonic transit time-resistivity anomalies in the Jonah area. (A) Jonah field well, McMurry Oil Stud Horse Butte 15-28 (SW SE Sec. 28, T29N, R108W), with less than 300 ft (90 m) between the log anomaly and the top of overpressure; (B) a well outside the overpressured area, HS Resources Holmes Federal 5-1 (SW NW Sec. 1, T27N, R109W), with more than 3800 ft (1160 m) between log anomaly and the top of overpressure. GR = gamma-ray log; R_{sh} = filtered deep resistivity in shales; DT_{sh} = filtered sonic transit time in shales; T_{gas} = mud-log total gas curve.

periods of relative uplift, during which substantial amounts of sediment were eroded. If an interval of strata in one of these basins becomes overpressured at any time in its burial history because of disequilibrium compaction and then is subsequently uplifted, the present-day pressure may not accurately reflect the net state of compaction.

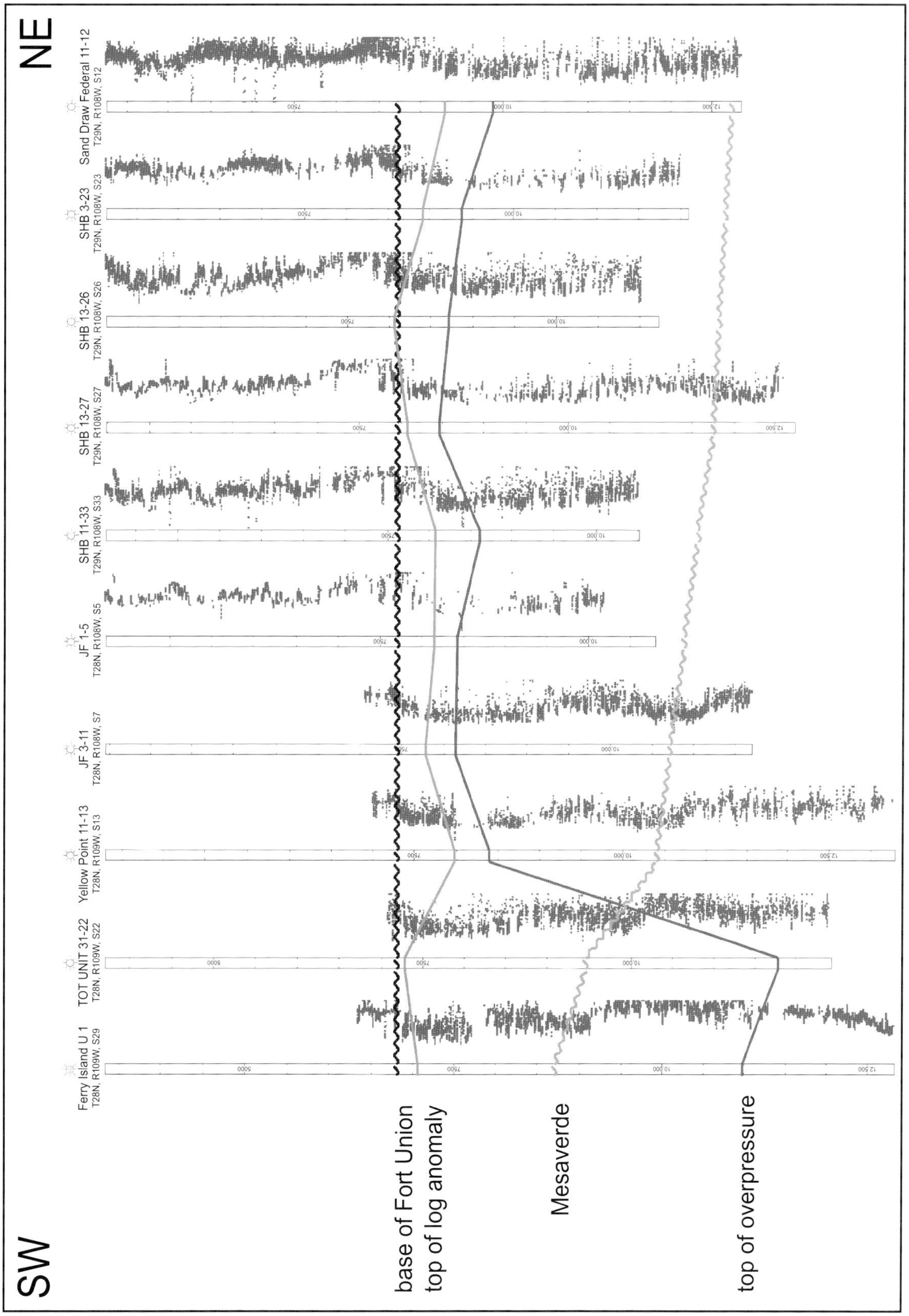

Figure 7. Southwest–northeast cross section across Jonah field showing position of resistivity anomaly relative to the top of overpressure. A large-scale version of this cross section is included on the CD-ROM with this publication as Appendix E. JF = Jonah Federal. SHB = Stud Horse Butte.

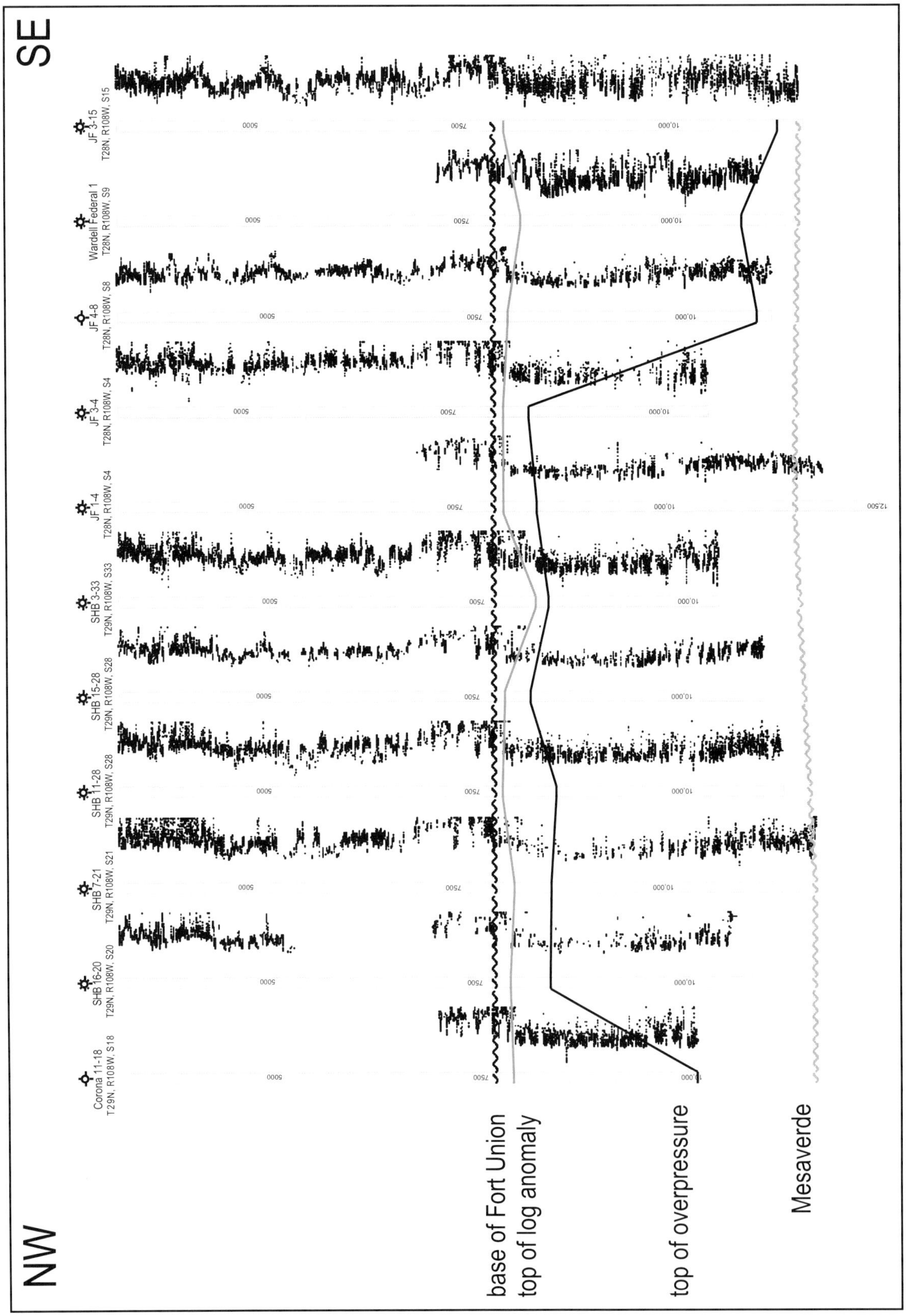

Figure 8. Northwest–southeast cross section across Jonah field showing position of resistivity anomaly relative to top of overpressure. A large-scale version of this cross section is included on the CD-ROM with this publication as Appendix E. JF = Jonah Federal. SHB = Stud Horse Butte.

Total pore pressure decreases with uplift and erosion because the trapped fluids cool and the pore system dilates as the net overburden stress is reduced. Leakage of the overpressured fluids can occur if faults and fractures are activated and form permeable pathways for fluids to migrate away from the overpressured interval. In a leaky section, the top of the overpressure will drop stratigraphically as the interval is uplifted.

Sonic and resistivity logs in the Rocky Mountain region probably respond to undercompaction of the sediments in an overpressured section just as they do in younger basins. Where overpressure results from disequilibrium compaction, the sonic-log and resistivity-log anomalies will mark the top of the overpressure during the compaction phase. The undercompacted rock fabric is an irreversible change that does not diminish as the section is uplifted. The only way to compact the rocks to the degree expected from a normal compaction gradient and thereby erase the undercompaction signature would be to rebury them to their former maximum depth in an open state, where pore fluids can freely escape. Therefore, in a section that has undergone relative uplift and fluid leakage during the uplift phase, the top of the present-day overpressure will lie somewhere below the log anomalies that mark the top of the overpressure during the burial and compaction phase. This decoupling of present overpressure with the position of the log anomaly is exactly what we observed in the area surrounding Jonah field.

Rathbun (1968) described conductivity anomalies in the northern Green River basin that he attributed to changes in pore pressure, and these were used as mappable parameters. Evers and Ezeanyim (1983) developed a set of calibration charts to predict pore pressure from conductivity and sonic transit time anomalies and presented a comparison of the predicted and actual pressures for the Wagon Wheel 1 well on the Pinedale anticline. They concluded that the method was accurate in this area to within ±0.03 psi/ft (0.679 MPa/km) of depth, or 300 psi (2.069 MPa) at 10,000 ft (3000 m). Subsequently, Prensky (1986), in a study of well-log responses in the northern Green River basin, found a poor correlation between the top of the conductivity anomaly in the Pinedale area and the top of the overpressure estimated from mud-weight data. Noting the close proximity of the conductivity anomaly in wells on the Pinedale anticline with the Cretaceous–Tertiary boundary and with a major change in GR intensity at approximately the same depth, Prensky (1986) concluded that the conductivity anomaly was a stratigraphic phenomenon related to a change in bulk mineralogy near the base of the Fort Union. He further concluded that shale conductivity and shale transit time plots would not be useful for pore-pressure prediction in the Green River basin because of complex changes in lithology through time, the lower salinity of pore waters compared to the Gulf Coast basin where the techniques were developed, and differences in the mechanism presumed to be responsible for overpressuring (hydrocarbon generation versus disequilibrium compaction). Prensky (1986) did not offer any explanation for the observation made by Rathbun (1968), which this study confirms, that the conductivity anomaly seems to cross stratigraphy from east to west, until it is present near the base of the Hilliard Shale on the west flank of the basin and in the La Barge platform area.

Although we concur with Prensky's (1986) observation that the resistivity and sonic anomalies in the Pinedale area roughly coincide with the Cretaceous–Tertiary boundary, it is not clear that this is a result of a change in shale mineralogy. Equally likely is that the sandstones and shales of the Tertiary are poor seals and had higher fluid transmissibility during compaction, consequently follow a normal compaction-versus-depth curve. The sandstones and shales in the Lance and at least part of the underlying Mesaverde, however, apparently had poor transmissibility and are undercompacted relative to the shallower, normally compacted, and normally pressured Tertiary units. Thus, the log anomalies follow the stratigraphic boundary, at least in this immediate area, because of a fluid-transmissibility contrast in the major units, which may or may not be related to mineralogical differences.

Surdam et al. (1997, their figure 12) proposed another explanation for the sonic-log anomaly, wherein relatively small levels of gas saturation in the shales can generate in a significant increase in sonic transit times. They cited laboratory data showing that as little as 5% gas saturation in the matrix porosity can result in a strong decrease in the compressional wave velocity (Timur, 1987). In this model, the top of the overpressure coincides with the onset of gas generation and results in partial gas saturation throughout the shales. The principal problem with this hypothesis is that an increase in gas saturation in the shales should result in an increase in average shale resistivity and not a decrease. Studies of mature source rock shales invariably document a major increase in resistivity that coincides with the top of hydrocarbon generation because pore waters are displaced by nonconductive hydrocarbons (e.g., Meyer and Nederlof, 1984; Schmoker and Hester, 1990), which is the opposite of what we observed in the Lance section in and around Jonah. Additionally, available vitrinite reflectance data at Jonah indicate that shales in the Lance are immature and have not reached the onset of gas generation (Bowker and Robinson, 1997; Coskey, 2004).

Jonah Field: Is It Really Just a Paleopressure Remnant?

The presence of a sonic-log and resistivity-log anomaly throughout the area surrounding Jonah field suggests that the Lance section is undercompacted and, at its maximum burial depth, was overpressured over a broad area of the northern Green River basin. Gas shows and marginal gas production from nearby wildcats further suggest that the Lance section contains some gas, but it is normally pressured or underpressured outside the field boundaries. These two observations, in combination, lead us to propose the following model for the origin of Jonah field.

The top of the overpressure was widely established as a regional surface during the early to middle Tertiary. The position of this surface coincides with the anomaly we see today on sonic and resistivity logs. The exact timing and burial history of the northern Green River basin is poorly constrained, in large part because portions of the section are now missing and therefore cannot be directly dated. Maximum overburden was probably attained significantly after deposition of Paleocene strata but before the middle Miocene. Coskey (2004), for example, suggests that most of the removed section is Wasatch, and erosion begins at about 5 Ma. The stratigraphic position of the top of the regional overpressure was controlled by the low permeability of the Lance section, which is composed of a heterolithic assemblage of low-permeability sandstones and shales that impeded fluid flow out of the section during burial. The sandstones of the lower Fort Union in the Jonah area have good lateral continuity, are regionally porous and permeable, and allowed fluids expelled during compaction to escape the system. Consequently, the Fort Union is currently normally pressured, exhibits high water saturations, and has probably always been normally pressured.

At its time of maximum burial, at depths 3000–5000 ft (910–1500 m) greater than today, the entire Lance section was receiving gas charge from deeper, more mature source rocks (Coskey, 2004). The Lance itself is thermally immature in the Jonah area (Warner, 1998; Coskey, 2004). Gas migrated vertically through a regional system of joints and fractures that must have been sufficiently extensive to contact a high percentage of the reservoir sand bodies. The formation was somewhat overpressured because the Lance had lower transmissibility than the overlying strata and, consequently, was in a state of disequilibrium compaction, resulting in the sonic-log and resistivity-log anomalies found today both within and around the field.

The lithostatic overburden decreased in direct proportion to the rate and amount of erosion during late Tertiary uplift. The porosity of the sandstones increased slightly as the pore system dilated with reduced confining stress and the section cooled with uplift, resulting in contraction of the trapped gases. Total reservoir pore pressure decreased, although the pore-pressure gradient may have actually increased with relative uplift because of the high compressibility of gas (Figure 9) (Katahara and Corrigan, 2002). Another factor to consider is that during uplift, erosion, and cooling, the top of the active hydrocarbon generation window falls away from the Lance in rough proportion to the amount and rate of erosion. It is unlikely that active hydrocarbon generation was able to contribute to overpressured conditions in the Lance once uplift commenced.

The sharp difference in pressure gradients across the field boundaries, with similar reservoir rock properties and identical burial histories within and outside the field, can be explained by a difference in the rate of fluid leakage out of the Lance section during uplift. The Jonah compartment appears to be very tightly sealed and has retained its gas charge during uplift. Leakage of fluids out of Jonah was limited by extremely low-permeability seals on three sides of the field. The major fault zones on the south and west flanks of Jonah form lateral sealing surfaces that are capable of maintaining a pressure differential greater than 1000 psi (6.9 MPa) over short distances. The top of the gas accumulation at Jonah coincides with the shaly, low net-to-gross Unnamed unit. This apparently sealing interval separates overpressured, gas-charged Lance sandstones from water-saturated and normally pressured Fort Union sandstones.

These three major sealing surfaces, two fault systems, and a top-sealing shale, form a structural trap similar in geometry to the bow and decking of a ship. Because the bow is fortuitously pointed directly updip, fluids were unable to leak out of the trap as the system was uplifted. The top-sealing shale, however, is truncated by post-Lance erosion a short distance to the south and west of Jonah (Coskey, 2004). If the truncation edge of the Unnamed Tertiary was within the present field area, anywhere north and east of the intersection of the two major bounding faults, a large part of the accumulation would have likely leaked upward into the Fort Union, and the field would be much smaller or might not even exist.

In the area surrounding Jonah, gas freely escaped upward into the overlying Fort Union section as the interval was uplifted. The exact mechanism and pathways for leakage are speculative, but it is likely that gas leaked though a system of regional joints and fractures. This regional fracture system may very well be identical to that inside the field, but a key difference is that outside the field, there are no lateral or top seals to prevent gas from migrating

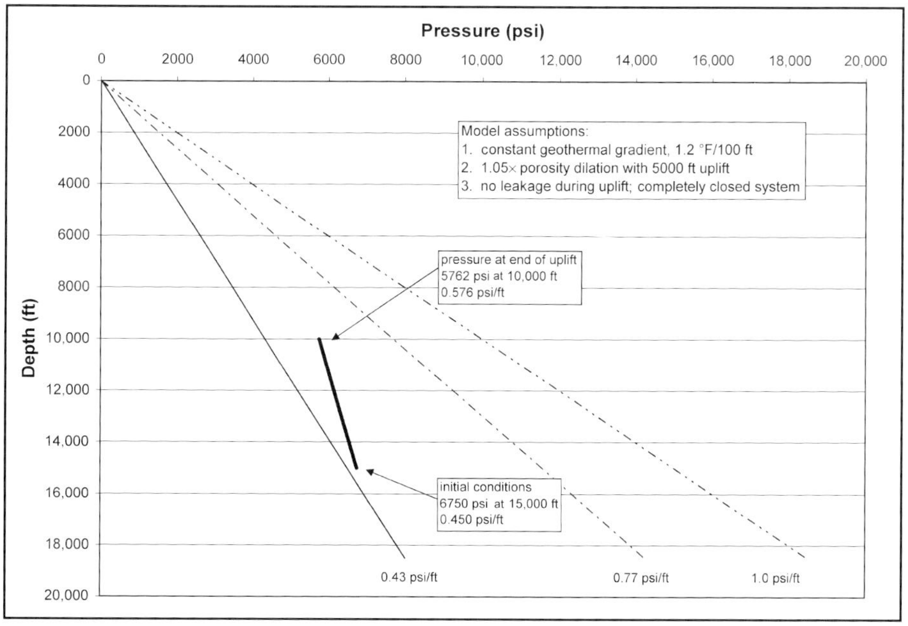

Figure 9. Model calculations of 5000 ft (1500 m) of relative uplift of the Lance show a net reservoir pressure decrease of approximately 15%, but the pore-pressure gradient relative to the surface actually increases from 0.45 to 0.58 psi/ft (10–13 MPa/km).

out of the Lance into the Fort Union. The Lance sandstones outside Jonah still retain some gas saturation because minimal free water was available to be imbibed as the section was uplifted, but they have lost a large portion of their charge and yield much weaker gas shows during drilling.

Jonah field is a tightly sealed and overpressured compartment that stands high above the surrounding top of the pressure surface, much like a mesa above an eroded landscape. The top of the overpressure surface surrounding the field, established in the early Tertiary at maximum burial depth, has been dropping since uplift and erosion began in the late Miocene or Pliocene. Because Jonah field owes its origin to the preservation of overpressure conditions during uplift, and maintaining this state of overpressure is a direct result of the unique sealing conditions at Jonah, exploration for additional overpressured accumulations of this kind should focus on sealing and trapping conditions that will prevent leakage. This is in stark contrast to the prevailing exploration philosophy for basin-centered gas sweet spots, which emphasizes finding fractures and areas of enhanced reservoir quality favorable for gas leaking upward into shallower horizons.

CONCLUSIONS

The Lance Formation consists of highly discontinuous sandstones in a shale matrix. Fluids contained in the Lance were unable to freely escape the section during rapid burial, and as a consequence, the fine-grained sediments are undercompacted relative to their former maximum burial depths. Sonic and resistivity logs through the Lance and Mesaverde retain a strong signature of reduced velocity and reduced resistivity directly related to undercompaction during burial. The top of the anomalous sonic velocity-versus-depth and resistivity-versus-depth trends generally coincides with the top of the shaly Lance section and is everywhere beneath the more continuous and permeable Fort Union sandstones.

Coincidentally, the top of the anomalous velocity and resistivity profiles corresponds, within a few hundred feet, to the top of the overpressured gas in Jonah field, leading to the idea that the geophysical signatures might be directly related to the presence of overpressure. Regional examination of logs, however, demonstrates that the anomalies extend far beyond the field limits and persist at roughly the same stratigraphic position, clearly demonstrating that the anomaly is decoupled from present-day pressure conditions.

The best explanation for the similarity of velocity-resistivity profiles inside and outside Jonah is that they record the former top-of-pressure conditions attained at maximum burial, and the signature was frozen into the rocks during subsequent exhumation. Jonah field is thus an anomalous remnant of former regional overpressure conditions that were established during maximum burial conditions in the early or middle Tertiary.

The origin of Jonah field is therefore related to relative rates of fluid leakage out of the Lance section during relative uplift. Jonah field is tightly sealed on two sides by impermeable sealing faults and is sealed on the top by a thick, low-permeability shaly interval. Outside Jonah field, fluids were able to leak updip through fractures and escaped the system. Where the fluids were unable to escape, a deep, overpressured system was able to survive uplift to much shallower depths. Thus, a multi-tcf gas accumulation in the Greater Green River basin owes its origin to the preservation of gas charge and overpressure over millions of years of regional uplift and erosion.

ACKNOWLEDGMENTS

The authors acknowledge the contributions of numerous colleagues and clients we have worked with over the years on the Jonah problem, many of whom contributed substantially to the development of the ideas presented in this chapter. These include Alan Byrnes of the Kansas Geological Survey, and John Webb of The Discovery Group, with whom we first started working on the petrophysical aspects of Jonah back in 1994; Dennis Storhaug and Denise Delozier, both formerly with Snyder Oil Company; Kent Bowker and Deane Foss, both formerly with Chevron U.S.A.; Dean DuBois, EnCana U.S.A.; Jim Folcik, EOG Resources; and Phil Winner, formerly with HS Resources. We especially thank Keith Shanley and John Robinson for uncounted hours of discussion on the dilemma of the origin of the pressure conditions around Jonah and for their support in the preparation of this chapter.

REFERENCES CITED

Athy, L. F. 1930, Density, porosity, and compaction of sedimentary rocks: AAPG Bulletin, v. 14, p. 1–24.

Barker, C., 1990, Calculated volume and pressure changes during the thermal cracking of oil to gas in reservoirs: AAPG Bulletin, v. 74, p. 1254–1261.

Battin, D. P., and S. R. Clark, 1994, Jonah field, a basin-centered gas field in the northern Green River basin, Sublette County, Wyoming (abs.): AAPG Annual Convention Abstracts, v. 3, p. 101.

Billingsley, R. L., and L. W. E. Reinert, 1994, The Wamsutter sweet spot: A continuing enigma in tight formation gas (abs.): AAPG Annual Convention Abstracts, v. 3, p. 104–105.

Bowker, K. A., and J. W. Robinson, 1997, Jonah field: A shallow sweet spot in the basin-centered gas accumulation of the northern Green River basin (abs.): AAPG Bulletin, v. 81, p. 1219–1220.

Burrus, J., 1998, Overpressure models for clastic rocks, their relation to hydrocarbon expulsion: A critical reevaluation, *in* B. E. Law, G. F. Ulmishek, and V. I. Slavin, eds., Abnormal pressures in hydrocarbon environments: AAPG Memoir 70, chapter 3, p. 35–63.

Coskey, R. J., 2004, Burial-history modeling of the Jonah field area: An unusual and possibly unique gas accumulation in the Green River basin, Wyoming, *in* J. W. Robinson, and K. W. Shanley, eds., Jonah field: Case study of a tight-gas fluvial reservoir: AAPG Studies in Geology 52, p. 93–126

Dickinson, W. W., 1989, Analysis of vitrinite maturation and Tertiary burial history, northern Green River basin, Wyoming: U.S. Geological Survey Bulletin, n. 1886-F, p. F1–F13.

Evers, J. F., and R. Ezeanyim, 1983, Prediction of abnormal pressures in Wyoming sedimentary basins using well logs: Society of Petroleum Engineers 1983 Rocky Mountain Regional meeting, Salt Lake City, Utah, SPE Paper 11859, 12 p.

Foster, J. B., and H. E. Whalen, 1966, Estimation of formation pressures from electrical surveys— Offshore Louisiana: Journal of Petroleum Technology, SPE Paper 1200, February 1966, p. 165–171.

Ham, H. H., 1966, A method of estimating formation pressures from Gulf Coast well logs: Transactions of the Gulf Coast Association of Geological Societies, v. 16, p. 185–197.

Hottman, C. E., and R. K. Johnson, 1965, Estimation of formation pressures from log-derived shale properties: Journal of Petroleum Technology, SPE Paper 1110, June 1965, p. 717–722.

Hunt, J. M., J. K. Whelan, L. B. Eglinton, and L. M. Cathles, III, 1998, Relation of shale porosities, gas generation, and compaction to deep overpressures in the U.S. Gulf Coast, *in* B. E. Law, G. F. Ulmishek, and V. I. Slavin, eds., Abnormal pressures in hydrocarbon environments: AAPG Memoir 70, chapter 5, p. 87–104.

Jiao, Z. S., and R. C. Surdam, 1997, Characteristics of anomalously pressured Cretaceous shales in the Laramide basins of Wyoming, *in* R. C. Surdam, ed., Seals, traps and the petroleum system: AAPG Memoir 67, chapter 14, p. 243–253.

Katahara, K. W., and J. D. Corrigan, 2002, Effect of gas on poroelastic response to burial or erosion, *in* A. R. Huffman and G. L. Bowers, eds., Pressure regimes in sedimentary basins and their prediction: AAPG Memoir 76, p. 73–78.

Law, B. E., 1984, Relationships of source-rock, thermal maturity, and overpressuring to gas generation and occurrence in low-permeability Upper Cretaceous and lower Tertiary rocks, Greater Green River basin, Wyoming, Colorado, and Utah, *in* J. Woodward, F. F. Meissner, and J. L. Clayton, eds., Hydrocarbon source rocks of the greater Rocky Mountain region: 1984 Rocky Mountain Association of Geologists Guidebook, p. 469–490.

Law, B. E., 2002, Basin-centered gas systems: AAPG Bulletin, v. 86, p. 1891–1919.

Law, B. E., and W. W. Dickinson, 1985, Conceptual model for origin of abnormally pressured gas accumulations in low-permeability reservoirs: AAPG Bulletin, v. 69, p. 1295–1304.

Law, B. E., and R. C. Johnson, 1989, Structural and stratigraphic framework of the Pinedale anticline, Wyoming, and the multiwell experiment site, Colorado: U.S. Geological Survey Bulletin, n. 1886-B, p. B1–B11.

Law, B. E., and C. W. Spencer, 1998, Abnormal pressure in hydrocarbon environments, *in* B. E. Law, G. F. Ulmishek, and V. I. Slavin, eds., Abnormal pressures in hydrocarbon environments: AAPG Memoir 70, chapter 1, p. 1–11.

Magara, K., 1976, Thickness of removed sedimentary rocks, paleopore pressure, and paleotemperature, southwestern part of Western Canada basin: AAPG Bulletin, v. 60, p. 554–565.

Masters, J. A., 1979, Deep basin gas trap, western Canada: AAPG Bulletin, v. 63, p. 152–181.

Masters, J. A., 1984, ed., Elmworth— Case study of a deep basin gas field: AAPG Memoir 38, 316 p.

MacGregor, J. R., 1965, Quantitative determination of reservoir pressures from conductivity log: AAPG Bulletin, v. 49, p. 1502–1511.

McPeek, L. A., 1981, Eastern Green River basin— A developing giant gas supply from deep, overpressured Upper Cretaceous sandstones: AAPG Bulletin, v. 65, p. 1078–1098.

Meissner, F. F., 2000, Causes of anomalous deep basin fluid pressures in Rocky Mountain basins and their relation to regional gas accumulation: Rocky Mountain Association of Geologists, Basin-centered Gas Symposium, October 6, 2000, Denver, Colorado.

Meyer, B. L., and M. H. Nederlof, 1984, Identification of source rocks on wireline logs by density/resistivity and sonic transit time/ resistivity crossplots: AAPG Bulletin, v. 68, p. 121–129.

Montgomery, S. L., and J. W. Robinson, 1997, Jonah field, Sublette County, Wyoming: Gas production from overpressured Upper Cretaceous Lance sandstones of the Green River basin: AAPG Bulletin, v. 81, p. 1049–1062.

Osborne, M. J., and R. E. Swarbrick, 1997, Mechanisms for generating overpressure in sedimentary basins: A reevaluation: AAPG Bulletin, v. 81, p. 1023–1041.

Prensky, S. E., 1986, Geologic implications of large-scale trends in well-log response, northern Green River basin, Wyoming: Transactions of the Society of Professional Well Log Analysts 27th Annual Logging Symposium, Paper EEE, 24 p.

Rathbun, F. C., 1968, Abnormal pressures and conductivity anomaly, northern Green River basin, Wyoming: Society of Petroleum Engineers 43rd Annual Meeting, Houston, Texas, SPE Paper 2205, 8 p.

Robinson, J. W., 2000, Discovery of Jonah field, Sublette County, Wyoming: The Mountain Geologist, v. 37, no. 3, p. 135–143.

Sauvageau, K. A., 1984, A proven technique for liner cementing in Greater Green River basin, Wyoming: Journal of Petroleum Technology, SPE Paper 11418.

Schmoker, J. W., and T. C. Hester, 1990, Formation resistivity as an indicator of oil generation— Bakken Formation of North Dakota and Woodford Shale of Oklahoma: The Log Analyst, v. 31, no. 1, p. 1–9.

Spencer, C. W., 1983, Geologic aspects of tight-gas reservoirs in the Rocky Mountain region, *in* Society of Petroleum Engineers/Department of Energy Symposium on low-permeability reservoirs: SPE paper 11647, p. 399–408; *also in* Journal of Petroleum Technology, v. 37, p. 1308–1314 (1985).

Spencer, C. W., 1987, Hydrocarbon generation as a mechanism for overpressuring in Rocky Mountain Region: AAPG Bulletin, v. 71, p. 368–388.

Surdam, R. C., 1997, A new paradigm for gas exploration in anomalously pressured "tight-gas sands" in the Rocky Mountain Laramide basin, *in* R. C. Surdam, ed., Seals, traps, and the petroleum system: AAPG Memoir 67, chapter 17, p. 283–298.

Surdam, R. C., Z. S. Jiao, and R. S. Martinsen, 1994, The regional pressure regime in Cretaceous sandstones and shales in the Powder River basin, *in* P. J. Ortoleva, ed., Basin compartments and seals: AAPG Memoir 61, chapter 15, p. 213–233.

Surdam, R. C., Z. S. Jiao, and H. P. Heasler, 1997, Anomalously pressured gas compartments in Cretaceous rocks of the Laramide basins of Wyoming: A new class of hydrocarbon accumulation, *in* R. C. Surdam, ed., Seals, traps, and the petroleum system: AAPG Memoir 67, chapter 12, p. 199–222.

Surdam, R. C., J. Robinson, Z. S. Jiao, and N. K. Boyd, III, 2001, Delineation of Jonah field using seismic and sonic velocity interpretations, *in* D. Anderson, J. W. Robinson, J. E. Estes-Jackson, and E. B. Coalson, eds., Gas in the Rockies: 2001 Rocky Mountain Association of Geologists Guidebook, p. 189–208.

Swarbrick, R. E., and M. J. Osborne, 1998, Mechanisms that generate abnormal pressures: An overview, *in* B. E. Law, G. F. Ulmishek, and V. I. Slavin, eds., Abnormal pressures in hydrocarbon environments: AAPG Memoir 70, chapter 2, p. 13–34.

Swarbrick, R. E., M. J. Osborne, and G. S. Yardley, 2002, Comparison of overpressure magnitude resulting from the main generating mechanisms, *in* A. R. Huffman and G. L. Bowers, eds., Pressure regimes in sedimentary basins and their prediction: AAPG Memoir 76, chapter 1, p. 1–12.

Timur, A., 1987, Acoustic logging, *in* H. Bradley, ed., Petroleum Engineering Handbook: Society of Petroleum Engineers, chapter 51, p. 1–52.

Vega, V. H., and W. B. Hanson, 2001, Jonah field area: Geophysical imaging of trap boundaries and reservoir heterogeneity, *in* J. W. Robinson and K. W. Shanley, co-convenors: Tight gas fluvial reservoirs: A case history from the Lance Formation, Green River basin, Wyoming: AAPG Short Course Notes 2, AAPG Annual Meeting, Denver, 2 June 2001.

Warner, E. M., 1997a, A structural model for shallower emplacement of gas in overpressured basin centers, Jonah field as a previously undetected abnormally pressured structural trap (abs.): AAPG Bulletin, v. 81, p. 1236.

Warner, E. M., 1997b, Structural geology of Jonah field as determined from subsurface geology and 3-D seismic (abs.): AAPG Bulletin, v. 81, p. 1236.

Warner, E. M., 1998, Combining a basin-centered accumulation with a structural trap, Jonah field: A pressure compartment defined by subsurface geology and 3-D seismic, *in* Developing a better understanding of basin-centered gas plays: Gas Research Institute GRI 98/ 0119, p. 208–228.

CHAPTER 9

Basin-centered Gas Systems and the Jonah Field

B. E. LAW
Pangea Hydrocarbon Exploration, LLC, Lakewood, Colorado

C. W. SPENCER
U.S. Geological Survey, Lakewood, Colorado, Retired

ABSTRACT

The giant Jonah gas field, located in western Wyoming, is a gas chimney rooted in a regionally pervasive, direct-type, basin-centered gas accumulation (BCGA). The field is an excellent example of a structural sweet spot in a BCGA. Basin-centered gas systems (BCGSs), of which BCGAs are products, are potentially one of the more economically important, unconventional gas systems in the world; in the United States, they contribute as much as 17% of the total annual gas production. These regionally pervasive gas accumulations are different from conventional accumulations in several respects. The BCGAs associated with BCGSs are typically characterized by regionally pervasive reservoirs that are gas saturated, abnormally (high or low) pressured, commonly lack a downdip water contact, and have low-permeability reservoirs. The accumulations range from single, isolated reservoirs a few feet thick to multiple, stacked reservoirs several thousand feet thick.

Two types of BCGSs are recognized: a direct type, characterized by having gas-prone source rocks, and an indirect type, characterized by having liquid-prone source rocks. During the burial and thermal histories of these systems, the source rock differences between the two types of BCGSs result in strikingly different characteristics. Based on these criteria, gas in the Jonah field is interpreted to have been sourced from gas-prone, type III kerogen and is therefore a direct type of BCGA.

INTRODUCTION

A petroleum system as defined by Magoon and Dow (1994, p. 3) "... includes all the elements and processes needed for an oil and gas accumulation to exist." In Magoon and Dow's definition, the elements include source rock, reservoir rock, seal rock, and overburden rock. Relevant processes include trap formation and the generation, expulsion, migration, and accumulation of petroleum. A basin-centered gas system (BCGS) contains all of these components; however, the magnitude and role of some of the components interact to form a unique type of hydrocarbon accumulation.

Of all the different types of unconventional gas systems, none have been more poorly defined than BCGSs. The problem of definition has led to misconceptions that have, in some cases, impeded exploration efforts. It is uncertain when regionally pervasive gas accumulations, like basin-centered gas accumulations (BCGAs), first became known; however, Silver (1950) alluded to pervasive gas accumulations in Cretaceous rocks in the San Juan basin of New Mexico and Colorado. Silver (1950) also recognized the gas-saturated nature of the reservoirs and the downdip absence of water. Later, nuclear stimulation experiments conducted in the United States from 1967 to 1973 seem to have implied knowledge of the presence of regionally pervasive, gas-charged reservoirs, although there are no geologic reports confirming this assumption. Nuclear detonations conducted in Cretaceous rocks in the San Juan and Piceance basins of New Mexico and Colorado were unsuccessful, and eventually, because of concerns about environmental and radioactive contamination issues, the tests were abandoned (Randolph, 1973, 1974a, b, c).

Jonah Field: Case Study of a Giant Tight-Gas Fluvial Reservoir, J. W. Robinson and K. W. Shanley, eds.: AAPG Studies in Geology 52 and Rocky Mountain Association of Geologists 2004 Guidebook

The first published, unmistakable reference to this type of gas accumulation was by Masters (1979). In his paper, Masters identified the basic concepts of BCGAs, referring to them as "deep basin gas." He provided several defining characteristics of gas-saturated reservoirs in the deep basin of Alberta, Canada, and in the San Juan basin of New Mexico and Colorado as examples of such accumulations. Later, in a publication edited by Masters (1984), the various geologic aspects of the "deep-basin" gas accumulation in the Elmworth field of Alberta, Canada, were provided. Other significant papers concerning BCGAs include those by Law et al. (1979, 1980); Law (1984) in the Greater Green River basin of Wyoming, Colorado, and Utah; McPeek (1981) in the Great Divide basin of Wyoming; and Brown et al. (1986) in the Piceance basin of Colorado. Spencer (1985, 1989) and Law and Spencer (1993) described many of the attributes common to BCGAs, and Law (2002) recognized direct and indirect types of BCGAs. Although reservoirs referred to as "tight" may include conventionally trapped gas, several examples of "tight-gas reservoirs" in the United States are provided in a volume edited by Spencer and Mast (1986). Finley (1984) and Dutton et al. (1993) also described many additional low-permeability reservoirs in the United States.

It is uncertain when the term "basin-centered gas accumulations (BCGAs)" came into use; however, the first published references to the term appeared in papers by Rose et al. (1986) in a study of gas accumulations in the Upper Cretaceous Trinidad Sandstone of the Raton basin and by Brown et al. (1986) in a study of the Upper Cretaceous Corcoran, Cozette, and Rollins sandstones in the Piceance basin of Colorado. It is likely, however, that the term "basin-centered gas accumulations" had been informally used by United States industry people prior to this time.

The name "tight-gas sands" has been widely used to describe BCGAs for many years, and many explorationists still use the term. In many cases, tight-gas sands is an appropriate term; however, the term is somewhat ambiguous and may include gas accumulations that are trapped as conventional, buoyant accumulations. The use of "deep basin gas" (Masters, 1979) has some problems also because all BCGAs do not occur at great depths. For example, much of the gas production in the San Juan basin is from a BCGA at depths as shallow as 2500 ft (760 m). The term "deep basin gas" is more recently used to define gas accumulations deeper than 15,000 ft (4600 m) (Dyman et al., 1997); it is an economic definition and is not based on geologic processes. Finally, the name "continuous gas accumulation" (Schmoker, 1995), although accurately portraying the pervasive nature of BCGAs, is too broad and includes such gas systems as coalbed methane and shale gas. In the absence of other suitable names, the name basin-centered gas accumulation is used here.

In light of the extremely large gas resources contained in BCGAs and the significance of Jonah field as an exploration model, the objectives of this chapter are to provide an overview of BCGAs; expand on the nature of BCGAs previously published by Masters (1979), Spencer (1985, 1989), Law and Spencer (1993), and Law (2002); and speculate on the nature of the gas accumulation in the Jonah field, located in the Greater Green River basin of western Wyoming (Figure 1). To accomplish these objectives, BCGAs are discussed in the context of a petroleum system.

BASIN-CENTERED GAS SYSTEM DEVELOPMENT

Two types of BCGSs exist: a direct type and an indirect type (Law, 2002). The attributes of these two types of systems are provided in Table 1. Direct and indirect types of BCGSs are distinguished on the basis of source rock composition: direct BCGSs have a gas-prone source rock, and an indirect BCGS has an oil-prone source rock. These fundamental differences, gas-prone versus oil-prone source rocks, lead to the development of different characteristics, as shown in Table 1. In addition to the two basic types of systems, there may be hybrid systems where source rocks composed of gas- and liquid-prone organic matter, such as humic, type III kerogen, have contributed to the development of a BCGA.

The developmental history of a BCGS may be viewed as four reservoir-pressure cycles. As a consequence of the dynamic nature of geologic processes and the response to those processes, the phases discussed here are geologically ephemeral. Figure 2 is a diagrammatic representation showing these pressure phases and the development of direct and indirect BCGSs. F. F. Meissner (1978, personal communication) and Law and Dickinson (1985) discussed these phase changes for gas accumulations in low-permeability (tight) reservoirs. The development of direct and indirect BCGSs has been described by Law (2002).

From an exploration perspective, the ability to identify the various phases becomes important through the observation that underpressured systems (phase III) are generally more commercial throughout their areal extent than overpressured systems (phase II). For example, the Upper Cretaceous underpressured BCGA in the San Juan basin (phase III) is, for the most part, productive throughout the BCGA, whereas the Upper Cretaceous overpressured BCGA in the Green River basin is not productive throughout the BCGA; the presence of a structural or stratigraphic sweet spot is necessary for commercial rates of production in overpressured BCGAs.

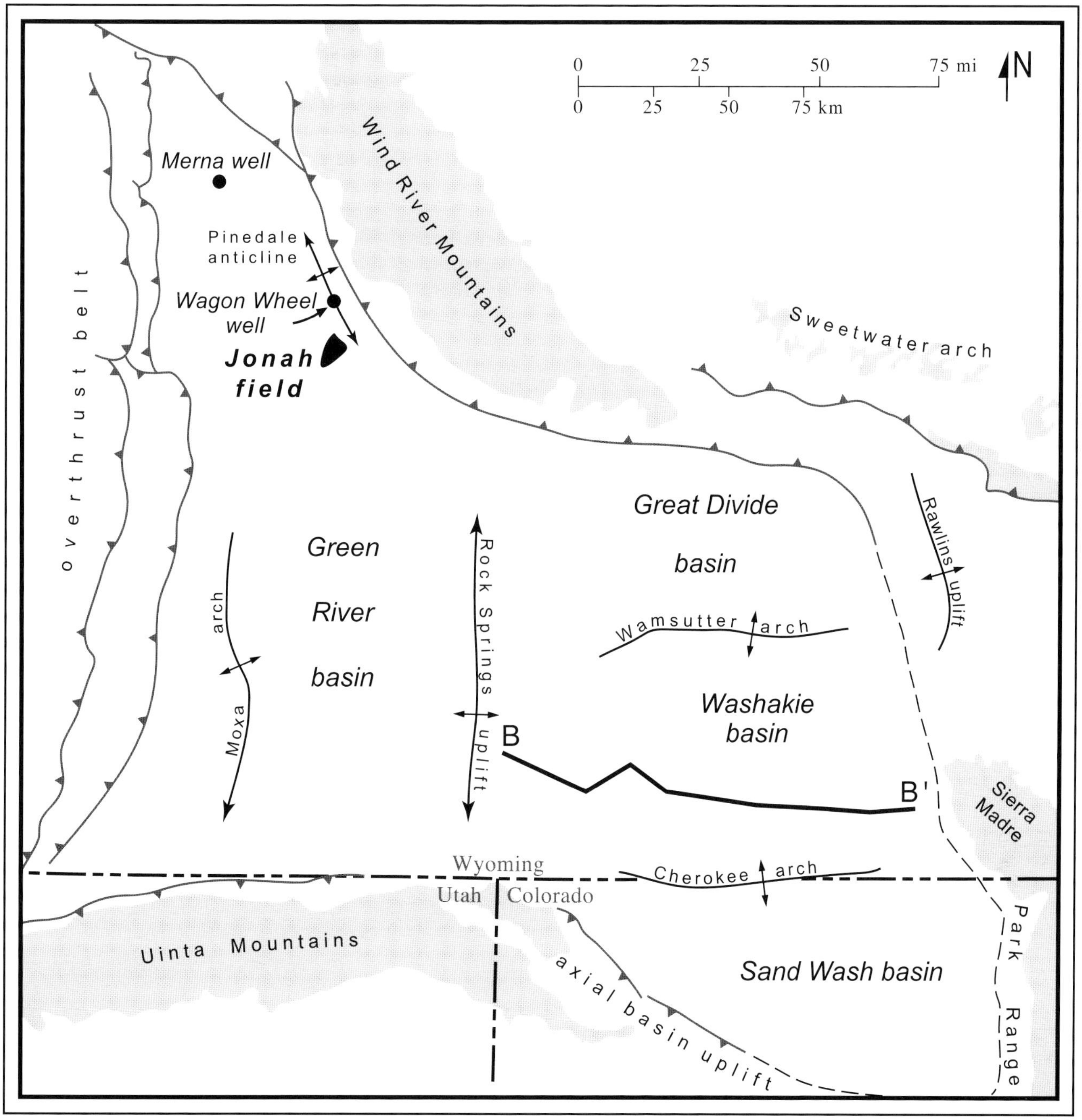

Figure 1. Map of Greater Green River basin showing the location of the Jonah field and major structural elements in the basin (modified from Law, 2002).

Phase I

Direct and Indirect Systems

During the early burial and thermal histories of direct and indirect systems, the reservoirs are normally pressured, and the fluid phase in the pore system is 100% water saturated. The compaction of framework grains during this phase is an important process. The defining processes for each system, however, are different. For direct systems, phase I terminates with the initiation of thermal gas generation from a source rock, whereas the termination of phase I in indirect systems occurs after migration and entrapment of oil and gas and the initiation of thermal cracking of oil to gas (Figure 2). Therefore, the composition of

Table 1

Attributes of Direct and Indirect Basin-centered Gas Systems (Law, 2002).

Type	Source Rocks	Reservoir In-situ Permeability (md)	Hydrocarbon Migration Distance	Reservoir Pressure	Pressure Mechanism	Seal	Seal Quality	Nature of Upper Boundary	Thermal Maturity Top of BCGA	Occurrence
Direct	gas-prone, type III kerogen	<0.1	short	overpressure and underpressure	hydrocarbon generation	capillary	variable	cuts across stratigraphy	>0.7% R_o	downdip from water
Indirect	liquid-prone, types I and II kerogens	<0.1	short and long	overpressure and underpressure	thermal cracking of oil to gas	lithologic and capillary	good	bedding parallel	highly variable	downdip from water

migrating hydrocarbons in direct systems would have been predominantly gas, with lesser amounts of liquids, and the composition of hydrocarbons in indirect systems would have been oil and gas.

Law and Spencer (1998) suggested that in the early burial stages of a BCGA sequence, prior to the development of a recognizable BCGA, it is conceivable that in some depositional settings of rapid sedimentation, compaction disequilibrium may have been the initial overpressuring mechanism. In this scenario, the pressuring fluid phase is water. However, as the sequence experiences further burial and higher temperatures, the compaction disequilibrium pressure mechanism may be replaced by hydrocarbon generation and the development of abnormally high pressures characterized by pore fluids composed of gas and little or no water. A possible example of the transition of pressure mechanisms from compaction disequilibrium to hydrocarbon generation may be present in Miocene and Pliocene rocks in the Beckes basin (Spencer et al., 1994) and the Makü Trench of Hungary. In these areas, Miocene and Pliocene rocks are overpressured and possess many of the distinguishing characteristics of a BCGA. The overpressures in Miocene rocks appear to be caused by hydrocarbon generation, whereas overlying, overpressured Pliocene rocks appear to be in a transitional pressure phase between compaction disequilibrium and hydrocarbon generation. In this case, knowledge of pore-fluid composition (mainly gas or mainly water) in the Pliocene sequence would offer considerable insight in resolving the problem.

Phase II

Direct Systems

Direct systems require gas-prone source rocks and low-permeability reservoirs in close proximity to each other. As the sequence of source and reservoir rocks undergoes burial and exposure to increasing temperatures, a point in time is reached where the source rocks begin to generate gas (Figure 2). Concomitant with increased gas generation, expulsion, and migration, gas begins to enter adjacent, water-wet sandstones. Because these sandstones have low permeability, the rate at which gas is generated and accumulated in reservoirs is greater than the rate at which gas is lost. Eventually, the capillary pressure of the water-wet pores is exceeded, forcibly expelling free, mobile water from the pore system, resulting in the development of an overpressured, gas-saturated reservoir with little or no free water. Examples of BCGA systems exhibiting this overpressured phase include the Greater Green River (Law, 1984), Wind River (Johnson et al., 1996), Big Horn (Johnson et al., 1999), and Piceance basins

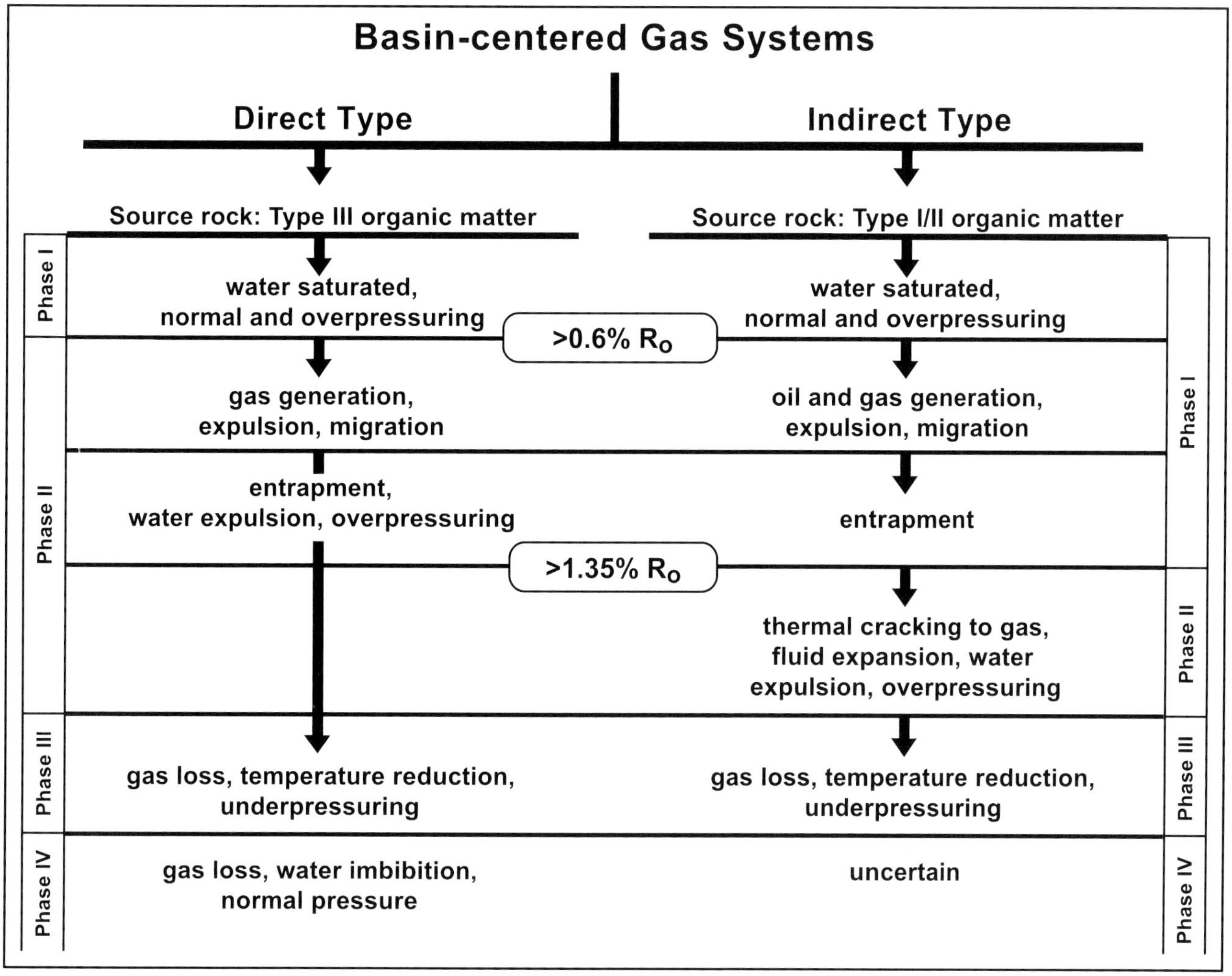

Figure 2. Schematic diagram showing evolution of direct and indirect basin-centered gas systems. Evolutionary phases are shown along the side of each system (modified from Law, 2002).

(Johnson et al., 1987) in the Rocky Mountain region of the United States, as well as the Taranaki basin, New Zealand.

Indirect Systems

In contrast to direct systems, indirect systems have liquid-prone source rocks (Figure 2). Reservoir quality in indirect systems is assumed to be better than in direct systems. In this case, oil and gas are generated and expelled and migrate to reservoirs where they accumulate in structural and stratigraphic traps as discrete, buoyant accumulations with downdip water contacts. With subsequent burial and exposure to higher temperatures, the accumulated oil undergoes thermal cracking to gas, accompanied by a significant increase of fluid volume and pressures (Barker, 1990). The level of thermal maturity at which oil is transformed to gas is commonly thought to be 1.35% R_O; however, there is some evidence, as discussed in a later section, indicating that the transformation may occur at higher levels of thermal maturity. Alternatively, gas derived from thermally cracked oil in a source rock may subsequently be expelled and migrate to low-permeability reservoirs (Garcia-Gonzales et al., 1993a, b; MacGowan et al., 1993; Hunt, 1996). Under these conditions of changing fluid volume and pressure, the capillary pressure of the water-wet pore system is exceeded, and like pore pressures in direct systems, the high pressures forcibly expel mobile, free water from the pore system, replacing water with gas, and the development of an overpressured BCGA ensues. The distinction between a BCGA charged by gas transformed from oil in a reservoir versus gas transformed from oil in a source rock is not readily apparent and may require a geochemical evaluation of the correlation between source rock and gas.

Phase III

If direct and indirect systems reach an overpressured state in phase II, then the processes involved in the transition to phase III are identical for both systems. Phase III of BCGS history occurs when the overpressured phase of direct and indirect systems evolves into underpressured conditions (Figure 2). Both systems, subsequent to the phase II history of overpressure, may experience a period of uplift and erosional unloading and/or heat-flow perturbations. During or subsequent to these burial-history disruptions, some gas is lost from the accumulation by diffusion and gas-phase migration, and the overpressured gas reservoirs are subjected to reduced temperatures. Loss of gas volume, in conjunction with reduced temperatures, will effectively result in the development of an underpressured BCGA (F. F. Meissner, 1978, personal communication; Law and Dickinson, 1985). Meissner (2000) emphasizes the importance of gas loss over temperature reduction as the dominant process during this pressure transition. Some examples of underpressured, phase III direct systems include Cretaceous rocks in San Juan, Raton, and Denver basins, and examples of phase III, underpressured, indirect systems include Lower Silurian reservoirs in the Appalachian basin, Ordovician reservoirs in the Risha area of eastern Jordan, and Cambrian and Ordovician reservoirs in the Ahnet basin of Algeria.

Phase IV

Phase IV (Figure 2) is theoretical and may be more applicable to direct systems than indirect systems because of the perceived, more effective nature of a lithologic seal in indirect systems as opposed to a capillary seal in direct systems. During phase IV, continued gas loss from capillary seals in BCGAs is accompanied by water slowly reentering underpressured, gas-bearing reservoirs. Under these conditions, F. F. Meissner (1978, personal communication) and Law and Dickinson (1985) hypothesized that the underpressured, gas-bearing reservoirs will eventually evolve into normally pressured, water-bearing reservoirs.

SYSTEM ELEMENTS AND PROCESSES

Source Rocks

Source rock composition is the fundamentally most important element distinguishing direct from indirect BCGSs and sets the stage for all subsequent differences between the two systems. The source rocks for direct BCGSs are most commonly humic-type coal beds and carbonaceous shale, such as those that occur in Cretaceous rocks in most Rocky Mountain basins or Carboniferous rocks in Europe. Source rocks for indirect BCGSs are hydrogen-rich shales, such as those in the Ordovician shales in the Appalachian basin or in Silurian shales in the Middle East and North Africa. Garcia-Gonzales et al. (1993a, b), MacGowan et al. (1993), and Surdam (1997) concluded that some of the coal beds in the Greater Green River basin of Wyoming (Upper Cretaceous Almond coal beds) have generated liquid hydrocarbons that were subsequently thermally cracked to gas while still in the coal beds. They further speculated that because of the increased fluid volume associated with the oil-to-gas transformation, high pressures created fractures in the coal beds, facilitating the expulsion of gas. The gas then migrated and accumulated in adjacent, low-permeability reservoirs.

Based on source rock and gas analyses, Law (1984) concluded that all or most of the gas in low-permeability reservoirs in the Greater Green River basin was sourced from humic, type III kerogen contained in coal beds and carbonaceous shale in several coal-bearing Upper Cretaceous intervals. It is important to note that virtually all coal beds, including humic coals, contain some hydrogen-rich macerals capable of generating liquid hydrocarbons. The relative contribution of gas to BCGA reservoirs from these different macerals is uncertain. However, if oil generated from more liquid-prone source rocks was an important contributor, one would expect to observe oil accumulations around the margins of the basin where source rocks were in the oil window. The Greater Green River basin, however, is a gas-prone province with very few oil fields. In the Greater Green River basin, it seems likely that the BCGA gas is dominantly sourced directly from gas-prone, humic coal beds in the Lance, Almond, and Rock Springs formations, with a minor contribution from the cracking of oil to gas in the Almond Formation in the very deepest parts of the Great Divide and Washakie basins, where levels of thermal maturity may be sufficiently high.

Reservoir Rocks

Gas-charged reservoirs in direct and indirect BCGSs are regionally pervasive, commonly encompassing several thousand square miles and may consist of single, isolated reservoirs a few feet thick or vertically stacked reservoirs several thousand feet thick. Multiple, stacked reservoirs are common in direct BCGSs, whereas single, discrete reservoirs are common in indirect BCGSs. Direct and indirect reservoirs always exhibit low porosity (<13%) and low, in-situ, permeability (<0.1 md) (Spencer, 1985, 1989). They are composed of sandstone, siltstone, and, to a much lesser degree, carbonates; the only occurrence of a BCGA carbonate reservoir we know of is in the Sichuan basin of China (Da-jun and Yun-ho, 1994). The environments of deposition of BCGA reservoirs range from marine to nonmarine. Reservoirs are gas saturated, with little or no producible water and are downdip from water-bearing reservoirs, a reversal of conditions found in conventional gas systems (Masters, 1979; Law, 1984; Spencer, 1985, 1989).

BCGS reservoirs can be divided into two groups: lenticular and blanket reservoirs (Finley, 1984; Spencer, 1985, 1989). Lenticular reservoirs, such as small, fluvial-channel sandstones, typically have a limited reservoir volume and very low permeability. In contrast, blanket-like reservoirs, such as braided-stream, delta-front, and eolian sandstones, have very large reservoir volumes and generally exhibit relatively better permeability than lenticular reservoirs. The distinction between these types of reservoirs becomes important when attempting to distinguish between gas- and water-bearing reservoirs and is an important factor in the design of drilling and completion programs.

Interbedded, water-bearing reservoirs are common in thick, vertically stacked, direct BCGA reservoirs. For example, the Upper Cretaceous Ericson Sandstone of the Mesaverde Group in western Wyoming is a water-bearing reservoir interbedded with a regionally pervasive BCGA. Additional examples in western Wyoming of interbedded, water-bearing reservoirs have been observed by various operators in the Upper Cretaceous Frontier and Blair formations, Almond sandstone, and Lewis Shale. The Rollins and Trout Creek members of the Mesaverde Formation in the Piceance basin of Colorado are blanket reservoirs that are water bearing in the deep part of the basin (Johnson et al., 1987). Water-bearing reservoirs in the BCGA in the Elmworth field in the Alberta basin of Canada also exist. The occurrence of water in thick, BCGA sequences is also possible through the introduction of water along fractures and faults.

Seals

Where detailed work has been conducted in direct BCGAs, gas-saturated reservoirs grade vertically as well as updip into water- and gas-bearing intervals that, in turn, grade into normal-pressured, water-bearing reservoirs (Figure 3). In indirect BCGAs, such as the Lower Silurian Clinton-Medina in the Appalachian basin, gas-saturated reservoirs grade updip into transitional, water- and gas-bearing zones (Ryder and Zagorski, 2003); however, vertical transition zones across bed boundaries do not occur, and there is an abrupt, distinct boundary between abnormally pressured, gas-bearing reservoirs and normally pressured, water-bearing reservoirs (Figure 3). The nature of these bounding surfaces is related to seal integrity. Seals in BCGAs range from lithologic, such as those associated with shales or evaporites, to relative permeability or water block seals, referred to here as capillary-pressure seals. Capillary-pressure seals generally occur in reservoirs with very small pore throats and two or more fluid phases present. Under these conditions, the permeability to each fluid phase is effectively reduced. Top seals in indirect BCGA are most commonly lithologic seals, whereas top seals and updip seals in direct BCGAs are most commonly related to capillary-pressure seals. Indirect systems may have dual seals in which the top seal is lithologic and the updip seal is a capillary-pressure seal.

It is generally accepted that gas can diffuse through nearly all seals given sufficient time; evaporite seals may be the exception. Studies of the origin and history of gas generation, overpressure, and pressure decay attest to the ephemeral nature of gas in reservoirs (F. F. Meissner, 1978, personal communication; Law and Dickinson, 1985). If the perception of a leaky seal over significantly long periods of time is correct, then one might expect to see a predominance of BCGAs in rocks that have experienced the formation of a BCGA within a few tens of million years. In a more general sense, one would expect to observe a higher frequency of BCGAs in younger rocks than in older rocks. Observational data are skewed toward Cretaceous systems, largely because most of the work conducted on BCGAs has been in Cretaceous and Tertiary rocks.

Hydrocarbon Generation, Expulsion, and Migration

A large body of literature concerning hydrocarbon generation, expulsion, and migration (e.g., Hunt, 1996) is available. The following discussion is of a general nature because of the variability of source rock quality and kinetics. As depicted in Figure 2, in both direct and indirect BCGSs, the generation of hydrocarbons from source rocks occurs at levels of thermal maturity exceeding 0.6% vitrinite reflectance (Hunt, 1996). According to Meissner (1984), thermal generation of gas from humic coal beds begins at a vitrinite reflectance of 0.73%. Peak generation may occur at levels of thermal maturity between 0.8 and 0.9% R_o (Tissot and Welte, 1984). Measured levels of thermal maturity at the top of direct BCGAs provide some insight into the question of peak gas generation. In the

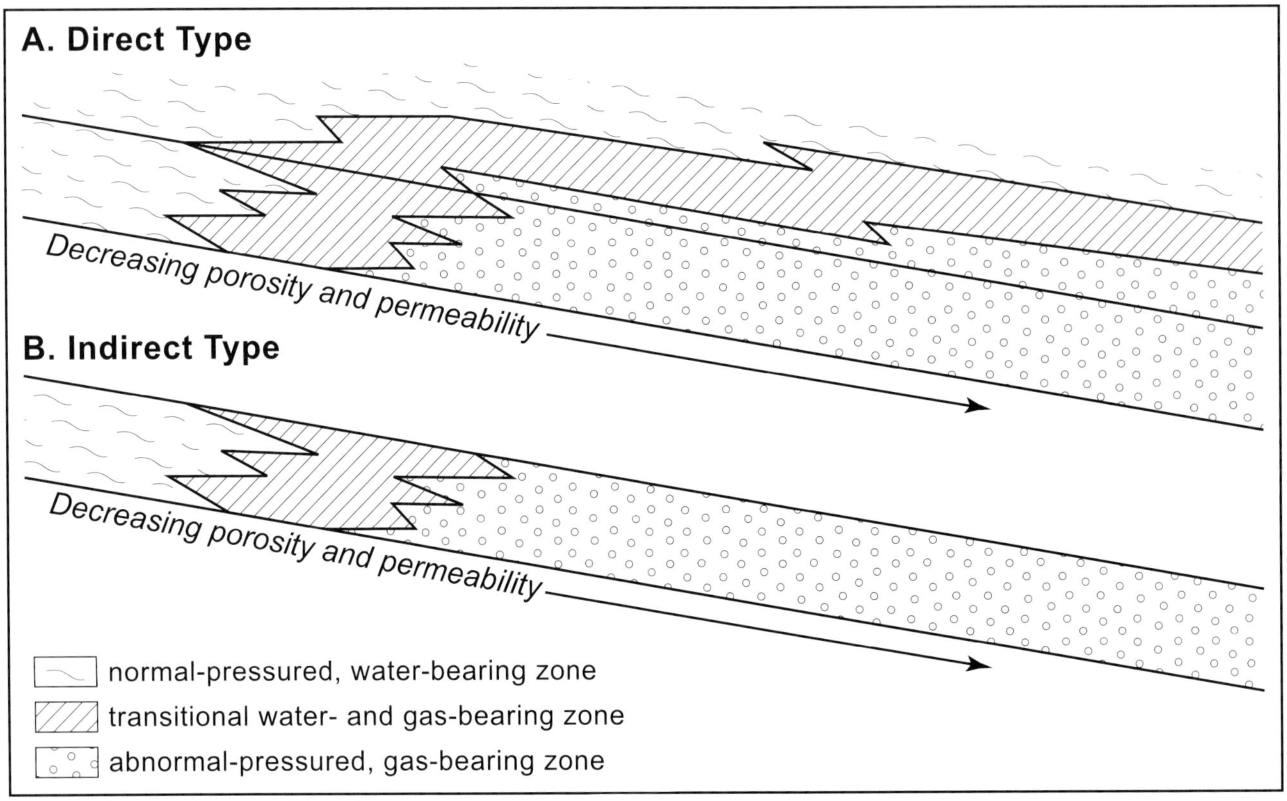

Figure 3. Diagrammatic illustrations showing normal-pressured and water-bearing zones, water- and gas-bearing transitional zones, and abnormally pressured and gas-bearing zones for direct (A) and indirect (B) BCGA (modified from Law, 2002).

Greater Green River basin, measured levels of thermal maturity at the top of direct BCGAs range from 0.7 to 0.9% R_o (Law, 1984). This would suggest that the source beds for the gas must have levels of thermal maturity equal to or greater than 0.7–0.9% R_o.

The level of thermal maturity marking the transformation of oil to gas in indirect systems (initiation of phase II in Figure 2) is uncertain. Conventional wisdom indicates that the thermal cracking of oil to gas occurs at a vitrinite reflectance of about 1.35%. Price (1997) questioned this value and concluded that the transformation of oil to gas occurred at much higher levels of thermal maturity. More recent kinetic studies by Tsuzuki et al. (1999) using hydrous-pyrolysis experiments also suggest that oil is stable over higher levels of thermal maturity than previously thought. Applying these kinetic parameters to burial history curves in the United States Gulf Coast indicate that oil cracking to gas starts at vitrinite reflectance values of 1.75% (M. D. Lewan, 2002, personal communication). Petersen (2002), while conducting hydrous-pyrolysis experiments with type III kerogens, concluded that thermal cracking of oil to gas is coincident with a vitrinite reflectance of 1.8%.

In general, hydrocarbon migration distances in direct BCGSs are short, perhaps on the order of a few hundred feet or less. The exception to short hydrocarbon migration distances may occur in cases where the regional top of a BCGA has been locally modified by faulting and fracturing. These faults and associated fractures facilitate vertical migration of gas along faults and fractures for distances far greater than a few hundred feet, such as in the Jonah field of western Wyoming, which is discussed in a later section.

In indirect BCGSs, hydrocarbon migration distances may be highly variable, similar to migration distances in conventional gas systems. Approximately 1000 ft (300 m) of vertical migration is proposed for the Clinton-Medina BCGA (Ryder and Zagorski, 2003). In direct BCGSs, gas is the dominant migrating hydrocarbon phase, and in indirect BCGSs, both oil and/or gas may be expected to be the migrating fluid phases.

Trap Formation

The development of a trap in the conventional sense of a structural or stratigraphic trap is an important process in a petroleum system. In a direct BCGS system, however, it is of secondary importance, whereas in indirect systems, it is very important, because indirect accumulations are formed from the thermal destruction of previously accumulated oil. In direct systems, the tops of gas accumulations cut across structural and stratigraphic boundaries (Law, 1984; Law and Spencer, 1993; Spencer, 1985) and are therefore not normally dependent on the development of structural or stratigraphic traps. An exception to this occurs in the Jonah field of western Wyoming, which is discussed in a later section. Indirect BCGAs, however, require the presence of a structural or stratigraphic trap during phase II of development.

In the development of indirect BCGAs, a conventional structural or stratigraphic trap is necessary for the accumulation of oil and gas, much the same way as oil and gas accumulate in conventional, buoyancy-driven accumulations. The development of indirect BCGAs occurs at a later burial stage than direct systems, when conventionally accumulated oil is thermally cracked to gas, accompanied by a significant increase in pore-fluid volume and pore pressure. Oil, however, does not always accumulate in discrete accumulations and may be disseminated throughout a reservoir. However, the amount of oil present in a disseminated accumulation is probably insufficient to develop pore pressures high enough to form a BCGA during the thermal conversion of oil to gas. Thus, the formation of a suitable trap and the temporal relationships among trap formation, gas generation, expulsion, migration, and entrapment are critical processes in indirect systems.

CLASSIFICATION OF THE JONAH FIELD

Entrapment of gas in the Jonah field (Figures 1, 4) has been variously referred to as a structural and/or stratigraphic trap (Montgomery and Robinson, 1997; Warner, 2000; Hanson et al., 2004). Although a discussion of the origin of the field is beyond the scope of this report, Bowker and Robinson (1997), Warner (1998, 2000), and Law (2002) have indicated that the gas in the field migrated upward along faults and fractures into the productive Lance Formation reservoirs. Alternatively, Cluff and Cluff (2001) have interpreted the Jonah field to be a remnant of a larger, much shallower BCGA than presently identified. The field, in our opinion, is a fault-bounded, gas chimney rooted in a regionally pervasive BCGA. We interpret the field to have formed as a consequence of structural breaching of the regionally pervasive BCGA in the Greater Green River basin. The resulting faults and fractures then facilitated the vertical migration of gas into stratigraphically higher positions (more than 2500 ft [760 m]) than adjacent rocks (Montgomery and Robinson, 1997; Warner, 2000).

The classification of the Jonah field as a direct or indirect type of BCGA largely depends on the nature of the source rock. Unfortunately, there is very little detailed source rock data available for the Jonah field. Therefore, we have used the gas compositional data reported by Hanson et al. (2004) and source rock and gas compositional data from the nearby Pinedale anticline (Law, 1984). Hanson et al. (2004) concluded that coal beds and carbonaceous shales in the lower part of the Lance Formation and upper part of the Mesaverde Group are the principal source rocks. They further state that these source rocks are composed of humic, type III kerogen, a conclusion similar to that of Law (1984) for the entire Greater Green River basin BCGA. Hanson et al. (2004) further concluded that 90% of the gas is derived from the thermal cracking of oil. Source rock data from two wells on the nearby Pinedale anticline (Law, 1984) do not support that conclusion.

In the Pinedale area, the composition of organic matter in the Lance and Mesaverde is composed predominantly of vitrinite and inertinite macerals (>75%), having hydrogen indices (HI) less than 120 mg of hydrocarbon per gram of carbon (mg HC/g C) (Law, 1984). Although those samples with HI values less than 120 mg HC/g C are from rocks with vitrinite reflectance values greater than 0.7% and do not reflect the original HI values in those rocks, they are, in our opinion, indicative of kerogen with limited oil-generative capability. In addition, the observed levels of thermal maturity in the Lance and Mesaverde in the Jonah field, as shown by vitrinite reflectance profiles from two wells (Warner, 1998), are less than 1.0% R_o, values far too low to have transformed oil to gas. Therefore, given that gas in the Jonah field is most likely from humic, gas-prone, type III kerogen and may contain some gas from thermally cracked oil, we believe the Jonah field is best interpreted as a direct type of BCGA.

SUMMARY

Two types of BCGSs are recognized: a direct type, distinguished by having a gas-prone source rock, and an indirect type, having a liquid-prone source rock. Although the two systems have several similar attributes, the fundamental differences between the systems, gas-prone versus liquid-prone source rocks, result in some dissimilar attributes.

Based on the characteristics of direct and indirect types of BCGAs, the Jonah field is interpreted to be a direct type.

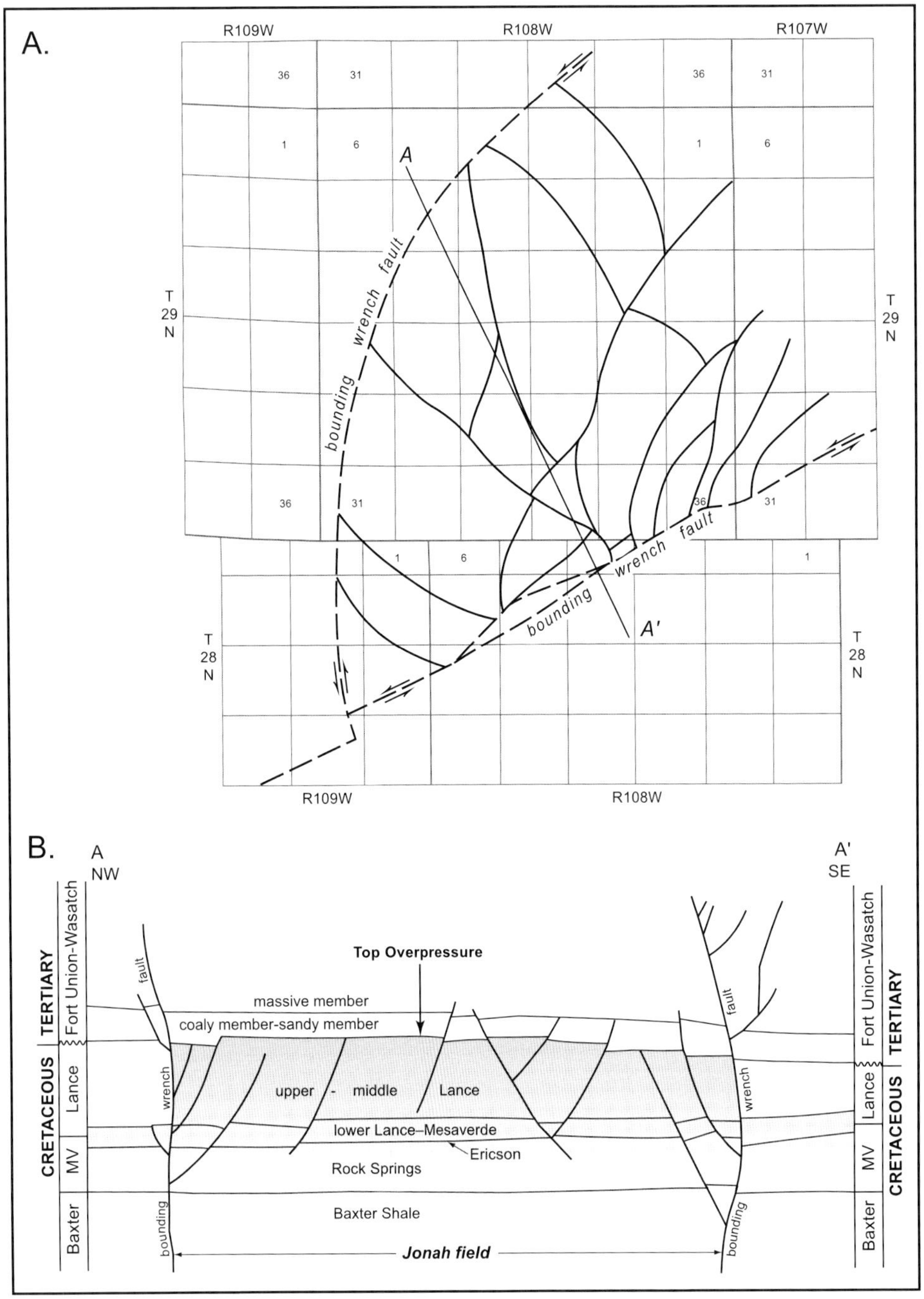

Figure 4. Map of Jonah field (A) showing location of major faults and cross section AA′ (B) through Jonah field. Shaded areas represent overpressured, gas-saturated reservoirs. Relief on top of overpressured reservoirs outside to inside the field ranges from 2500 to 3000 ft (760 to 910 m). Location of Jonah field shown in Figure 1. Figure was modified from Warner (1998). MV = Mesaverde.

The determination of the specific type of BCGA is strictly based on the nature of the dominant source rock. Available data indicate that the most likely source rocks are coal and carbonaceous shale in the lower Lance Formation and the upper part of the Mesaverde Group, and these source rocks are composed of a humic, gas-prone, type III kerogen. Some uncertainty remains, however, regarding the proportions of gas contributed directly from the source rocks versus the amount of gas that may have originated as a consequence of thermal cracking of oil generated from the small amounts of hydrogen-rich components associated with the humic organic matter.

ACKNOWLEDGMENTS

This chapter draws on the publications of many workers and organizations. We were employed by the U.S.

Geological Survey (USGS) in 1977 when the U.S. Department of Energy (DOE) requested that the USGS design a program to study the then poorly known characteristics and resources in tight-gas sands. We thank the DOE for supporting the subsequent USGS program. Without this support, the USGS research would probably have been done on a very small scale. DOE managers assisting the research include Keith Westhusing, Karl Frohne, and William Gwilliam.

Many USGS researchers contributed to the program, including R. C. Johnson, T. M. Finn, D. D. Rice, T. D. Fouch, V. F. Nuccio, C. W. Keighin, J. Pittman, R. M. Pollastro, J. L Ridgley, C. J. Wandrey, and R. T. Ryder. Special thanks are extended to W. J. Barrett, F. J. Barrett, Terry Barrett, C. A. Brown, W. B. Hanson, Greg Anderson, D. P. Battin, J. W. Robinson, and Ira Pasternak for their help in obtaining subsurface data and useful discussions.

Reviews by Barry C. McBride, Mark A. Kirschbaum, and an anonymous reviewer provided many helpful suggestions and corrections.

REFERENCES CITED

Barker, C., 1990, Calculated volume and pressure change during the thermal cracking of oil to gas in reservoirs: AAPG Bulletin, v. 74, p. 1254–1261.

Bowker, K. A., and J. W. Robinson, 1997, Jonah field: A shallow sweetspot in the basin-centered gas accumulation of the northern Green River basin, Wyoming: AAPG Bulletin, v. 81, p. 1219.

Brown, C. A., T. M. Smagala, and G. R. Haefele, 1986, Southern Piceance basin model— Cozette, Corcoran and Rollins sandstones, *in* C. W. Spencer and R. F. Mast, eds. Geology of tight gas reservoirs: AAPG Studies in Geology 24, p. 207–219.

Cluff, R. M., and S. G. Cluff, 2001, Overpressure determination from sonic and resistivity log anomalies, Jonah field, northern Green River basin, Wyoming, *in* J. W. Robinson and K. W. Shanley, eds., Tight gas fluvial reservoirs: A case history from the Lance Formation, Green River basin, Wyoming: AAPG Annual Meeting Short Course Notes 2.

Da-jun, P., and L. Yun-ho, 1994, Genetic mechanism of abnormal pressure, pressure seals and natural gas accumulation in carbonate reservoirs, Sichuan Basin, *in* Abnormal pressures in hydrocarbon environments: Proceedings, AAPG Research Conference, June 8–10, 1994, Golden, Colorado, unpaginated.

Dutton, S. P., S. J. Clift, D. S. Hamilton, H. S. Hamlin, T. F. Hentz, W. E. Howard, M. S. Akhter, and S. E. Laubach, 1993, Major low-permeability sandstone reservoirs in the continental United States: Texas Bureau of Economic Geology Report of Investigations 211, 221 p.

Dyman, T. S., D. D. Rice, and P. A. Westcott, eds., 1997, Geologic controls of deep natural gas resources in the United States: U.S. Geological Survey Bulletin, v. 2146, 239 p.

Finley, R. J., 1984, Geology and engineering characteristics of selected low-permeability gas sandstones: A national survey: Texas Bureau of Economic Geology Report of Investigations 138, 220 p.

Garcia-Gonzales, M., D. B. MacGowan, and R. C. Surdam, 1993a, Coal as a source rock of petroleum and gas— A comparison between natural and artificial maturation of the Almond Formation coals, Greater Green River basin in Wyoming, *in* D. G. Howell, ed., The future of energy gases: U.S. Geological Survey Professional Paper 1570, p. 405–437.

Garcia-Gonzales, M., D. B. MacGowan, and R. C. Surdam, 1993b, Mechanisms of petroleum generation from coal, as evidenced from petrographic and geochemical studies: Examples from Almond Formation coals in the Greater Green River basin, *in* B. Strook and S. Andrew, eds., Wyoming Geological Association Field Conference Guidebook, p. 311–323.

Hanson, W. B., V. Vega, and D. Cox, 2004, Structural geology, seismic imaging, and genesis of the giant Jonah gas field, Wyoming, U.S.A., *in* J. W. Robinson and K. W. Shanley, eds., Jonah field: Case study of a tight-gas fluvial reservoir: AAPG Studies in Geology 52, p. 61–92.

Hunt, J. M., 1996, Petroleum geochemistry and geology, 2d ed.: New York, W.H. Freeman and Co., 743 p.

Johnson, R. C., R. A. Crovelli, C. W. Spencer, and R. F. Mast, 1987, An assessment of gas resources in low-permeability sandstones of the Upper Cretaceous Mesaverde Group, Piceance basin, Colorado: U.S. Geological Survey Open-file Report 87-357, 165 p.

Johnson, R. C., T. M. Finn, R. A. Crovelli, and R. H. Balay, 1996, An assessment of in-place gas resources in low-permeability Upper Cretaceous and lower Tertiary sandstone reservoirs, Wind River basin, Wyoming: U.S. Geological Survey Open-file Report 96-264, 67 p.

Johnson, R. C., R. A. Crovelli, B. G. Lowell, and T. M. Finn, 1999, An assessment of in-place gas resources in the low-permeability basin-centered gas accumulation of the Big Horn basin, Wyoming and Montana: U.S. Geological Survey Open-file Report 99-315A, 123 p.

Law, B. E., 1984, Relationships of source rocks, thermal maturity, and overpressuring to gas generation and occurrence in low-permeability Upper Cretaceous and lower Tertiary rocks, Greater Green River basin, Wyoming, Colorado, and Utah, *in* J. Woodward, F. F. Meissner, and J. L. Clayton, eds., Hydrocarbon source rocks of the greater Rocky Mountain region: Rocky Mountain Association of Geologists Guidebook, p. 469–490.

Law, B. E., 2002, Basin-centered gas systems, *in* B. E. Law and J. B. Curtis, eds., Unconventional petroleum systems: AAPG Bulletin, v. 86, p. 1891–1919.

Law, B. E., and W. W. Dickinson, 1985, A conceptual model for the origin of abnormally pressured gas accumulations in low-permeability reservoirs: AAPG Bulletin, v. 69, p. 1295–1304.

Law, B. E., and C. W. Spencer, 1993, Gas in tight reservoirs— An emerging source of energy, *in* D. G. Howell, ed., The future of energy gases: U.S. Geological Survey Professional Paper 1570, p. 233–252.

Law, B. E., and C. W. Spencer, 1998, Abnormal pressure in hydrocarbon environments, *in* B. E. Law, G. F. Ulmishek, and V. I. Slavin, eds., Abnormal pressures in hydrocarbon environments: AAPG Memoir 70, p. 1–11.

Law, B. E., C. W. Spencer, and N. H. Bostick, 1979, Preliminary results of organic maturation, temperature, and pressure studies in the Pacific Creek area, Sublette County, Wyoming, *in* 5th Department of Energy Symposium on Enhanced Oil and Gas Recovery and Improved Drilling Methods— Oil and gas recovery: Tulsa, Oklahoma, Petroleum Publishing Co., v. 3, p. K2/1–K2/13.

Law, B. E., C. W. Spencer, and N. H. Bostick, 1980, Evaluation of organic maturation, subsurface temperature, and pressure with regard to gas generation in low-permeability Upper Cretaceous and lower Tertiary strata in the Pacific Creek area, Sublette County, Wyoming: The Mountain Geologist, v. 17, no. 2, p. 23–35.

MacGowan, D. B., M. Garcia-Gonzales, D. R. Britton, and R. C.

Surdam, 1993, Timing of hydrocarbon generation, organic-inorganic diagenesis, and the formation of abnormally pressured gas compartments in the Cretaceous of the Greater Green River basin: A geochemical model, *in* S. Strook and S. Andrew, eds., Wyoming Geological Association Field Conference Guidebook, p. 325–357.

Magoon, L. B., and W. G. Dow, eds., 1994, The petroleum system— From source to trap: AAPG Memoir 60, 655 p.

Masters, J. A., 1979, Deep basin gas trap, western Canada: AAPG Bulletin, v. 63, p. 152–181.

Masters, J. A., ed., 1984, Elmworth— Case study of a deep basin gas field: AAPG Memoir 38, 316 p.

McPeek, L. A., 1981, Eastern Green River basin— A developing giant gas supply from deep, overpressured Upper Cretaceous sandstones: AAPG Bulletin, v. 65, p. 1078–1098.

Meissner, F. F., 1984, Cretaceous and lower Tertiary coals as sources for gas accumulations in the Rocky Mountain area, *in* J. Woodward, F. F. Meissner, and J. L. Clayton, eds., Hydrocarbon source rocks of the greater Rocky Mountain region: Rocky Mountain Association of Geologists Guidebook, p. 401–431.

Meissner, F. F., 2000, Causes of anomalous deep basin fluid pressures in Rocky Mountain basins and their relation to regional gas accumulation, *in* Basin-center Gas Symposium: Rocky Mountain Association of Geologists Short Course Notes, 11 p.

Montgomery, S. L., and J. W. Robinson, 1997, Jonah field, Sublette County, Wyoming: Gas production from overpressured Upper Cretaceous Lance sandstones of the Green River basin: AAPG Bulletin, v. 81, p. 1049–1062.

Petersen, H. I., 2002, A re-consideration of the "oil window" for humic coal and kerogen type III source rocks: Journal of Petroleum Geology, v. 25, no. 4, p. 407–432.

Price, L. C., 1997, Minimum thermal stability levels and controlling parameters of methane, as determined by C_{15}+ hydrocarbon thermal stabilities, *in* T. S. Dyman, D. D. Rice, and P. A. Westcott, eds., Geologic controls of deep natural gas resources in the United States: U.S. Geological Survey Bulletin, v. 2146, p. 139–176.

Randolph, P. L., 1973, Nuclear stimulation of gas fields: Canadian Gas Association National Technical Conference, Calgary, Alberta, Canada, October 17–19, 1973, 21 p.

Randolph, P. L., 1974a, Massive stimulation effort may double U.S. gas supply: Part 1: World Oil, p. 33–36.

Randolph, P. L., 1974b, Massive stimulation effort may double U.S. gas supply: Part 2: World Oil, p. 65–71.

Randolph, P. L., 1974c, Massive stimulation effort may double U.S. gas supply: Part 3: World Oil, p. 131–134.

Rose, P. R., J. R. Everett, and I. S. Merin, 1986, Potential basin-centered gas accumulation in Cretaceous Trinidad Sandstone, Raton basin, Colorado, *in* C. W. Spencer and R. F. Mast, eds., Geology of tight gas reservoirs: AAPG Studies in Geology 24, p. 111–128.

Ryder, R. T., and W. A. Zagorski, 2003, Nature, origin, and production characteristics of the Lower Silurian regional oil and gas accumulation, central Appalachian basin, United States: AAPG Bulletin, v. 87, p. 847–872.

Schmoker, J. W., 1995, Method for assessing continuous-type (unconventional) hydrocarbon accumulations, *in* D. L. Gautier, K. I. Takahashi, and K. L. Varnes, eds., 1995 National assessment of United States oil and gas resources— Results, methodology, and supporting data: U.S. Geological Survey Digital Data Series 30 (CD-ROM).

Silver, C., 1950, The occurrence of gas in the Cretaceous rocks of the San Juan basin, New Mexico and Colorado: New Mexico Geological Society First Field Conference, San Juan basin, p. 109–123.

Spencer, C. W., 1985, Geologic aspects of tight gas reservoirs in the Rocky Mountain region: Journal of Petroleum Technology, July issue, p. 1308–1314.

Spencer, C. W., 1989, Review of characteristics of low-permeability gas reservoirs in western United States: AAPG Bulletin, v. 73, p. 613–629.

Spencer, C. W., and R. F. Mast, 1986, eds., Geology of tight gas reservoirs: AAPG Studies in Geology 24, 299 p.

Spencer, C. W., A. Szalay, and E. Tatar, 1994, Abnormal pressure and hydrocarbon migration in the Bekes basin, *in* P. G. Teleki, R. E. Mattick, and J. Kokai, eds., Basin analysis for oil and gas exploration: A case study from Hungary: Netherlands, Kluwer Academic Publishers, p. 201–219.

Surdam, R. C., 1997, A new paradigm for gas exploration in anomalously pressured "tight gas sands" in the Rocky Mountain Laramide basins, *in* R. C. Surdam, ed., Seals, traps, and the petroleum system: AAPG Memoir 67, p. 283–298.

Tissot, B. P., and D. H. Welte, 1984, Petroleum formation and occurrence, 2d revised ed.: Berlin, Springer-Verlag, 699 p.

Tsuzuki, N., N. Takeda, M. Suzuki, and K. Yokoi, 1999, The kinetic modeling of oil cracking by hydrothermal pyrolysis experiments: International Journal of Coal Geology, v. 39, p. 227–250.

Warner, E. M., 1998, Structural geology and pressure compartmentalization of Jonah field, Sublette County, Wyoming, *in* R. M. Slatt, ed., Compartmentalized reservoirs in Rocky Mountain basins: Rocky Mountain Association of Geologists Guidebook, p. 29–46.

Warner, E. M., 2000, Structural geology and pressure compartmentalization of Jonah field based on 3-D seismic data and subsurface geology, Sublette County, Wyoming: The Mountain Geologist, v. 37, p. 15–30.

CHAPTER 10

Fluvial Reservoir Description for a Giant, Low-permeability Gas Field: Jonah Field, Green River Basin, Wyoming, U.S.A.

KEITH W. SHANLEY[1]
BP America/Upstream Technology Group

ABSTRACT

The Lance Formation at Jonah field comprises a thick succession, locally more than 3500 ft (1100 m), of low-permeability, fine-grained alluvial sandstones. These alluvial sandstones were deposited by modest rivers in a rapidly subsiding basin in the northern portion of the Greater Green River basin of southwest Wyoming. Stacking of these alluvial sand bodies has produced an extremely heterogeneous reservoir in which reservoir sand bodies are from 9 to 15 ft thick (3 to 5 m) and approximately 200–700 ft wide (60–210 m P_{50} values). Because of the low net/gross of the overall Lance Formation (10–35% overall, 40–80% locally) and the comparatively small sand bodies, it is most likely that wells drilled on 40-ac (0.16-km^2) density (1320 ft [402 m] closest spacing between wells) will result in more than 75% reservoir additions and less than 25% rate acceleration. It is also quite likely that wells drilled on increased density, less than 40 ac (0.16 km^2), will result in significant reserve additions as well.

Because the Lance Formation at Jonah field is dominated by intercalated, relatively thin alluvial sand bodies and alluvial-plain deposits, much of the reservoir-bearing interval is at or below seismic tuning. As a result, despite the presence of a high-quality, three-dimensional seismic survey, the reservoir cannot be consistently imaged or described from seismic data. In this chapter, a description of the reservoir at Jonah field is developed based on a sedimentological description of the available cores and the integration of well-log data. These data are then used to develop quantitative estimates of the sizes of fluvial systems that resulted in the sand bodies found in Jonah field as well as estimates of the lateral extents of the sand bodies themselves. Traditional techniques for estimating sand body geometry are compared with a probabilistic approach based on numerous analog sand body studies. For subsurface decision making, a probabilistic approach more accurately captures the range of likely outcomes than the more traditional approach of attempting to derive single-point estimates of fluvial dimensions.

GENERAL BACKGROUND AND FIELD OVERVIEW

Jonah field is a fault-bounded, structural trap located in the northwestern part of the Green River basin, Sublette County, Wyoming (e.g., DuBois et al., 2004; Hanson et al., 2004). The field produces gas from overpressured, lenticular, fluvial deposits of the Late Cretaceous (Maastrichtian) Lance Formation (Figures 1, 2). As of July 2003, Jonah field has produced approximately 848.6 bcf gas, 8.6 million bbl oil, and 2.6 million bbl water from approximately 504 completed wells; daily production is approximately 697 mmcf gas, 6.0 thousand bbl, and 2.4 thousand bbl water (July 2003 production data reported to Wyoming Oil and Gas Conservation Commission). The Lance Formation reservoir interval may contain more than 100 bcf gas per section (640 ac; 2.60 km^2), with total field gas-in-place estimates ranging to as much as 8.3 tcf gas (DuBois et al., 2004). Total recoverable reserves from the Lance Formation may well exceed 2.5 tcf gas. In addition to gas resources in the Lance Formation, additional resource

[1]*Present address*: Consultant, Littleton, Colorado, U.S.A.

Jonah Field: Case Study of a Giant Tight-Gas Fluvial Reservoir, J. W. Robinson and K. W. Shanley, eds.: AAPG Studies in Geology 52 and Rocky Mountain Association of Geologists 2004 Guidebook

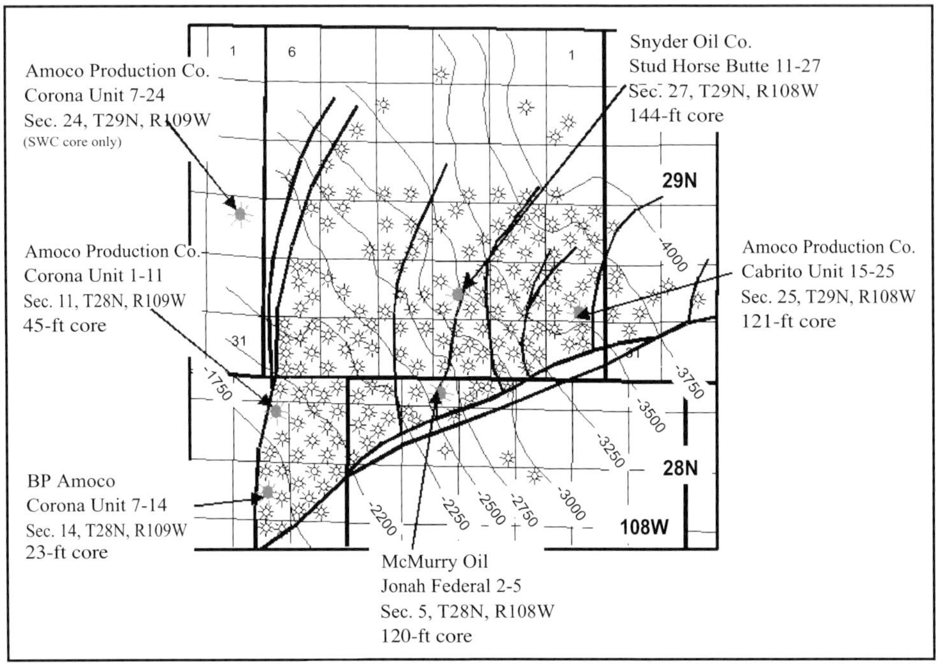

Figure 1. Generalized structure map of the Jonah field area, Sublette County, Wyoming, illustrating the locations of cored wells discussed in this chapter.

potential may also exist in the middle Cretaceous (Turonian) Frontier Formation, the Late Cretaceous (Campanian) Ericson and Rock Springs formations, and the Paleocene Fort Union Formation. Overpressuring in the field ranges between 0.55 and 0.59 psi/ft (0.012 and 0.013 MPa/m) and may range as much as 0.70 psi/ft (0.016 MPa/m) as suggested by some stimulation data. At the time this book was being compiled (Fall 2003), field spacing allowed for two wells per 80-ac (0.32-km^2) spacing unit, effectively allowing operators to drill wells on a 40-ac (0.16-km^2) well spacing (1320 ft [402 m] between closest wells). This phase of field development is nearing completion, and pilot projects designed to explore the feasibility of 20-ac (0.08-km^2) spacing are underway by both EnCana and BP America, the two dominant operators in Jonah field.

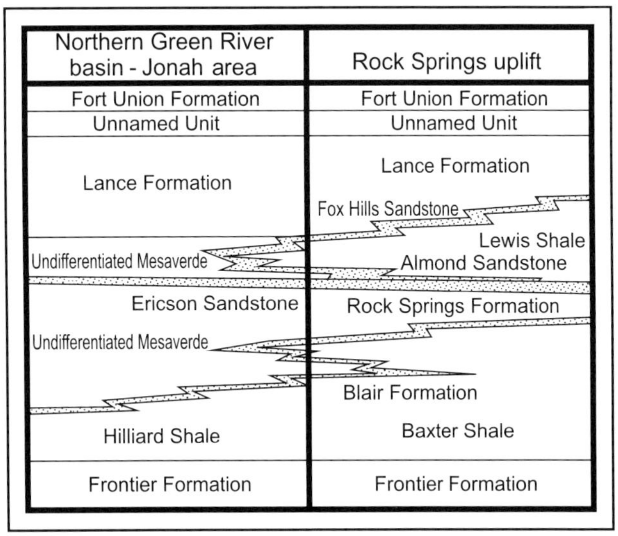

Figure 2. Regional-stratigraphic column for the Greater Green River basin, Wyoming (modified from Law, 1984).

Circumstances related to the discovery and development of Jonah field have been previously described by Robinson (2000), Robinson et al. (1995, 1996), Montgomery (1997), and Montgomery and Robinson (1997) and are summarized in Robinson (2004). These aspects of the field history are not dealt with any further in this chapter. This chapter results from an in-house project performed for Amoco Production Company and its successor BP America. The project was undertaken in late 1995 and early 1996 to better understand the general nature of the reservoir at Jonah field. As drilling matured, the project was extended to examine reservoir heterogeneity, sand body geometry, and connectivity issues in support of optimal well-spacing decisions (e.g., Shanley, 2001).

Stratigraphic Overview

The primary reservoir at Jonah field lies within overpressured Late Cretaceous (Campanian to Maastrichtian) fluvial sandstones of the Lance Formation that lie at depths ranging between 7800 and 12,000 ft (2400 and 3700 m). The Lance Formation is broadly regarded as facies equivalent to progradational, nearshore deposits to the east that are assigned to the Fox Hills Formation (Figure 2). In the vicinity of Jonah field, the Lance Formation exceeds 3500 ft (1100 m) in thickness (Cluff and Murphy, 2004); however, productive intervals (net pay) generally range between 100 and 750 ft (30 and 230 m) per well (Figure 3). Despite the internal complexities in the Lance Formation, several attempts have been made by field operators to recognize a more detailed internal stratigraphy. An early effort to subdivide the Lance Formation primarily on the basis of net/gross trends resulted in approximately eight informal subdivisions (top Lance, upper Lance, Lance B, Lance B2, middle Lance, Jonah, Yellow Point, and Wardell; see Figure 3). More recently, however, increasing well control has led to the realization that net/gross trends recognized on more sparsely spaced data are, in fact, difficult to reconcile on more closely spaced data. For some operators, this has resulted in substantially fewer stratigraphic subdivisions as the field matured (e.g., DuBois et al., 2004).

Compositional data from the Lance Formation are sparse; however, samples from Jonah field are similar to more extensive analyses conducted in the Pinedale anticline to the north (Law and Spencer, 1989). These data suggest that sandstones in the Lance Formation are quartz- and rock fragment-rich, and feldspar-poor lithic arenites; quartz composition ranges from 50 to 60%, rock fragments range between 40 and 50%, feldspar comprises between 0 and 5%, and clay comprises between 5 and 8%. These compositional data are consistent with the interpretation of a sediment source to the west (possibly the Pennsylvanian Phosphoria Formation) in the fold and thrust belt and regional west-to-east sediment transport. Previous interpretations of depositional environments suggest alternating braided and meandering rivers to be responsible for the sandstone reservoirs, with the degree of channel sand body stacking reflecting changes in sediment supply and local accommodation. In general, net/gross ranges from 10 to 35%, with locally 40–80% net/gross in the Jonah subinterval of the Lance Formation.

RESERVOIR PROPERTIES

Data Sources

Despite the large number of wells in Jonah field, at the time of this study, there were relatively few conventional cores with which to calibrate the reservoir description. It was therefore necessary to include some cores taken on the nearby Pinedale anticline located to the north-northeast to supplement the core data within Jonah field proper. Since this reservoir description was performed, additional cores have been taken in both Jonah field and in nearby portions of the Pinedale anticline, which have been made available to the author. These additional sources of data are not presented in this chapter; however, the features described in this reservoir description are consistent with this additional core information. The reservoir description presented here is based on examination of slightly less than 1100 ft (340 m) of core taken from six different wells. The cored wells include (Figure 1)

(1) McMurry Oil, Jonah Federal 2-5 (SE Sec. 5, T28N, R108W), 120-ft (37 m) core
(2) Amoco Production Co., Corona Unit 1-11 (NE Sec. 11, T28N, R109W), 45-ft (14-m) core
(3) BP Amoco, Corona Unit 7-14 (NE Sec. 14, T28N, R109W), 23-ft (7-m) core
(4) Amoco Production Co., Cabrito Unit 15-25 (SE Sec. 25, T29N, R108W), 121-ft (37-m) core
(5) Snyder Oil Co., Stud Horse Butte 11-27 (SW Sec. 27, T29N, R108W), 144-ft (44-m) core
(6) El Paso Natural Gas, Wagon Wheel 1 (SE Sec. 5, T30N, R108W), 607-ft (185-m) core

In addition to the conventional cores described above, porosity, permeability, and textural data were available from rotary sidewall cores taken in the Corona 7-24 (NE Sec. 24, T29N, R109W). With the exception of the El Paso Natural Gas, Wagon Wheel 1 well (Sec. 5, T30N, R108W), all the above-listed wells are within the immediate Jonah field area. The Wagon Wheel core is significant in terms of the sheer amount of core that is available from a single well and the perspective it provides for the Pinedale anticline area. Core descriptions for the Cabrito 15-25, Stud Horse Butte 11-27, and the Wagon Wheel 1 are included as Appendices to this chapter.[a]

Reservoir Property Overview

Reservoir properties for the Lance Formation at Jonah field are more thoroughly discussed by Yin et al. (2001) and Cluff and Cluff (2004). The material presented here is only intended to serve as a broad overview of the Lance Formation at Jonah field. Core porosity in the Lance Formation ranges from 2 to 14%, with the reservoir rock porosity ranging from approximately 8 to 14%. Permeability in the Lance Formation ranges from less than 1 to

[a] Appendices 1 through 7, referenced throughout this chapter, can be found in Appendix D on the CD-ROM accompanying this book.

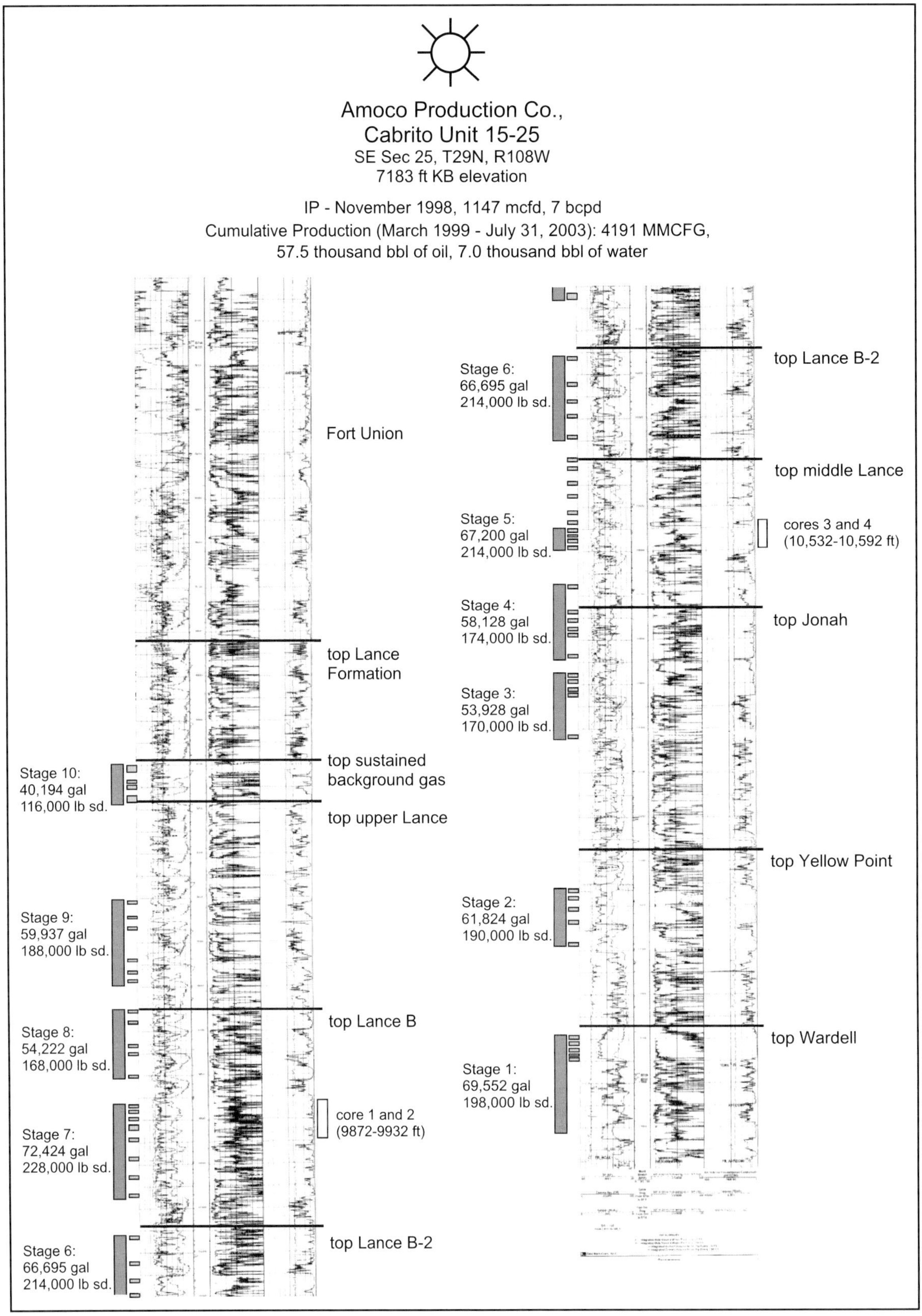
Amoco Production Co.,
Cabrito Unit 15-25
SE Sec 25, T29N, R108W
7183 ft KB elevation
IP - November 1998, 1147 mcfd, 7 bcpd
Cumulative Production (March 1999 - July 31, 2003): 4191 MMCFG,
57.5 thousand bbl of oil, 7.0 thousand bbl of water
Fort Union
top Lance Formation
top sustained background gas
Stage 10: 40,194 gal 116,000 lb sd.
top upper Lance
Stage 9: 59,937 gal 188,000 lb sd.
top Lance B
Stage 8: 54,222 gal 168,000 lb sd.
Stage 7: 72,424 gal 228,000 lb sd.
core 1 and 2 (9872-9932 ft)
top Lance B-2
Stage 6: 66,695 gal 214,000 lb sd.
Stage 6: 66,695 gal 214,000 lb sd.
top Lance B-2
top middle Lance
Stage 5: 67,200 gal 214,000 lb sd.
cores 3 and 4 (10,532-10,592 ft)
Stage 4: 58,128 gal 174,000 lb sd.
top Jonah
Stage 3: 53,928 gal 170,000 lb sd.
top Yellow Point
Stage 2: 61,824 gal 190,000 lb sd.
top Wardell
Stage 1: 69,552 gal 198,000 lb sd.

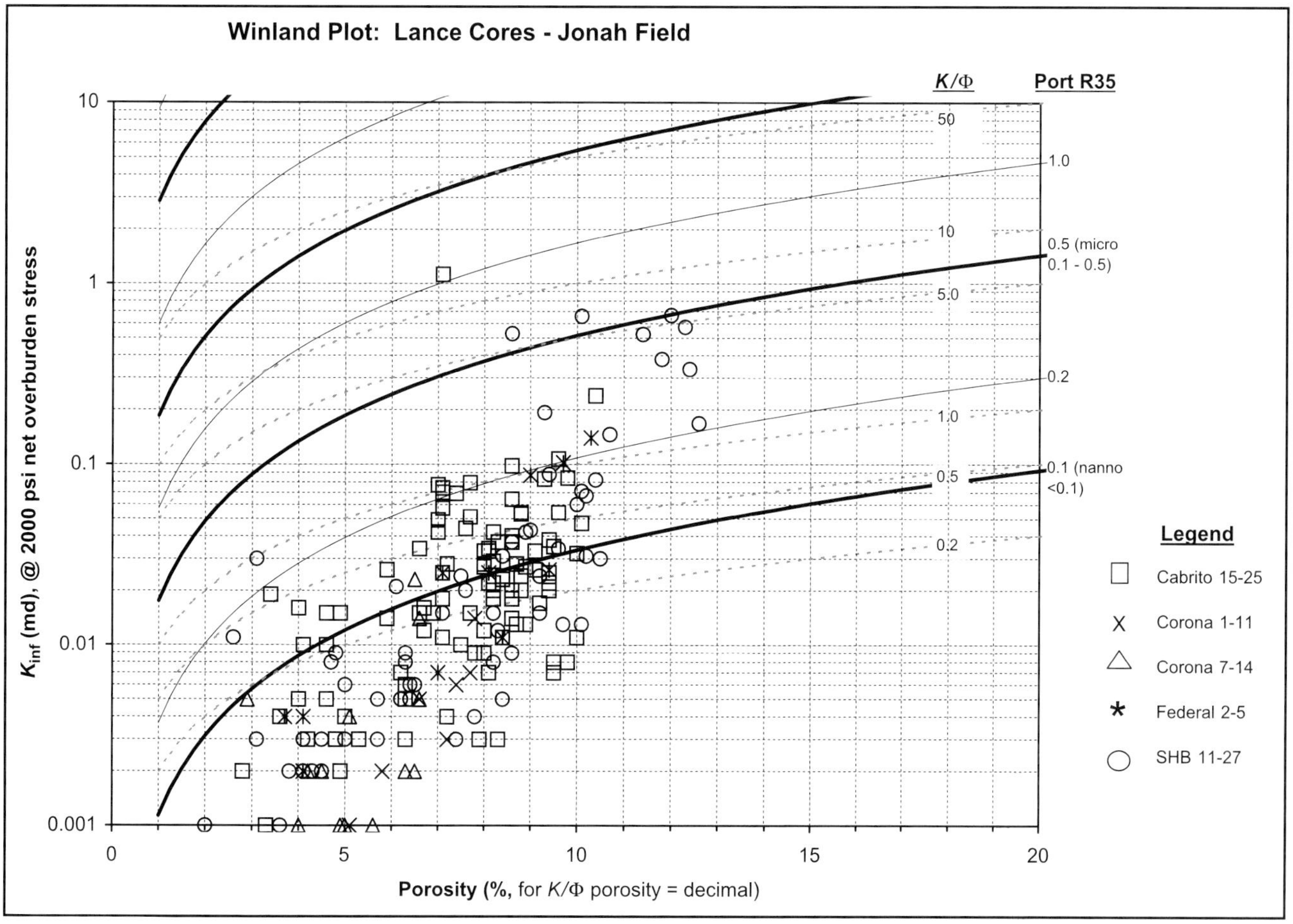

Figure 4. Crossplot of porosity and permeability from cores taken in the Lance Formation in Jonah field. These data were collected at an equivalent net overburden of 2000 psi. Shown on the crossplot are Winland R35 pore-throat ("port") contours as well as K/Φ contours of reservoir quality.

more than 1000 μd. Samples with porosity more than 8% generally have permeabilities between 10 and 100 μd (Figure 4). In-situ permeabilities range from 3 to 20 μd (reported in Robinson et al., 1995). Water saturations in productive sandstones in the field are generally less than 35% (Montgomery and Robinson, 1997). Net pay thickness ranges from 100 to 750 ft/well (30 to 229 m/well), and gas-in-place estimates exceed 100 bcf/section (640 ac; 2.60 km^2). Following an early phase of field development at 160-ac (0.65 km^2) (2640 ft [805 m] closest spacing between wells) and 80-ac (0.32-km^2) spacing (1867 ft [569 m] closest spacing between wells), operators were granted the flexibility to drill a second well in an 80-ac (0.32-km^2) spacing unit effectively allowing for 40-ac (0.16-km^2) infill (1320 ft [402 m] closest well spacing). It has been variously estimated that drilling at 40 ac/well (0.16 km^2/well) results in more than 75% reserve additions and less than 25% rate acceleration. Subsequent studies of reservoir architecture may suggest that portions of the field will require 20-ac (0.08-km^2) development to fully access the in-place resource in an efficient manner.

Reservoir Rock Quality

An important part of the reservoir description is to determine the relative efficiency of various reservoir elements in terms of productive flow capacity. Porosity (Φ) and permeability (K) are standard components of many

◄

Figure 3. Type log for the Jonah field area for this study. The well is Amoco Production Company, Cabrito Unit 15-25, SE Sec. 25, T29N, R108W, Sublette County, Wyoming. Cored intervals are discussed in the text; perforations, stimulation stages, and informal stratigraphic subdivisions of the Lance Formation are shown.

engineering flow-performance equations. The K/Φ ratio, therefore, reflects rock quality in terms of flow efficiency and is used as a point of comparison between various sedimentological attributes (e.g., Hartmann and Beaumont, 1999). In Jonah, Pinedale, and much of the surrounding area, the Lance Formation is dominated by fine- and very fine-grained sandstones (Figure 5). Cores such as the Jonah Federal 2-5 and the Old Road 1 (see Webb et al., 2004) contain scattered intervals with relatively coarse-grained material; however, their overall importance in the Lance Formation in the northern Green River basin is uncertain. In the cores examined in Jonah and Pinedale, less than 5% of the cored rock contains grain sizes larger than upper fine (Figure 5). Examination of rock types, as determined by dominant grain size, suggests a strong relationship between grain size and reservoir quality as described by K/Φ ratios (Figure 6). Medium-grained, sand-sized rocks have superior reservoir quality; however, they do not make up a significant proportion of the Lance Formation at Jonah field. Although the data thus far would indicate that the contribution of coarser grained lithologies to be relatively minor, anecdotal evidence from adjacent producing areas on the nearby Pinedale anticline suggests that some anomalously high producing zones may have encountered a substantially improved rock type. To further address the potential role of these coarser grained lithologies, additional data in the form of sidewall cores will be required. When reservoir quality is compared to sedimentary fabric (Figure 7), it appears that cross-bedded sandstones generally provide improved reservoir properties. The change in fabric from cross-bedded sandstones to convolute- and massive-bedded sandstones, which dominate the channel-fill deposits, results in a significant loss in reservoir quality. When reservoir quality is compared to major depositional facies (Figure 8), it is clear that the channel-fill sandstones are superior in terms

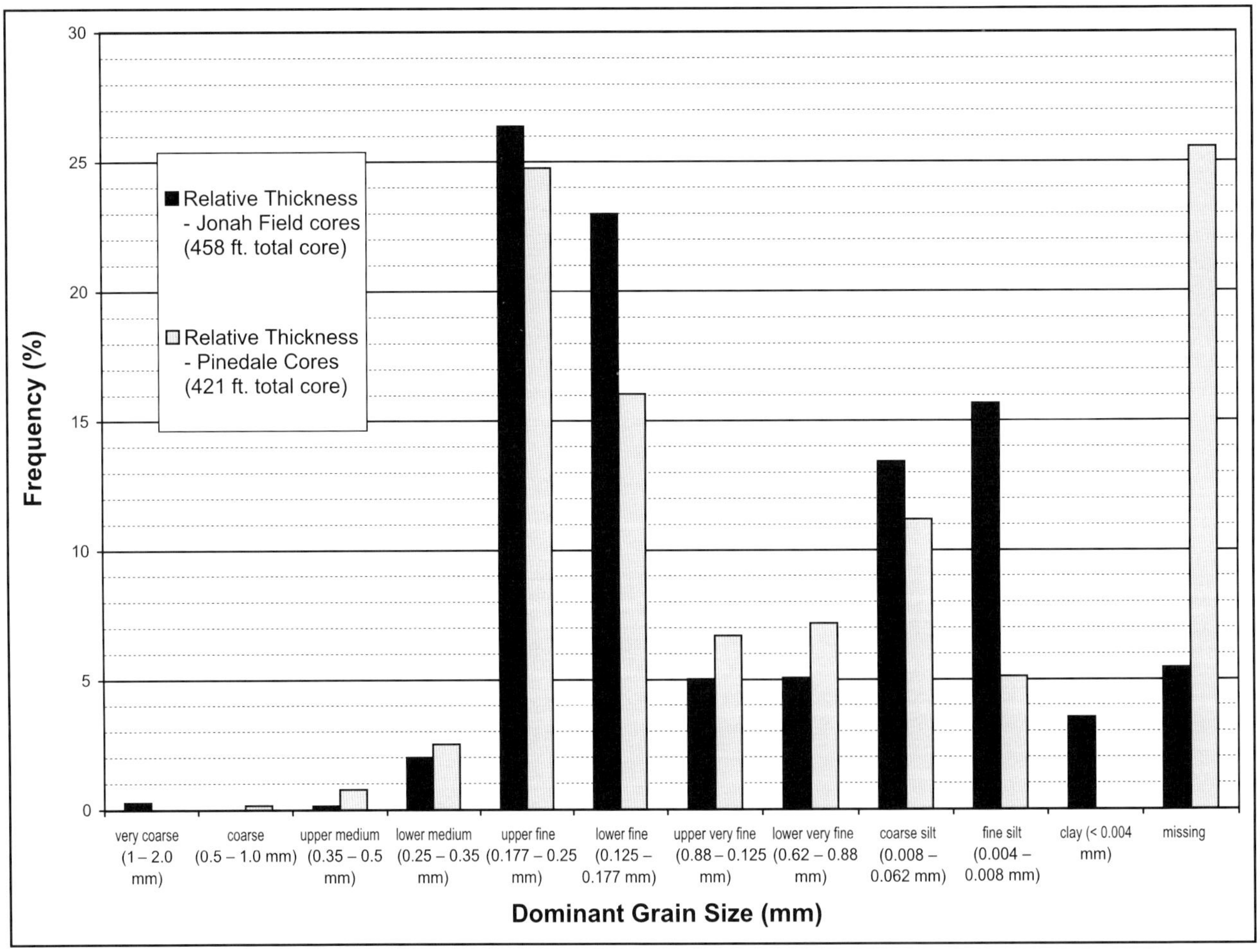

Figure 5. Relative frequency plot illustrating grain-size variation seen in cores in Jonah and adjacent areas of Pinedale anticline. The dominant grain size is very fine grained (0.125–0.250 mm; 0.005–0.010 in.).

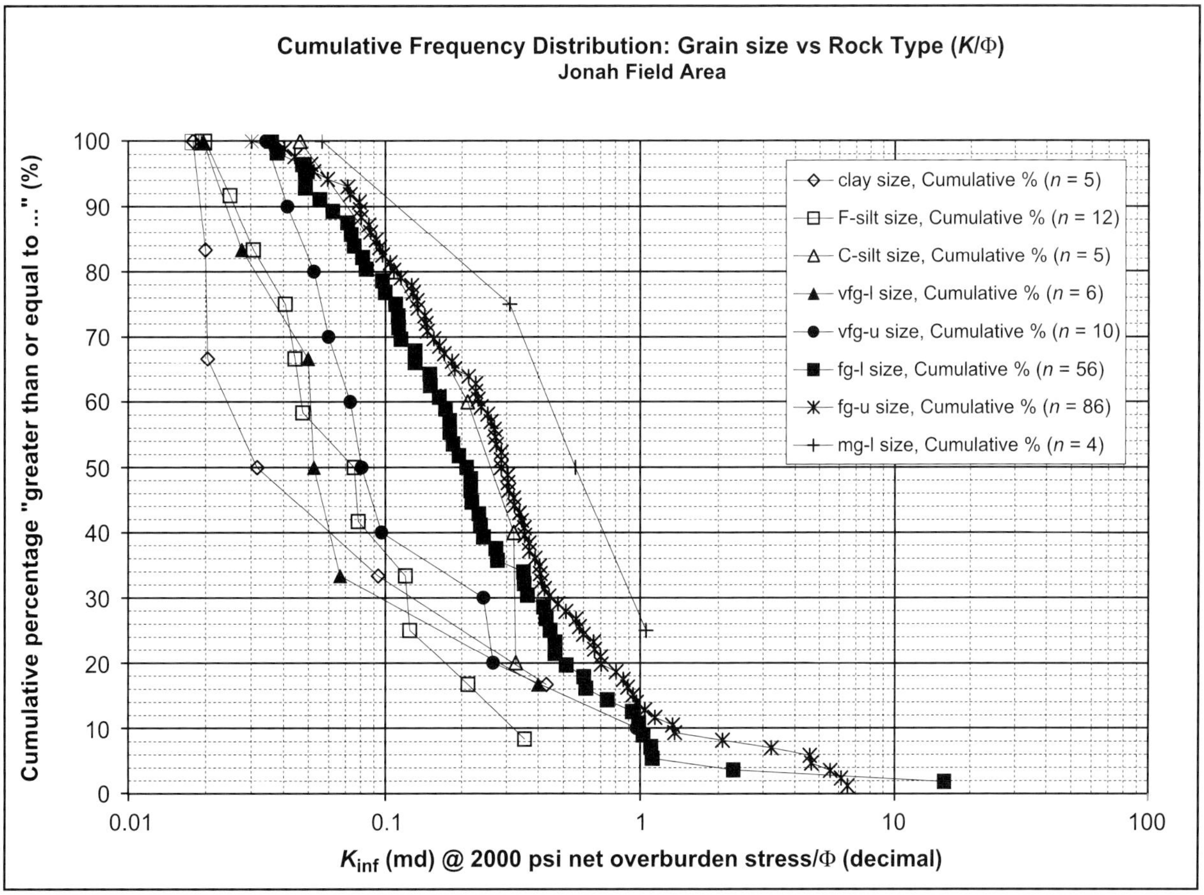

Figure 6. Cumulative frequency distribution illustrating variations in reservoir quality as a function of grain size. Reservoir quality is almost a direct function of grain size, with the best reservoir quality occurring in the medium-grained sandstone. Comparison with Figure 5, however, indicates that these coarser grained sandstones comprise a relatively minor portion of the Lance Formation as seen at Jonah field. F-silt = fine silt; C-silt = coarse silt; vfg-l = lower very fine-grained sand; vfg-u = upper very fine-grained sand; fg-l = lower fine-grained sand; fg-u = upper fine-grained sand; mg-l = lower medium-grained sand.

of reservoir rock quality and should dominate reservoir performance. It should also be noted that relatively thin crevasse-splay sandstones also have favorable reservoir properties, and although they are unlikely to form aerially extensive reservoirs, they may provide connectivity between channel-fill sandstones.

RESERVOIR DESCRIPTION

Reservoir Sedimentology

Cores were generally described at a scale of 1:24, allowing for detailed variations in sedimentary fabric to be captured. On the basis of observations made from cores, sedimentary facies in the Lance Formation at Jonah field were combined into five major depositional facies: These depositional facies associations are channel-fill deposits, abandonment-fill deposits, crevasse-splay deposits, rooted flood-plain deposits, and flood-plain lake deposits.

Channel-fill Deposits

Description. The primary reservoir at Jonah occurs in fine-grained, trough cross-bedded and convolute-bedded sandstones that are stacked in either "single-story" or "multistory" units (Appendices 1–7; Figures 9–18). Examination of single-story units (e.g., Cabrito Unit 15-25, 9937–9927.5 ft [3029–3026 m], Appendix 1, p. 3; Stud Horse Butte 11-27, 8928–8921 ft [2721–2719 m], Appendix 3, p. 1–2) allows a glimpse of the full range of sedimentary features that may be found in this facies. Contacts with underlying facies are generally sharp to erosional and are commonly lined with shale clasts (Figures 9–11). These shale-clast lags locally exceed 2 ft

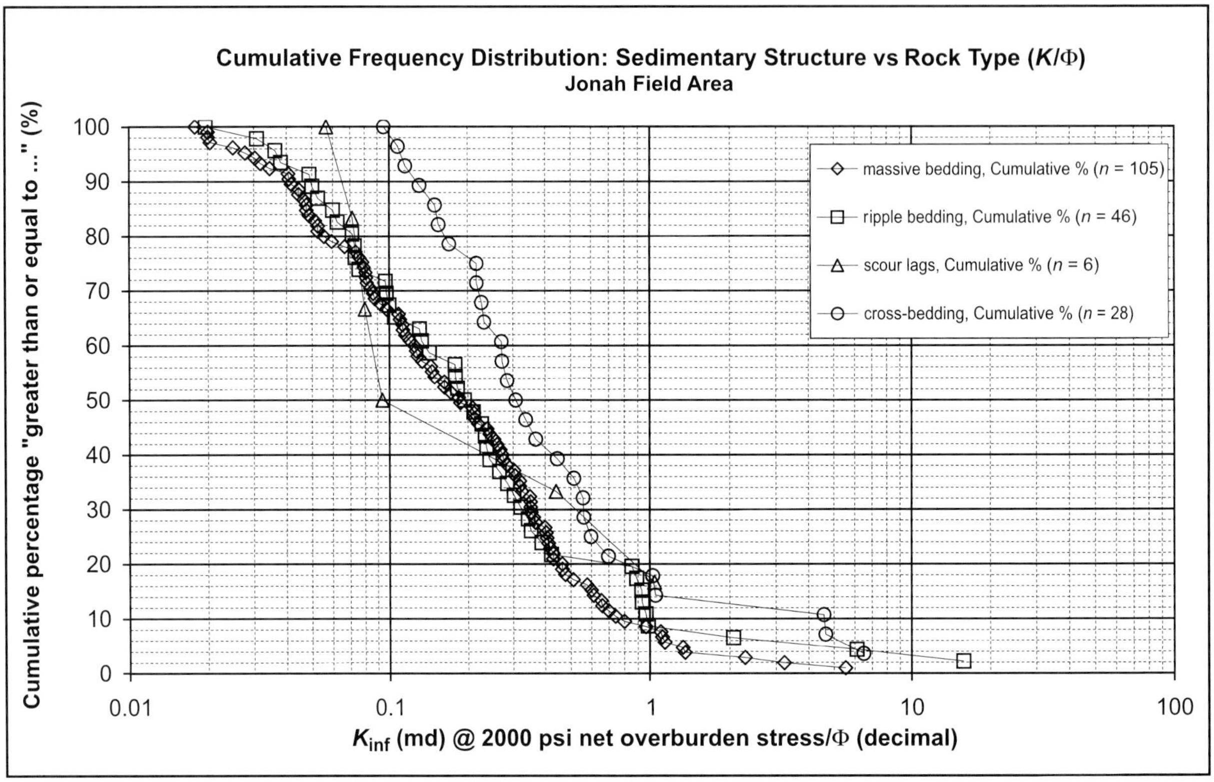

Figure 7. Cumulative frequency distribution illustrating variations in reservoir quality as a function of sedimentary structures and fabric. These data indicate that cross-bedded sandstones have superior reservoir quality. The deformation of cross-bedded sands to more massive and convolute-bedded sands results in a significant loss in reservoir quality.

(0.6 m) in thickness; however, intervals as much as 4 ft (1.2 m) may contain abundant shale clasts. In places, these clay-clast lags are sufficiently thick, and the concentration of shale clasts are sufficiently abundant that they are seen on gamma-ray logs as high-radioactivity zones; these are easily misinterpreted as shales. Caution must be exercised during the well-log interpretation of "shales" in thick, multistory sands; many are undoubtedly shale-clast conglomerates with very limited lateral extent.

Sedimentary structures in channel-fill deposits are dominated by trough cross-beds (Figure 16) that, in single-story units, pass upward to ripple lamination (Figure 17). Convolute and massive bedding is common in these deposits (Figures 12–15) and, in some cores, are the most common sedimentary structures. In many cases, convolute bedding can be identified as having disturbed primary trough cross-beds or ripple lamination.

The vast majority of channel-fill deposits in Jonah field and in the Pinedale anticline region suggests a fine-grained depositional system and resultant reservoir. However, discrete intervals of core are locally medium- to coarse-grained sandstone and contain lags of coarse-grained sandstone to pebbly material (i.e., McMurry Oil Co., Jonah Federal 2-5, core 3; American Hunter, Old Road Unit 1; see Webb et al., 2004). The overall volume and, hence, significance of these coarser grained strata remain poorly understood at present. This coarser grained lithofacies was first observed at the Old Road Unit 1 outside of Jonah field and was later observed in one of the early cores taken in Jonah field at the Jonah Federal 2-5. Early descriptive efforts at Jonah field accorded this coarser grained lithofacies considerable significance. With the benefit of additional cores in Jonah field and subsequent core description efforts in the Pinedale anticline area, the coarser grained lithofacies, such as that seen in the Jonah Federal 2-5, is now thought to be relatively uncommon.

Channel-fill deposits may overlie any of the other five depositional facies associations recognized at Jonah. In cases where preservation is more complete (i.e., single-story sand bodies), the channel-fill deposits generally grade up to either abandonment-fill deposits or rooted flood-plain deposits.

Stacking Patterns. Examination of cores, and subsequent calibration to wire-line logs, indicates that the majority of sandstones in the Lance are multistory in nature

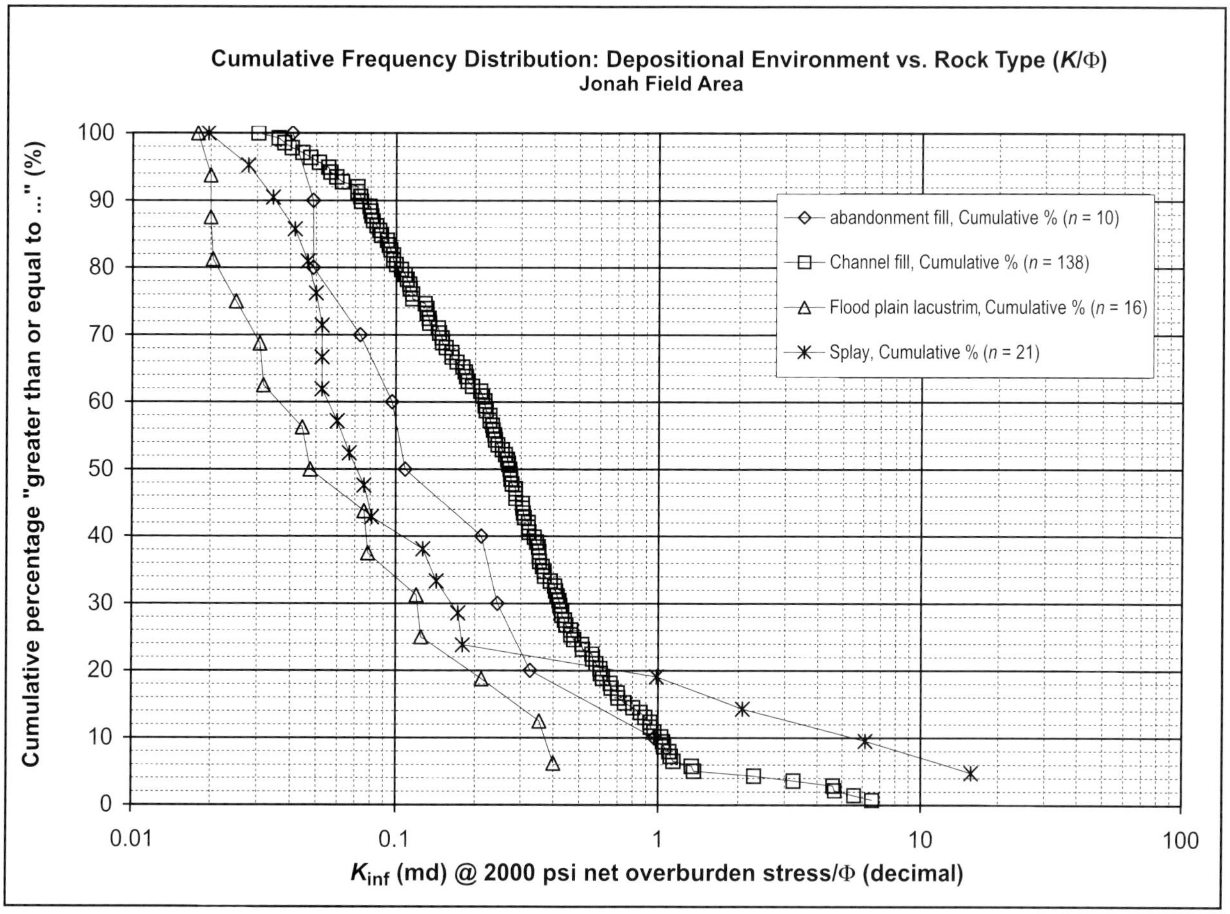

Figure 8. Cumulative frequency distribution illustrating variations in reservoir quality as a function of interpreted depositional environment. Channel-fill sandstones are clearly the dominant reservoir type at Jonah field. These data also illustrate that some crevasse-splay sandstones may have superior reservoir quality. Although these crevasse-splay sandstones are not expected to have broad aerial extent, they may provide connectivity between channel-fill sandstones.

(Appendices 1–7). The interpretation presented here suggests that the multistory sandstones are developed through the amalgamation of relatively small, single-story sand bodies as the result of channel-belt migration. Single-story sandstones range in thickness from approximately 9 to 15 ft (3 to 5 m) with many multistory sandstones exceeding 50 ft (15 m) in thickness. The various stacking patterns are indicated on the core descriptions provided in Appendices 1–7.

Interpretation. Channel-fill deposits reflect deposition in primarily meandering to perhaps anastomosed rivers. The relatively fine grain size, the common occurrence of fining-upward grain-sized trends in single-story sand bodies, the common occurrence of ripple lamination, well-developed flood-plain deposits, and the lack of planar-tabular cross-beds all suggest a rather modest, low-energy fluvial system similar, perhaps, to the small, meandering rivers of the upper high plains of the United States. The interpretation of either meandering or anastomosed river deposits at this stage of analysis is to allow for the possibility of either single-channel or multichannel systems. Multistory sand bodies reflect vertical and lateral stacking of these low-energy channel deposits as opposed to dramatic alternations in the style of the fluvial system itself and the development of larger sand bodies. Analysis of the thickness trends seen in core suggests that these rivers had mean bankfull depths of approximately 6–10 ft (2–3 m), having maximum depths of 10–17 ft (3–5 m). These channels likely had widths between 80 and 280 ft (24 and 85 m) that resulted in sandstone widths between 160 and 2340 ft (50 and 710 m) through lateral migration. These dimensional data are discussed more fully elsewhere in this chapter and are presented in Table 1A–C.

Figure 9. Core photograph of shale-clast conglomerates from the basal scour surface of a multistory, channel-fill deposit from the Snyder Oil Company, Stud Horse Butte 11-27 well. Core depth = 10,431 ft (3179 m). A description of this core is contained in Appendix 4.

Abandonment-fill Deposits

Description. Abandonment-fill deposits are composed of ripple-laminated (Figures 17, 18), very fine-grained, and fine-grained sandstones and siltstones that overlie channel-fill deposits (Figure 19) (e.g., Cabrito Unit 15-25, 10,593.5–10,590.4 ft [3228.9–3228.0 m], Appendix 2, p. 2; Cabrito Unit 15-25, 9932–9927.5 ft [3027–3026 m], Appendix 1, p. 3). Scattered shale clasts, abundant organic material, and contorted bedding are common. Inclined, ripple-laminated bedding has also been observed (e.g., Stud Horse Butte 11-27, 8922–8925 ft [2719–2720 m], Appendix 3, p. 1–2). Abandonment-fill deposits range in thickness from 2 to 5 ft (0.6 to 1.5 m) and are gradational with underlying channel-fill deposits.

Interpretation. These strata reflect progressive abandonment of active channels. Fluvial avulsion results in reduced flow within channels, leading to lower shear stresses and lower flow-regime deposition; this results in a predominance of ripple lamination and preservation of organic material. Inclined bedding, where preserved, reflect the upper part of lateral accretion sets associated with point bars in meandering rivers. The abandonment-fill deposits represent the final infill of channel morphology and, as a result, are only preserved in the upper part of single-story sand bodies or the upper part of multistory sand bodies. The finer grain size and heterolithic lithologies suggest that the abandonment-fill deposits do not contribute significantly to overall reservoir quality and performance.

Crevasse-splay Deposits

Description. Crevasse-splay deposits are comprised of thin, climbing-ripple–laminated, ripple-laminated, and parallel-laminated, very fine-grained, and fine-grained sandstones (Figures 15, 18) (e.g., Cabrito Unit 15-25, 10,538.4–10,535 ft [3212.1–3211 m], Appendix 2, p. 1; Stud Horse Butte 11-27, 8921–8915 ft [2719–2717 m],

Figure 10. Core photograph of shale-clast conglomerates from the basal scour surface of a multistory channel-fill deposit from the El Paso Natural Gas, Wagon Wheel 1 well. Core depth = 10,175 ft (3101 m). A description of this core is contained in Appendix 6.

Figure 11. Core photograph of shale clasts in a fine-grained, ripple-laminated sandstone defining a scour surface in a multistory, channel-fill complex from the El Paso Natural Gas, Wagon Wheel 1 well. Core depth = 10,172 ft (3100 m). A description of this core is contained in Appendix 6.

Appendix 3, p. 1; Stud Horse Butte 11-27, 10,956.5–10,950 ft [3339.5–3338 m], Appendix 5, p. 1; Stud Horse Butte 11-27, 10,980–10,974 ft [3347–3345 m], Appendix 5, p. 2). These sandstones commonly have a sharp base and fine upward. Typical thicknesses for these sandstones range from 2 to 4 ft (0.6–1.2 m). These deposits are commonly overlain by rooted flood-plain deposits.

Interpretation. These strata reflect crevasse-splays that developed on the flood plain adjacent to channel deposits. Within the crevasse-splay facies assemblage, both proximal and distal splays may be interpreted based on the amount of climbing-ripple lamination, overall thickness, basal contact, and degree of bioturbation and root modification. These strata are significant because they may facilitate communication (pressure and production) between thicker, more discontinuous channel deposits.

Rooted Flood-plain Deposits

Descriptions. Rooted flood-plain deposits are comprised of gray to greenish gray massive siltstones and fine siltstones. Organic debris, slickensides, and root traces are common throughout this facies association (Figure 20). These strata may range in thickness from a few feet to 15 ft (5 m) (e.g., Stud Horse Butte 11-27, 8941–8928 ft [2725–2721 m], Appendix 3, p. 2; Stud Horse Butte 11-27, 10,439–10,432 ft [3181–3179 m], Appendix 4, p. 1–2; Stud Horse Butte 11-27, 10,947–10,942 ft [3337–3335 m], Appendix 5, p. 1; Stud Horse Butte 11-27, 10,974–10,967 ft [3345–3343 m], Appendix 5, p. 2). Rooted flood-plain deposits commonly overlie flood-plain lake deposits, suggesting an evolutionary relationship.

Interpretation. Rooted flood-plain deposits reflect the development of a well-drained, vegetated flood plain. This environment was likely marginal to the active channels and provides the probable source for the abundant shale clasts in the channel deposits. These deposits are likely to have significant areal extent and may well correlate between wells over local areas. The association of rooted flood-plain and flood-plain lake deposits suggests that with increased sediment flux, flood-plain lakes gradually became vegetated and emergent. In those few cases

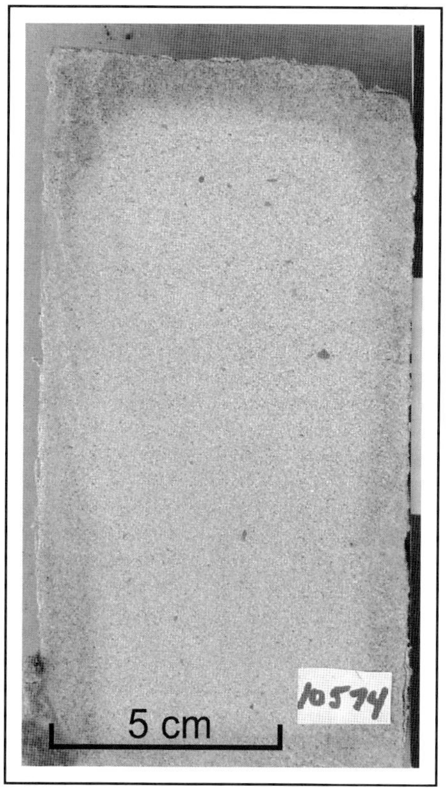

Figure 12. Core photograph of a massive, fine-grained sandstone from within a multistory channel-fill complex from the Amoco Production Company, Cabrito 15-25 well. Core depth = 10,574 ft (3223 m). A description of this core is contained in Appendix 2.

Figure 13. Core photograph of a convolute bedding in a fine-grained, multistory, channel-fill sandstone in the Amoco Production Company, Cabrito 15-25 well. Core depth = 9911 ft (3021 m). A description of this core is contained in Appendix 1.

where lake deposits are seen to overlie rooted flood plains, it is likely that channel avulsion resulted in isolated areas within the flood plain that were sediment starved and became subaqueous.

Flood-plain Lake Deposits

Description. Lacustrine deposits are comprised of horizontal-laminated to low-angle–laminated, very dark gray to greenish gray claystones and siltstones. Small-scale ripple lamination (pseudohorizontal lamination) is common (e.g., Stud Horse Butte 11-27, 10,950–10,947 ft [3338–3336 m], Appendix 5, p. 1). This facies association sparsely exceeds 3 ft (1 m) in thickness as determined by core examination. These lake deposits generally grade upward into massive and root-penetrated siltstones and claystones that are part of the rooted flood-plain facies association.

Interpretation. Flood-plain lake deposits reflect deposition in small lakes and isolated bodies of water located in the flood plain. The lack of root modification and the preservation of fine-scale lamination suggest deposition in a subaqueous setting. The geometry and occurrence of these deposits is unclear; however, given the fact that they appear sparsely in core, these lakes may be relatively small in areal extent. The distribution of these flood-plain lakes is thought to be essentially random within the flood plain. The upward transition to rooted flood-plain deposits reflects the gradual infilling, drying, and vegetation of these small bodies of water.

RESERVOIR GEOMETRY

A significant operational and financial decision at any oil and gas field concerns appropriate well spacing. Early identification of optimal well spacing allows financial value

Figure 14. Core photograph of a convolute bedding in a fine-grained, ripple-laminated sandstone from within a multistory channel-fill complex in the Snyder Oil Company, Stud Horse Butte 11-27 well. Core depth = 10,427 ft (3178 m). A description of this core is contained in Appendix 4.

Figure 15. Core photograph of a convolute bedding in a fine-grained sandstone interpreted as part of a crevasse-splay deposit in the El Paso Natural Gas, Wagon Wheel 1 well. Note the outline of fold structures highlighted by the concentration of organic debris. Core depth = 11,027 ft (3361 m). A description of this core is contained in Appendix 6.

(measured in terms of present value) to be maximized. From a technical perspective, the central issue surrounding well spacing concerns the degree to which close-spaced wells will drain a single reservoir body or will interfere with one another. In an attempt to aid in that decision process, data derived from core examination were used to develop estimates of both channel and sand body dimension. Attempts to reconstruct channel geometry from vertical data are an ongoing area of sedimentary research (e.g., Bridge, 1997; Bridge and Tye, 2000; Dalrymple, 2001), and there are significant differences of opinion as to how this problem should be addressed. Although high-quality, three-dimensional seismic surveys have been acquired across Jonah field, many of the reservoir-bearing sandstones are at or below seismic tuning thickness, substantially reducing the effectiveness of seismic data to consistently resolve reservoir-scale features across the field. Instead of providing a single solution concerning the reservoir geometry, the data derived from core examination were used as input data to a variety of techniques. Each solution was then compared, and the impact of that solution was considered in terms of both engineering and financial impact.

Reservoir description efforts have long attempted to characterize fluvial strata as either "braided" or "meandering." Outcrop studies over the last 15 yr, however, have shown conclusively that distinguishing truly meandering from braided morphologies is virtually impossible with vertical-based data (cores and well logs) alone (e.g., Bridge, 1985). From a practical standpoint, the desire to separate what are perceived as braided from meandering deposits is driven by the perception that meandering rivers will result in isolated reservoirs, whereas braided deposits will result in more sheetlike reservoirs. This perception reflects interpretations largely derived from the

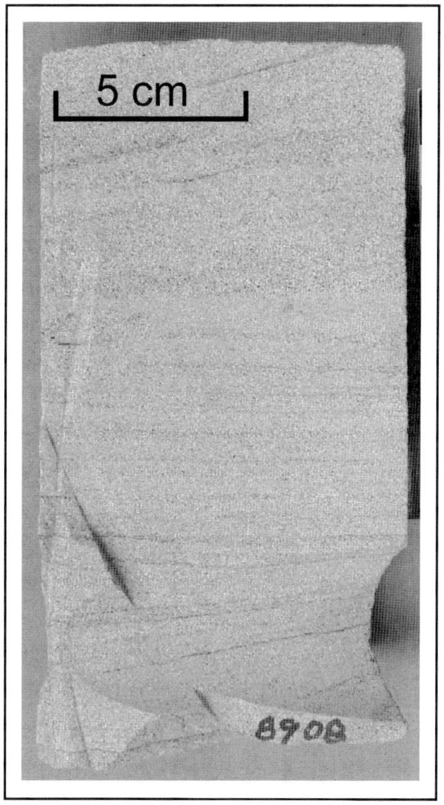

Figure 16. Core photograph of a trough cross-bedded, fine-grained sandstone from within a multistory channel-fill complex in the Snyder Oil Company, Stud Horse Butte 11-27 well. Core depth = 8908 ft (2715 m). A description of this core is contained in Appendix 3.

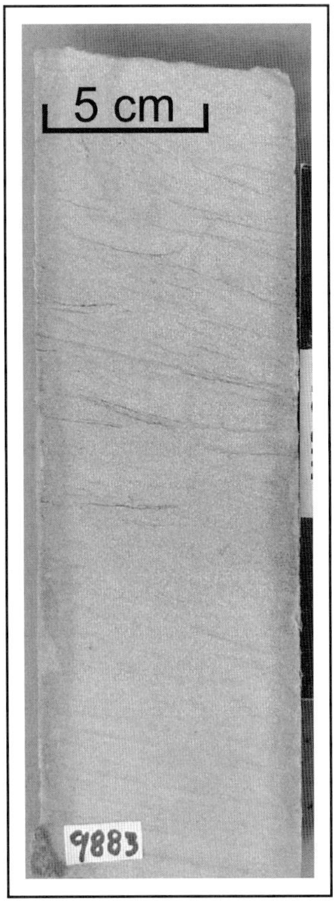

Figure 17. Core photograph of a fine-grained, ripple-laminated sandstone from within a multistory channel-fill complex from the Amoco Production Company, Cabrito Unit 15-25 well. Core depth = 9883 ft (3012 m). A description of this core is contained in Appendix 1.

Recognizing that there is no single solution when trying to determine sand body dimensions from purely vertical data, several techniques were examined and presented for purposes of comparison and to better understand the likely distribution of plausible results. It is the contention of the author that an understanding of the uncertainty associated with possible sand body geometries and the range in interconnectedness is of far greater value to those developing subsurface reservoirs than techniques that report a single, unique value of width-thickness.

To be useful, reservoir models must account for the three-dimensional geometry of reservoir bodies. In alluvial settings, historical focus has been on the development of two-dimensional aspect ratios (width-thickness) for alluvial sand bodies. Approaches to estimating sand body widths generally fall into one of three types of analyses: (1) numerical models based on theoretical considerations, (2) numerical estimates based on an understanding of modern fluvial systems, and (3) estimates derived from the study of outcrop analogs. In all cases, it should be noted study of modern fluvial analogs and their geomorphology. Although the geologic literature is replete with descriptions of the geomorphology of fluvial systems, modeling studies of fluvial systems and studies of sedimentary architecture strongly suggest that the interplay of subsidence, base level, and magnitude of sediment supply exerts a far greater control on the degree to which fluvial deposits are amalgamated or isolated than the many short-term processes commonly viewed in the study of modern analogs (e.g., Bridge and Leeder, 1979; Bridge and Mackey, 1993; Shanley and McCabe, 1994). Careful study of numerical models and large, expansive outcrops suggests that for systems with a net/gross of less than 40%, fluvial deposits are likely to be isolated from one another. For systems with a net/gross greater than 65–70%, fluvial deposits are likely to be amalgamated, and for systems between 40 and 70%, the degree of amalgamation is difficult to generalize.

Figure 18. Core photograph of a fine-grained, ripple-laminated, and climbing-ripple–laminated sandstone from a crevasse-splay deposit from the Snyder Oil Company, Stud Horse Butte 11-27 well. Core depth = 10,978 ft (3346 m). A description of this core is contained in Appendix 5.

Table 1-A

A) Summary of Techniques Used to Estimate Channel Depths and Channel Widths Given the Full-story Thicknesses Observed in Cores in the Jonah–Pinedale Area. References and Comments Relative to Each Technique are Included.

	Channel-depth Analysis								Channel-width Analysis					
	Maximum Bankfull Depth Estimate (h)		**10% Decompaction for True Maximum Bankfull Depth Estimate (h)**		**Mean Bankfull Depth Estimate, d_m (ft), $d_m = 0.57h$**		**Mean Bankfull Depth Estimate From Mean Cross-set Thickness, d_m (ft), $d_m = 16S_m - 34S_m$**		**Bankfull Channel Width (m) $W = 6.8h$(exp 1.54)**		**Bankfull Channel Width (m) $W = 8.88d_m$(exp 1.82)**		**Bankfull Channel Width (m) $W = 15.85d_m$(exp 1.58)**	
	(ft)	**(m)**	**(ft)**	**(m)**	**(ft)**	**(m)**	**(ft)**	**(m)**	**(ft)**	**(m)**	**(ft)**	**(m)**	**(ft)**	**(m)**
Source/ Reference Additional comment	from core examination		Ethridge and Schumm (1978) similar to Lorenz, 1985		Bridge and Mackey (1993) compare to modern river data sets		Bridge (1997) and Bridge and Tye (2000) uses mean cross-set thickness; cores suggest $S_m = 0.10–0.15$ m		Leeder (1973) almost everyone goes back to Leeder, 1973 for sinuous rivers $S_n > 1.7$ uses maximum bankfull estimate for h; equation input in meters		Bridge and Mackey (1993) $S_n > 1.0$ uses mean bankfull estimate for d_m; equation input in meters		Bridge and Mackey (1993) $S_n > 1.0$ uses mean bankfull estimates for d_m; similar results to Leeder, 1973; equation input in meters	
Full-story thickness														
	8	2.4	8.9	2.7	5.1	1.5			103.5	31.6	64.3	19.6	103.3	31.5
	9	2.7	10.0	3.0	5.7	1.7			124.1	37.8	79.6	24.3	124.5	37.9
	10	3.0	11.1	3.4	6.3	1.9			146.0	44.5	96.4	29.4	147.0	44.8
	12.5	3.8	13.9	4.2	7.9	2.4			205.9	62.7	144.8	44.1	209.1	63.7
	15	4.6	16.7	5.1	9.5	2.9			272.6	83.1	201.7	61.5	279.0	85.0
	20	6.1	22.2	6.8	12.7	3.9			424.6	129.4	340.5	103.8	439.5	134.0
	25	7.6	27.8	8.5	15.8	4.8			598.6	182.5	511.1	155.8	625.3	190.6
							5.2–16.7	1.6–5.1						

Table 1-B

B) Summary of Techniques Used to Estimate Sandstone Widths (Channel-belt Widths of Some Authors) from Estimates of Channel Depth and Width. The Nine Techniques Shown here are Commonly Cited in the Literature and Provide an Estimate of the Range of Possible Solutions for a Given Estimate of Channel Depth and Width.

			Calculation of Sandstone Body Width (various algorithms)							
			Meander-belt Width (m) $W_m = 65.6 \times h(\exp 1.57)$		**Meander-belt Width (m) $W_m = 7.44 \times W(\exp 1.01)$**		**Channel-belt Width (m) CBW = $W_m = 12.1 \times d_m(\exp 1.85)$**		**Channel-belt Width (m) CBW = $W_m = 0.95 \times d_m(\exp 2.07)$**	
			(ft)	**(m)**	**(ft)**	**(m)**	**(ft)**	**(m)**	**(ft)**	**(m)**
Source/ Reference			Collinson (1978) modified *Canadian Society of Petroleum Geologists Memoir* 5, p. 580		Lorenz et al. (1985) *AAPG Buletin*, v. 69, p. 713		Fielding and Crane (1987) *SEPM Special Publication* 39, p. 321–323		Fielding and Crane (1987) *SEPM Special Publication* 39, p. 323	
Additional comment			see Bridge and Mackey, 1993, p. 216, for modification to original Collinson (1978)		based on MWX site work in Mesaverde		best fit solution, case 2A in Fielding and Crane (1987)		meandering deposit fit, lower limit, case 1B in Fielding and Crane (1987)	
			use maximum bankfull depth estimate for *h*; equation input in meters		use Bridge and Mackey (1993) estimate of *W*; equation input in meters		use mean bankfull estimate for d_m (implied in text) from Bridge and Mackey (1993); equation input in meters		use mean bankfull estimate for d_m (implied in text) from Bridge and Mackey (1993); equation input in meters	
Full Story thickness	(ft)	(m)								
(see Table 1A)	8	2.4	872.3	265.9	795.7	242.5	88.7	27.0	7.7	2.3
	9	2.7	1049.4	319.9	960.3	292.7	110.3	33.6	9.8	3.0
	10	3	1238.2	377.4	1136.1	346.3	134.0	40.9	12.2	3.7
	12.5	3.8	1757.7	535.7	1622.1	494.4	202.5	61.7	19.3	5.9
	15	4.6	2340.2	713.3	2169.9	661.4	283.8	86.5	28.2	8.6
	20	6.1	3676.3	1120.5	3434.1	1046.7	483.2	147.3	51.1	15.6
	25	7.6	5218.7	1590.6	4902.9	1494.4	730.1	222.5	81.0	24.7

that sandstone body width is being estimated normal to the direction of sediment transport. Although there are several established methods for deriving estimates of sand body width from thickness, the estimation of the third dimension, sand body length ("downtransport"), has far greater uncertainty. Robinson (1994) examined sandy, braided fluvial deposits of the Morrison Formation in superbly exposed outcrops of Utah in the United States and provided one of the very few outcrop-based data sets that allowed length dimensions to be estimated as a function

Table 1-B

B) Summary of Techniques Used to Estimate Sandstone Widths (Channel-belt Widths of Some Authors) from Estimates of Channel Depth and Width. The Nine Techniques Shown here are Commonly Cited in the Literature and Provide an Estimate of the Range of Possible Solutions for a Given Estimate of Channel Depth and Width (cont.).

Calculation of Sandstone Body Width (various algorithms)

Channel-belt Width (m) CBW = $W_m = 64.6 \times d_m$(exp 1.54)		**Channel-belt Width (m) CBW = $W_m = 59.9 \times d_m$(exp 1.8)**		**Channel-belt Width (m) CBW = $W_m = 45.76 \times h$(exp 1.52)**		**Average Single-Sandstone Body Width (m) W_m = 12.1 × h(exp 1.87)**		**Average Single-Sandstone Body Width (m) $W_m = 6.0 \times h$(exp 1.88)**	
(ft)	**(m)**	**(ft)**	**(m)**	**(ft)**	**(m)**	**(ft)**	**(m)**	**(ft)**	**(m)**
Fielding and Crane (1987) *SEPM Special Publication* 39, p. 323		Bridge and Mackey (1993) *International Association of Sedimentologists Special Publication* 15, p. 216		Bridge and Mackey (1993) *International Association of Sedimentologists Special Publication* 15, p. 216		Robinson and McCabe (1997) *AAPG Bulletin*, v. 81, p. 1278–1280		Robinson and McCabe (1997) *AAPG Bulletin*, v. 81, p. 1278–1280	
meandering river fit, upper limit, case 2B in Fielding and Crane (1987)		table 2		table 2		sandy braided systems, Fig. 8b		muddy braided systems, Fig. 8a	
based on Collinson (1978) (maximum bankfull depth estimate), use maximum bankfull depth for d_m; equation input in meters; see later modification by Bridge and Mackey (1993)		one of several possible equations; use mean bankfull estimate for d_m; equation input in meters		one of several possible equations; use maximum bankfull estimate; decompacted for h; equation input in meters		based on Ph.D. study of Morrison Formation braided stream deposits in Utah; use maximum bankfull estimate; decompacted for h; equation input in meters		based on Ph.D. study of Morrison Formation braided stream deposits in Utah; use maximum bankfull estimate; decompacted for h; equation input in meters	
836.3	254.9	429.7	131.0	683.0	208.2	256.0	78.0	128.2	39.1
1002.6	305.6	531.1	161.9	816.9	249.0	319.1	97.3	160.0	48.8
1179.2	359.4	642.1	195.7	958.8	292.2	388.5	118.4	195.0	59.4
1662.8	506.8	959.4	292.4	1345.9	410.2	589.7	179.8	296.7	90.4
2201.8	671.1	1332.1	406.0	1775.8	541.3	829.3	252.8	418.0	127.4
3429.2	1045.2	2235.8	681.5	2749.7	838.1	1420.3	432.9	717.9	218.8
4835.4	1473.8	3340.9	1018.3	3860.1	1176.5	2155.7	657.1	1092.0	332.9

of sand body width. Robinson's (1994) work suggests sand body length to be a factor of approximately 1.5 times sand body width. Ebanks and Weber (1987) studied Pennsylvanian–age meandering river deposits in the shallow subsurface of Missouri in the United States in a shallow oil field where drilling had occurred on approximately 1–2-ac (0.004–0.008-km^2) density. Based on these data, Ebanks and Weber (1987) mapped discrete fluvial point-bar deposits that had "cross-flow" dimensions of approximately 1.4 times the "downflow" dimension.

Table 1-C

C) Summary of Probabilistic Estimates of Fluvial Sand Body Width for Given Ranges of Single-story Thicknesses. The ''1–5-m'' and ''More Than 5- to 10-m'' Thickness Curves Approximate the Range of Single-story Thicknesses Commonly Observed in Cores from the Lance Formation of Jonah Field. The ''All-thickness'' Curve Summarizes All Data (n = 595) and Provides a General Description of Probabilistic Width Distributions. The ''Jonah–Pinedale'' Specific Sort Summarizes the Distribution of Widths for Thicknesses of Between 2.5 and 5 m (8.2 and 16.4 ft).

	Probabilistic Calculation of Sandstone Body Widths					
	Sandstone Body Width, P_{90}		**Sandstone Body Width, P_{50}**		**Sandstone Body Width, P_{10}**	
	(ft)	**(m)**	**(ft)**	**(m)**	**(ft)**	**(m)**
Source/Reference Additional comment	P_{90} = minimum case; 90% of the data has widths greater than the value expressed here as the P_{90}; use max bankfull depth estimate to enter nomogram		P_{50} = median case; 50% of the data has widths greater than the value expressed here as the P_{50}; use maximum bankfull depth estimate to enter nomogram		P_{10} = maximum case; 10% of the data has widths greater than the value expressed here as the P_{10}; use max bankfull depth estimate to enter nomogram	
Thickness curve						
1–5 m thick (3.3–16.4 ft thick)	45.9	14.0	206.7	63.0	1476.4	450.0
more than 5 to 10 m thick (more than 16.4 to 32.8 ft thick)	213.3	65.0	705.4	215.0	5905.5	1800.0
''all-thickness'' curve	59.1	18.0	492.1	150.0	6233.6	1900.0
Jonah–Pinedale specific sort 2.5–5 m thick (8.2–16.4 ft thick)	65.6	20.0	246.1	75.0	2296.6	700.0
Estimation of Aerial Extent/Reservoir Sand (from Probabilistic Data, P_{10}, P_{50}, and P_{90})						
Area of an ellipse = (Π × long axis × short axis)/4	Area (ac)		Area (ac)		Area (ac)	
1–5 m thick (3.3–16.4 ft thick)	0.06		1.16		58.95	
more than 5 to 10 m thick (more than 16.4 to 32.8 ft thick)	1.23		13.46		943.21	
''All thickness'' curve	0.09		6.55		1050.92	
Jonah–Pinedale specific sort 2.5–5 m thick (8.2–16.4 ft thick)	0.12		1.64		142.65	
Comparison of Well-spacing and Probabilistic Width-thickness Data						
Well spacing (ac)	20	40	80	160	320	640
Closest wells (m)	284.5	402.3	569.0	804.6	1137.9	1609.3
Closest wells (ft)	933.4	1320.0	1866.8	2640.0	3733.5	5280.0
Approximate % sandstone bodies greater than or equal to well spacing (using 2.5–5-m Jonah–Pinedale sort)	17	14	12	9	8	7
Approximate % sandstone bodies greater than or equal to well spacing (using ''all-thickness'' curve)	35	30	24	18	16	13

Geometric Estimates

Table 1A–C present a comparison of several different techniques that have been used to derive estimates of sand body width. Some of these techniques follow well-tested approaches (e.g., Table 1B), whereas others rely on a compilation of outcrop data in a more probabilistic manner. For purposes of reservoir description and understanding the possible impact of variable well spacings, several methods of determining sand body widths were considered, and their results were compared to variable well spacing.

Core examination suggests that single-story sand bodies range in thickness from 9 to 15 ft (3 to 5 m) (Appendices 1–5). These single-story measurements serve as an estimate of the maximum bankfull channel depth. Decompaction of these single-story depths (Ethridge and Schumm, 1978; Lorenz, 1985) results in a modified maximum bankfull

Figure 19. Core photograph of a poorly laminated, root-penetrated carbonaceous siltstone from an abandoned channel-fill deposit in the El Paso Natural Gas, Wagon Wheel 1 well. Core depth = 10,204 ft (3110 m). A description of this core is contained in Appendix 6.

depth range of approximately 10–17 ft (3–5 m). Bridge and Mackey (1993) developed an algorithm to estimate the mean bankfull depth from the maximum depth estimate. Mean bankfull depths range from approximately 6 to 10 ft (2 to 3 m). Subsequent to the initial compilation of geometric estimates performed for this study, Bridge (1997) and Bridge and Tye (2000) presented a technique that allowed for the estimation of mean bankfull depth from an estimation of mean cross-set thickness. This approach suggests a range in mean bankfull depth of approximately 5.2–16.7 ft (2–5 m); this result overlaps somewhat with estimates derived from other techniques; however, it is skewed somewhat toward deeper mean bankfull depths (see Table 1A).

Armed with estimates of both mean and maximum channel depth, estimates of channel width can be developed (Leeder, 1973; Bridge and Mackey, 1993). Single-story sand bodies that are approximately 9 ft (3 m) in thickness were deposited by rivers that ranged in widths from approximately 80 to 125 ft (24 to 38 m), whereas single-story sand bodies that were approximately 15 ft (5 m) thick were deposited by rivers that ranged in widths from 200 to 280 ft (61 to 85 m) in width (see Table 1A). By all measures, these rivers are moderately small river systems and are typical of systems found in both coastal-plain and alluvial settings, such as the upper high plains of the United States.

As channels migrate across their flood plain, they create sand bodies that are greater in width than any individual channel deposit as seen at any given point in time. A wide variety of techniques have been developed to estimate sand body widths from limited vertical core and/or log data (e.g., Collinson, 1978; Lorenz et al., 1985; Fielding and Crane, 1987; Bridge and Mackey, 1993; Robinson, 1994; Robinson and McCabe, 1997; Bridge and Tye, 2000). Differences between the various algorithms presented by the authors mentioned lie in the best fit solutions derived from the particular data set individual authors were considering. For single-story sand bodies that are

Figure 20. Core photograph of a ripple-laminated siltstone in which lamination has been disrupted by root penetration and bioturbation in flood-plain deposits from the Snyder Oil Company, Stud Horse Butte 11-27 well. Core depth = 10,973 ft (3345 m). A description of this core is contained in Appendix 5.

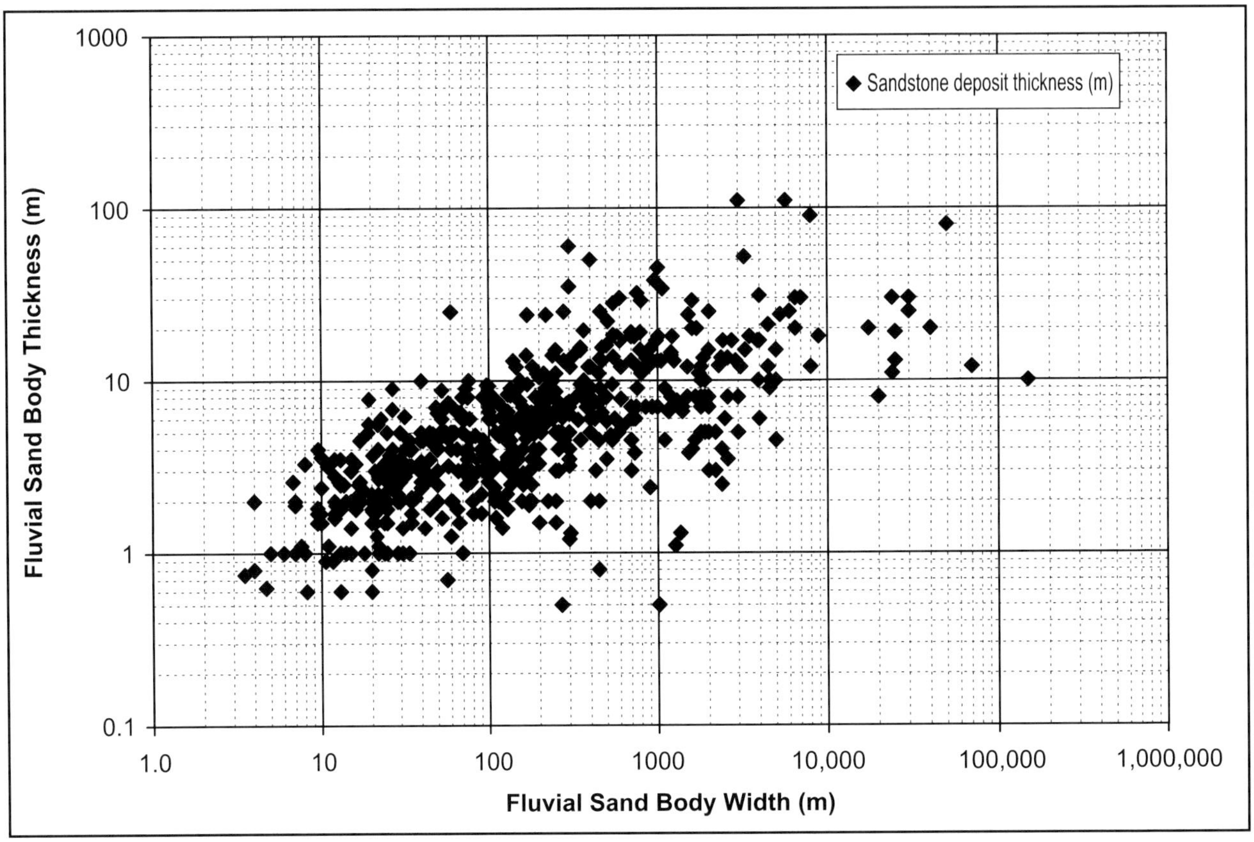

Figure 21. Log-log crossplot of fluvial sand body thickness versus fluvial sand body width compiled from a database of approximately 595 descriptions of fluvial geometry. These data are largely derived from the literature and represent well-constrained outcrop descriptions.

approximately 9 ft (3 m) in thickness, the range of calculated sand body widths range from approximately 9 to 1050 ft (3 to 320 m). For single-story sand bodies that are approximately 15 ft (5 m) in thickness, the range in calculated sand body widths range from approximately 28 to 2340 ft (9 to 710 m) (see Table 1B).

In addition to the more traditional methods of calculating sand body widths that are presented above, a probabilistic description that used sand body dimensions taken from the description of a wide variety of fluvial outcrops was also considered. Data of this type are commonly portrayed as log-log plots of width vs. thickness (e.g., Figure 21). Although these types of plots are useful, they do not easily convey the uncertainty associated with any given prediction of sand body geometry. To overcome this shortcoming, the same data shown in Figure 21 was portrayed as a series of width-thickness nomograms that describe the probabilistic range of sand body widths for a given range of thicknesses (Figure 22). In the case of the Jonah field reservoir description, specific distributions were constructed for the common range of single-story thicknesses observed in the core (see details in Figure 22). Portraying the data in terms of probability distributions provides a much better sense of the range of outcomes and their likelihood, as opposed to single-point calculations of width that are commonly used. Consideration of width-thickness data in a probabilistic sense suggests the following: when the entire fluvial database is considered ("all thickness curve"), sand bodies have a minimum (P_{90}) width of approximately 59.1 ft (18 m), a median (P_{50}) width of 492 ft (150 m), and a maximum (P_{10}) width of 6233 ft (1900 m). If only those sand bodies between 3.3 and 16.4 ft (the "1–5-m" nomogram line) are considered, the minimum (P_{90}) is approximately 46 ft (14 m), the median (P_{50}) is approximately 206.7 ft (63 m), and the maximum (P_{10}) is approximately 1476 ft (450 m). When a subset of the data that more closely approximate the thicknesses seen in the Lance Formation at Jonah and Pinedale are considered ("Jonah thickness 2.5–5.0-m"), the minimum (P_{90}) width is approximately 65.6 ft (20 m), the median (P_{50}) width is

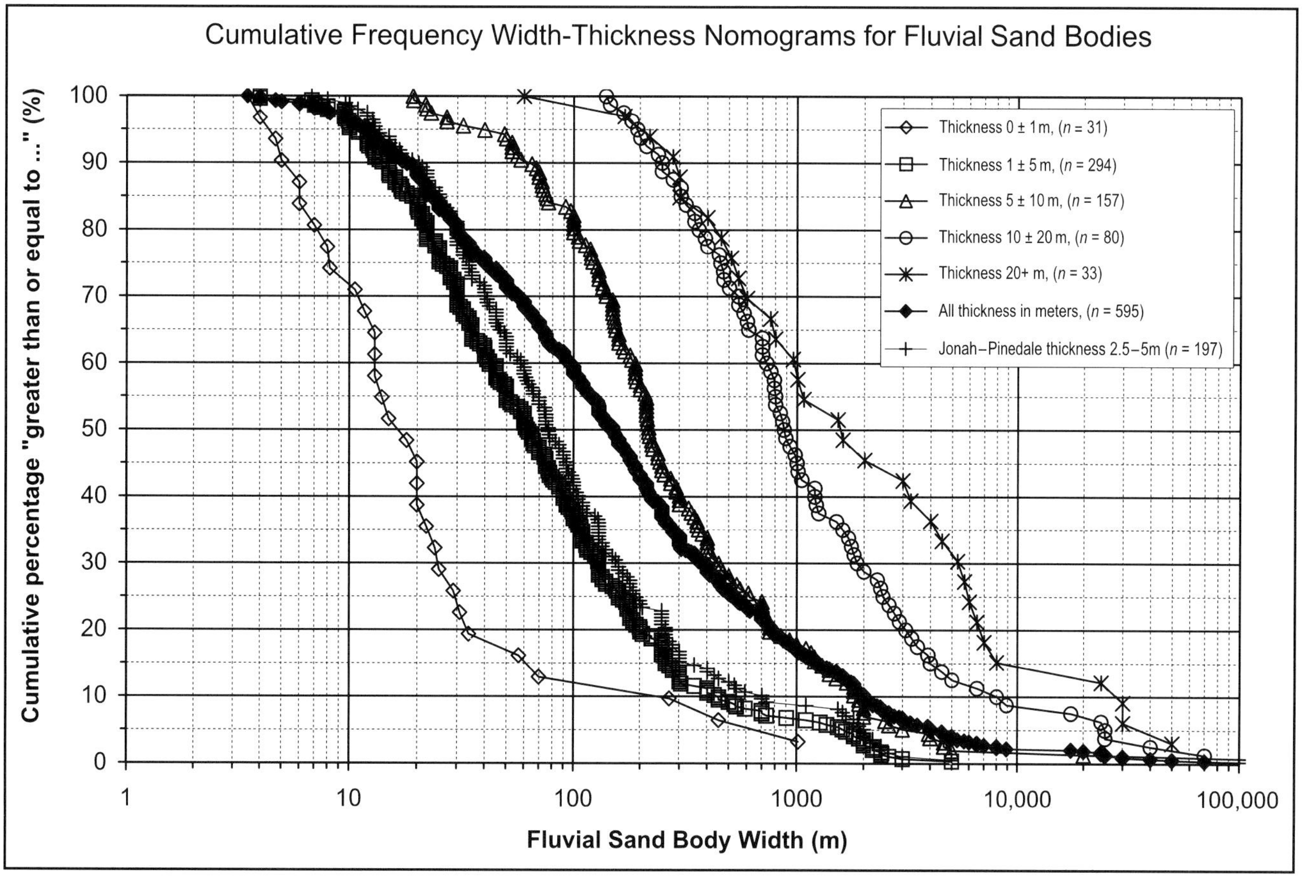

Figure 22. The same data presented in Figure 21 is illustrated as a series of "width-thickness" nomograms in which fluvial sand body width is portrayed for a "family" of thickness curves. In addition, a curve summarizing fluvial sand body widths for all data, irrespective of thickness, is shown as a cumulative frequency curve for thicknesses that are commonly observed in the Lance Formation at Jonah field. Representation of width-thickness data in this form allows for an improved understanding of uncertainty and the range of possible outcomes.

approximately 246 ft (75 m), and the maximum (P_{10}) width is approximately 2297 ft (700 m) (see Table 1C). Viewed another way, for sand body thicknesses of between 8.2 and 16.4 ft (2.5 and 5.0 m) and well spacings of 40 ac (0.16 km^2) (the original case considered for this reservoir description), approximately 14% of known sand bodies have widths equal to or greater than the closest well spacing of 1320 ft (402 m). At well spacing of 20 ac (0.08 km^2), 17% of known sand bodies have widths equal to or greater than the closest well spacing of 933 ft (284 m). As well spacing increases to 80, 160, 320, and 640 ac (0.32, 0.65, 1.3, and 2.60 km^2), the percentages fall to 12, 9, 8, and 7% respectively. Examination of the calculated sandstone body widths (Table 1B) suggests that at a well spacing of 40 ac (0.16 km^2) (closest wells at 1320 ft [402 m]), all estimations of the calculated widths for a 9-ft (3-m) single-story sandstone are less than the closest well spacing. If the sand body thickness is increased to 15 ft (5 m), then five of the nine approaches presented result in width estimates that are less than the closest well spacing at 40 ac (0.16 km^2).

The probabilistic data illustrated in Table 1C were also used to estimate the range in aerial extents of individual reservoir sand bodies, i.e., the reservoir building blocks. Using an aspect ratio of length to width equal to 1.5, areas were calculated (using the area of an ellipse) for the P_{10}, P_{50}, and P_{90} scenarios for each of the four cases considered (1–5-m case, 5- to 10-m case, all-thickness case, and Jonah sort case). Aerial extents for the "1–5-m," "5- to 10-m," "all-thickness," and the "Jonah" cases are approximately 1.1, 13.4, 6.5, and 1.6 ac (0.004, 0.05, 0.026, and 0.006 km^2), respectively. The aerial extents suggested here are for the individual reservoir sand bodies or the basic building blocks of the reservoir. Interwell communication and the compounding effect of thicker, multistory sand bodies would likely lead to wells locally contacting sandstones that are larger than the basic building block. Nevertheless, with individual wells attempting completion in 30–40 sandstone intervals per well (D. DuBois, 2003, personal communication), the estimates of the gas in place derived from a geometric analysis are directionally consistent with estimates of the gas in place per well derived from more traditional material balance techniques, suggesting that the estimates of sand body extent are appropriate.

The majority of the data presented here suggests that at 40-ac (16.2 ha) well spacing, the vast majority of sand bodies encountered in a vertical wellbore is smaller than the well spacing. An inherent problem in fully determining the size of sand bodies from limited vertical borehole data, however, is accounting for the connectivity of sand bodies in the interwell region. This problem is typically addressed through a variety of geostatistical techniques in which the possible solutions are conditioned to the well data, with the overall architecture being conditioned to net/gross trends in the Lance Formation. These techniques, however, commonly result in the larger sand bodies predicted by a distribution function always occurring in the interwell region. Developing an improved understanding of alluvial sand body geometry from vertical data alone is an active area of research. One of the more promising approaches to this problem is relating alluvial sand body geometry and connectivity to variations in vertical net/gross trends. Implicit to this approach is that sand body geometry is largely governed by the interaction of sediment flux and accommodation, and this suggests that net/gross might serve as a proxy for that accommodation-based interrelationship (e.g., Dalrymple, 2001). This approach has not been tried at Jonah; however, it is an approach that will likely provide some insights to sand body architecture.

Discussion

Consideration of the core data suggests that the vast majority of sand bodies seen in the Lance Formation in Jonah field have width dimensions significantly less than the closest well spacing at 40-ac (16.2 ha) drilling density. It is therefore most likely that sands seen in one wellbore will not be seen in an adjacent wellbore. An important issue, however, is the degree of connectivity between wellbores caused by sand bodies that lie within the interwell region and serve to connect sands in adjacent wellbores. Fluvial modeling studies (e.g., Bridge and Leeder, 1979; Bridge and Mackey, 1993) suggest that the degree of connectivity is heavily determined by the interaction of accommodation and sediment supply. At present, sufficient data is not available to comment on these variables at Jonah field. To develop some "operational comfort" concerning the degree of potential connectivity between wellbores, it is worth noting that, in general, the Lance has net/gross ratios of 10–35%, with higher net/gross ratios generally confined to the "Jonah" subinterval (40–80%). These data, coupled with the material presented in this chapter, suggest that well-to-well communication is quite limited at 40-ac (16.2 ha) spacing. In addition, limited reservoir engineering studies of closely spaced wells in the field suggest little to no well-to-well interference, and comparison of well logs from wells at varying degrees of offset distances suggest that at 40 ac (16.2 ha) (1320 ft; 402 m), well logs do not correlate well; at 20 ac (8.1 ha) (933 ft; 284 m), correlations have improved marginally; and at distances much less than 20 ac (8.1 ha), the degree of correlation improves noticeably. Finally, it is worth noting that in the Rullison field area in the Piceance basin of Colorado, the Williams Fork reservoir has a similar alluvial architecture to the Lance Formation, and drilling on both 20- and 40-ac (8.1 and 16.2 ha) spacing has seen little well-to-well interference.

ACKNOWLEDGMENTS

This chapter is based on a proprietary, in-house project performed for Amoco Production Company and its successor, BP America. Like all technology-unit projects, this work would not have been possible without the cooperation and support of the business units responsible for the Jonah asset. Within the business unit, Paul Boissevain, Chris Christensen, Dennis Cox, Gerald Craig, Lesley Evans, Bill Hanson, Roger Harris, Ira Pasternack, and Rick Ross were instrumental in supporting this project. Within the Upstream Technology Group, Allen Skorpen, Todd Stephenson, and Mike Bowman provided invaluable support for this project. John Robinson, formerly of McMurry Energy, has provided invaluable insights to Jonah field, the Green River basin in general, and to thoughts on alluvial architecture. Peter McCabe of the U.S. Geological Survey provided insights into alluvial architecture.

The manuscript was reviewed by Frank Ethridge at Colorado State University and Peter McCabe at the U.S. Geological Survey. Both reviewers provided thoughtful comments that resulted in substantial improvements to the chapter. I deeply appreciate both the time and effort expended by Frank and Peter. Finally, a debt of gratitude is owed to Bob and Sue Cluff of The Discovery Group,

Denver, Colorado. Bob and Sue have participated in numerous discussions about Jonah field over the last few years that have shaped my own thoughts. More recently, they have generously allowed me the use of their office, support and computer systems, etc. during completion of the manuscript. Without their help, this manuscript would not have been completed.

REFERENCES CITED

Bridge, J. S., 1985, Paleochannel patterns inferred from alluvial deposits: A critical evaluation: Journal of Sedimentary Petrology, v. 55, p. 579–589.

Bridge, J. S., 1997, Thickness of sets of cross strata and planar strata as a function of formative bed-wave geometry and migration, and aggradation rate: Geology, v. 25, p. 971–974.

Bridge, J. S., and M. R. Leeder, 1979, A simulation model of alluvial stratigraphy: Sedimentology, v. 26, p. 617–644.

Bridge, J. S., and S. D. Mackey, 1993, A theoretical study of fluvial sandstone body dimensions, *in* S. S. Flint and I. D. Bryant, eds., Geological modeling of hydrocarbon reservoirs: International Association of Sedimentologists Special Publication 15, p. 213–236.

Bridge, J. S., and R. S. Tye, 2000, Interpreting the dimensions of ancient fluvial channel bars, channels, and channel belts from wireline logs and cores: AAPG Bulletin, v. 84, p. 1205–1228.

Cluff, R. M., and C. M. Murphy, 2004, Stratigraphic Cross Sections through Jonah field, Sublette County, Wyoming, *in* J. W. Robinson and K. W. Shanley, eds., Jonah field: A case study of a tight-gas fluvial reservoir: AAPG Studies in Geology 52, accompanying CD-ROM, Appendix B.

Cluff, S. G., and R. M. Cluff, 2004, Petrophysics of the Lance sandstone reservoirs in Jonah field, Sublette County, Wyoming, *in* J. W. Robinson and K. W. Shanley, eds., Jonah field: A case study of a tight-gas fluvial reservoir: AAPG Studies in Geology 52, p. 215–242.

Collinson, J. D., 1978, Vertical sequence and sand body shape in alluvial sequences, *in* A. D. Miall, ed., Fluvial sedimentology: Canadian Society of Petroleum Geologists Memoir 5, p. 577–586.

Dalrymple, M., 2001, Fluvial reservoir architecture in the Statfjord Formation (northern North Sea) augmented by outcrop analogue statistics: Petroleum Geoscience, v. 7, p. 115–122.

DuBois, D. P., P. J. Wynne, T. M. Smagala, J. L. Johnson, K. D. Engler, and B. C. McBride, 2004, Geology of Jonah field, Sublette County, Wyoming, *in* J. W. Robinson and K. W. Shanley, eds., Jonah field: A case study of a tight-gas fluvial reservoir: AAPG Studies in Geology 52, p. 37–60.

Ebanks, W. J. Jr., and J. F. Weber, 1987, Shallow heavy-oil deposit in a Pennsylvanian fluvial sandstone reservoir, Eastburn field, Missouri, *in* R. F. Meyer, ed., Exploration for heavy crude oil and natural bitumen: AAPG Studies in Geology 25, p. 327–339.

Ethridge, F. G., and S. A. Schumm, 1978, Reconstructing paleochannel morphologic and flow characteristics: Methodology, limitations, and assessment, *in* A. D. Miall, ed., Fluvial sedimentology: Canadian Society of Petroleum Geologists Memoir 5, p. 703–721.

Fielding, C. R., and R. C. Crane, 1987, An application of statistical modeling to the prediction of hydrocarbon recovery factors in fluvial reservoir sequences, *in* F. G. Ethridge, R. M. Flores, and M. D. Harvey, eds., Recent developments in fluvial sedimentology: SEPM Special Publication 39, p. 321–327.

Hanson, W. B., V. Vega, and D. Cox, 2004, Structural geology, seismic imaging, and genesis of the giant Jonah gas field, Wyoming, U.S.A., *in* J. W. Robinson and K. W. Shanley, eds., Jonah field: A case study of a tight-gas fluvial reservoir: AAPG Studies in Geology 52, p. 61–92.

Hartmann, D. J., and E. A. Beaumont, 1999, Predicting reservoir system quality and performance, *in* E. A. Beaumont and N. H. Foster, eds., Exploring for oil and gas traps: AAPG Treatise of Petroleum Geology, chapter 9, p. 1–154.

Law, B. E., 1984, Relationship of source-rock, thermal maturity, and overpressuring to gas generation and occurrence in low-permeability Upper Cretaceous and lower Tertiary rocks, Greater Green River basin, Wyoming, Colorado, and Utah, *in* J. Woodward, F. F. Meissner, and J. L. Clayton, eds., Hydrocarbon source rocks of the greater Rocky Mountain region: Rocky Mountain Association of Geologists, p. 469–490.

Law, B. E., and C. W. Spencer, 1989, Geology of tight-gas reservoirs in the Pinedale anticline area, Wyoming and at the multiwell experiment site, Colorado: U.S. Geological Survey Bulletin, v. 1886, 260 p.

Leeder, M. R., 1973, Fluviatile fining-upward cycles and the magnitude of paleochannels: Geological Magazine, v. 110, p. 265–276.

Lorenz, J. C., 1985, Predictions of size and orientations of lenticular reservoirs in the 0Mesaverde Group, northwestern Colorado: Society of Petroleum Engineers Paper 13851.

Lorenz, J. C., D. M. Heinze, J. A. Clark, and C. A. Searls, 1985, Determination of widths of meander-belt sandstone reservoirs from vertical downhole data, Mesaverde Group, Piceance Creek basin, Colorado: AAPG Bulletin, v. 69, p. 710–721.

Montgomery, S. L., 1997, Jonah field: Expanding gas exploration and production in the Greater Green River basin: Petroleum Frontiers, v. 14, no. 2, 56 p.

Montgomery, S. L., and J. W. Robinson, 1997, Jonah field, Sublette County, Wyoming: Gas production from overpressured Upper Cretaceous Lance sandstones of the Green River basin: AAPG Bulletin, v. 81, p. 1049–1062.

Robinson, J. W., 1994, Fluvial architecture of the Salt Wash Member of the Morrison Formation in southern Utah: A model for reservoir heterogeneity in fluvial strata: Ph.D. Thesis 4418, Colorado School of Mines, Golden Colorado, 321 p.

Robinson, J. W., 2000, Discovery of Jonah field, Sublette County, Wyoming: The Mountain Geologist, v. 37, p. 135–143.

Robinson, J. W., 2004, Discovery of Jonah field, Sublette County, Wyoming, *in* J. W. Robinson and K. W. Shanley, eds., Jonah field: A case study of a tight-gas fluvial reservoir: AAPG Studies in Geology 52, p. 9–20.

Robinson, J. W., and P. J. McCabe, 1997, Sandstone body and shale body dimensions in a braided fluvial system: Salt Wash Sandstone Member (Morrison Formation), Garfield County, Utah: AAPG Bulletin, v. 81, p. 1267–1291.

Robinson, J. W., D. Delozier, and R. Finch, 1995, Integrated reservoir description and analysis of the Lance Formation, Jonah field, Sublette County, Wyoming, *in* Geological characterizations of the Cretaceous gas plays in the Greater Green River basin: Gas Research Institute/Department of Energy/Independent Petroleum Association of Mountain States sponsored Consortium for Emerging Gas Resources in the Greater Green River basin, p. 297–308.

Robinson, J. W., D. Delozier, and R. Finch, 1996, Integrated reservoir description and analysis of the Lance Formation, Jonah field, Sublette County, Wyoming (abs.): AAPG Bulletin, v. 80, p. 980.

Shanley, K. W. 2001, Fluvial reservoir description for a tight-gas giant: Jonah field, Green River basin, Wyoming, *in* J. W. Robinson and K. W. Shanley, eds., Tight-gas fluvial reservoirs: A case history from the Lance Formation, Green River basin, Wyoming: 2001 AAPG Annual Meeting, Denver, Colorado, June 2, 2001, AAPG Short Course 2, cosponsored by the Rocky Mountain Association of Geologists.

Shanley, K.W. and P. J. McCabe, 1994, Perspectives on the sequence stratigraphy of continental strata: AAPG Bulletin, v. 78, p. 544–568.

Webb, J. C., S. G. Cluff, C. M. Murphy, and A. P. Byrnes, 2004, Petrology and petrophysics of the Lance Formation (Upper Cretaceous), American Hunter, Old Road Unit 1, Sublette County, Wyoming, *in* J. W. Robinson and K. W. Shanley, eds., Jonah field: A case study of a tight-gas fluvial reservoir: AAPG Studies in Geology 52, p. 183–214.

Yin, P., S. Stookey, R. Surdam, 2001, Tight gas sands of the Lance Formation in Jonah field— The Jonah 2-5 core, *in* J. W. Robinson and K. W. Shanley, eds., Tight-gas fluvial reservoirs: A case history from the Lance Formation, Green River basin, Wyoming: 2001 AAPG Annual Meeting, Denver, Colorado, June 2, 2001, AAPG Short Course 2, cosponsored by the Rocky Mountain Association of Geologists.

CHAPTER 11

Petrology and Petrophysics of the Lance Formation (Upper Cretaceous), American Hunter, Old Road Unit 1, Sublette County, Wyoming

J. C. WEBB
The Discovery Group, Inc., Denver, Colorado, U.S.A.

S. G. CLUFF
The Discovery Group, Inc., Denver, Colorado, U.S.A.

C. M. MURPHY
The Discovery Group, Inc., Denver, Colorado, U.S.A.

A. P. BYRNES
Kansas Geological Survey, The University of Kansas, Kansas, U.S.A.

ABSTRACT

The American Hunter Old Road unit 1 was drilled to test multiple, stacked sandstones in lower Tertiary and Upper Cretaceous sandstones but was completed in 1981 as a water well producing from a shallow aquifer. Poor fluid recoveries from a drillstem test in the middle Lance Formation indicated a low-permeability reservoir with probable formation damage. Two cores, totaling 106 ft (32 m), were cut in the middle and lower Lance Formation. The cored intervals consist of thin- to medium-bedded, fine- to medium-grained sandstone and chert pebble conglomerate interbedded with silty mudstone and shale. Conglomeratic beds are cross-bedded and massive to normally and inverse graded. Sandstones are trough cross-bedded, massive and ripple laminated, and locally root mottled. Mudstones are rooted and burrowed. Conglomerate and sandstone were deposited in fining-upward genetic units by small meandering rivers. Mudstones were deposited in flood plain, swamp, lacustrine, and brackish to normal marine environments.

Sandstones consist of sublithic to lithic arenite and are cemented by moderate amounts of mixed-layer illite-smectite and quartz, sparse siderite, pyrite, ferroan calcite, and kaolinite. Feldspars are very sparse. Rock fragments are dominated by chert, with sparse limestone, dolomite, shaly and silty mudstone, reworked glauconite, and phosphate grains. Porosity consists of moderate to sparse, modified primary intergranular, clay-lined and clay-filled intergranular, secondary intergranular and grain-moldic pores, and natural fractures. Most natural fractures are filled by kaolinite but retain some permeability. Core-measured in-situ porosities range from 4 to 13%. In-situ Klinkenberg permeabilities range from 0.001 to 2.66 md. "Irreducible" water saturations, representing water saturations at gas column heights of approximately 750 ft (230 m), range from 21 to 72%, in close agreement with log-derived data, indicating that most of the potential reservoir sandstones are at or near irreducible water saturation. Relative gas permeability at irreducible water saturation ranges from 46 to 99% of the absolute permeability. Common clay-filled microporosity is responsible for low measured permeability, relatively high values of irreducible water saturation, and moderate susceptibility to formation damage.

Jonah Field: Case Study of a Giant Tight-Gas Fluvial Reservoir, J. W. Robinson and K. W. Shanley, eds.: AAPG Studies in Geology 52 and Rocky Mountain Association of Geologists 2004 Guidebook

INTRODUCTION

The American Hunter Old Road unit 1 spud in January and was drilled to a total depth of 12,422 ft (3790 m) in March 1981. It is located approximately 10 mi (16 km) south of the southern bounding fault of Jonah field (Figure 1). Drilled before the discovery well was drilled at Jonah field, the nearest production at the time from Tertiary or Upper Cretaceous rocks was at Pinedale field, more than 20 mi (30 km) to the north, and at Big Piney field, approximately 25 mi (40 km) to the west. The Old Road well was drilled as a basin-centered gas prospect as part of an exploration program in the northern Green River basin conducted by American Hunter Exploration, Ltd., Calgary, Canada, and was slated to test the Fort Union Formation (Paleocene), Lance and Mesaverde sandstones (Upper Cretaceous), and the Upper Cretaceous Hilliard (Baxter) Shale.

A drillstem test (DST) in the middle Lance (8344–8406 ft; 2540–2560 m) recovered water-cut mud and a trace of gas and was interpreted to indicate a low-permeability formation with probable formation damage (Johnson Schlumberger Computerized Data Analysis, 1981). Core analysis and logs in the DST interval indicated fair to good porosity (4–14%) but low average permeability (0.001–4.2 md, geometric average 0.076 md). Traces of oil were reported in the fluid-saturation data (Core Laboratories, Inc., 1981). Results of a second DST in the underlying Blair sandstone were less encouraging, and the well was eventually completed as a water supply well in October 1981. Mud weight at total depth was reported at 9.1 lb/gal (1090 kg/m^3). No other mud compositional data were reported. No mud logs are on file with the Wyoming Oil and Gas Conservation Commission. None of the potential reservoir zones in the well were fracture treated prior to abandonment.

A total of five cores were cut from what was originally identified as Lance, upper Mesaverde, and Blair formations. Our subsequent stratigraphic work indicates that

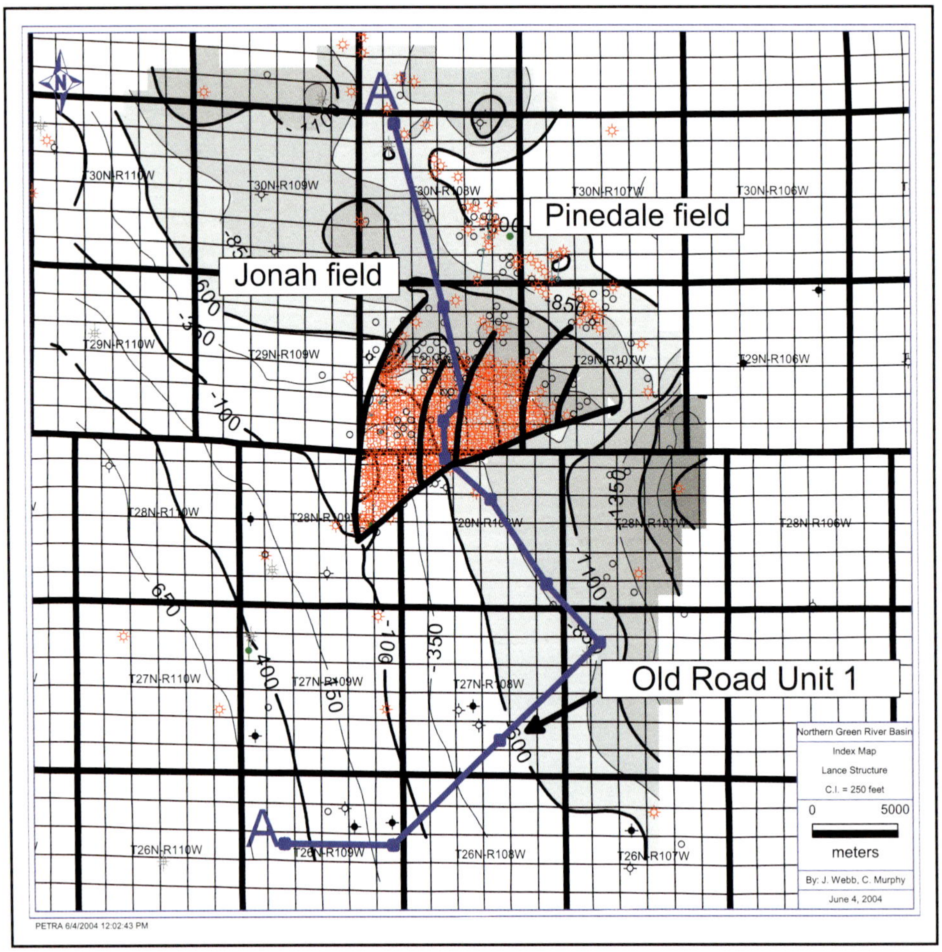

Figure 1. Map showing location of American Hunter Old Road unit 1, Sublette County, Wyoming, and the line of north-south cross section, northern Green River basin, Wyoming. Arrow indicates location of the American Hunter Old Road unit 1.

the uppermost cored interval (Lance) falls within the Unnamed Tertiary unit, below the base of the lower Fort Union, whereas the middle cored interval (originally reported as upper Mesaverde) is correlated as middle and lower Lance (Figure 2; Table 1). Petrographic data indicate that sandstones of the Unnamed unit are feldspathic and may be genetically related to the lower Fort Union. Further stratigraphic work is needed to reconcile this issue.

Cores 2 and 3 (8346–8452 ft [2544–2576 m], Figure 3) in the middle Lance have sedimentologic, petrographic, and petrophysical characteristics of a prospective low-permeability gas reservoir that has been conventionally cored, logged with modern logs, and drillstem tested (Figure 3). All of the core material from this well is stored at the U.S. Geological Survey Core Research Center (USGS CRC), Denver, Colorado, and is available to the public (USGS CRC E-894). Routine core analysis data were provided by the operator and are archived at the USGS CRC (Core Laboratories, Inc., 1981). Special core analysis (SCAL) data, comprising analysis of in-situ Klinkenberg gas permeability (k_{ik}), irreducible water saturation (S_{iw}), and effective and relative gas permeability at S_{iw}, were also obtained from samples of the Old Road unit 1. Copies of the SCAL data were provided to the USGS CRC as part of the sampling agreement (Byrnes, 1994). Petrographic data are based on analysis of 10 thin sections prepared from core samples of the middle and lower Lance (Table 2). Thin sections are archived at the USGS CRC. Three samples from the middle Lance were selected for x-ray diffraction (XRD) analysis (Table 2); four samples were selected from the middle Lance for scanning electron microscopy (SEM) analysis (Table 2). Additional supporting data are included in Appendix F on the CD-ROM accompanying this book.

LITHOFACIES

Eight lithofacies are present in the cored intervals and are described below.

Chert Pebble Conglomerate

Chert pebble conglomerates are light gray to medium olive gray, moderately to poorly sorted, clast supported, having cross-bedded to massive, inverse, and normally graded beds (Figure 4A). Moderately to poorly sorted sandy matrix is common. The conglomerate occurs in 1.5 ft (0.5 m) thick beds above sharp basal contacts in fining-upward sequences. Net thickness in the cored interval is approximately 3.5 ft (1 m). Conglomerates are sharply overlain by massive to cross-bedded, medium-grained sandstone and pebbly sandstone (Figure 3).

Medium-grained Sandstone and Pebbly and Granulitic Sandstone

Medium-grained sandstone and pebbly and granulitic sandstone are light gray, moderately to well sorted, massive to cross-bedded, with beds 1.5–2 ft (0.5–0.6 m) thick, and with angular shale lithoclasts locally present (Figure 4B). Net thickness in cored interval is 6 ft (2 m). Medium-grained sandstones are sharply overlain by fine sandstone and occur in the lower portions of fining-upward sequences (Figure 3).

Fine-grained Sandstone

Fine-grained sandstones are light gray, well sorted, planar tabular, trough, massive, and ripple cross-laminated with beds ranging from 1.5 to 4 ft (0.5 to 1.2 m) thick (Figure 4C). Net thickness in cored interval is 21 ft (6.3 m). This lithofacies is present in the middle to upper portions of fining-upward sequences and as discrete beds. Fine-grained sandstone is sharply overlain by conglomerate and medium sandstone or grades upward into bioturbated, fine-grained shaly sandstone (Figure 3).

Fine-grained Shaly Sandstone

Fine-grained shaly sandstones are moderate to dark olive gray, poorly sorted, massively bedded to strongly rooted, locally rippled, with beds ranging from 2 to 5 ft (0.6 to 1.5 m) thick (Figure 4D). Net thickness in the cored interval is 18 ft (5.5 m). Rooted, fine-grained, shaly sandstone grades into underlying sandstone or mudstone. Elsewhere, fine-grained shaly sandstone sharply overlies and is interbedded with ripple-laminated, bioturbated, very fine sandstone and mudstone. This lithofacies is present in the upper part of fining-upward sequences and is locally interbedded with shale and sandy and silty mudstone (Figure 3).

Very Fine Sandstone

Very fine sandstones are light gray to light olive gray, moderately to well sorted, locally shaly, low-angle planar cross-bedded, ripple laminated, having wavy and locally contorted bedding, and in 1–3 ft (0.3–1 m) thick beds (Figure 4E). Net thickness in cored interval is 5 ft (1.5 m).

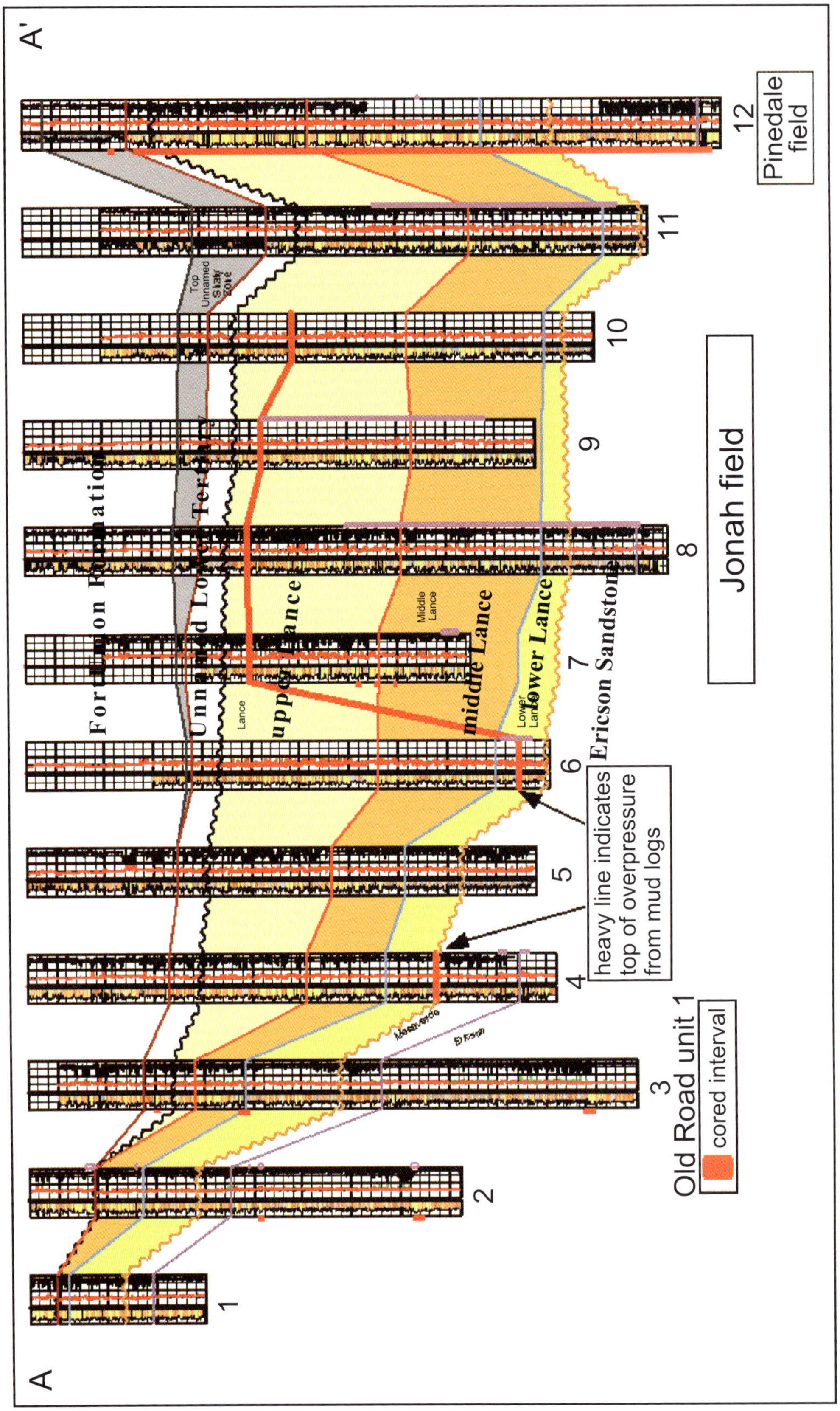

Figure 2. North–south structural cross section of the Lance Formation (Upper Cretaceous), the overlying unnamed interval, and lower Fort Union (Paleocene), northern Green River basin, Sublette County, Wyoming. Wells included in the cross section are listed in Table 1. Note the thinning of the Lance Formation from north to south and the truncation of the lower Lance and upper Lance by unconformities.

Table 1

List of Wells Included in Cross Section AA′.

Well Number	Operator	Lease	Section	Township, Range	County	State
1	Davis Oil Company	Government Mile Canyon 1	17	T26N, R109W	Sublette	Wyoming
2	American Hunter Exploration	Aspirin 1	13	T26N, R109W	Sublette	Wyoming
3	American Hunter Exploration	Old Road unit 1	27	T27N, R108W	Sublette	Wyoming
4	Snyder Oil Corporation	Sagebrush Federal 3-8	8	T27N, R107W	Sublette	Wyoming
5	McMurry Oil Company	Jonah State 3-36	36	T28N, R108W	Sublette	Wyoming
6	McMurry Oil Company	Jonah Federal 3-15	15	T28N, R108W	Sublette	Wyoming
7	McMurry Oil Company	Jonah Federal 2-5	5	T28N, R108W	Sublette	Wyoming
8	Snyder Oil Corporation	Stud Horse Butte 13-27	27	T29N, R108W	Sublette	Wyoming
9	McMurry Oil Company	Stud Horse Butte 7-27	27	T29N, R108W	Sublette	Wyoming
10	Snyder Oil Corporation	Stud Horse Butte 1-27	27	T29N, R108W	Sublette	Wyoming
11	Amoco Production Company	Stud Horse Butte 13-3	3	T29N, R108W	Sublette	Wyoming
12	El Paso Natural Gas Company	Wagon Wheel 1	5	T30N, R108W	Sublette	Wyoming

This lithofacies is present in the upper part of fining-upward sequences and is locally interbedded with bioturbated shale and sandy and silty mudstone (Figure 3).

Very Fine Shaly Sandstone

Very fine shaly sandstones are medium to light olive gray, moderately to poorly sorted, ripple and planar laminated, rooted, locally burrowed, contorted, and slumped, with beds 0.5–2 ft (0.2–0.6 m) thick (Figure 4F). Net thickness in cored interval is 10 ft (3 m). This lithofacies is present in the upper part of fining-upward sequences and is interbedded with shale, silty mudstone, and siltstone (Figure 3).

Shale, Silty and Sandy Mudstone, and Siltstone

Mudstone and siltstone are dark gray to medium olive gray, moderately to poorly sorted, rooted, and locally carbonaceous, having common plant impressions, massive bedding, and irregular, blocky to conchoidal fracture (Figure 4G). Shale and silty and sandy mudstone beds range from 0.5 to more than 10 ft (0.2 to more than 3 m) thick, with a net thickness of 31.5 ft (9.5 m). Shale and silty and sandy mudstone beds occur in the upper parts of fining-upward sequences and overlie very fine sandstone and shaly sandstone with sharp to gradational contact. Siltstones are planar and ripple laminated and rooted, with local contorted bedding. Siltstone beds range from 0.5 to 2 ft (0.2 to 0.6 m), having a net thickness of 10 ft (3 m). Siltstone overlies shale and silty and sandy mudstone with sharp to gradational contacts. Siltstone is interbedded with shale and silty and sandy mudstone and occurs in the upper part of fining-upward sequences (Figure 3).

Bioturbated Shale, Silty and Sandy Mudstone and Siltstone, and Shaly Sandstone

These are black, dark gray to medium olive gray, moderately to poorly sorted, planar laminated to bioturbated, and locally carbonaceous, with sparse plant impressions and moderate sand-filled burrows, and blocky to planar fracture (Figure 4H). Bioturbated shale and silty and sandy mudstone beds range from less than 0.25 to 1 ft (0.07 to 0.3 m) thick, with a net thickness of 3.5 ft (1.1 m). Bioturbated shale and mudstone overlie very fine sandstone and shaly sandstone with sharp contact. Bioturbated shale and silty and sandy mudstone beds occur in the upper parts of fining-upward sequences and are interbedded with rooted and bioturbated siltstone (Figure 3). Siltstones and shaly sandstones are bioturbated, and planar and ripple

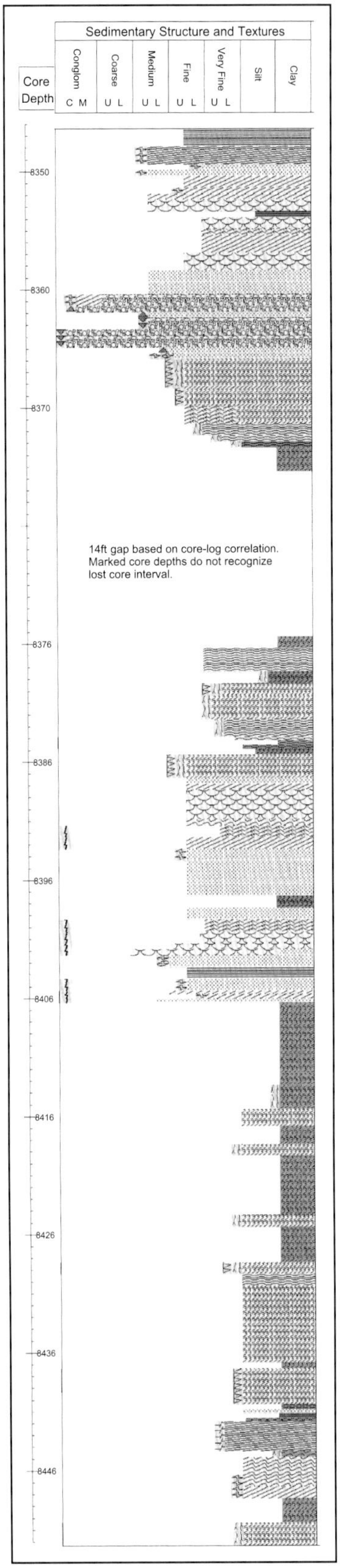

laminated, with locally contorted bedding. Siltstone beds range from 0.5 to 2 ft (0.2 to 0.6 m), with a net thickness of 4 ft (1.2 m). Bioturbated siltstone and shaly sandstone is interbedded with shale, silty and sandy mudstone and rooted siltstone, and very fine shaly sandstone. Basal contacts are gradational, and upper contacts with rooted facies are gradational (Figure 3).

LITHOFACIES DISTRIBUTION

The cored interval can be subdivided into three lithologic units: (1) an upper sandstone consisting of lower fine to upper medium sandstone and conglomerate and sparse upper very fine sandstone (8344–8372 ft; 2543–2552 m); (2) a lower sandstone consisting of upper very to lower fine sandstone and sparse lower medium sandstone (7385–8406 ft; 2251–2562 m); and (3) mudstone, dominated by shale, silty mudstone, and siltstone, with sparse very fine sandstone (8307–8452 ft; 2532–2576 m). The upper and lower sandstones are separated by 3 ft (1 m) of mudstone and 14 ft (4.6. m) of lost core (Figure 3) and comprise the basal portion of the middle Lance Formation. Regional stratigraphic work suggests that the lower sandstone overlies an unconformity. The mudstone in the lowermost interval represents the top of the lower Lance.

ENVIRONMENT OF DEPOSITION

Sharp basal contacts, strongly fining-upward character, interbedded mudstone and carbonaceous mudstones, rooted intervals, and poorly developed paleosols indicate that the sandstones of the middle Lance were deposited in meandering fluvial channels and sparse crevasse splays (Figure 3). These features are similar to those reported for cores in the Lance Formation at Jonah field (Shanley, 2001). Sandstone bodies consist of stacked deposits of amalgamated point bars and channel levees. Conglomerates and medium-grained sandstones were deposited in the lower to middle portions of point bars. Fine- to very fine-grained sandstones were deposited in upper point bars, channel levees, and crevasse splays. Soft sediment deformation, including contorted and slumped beds, was observed in very fine- to fine-grained sandstone. These features are characteristic of collapsed point-bar and levee

◄

Figure 3. Core description for the middle and lower Lance Formation, American Hunter Old Road unit 1, Sublette County, Wyoming.

deposits caused by falling water level and channel migration. Overbank splays are encased in mudstone, are finer grained, and are thinner bedded than the point-bar sandstones. Subaerial exposure resulted in the development of rooted zones, incipient paleosols, the introduction of detrital clay by mechanical infiltration in conglomerate and medium sandstone, and bioturbation (primarily because of plant roots) in sandstones. This is especially prominent in the levee and splay sandstones. Thin, poorly developed paleosols and rooted zones indicate intermittent subaerial exposure of sediments deposited in overbank and floodplain environments. Massively bedded silty mudstone with conchoidal fracture and sparse carbonized plant impressions was deposited in abandoned channel (clay plug) and lacustrine environments.

Mudstones and shaly sandstones in several intervals are burrowed (Figure 5). Burrows have been tentatively identified as *Teichichnus* sp., *Planolites* sp., *Palaeophycus*(?) sp., and *Helminthopsis* sp. (R. Martinson, 2003, personal communication). In addition, probable *Thalasinoides* sp. and *Skolithos* sp. are present (Figure 5). These trace fossils occur in open-marine, upper offshore to brackish environments (Pemberton et al., 2001). *Helminthopsis* sp. are characteristic of deeper water, including lower to upper offshore environments, but has been reported from low-energy, fine-grained marine bay environments (Pemberton et al., 2001). Rooting overprints some of these occurrences. The overlying thin, sharp-based, fining-upward sequences associated with burrowed mudstone suggest that transition from brackish or marginal marine environments to nonmarine environments may have occurred under low-energy conditions, such as in marsh, lagoonal, or marine embayment settings. Associated ripple-laminated, rooted very fine sandstone, shaly sandstone, and siltstone may represent low-energy tidal channel and lagoonal shorelines.

PETROGRAPHY AND MINERALOGY

Conglomerates and conglomeratic sandstones are composed of well-rounded chert pebbles and sparse sandstone pebbles (maximum pebble diameter <1 cm), with sparse sandy matrix (in clast-supported conglomerate) to moderate amounts of fine- to medium-grained sand in pebbly sandstone (Figure 6). The conglomerates exhibit a bimodal sorting, with both the pebbles and interstitial sand grains being moderately to poorly sorted. Chert pebbles exhibit a wide variety of textures, including ghost textures from lime grainstone, skeletal packstone, spiculitic and probable bedded chert, dolomitic limestone, and phosphatic chert (Figure 6B). The chert pebbles were derived from older Paleozoic sediments, most likely the Phosphoria Formation and Madison Limestone. Extremely sparse volcanic rock fragments (altered to bentonite) are also present in the conglomerate. Based on textural and point count analysis, the conglomerates are classified as sandy sedlithrudites. Both samples of conglomerate plot very near to the lithic pole of the Q:F:L ternary diagram in Figure 7. Sedimentary rock fragments (primarily chert) are the most abundant lithic grains in the conglomerates (Table 2). In a ternary plot of volcanic + plutonic:metamorphic:sedimentary (V + P:M:S) composition of the lithic fraction, the conglomerates plot closest to the sedimentary rock fragment pole (Figure 8).

Moderate to common amounts of authigenic clay, quartz overgrowths, and sparse siderite, pyrite, and ferroan calcite cement the conglomerates (Table 2). Clay cement lines and partly fills intergranular areas (Figure 6C, D) and is composed of mixed-layer illite-smectite (ML-IS), kaolinite (Figure 9A, B), and local chlorite (Figure 9E). Illite was detected during XRD analysis of the clay fraction but was not observed as a discrete phase in either thin-section petrography or SEM analysis. A possible exception might be the fibrous projections on ML-IS crystals that were observed during SEM analysis of sandstone samples.

Porosity in the conglomerates and conglomeratic sandstone consists almost entirely of clay-filled intergranular microporosity (Figure 6D, E). The flaky to filamentous crystal morphology and high specific surface area of the clay cement (in particular ML-IS clay) create microporosity by subdividing intergranular pores. Only traces of intergranular mesoporosity (Figure 6C) and fractures in chert grains are present (Figure 6F). Estimated mesoporosity (open intergranular and moldic pores) from point-count data ranges from trace to 1% of the entire rock volume in the conglomerates. Estimated microporosity (clay-filled intergranular and intragranular pores) ranges from 6 to 8% (Table 2).

Sandstones range from very fine to medium grained and are moderately to well sorted, with subangular to subrounded grains (Figures 10, 11). Fine-grained sandstone and fine-grained shaly sandstone comprise the bulk of the sandy lithofacies in the core. Angular to rounded shale lithoclasts (as much as 1.2 in. [3 cm]) are locally common (Figure 3). Plagioclase and potassium feldspar are conspicuously absent, with the exception of a sample from a crevasse-splay sandstone (8445 ft; 2574 m) in the lower, mudstone-dominated portion of the core (Table 2). Samples from the upper sandstone (8344–8372 ft; 2543–2552 m) are more coarse grained and contain more rock fragments than samples from the lower sandstone (Table 2; Figure 3). Samples from the upper sandstone are classified

Table 2

Petrographic Data for Middle and Lower Lance Sandstones, Old Road Unit 1, Sublette County, Wyoming.*

Sample Number	Core Depth (ft)	Log Depth (ft)	Helium Porosity**	Air Permeability (md)	Grain Density** (g/cm^3)	Water Saturation**	Member or Formation	XRD Sample	SEM Sample	Grain Size	Sorting	Detrital Matrix
1	8349.7	8347.7	9.4	0.19	2.66	46.6	middle Lance			UF/LM	well	0.3
2	8354.5	8352.5	11.3	0.140	2.67	41.5	middle Lance	XRD	SEM	UVF/LF	well	0.0
3	8360.8	8358.8	8.7	1.500	2.67	62.7	middle Lance		SEM	M/sm Pbl	very poor	0.7
4	8362.0	8360.0	8.8	0.160	2.66	56.2	middle Lance	XRD	SEM	L/UM/sm Pbl	poor	0.0
5	8363.9	8361.9	6.0	0.230	2.66	57.4	middle Lance			UVF/sm Pbl	very poor	3.0
6	8390.7	8404.7	13.8	1.300	2.63	51.5	middle Lance	XRD	SEM	L/UF	well	0.0
7	8394.8	8408.8	12.1	0.85	2.65	43.0	middle Lance			UVF-LM	moderate	0.0
8	8399.5	8413.5	11.0	0.30	2.64	31.7	middle Lance			L/UF	moderately well	0.0
9	8404.7	8418.7	7.3	0.21	2.64	35.2	middle Lance			UVF-UF	moderate	0.0
10	8445.0	8459.0	6.0	0.01	2.69	82.6	lower Lance			UVF-LF	poor	1.0

	Cements								Porosity Types			
Sample Number	Siderite	Pyrite	Quartz Overgrowth	Ferroan Calcite	Ferroan Dolomite	ML/Chl clfBP	Kaolinite Intergranular	Kaolinite-filled Fracture	Intergranular	Moldic	Fracture	Intragranular
1	0.0	0.0	15.0	0.0	0.0	5.0	0.0	0.0	1.3	0.7	0.3	2.2
2	0.0	0.0	5.7	0.0	0.3	10.3	1.0	0.0	2.0	0.0	0.0	1.0
3	0.0	1.0	1.3	0.0	0.0	4.0	3.0	0.0	0.3	0.0	0.0	3.0
4	0.0	0.0	3.3	5.0	0.0	5.7	2.0	0.0	1.3	0.0	0.0	1.2
5	0.3	0.7	0.0	0.0	0.0	4.0	0.0	0.0	0.0	0.0	0.0	1.3
6	0.0	0.0	2.7	0.0	0.0	16.0	0.7	0.0	4.7	1.0	0.0	1.0
7	0.3	0.0	13.3	0.0	0.0	13.0	0.0	0.0	6.7	1.0	0.0	1.0
8	0.3	0.0	9.7	0.0	0.0	9.3	0.3	1.7	9.0	1.3	0.0	2.3
9	0.0	0.0	12.7	0.0	0.0	10.7	0.3	0.0	5.3	1.0	0.0	2.5
10	1.0	0.0	1.7	6.0	0.0	13.0	0.0	0.0	0.0	0.0	0.0	1.0

Table 2

Petrographic Data for Middle and Lower Lance Sandstones, Old Road Unit 1, Sublette County, Wyoming (cont.).

	Porosity Types (cont.)		Porosity Summary							Framework Population		
Sample Number	Clay-filled Intergranular	Micro Intragranular	Mesoporosity	Microporosity	Fracture	Total Porosity	Mesoporosity of total porosity (%)	Microporosity of total porosity (%)	Fracture of total porosity (%)	Quartz	Feldspar	Lithic
1	2.5	0.4	2.0	5.1	0.3	7.5	26.8	68.7	4.5	75.4	0.0	24.6
2	5.2	2.3	2.0	8.4	0.0	10.4	19.2	80.8	0.0	71.5	0.4	28.1
3	3.6	1.1	0.3	7.7	0.0	8.0	4.2	95.8	0.0	13.6	0.0	86.4
4	3.8	0.9	1.3	5.9	0.0	7.3	18.4	81.6	0.0	52.0	0.8	47.2
5	2.3	0.3	0.0	3.9	0.0	3.9	0.0	100.0	0.0	13.4	0.0	86.6
6	8.0	0.4	5.7	9.4	0.0	15.1	37.6	62.4	0.0	85.3	0.0	14.7
7	6.5	0.2	7.7	7.7	0.0	15.3	50.0	50.0	0.0	90.8	0.0	9.2
8	4.7	0.2	10.3	7.2	0.8	18.3	56.4	39.1	4.5	86.8	0.0	13.2
9	5.5	0.6	6.3	8.6	0.0	14.9	42.5	57.5	0.0	83.8	0.0	16.2
10	6.6	0.2	0.0	7.8	0.0	7.8	0.0	100.0	0.0	50.0	1.8	48.2

	Lithic Population (V:M:S)					Textural Ratios						
Sample Number	Sedimentary	Plutonic	Volcanic	Quartzose Metamorphic	Micaceous Metamorphic	Ductile Grains	Framework	Matrix	Cement	Mesoporosity	Volume Pebbles	Intergranular Volume
1	86.0	0.0	5.3	7.0	1.8	3.9	77.3	0.3	20.0	2.3	0.0	26.7
2	60.3	0.0	30.9	8.8	0.0	8.3	80.7	0.0	17.3	2.0	0.0	29.0
3	97.4	0.0	1.8	0.9	0.0	10.0	89.7	0.7	9.3	0.3	64.0	14.3
4	79.3	0.0	18.1	2.6	0.0	10.5	82.7	0.0	16.0	1.3	0.0	23.0
5	97.4	0.0	1.7	0.9	0.0	3.3	92.0	3.0	5.0	0.0	73.3	12.0
6	51.5	0.0	33.3	12.1	3.0	2.2	75.0	0.0	19.3	5.7	0.0	39.3
7	61.1	0.0	16.7	22.2	0.0	2.0	65.7	0.0	26.7	7.7	0.0	46.3
8	63.0	0.0	18.5	14.8	3.7	2.9	68.3	0.0	21.3	10.3	0.0	37.7
9	61.8	0.0	35.3	2.9	0.0	3.8	70.0	0.0	23.7	6.3	0.0	39.7
10	52.8	0.0	37.7	4.7	4.7	14.7	77.3	1.0	21.7	0.0	0.0	35.7

* All data are in percent unless otherwise specified.
** Core analysis data from Core Laboratories, Inc., file RP-2-6443, all data are conventional core analysis, measured at ambient conditions.

Figure 4. Representative close-up core photographs from the American Hunter Old Road unit 1 for the lithofacies described in text: (A) poorly sorted, clast-supported chert pebble conglomerate with medium-grained sandy matrix, 8360.8 ft (2548.4 m); (B) medium-grained, well-sorted, trough cross-bedded sandstone with angular mudstone rip-up lag, 8351.4 ft (2545.5 m); (C) fine-grained, ripple-laminated sandstone, 8392 ft (2558 m); (D) fine-grained shaly sandstone, 8355 ft (2547 m); (E) very fine-grained, well-sorted, ripple-laminated sandstone, 8354 ft (2546 m); (F) very fine-grained, moderately sorted, rooted shaly sandstone, 8380 ft (2554.2 m); (G) silty mudstone with carbonaceous plant fragments, 8384.7 ft (2555.7 m); (H) bioturbated mudstone and laminated silty mudstone, 8353 ft (2546 m).

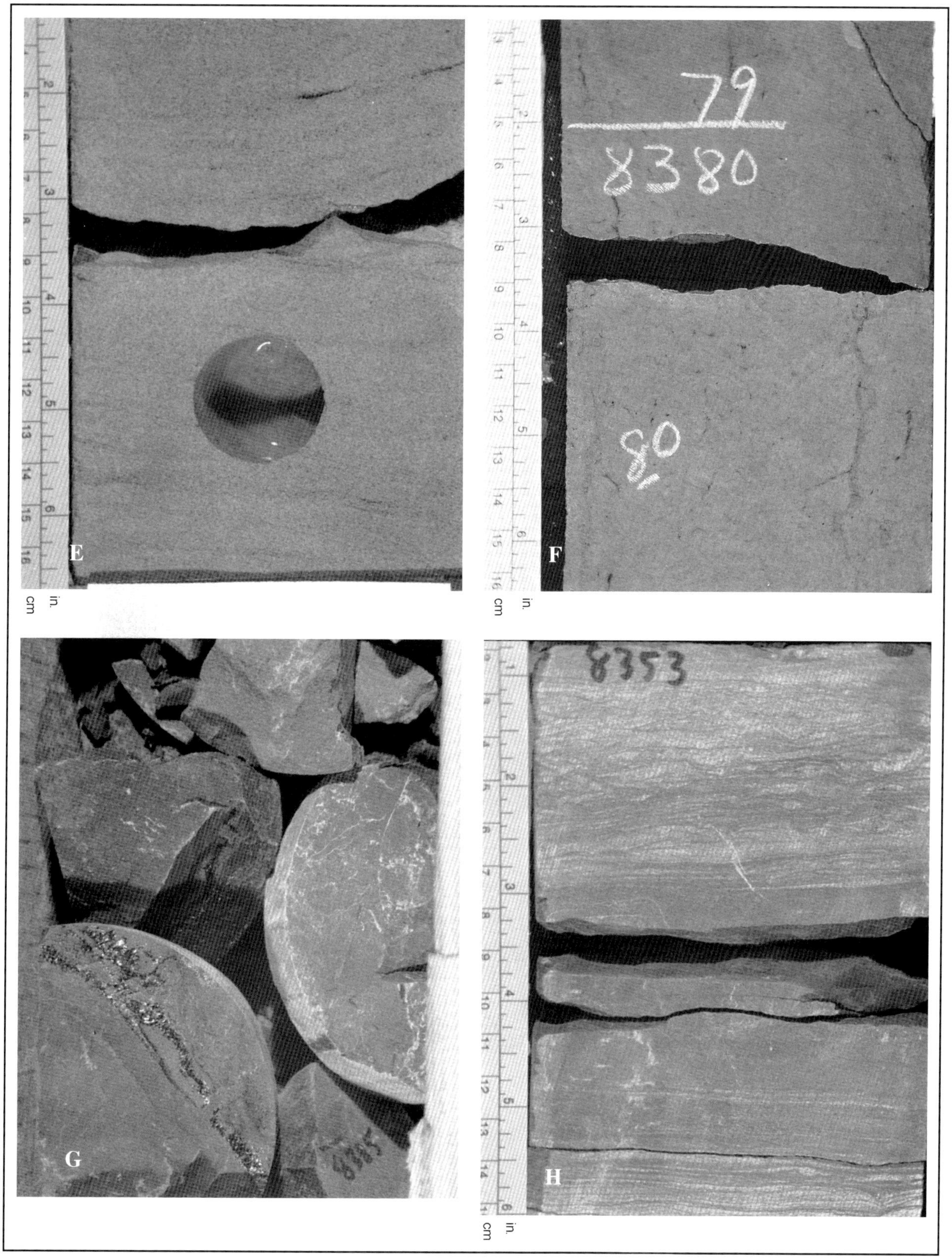

Figure 4. (cont.).

A
B
C
D
cm
in.

as litharenite, whereas those from the lower sandstone are classified as sublitharenite (Figure 7).

Detrital grains consist of common monocrystalline quartz with moderate to sparse amounts of chert, carbonate and phosphatic rock fragments, altered volcanic rock fragments, polycrystalline quartz, and quartzose and micaceous rock fragments of probable metamorphic origin (Table 2). In a ternary plot of V + P:M:S composition of the lithic fraction in sandstones, the samples in the upper sandstone plot closest to the sedimentary rock fragment pole (Figure 8), primarily because of an abundance of chert grains. The samples from the lower sandstone contain more altered volcanic and metamorphic grains (Table 2). The detrital composition and porosity distribution of the upper sandstone are more typical of sandstones in the Lance Formation. Petrographic data for samples from a core of the Lance from Jonah field, the McMurry Jonah Federal 2-5, shown on cross section AA′ (Figure 2), were published by Yin et al. (2001). When plotted on a Q:F:L ternary diagram, the data fall in the middle to upper portion (quartzose) of the litharenite field (Figure 12). Petrographic data from a Lance sandstone in core from the El Paso Natural Gas 1 Wagon Wheel at Pinedale field, also shown on cross section AA′, confirm this observation (Dincau, 1989).

Moderate to common amounts of authigenic clay, quartz overgrowths, and sparse amounts of siderite, pyrite, and ferroan calcite cement the sandstones (Table 2). Quartz overgrowths line open intergranular and moldic pores (Figure 10A, D). Clay cement lines and partly fills intergranular areas (Figure 10E, F). The clay cement consists dominantly of ML-IS (Figure 9C, D) and sparse to trace amounts of kaolinite (Figure 9C) and chlorite. The mixed-layer clay has low percentages of expandable layers (ranging from 20 to 40% smectite; Table 3), estimated by techniques described by Reynolds and Hower (1970). Scanning electron microscopy analysis reveals the pore lining and pore-throat occluding clay, with irregular flaky to filamentous crystal morphology typical of ML-IS authigenic cement (Figure 9F). In the lower sandstone, particles of clay are oriented parallel to detrital grain surfaces (Figure 11C, D) and indicate that this clay cement was deposited by mechanical infiltration of turbid water, prior to significant burial of the sediment (Walker et al., 1978). Scanning electron microscopy analysis of one of these samples indicates that this clay has recrystallized to ML-IS (Figure 9F and G). Where ML-IS and infiltrated clay cement line intergranular pores, quartz overgrowth cement is almost entirely absent (Figures 10E, 11E).

Kaolinite has locally replaced detrital grains (Figures 10E, 11G). These unstable grains may have been potassium feldspar, but no intermediate stages of this process were observed. Ferroan calcite cement locally fills intergranular pores that are lined by quartz overgrowths and, in the sample from a crevasse splay (8445 ft; 2574 m), has replaced detrital potassium feldspar grains.

Porosity consists of a mixture of sparse to moderate modified primary mesopores, sparse to moderate amounts of microporous clay-filled intergranular pores, sparse moldic pores, and very sparse natural fractures (Figure 13A, B). Primary intergranular pores are lined by quartz overgrowths, ML-IS, kaolinite, and local chlorite clay cement (Figures 9C, D; 10D, E; 11C, D). The clay cements also completely fill intergranular pores and moldic pores (Figure 10G).

Partial to complete dissolution of detrital grains and intergranular cement has resulted in the creation of secondary porosity (Figures 9, 10). Recognition criteria include inhomogeneous packing, detrital grain-sized voids, relics of former detrital grains in moldic pores, and relics of intergranular cement in intergranular voids (Schmidt and McDonald, 1979). Based on dissolution relics and preserved lithic and feldspar grains, the grains that were dissolved to form moldic pores included carbonate rock fragments, plagioclase, and potassium feldspar. The presence of grain-sized moldic pores and clots of kaolinite indicates that small amounts of chemically unstable grains have been removed by dissolution, resulting in a slightly more mature (more quartzose) Q:F:L ratio than in the original sediment. However, reconstruction of the detrital modes does not alter the original ternary Q:F:L classification.

Natural fractures are present in the sandstones within the cored interval (Figure 13A, B). They are lined by quartz overgrowths (Figure 13B, C) but are more typically filled by kaolinite cement (Figure 13A, D). Kaolinite-filled fractures have retained porosity. In thin section,

◄

Figure 5. Representative close-up core photographs for bioturbated lithofacies of the Lance Formation in the American Hunter Old Road unit 1: (A) bioturbated mudstone and wavy laminated silty mudstone, with *Planolites*, *Teichichnus*, and *Asterosoma* burrow. The upper portion of the core slab is moderately burrowed; the lower portion is heavily rooted, 8381 ft (2554.5 m). (B) Bioturbated silty mudstone and very shaly sandstone, exhibiting escape structure. Note the carbonaceous and argillaceous root structure at the top of the core slab, 8371 ft (2551.5 m). (C) Bioturbated shaly sandstone, heavily rooted and containing rhizoliths, 8437 ft (2571.6 m). (D) Very shaly sandstone, heavily burrowed, with *Planolites* sp. and other unidentifiable burrows, 8382 ft (2554.8 m).

blue-dyed epoxy has penetrated the kaolinite cement, indicating that microporosity is present in the kaolinite fracture filling (Figure 13D). Linear features that are locally present in some samples represent healed fractures (Figure 13E).

COMPACTION

Compaction is moderate, with long grain contacts being predominant. Curved contacts are common in the conglomeratic intervals, where the clast-supported fabric has

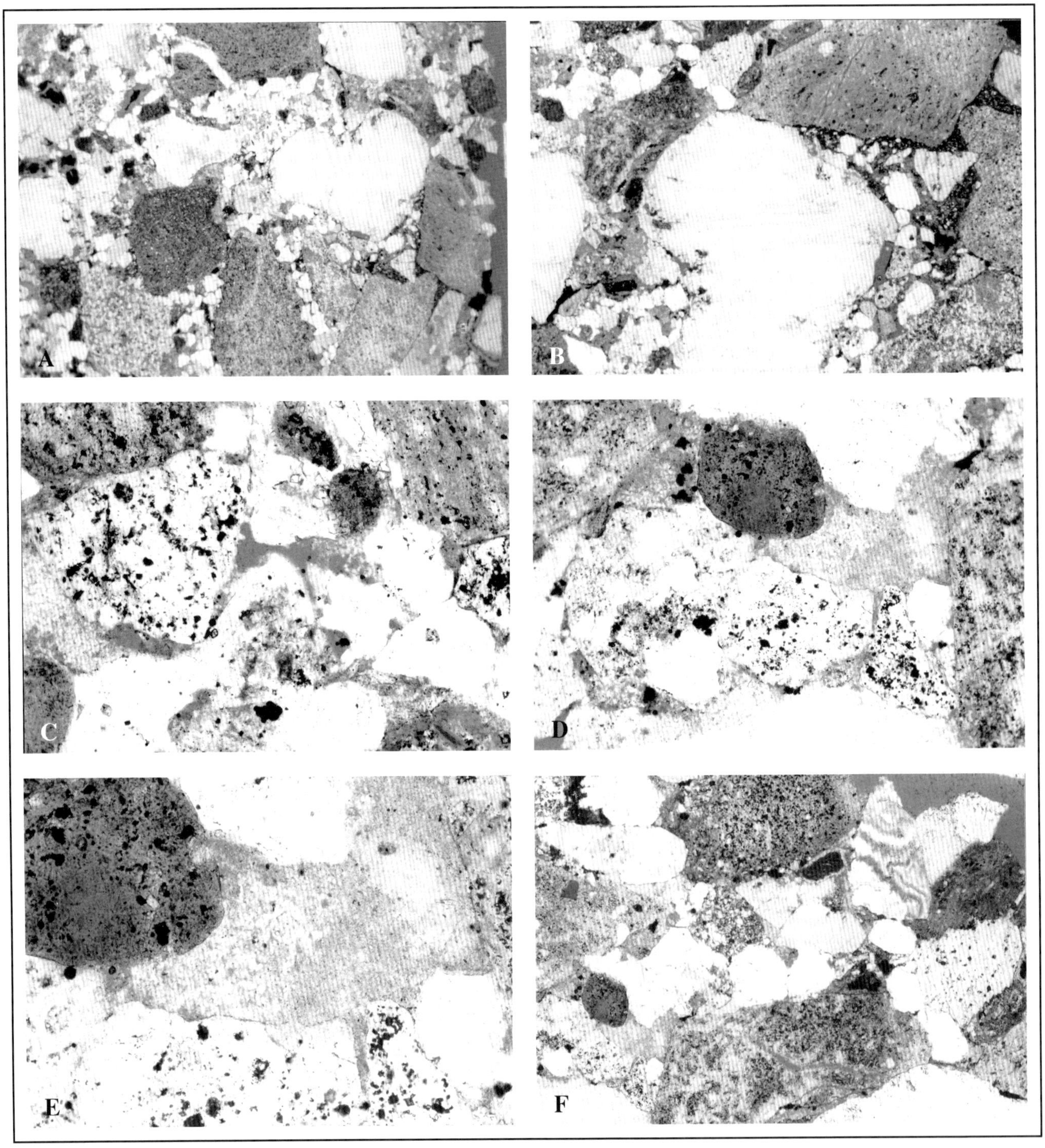

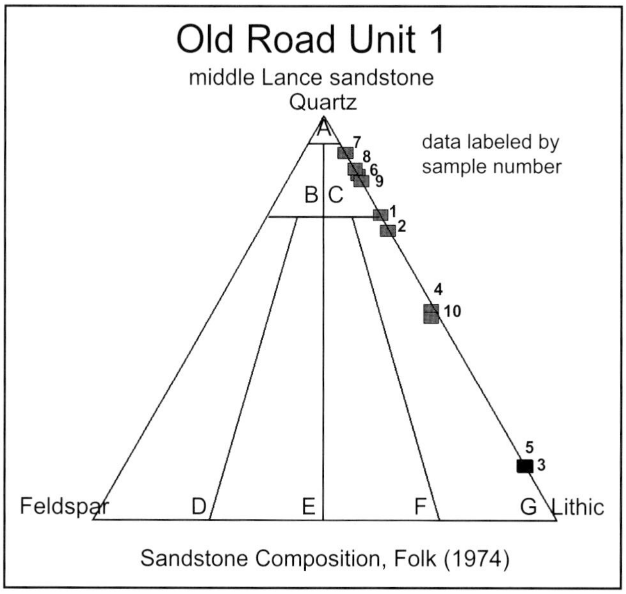

Figure 7. A ternary diagram illustrating detrital composition of sandstone and conglomerate samples, based on ratios of detrital quartz, feldspar, and lithic fragments (Q:F:L). Sandstone clans are designated according to the classification of Folk (1974). Samples in field C are sublitharenite; samples in field G are litharenite.

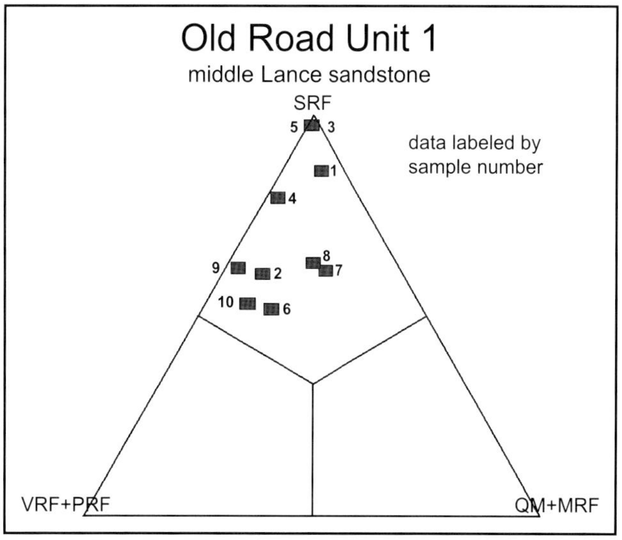

Figure 8. A ternary diagram illustrating the composition of lithic fragments in sandstone and conglomerate samples, based on the ratio of volcanic plus plutonic, quartzose and micaceous metamorphic, and sedimentary rock fragments (V:M:S). Lithic grain populations are designated according to the classification of Folk (1974).

responded to overburden pressure. Sutured to stylolitic contacts and fractured chert grains are locally present in clast-supported conglomerates. Intergranular volume (porosity plus volume of intergranular cement from point count) for conglomerates ranges from 12 to 14%. The low intergranular volume is primarily caused by the poor sorting of the conglomerates instead of excessive compaction. Intergranular volume for sandstone samples ranges from 23 to 46%. This indicates that as much as 15% of the original volume of some sandstone samples has been lost because of compaction of the unconsolidated sand (Beard and Weyl, 1973). Intergranular volumes for samples of the upper sandstone (23–29%) are lower than those of the lower sandstone (38–46%). This indicates an extremely small loss of porosity because of compaction in the lower sandstone (2–8%), and in fact, for one sample (8394.8 ft; 2558.7 m), it suggests an increase in intergranular volume beyond the typical intergranular volume for uncompacted sediments (Beard and Weyl, 1973). Ductile

◄

Figure 6. Photomicrographs illustrating characteristics of sandy conglomerate lithofacies, American Hunter Old Road unit 1, 8360.8 ft (2548.4 m). (A) Clast-supported chert pebble conglomerate with well-sorted, medium-grained sandy matrix. Long dimension of view is 12.23 mm (0.48 in.). Plane light, 10×. (B) Clast-supported chert pebble conglomerate with well-sorted, medium-grained sandy matrix. The sandy matrix has significantly reduced the porosity of this lithofacies. Intergranular mesoporosity is sparse. Compaction has resulted in open and partly cemented fractures in the chert pebbles. Long dimension of view is 6.1 mm (0.24 in.). Plane light, 20×. (C) Clast-supported chert pebble conglomerate with sparse intergranular mesopores. The neighboring intergranular areas are filled by clay cement, resulting in poor connectivity and low permeability. Long dimension of view is 1.2 mm (0.047 in.). Plane light, 100×. (D) Intergranular areas are lined by mixed-layer illite-smectite and are filled by kaolinite cement. The abundant clay cement causes low permeability and high irreducible water saturation. Long dimension of view is 0.6 mm (0.024 in.). Plane light, 200×. (E) Microporosity is present in the clay cement. Mixed-layer illite-smectite precipitated prior to pore-filing kaolinite cement. Long dimension of view is 0.6 mm (0.024 in.). Plane light, 200×. (F) Fractures in crushed chert pebbles (lower portion of the view) can enhance permeability and connectivity of the pore system in the clast-supported chert pebble conglomerate. Long dimension of view is 12.23 mm (0.48 in.). Plane light, 10×.

A
0.31kx
30kv 112
B
30kv 113
C
30kv 113
D
3.0kx
30kv 114
E
30kv 108
F
133

8390.7 ft 3000 ×, bar on left = 1 μm

Figure 9. (cont.).

grains are sparse in the conglomerates and sandstones. Therefore, deformation of ductile grains is responsible for only a minor proportion of porosity loss.

POROSITY DISTRIBUTION

Based on helium porosimetery, total porosity in these rocks ranges from 5 to 14%. Detailed petrography, SEM, and XRD analysis reveal that mesoporosity (intergranular, secondary intergranular, and moldic pores) comprises from 0 to 56% of the total porosity (Table 2). Microporosity (clay pore linings and fillings and intragranular porosity) is a large proportion of the total porosity of the middle Lance sandstone, ranging from 39 to 100% of the total porosity (Table 2). A ternary plot of the relative proportions of porosity types (intergranular mesoporosity:moldic: microporosity) illustrates the predominance of microporosity for samples from both the Lance sandstone and conglomerate (Figure 14).

THERMAL MATURITY: VITRINITE REFLECTANCE

Vitrinite reflectance (%R_O) analyses for two samples from the cored interval were performed by G. J. Jansen of the Rocky Mountain Coal Petrography, Golden, Colorado, and are reported in the well data files at the U.S. Geological Survey Core Research Center (Jansen, 1994). The sample at 8384 ft (2555 m) exhibited a reflectance of 0.52% R_O, and the sample from 8407 ft (2562 m) exhibited a reflectance of 0.64% R_O. These data indicate that the cored interval reached the upper portion of the oil generation window, but did not reach the threshold value of 0.8% R_O that has been linked to peak generation of natural gas and development of overpressure in the greater Green River basin (Pawlewicz, et al., 1986). Other types of kerogen were not identified, suggesting that the organic matter preserved in the samples from the Old Road core is gas-prone. Analyses of Mesaverde rocks in the greater Green River basin document the widespread occurence of gas-prone humic, type III kerogen (Law, 1984).

◄

Figure 9. Scanning electron micrographs of sandstone samples from the middle Lance Formation, American Hunter Old Road unit 1, Sublette County, Wyoming. (A) Kaolinite occupies a framework grain-size area, indicating either direct replacement of the unstable grain or selective filling of a grain moldic pore. Microporosity is present in the kaolinite cement. Sample depth = 8360.8 ft (2548.4 m), 310× magnification. (B) Kaolinite exhibits typical booklet crystal morphology that is the result of precipitation as cement. Kaolinite cement reduces effective porosity and occludes intergranular pore spaces. The increased specific surface area of the kaolinite cement results in elevated irreducible water saturation. Sample depth = 8360.8 ft (2548.4 m), 1500× magnification. (C) Intergranular pore lined by mixed-layer illite-smectite (left center) and quartz overgrowth cement (upper left) and subsequently filled by kaolinite cement (right center). Precipitation of the quartz cement has reduced intergranular porosity and constricted pore throats. Precipitation of the clay mineral cement has resulted in the formation of microporosity. Sample depth = 8354.5 ft (2546.5 m), 1500× magnification. (D) Pore-lining mixed-layer illite-smectite (ML-IS) exhibits typical flaky and filamentous crystal morphology that is the result of precipitation as cement. Mixed-layer illite-smectite cement reduces effective intergranular porosity and occludes pore throats. The high specific surface area of ML-IS results in elevated irreducible water saturation. Sample depth = 8354.5 ft (2546.5 m), 3000× magnification. (E) Pore-lining iron chlorite exhibits typical bladed rosette crystal morphology that is the result of precipitation as cement. Chlorite cement reduces effective intergranular porosity and occludes pore throats. The high specific surface area of chlorite cement results in elevated irreducible water saturation. Sample depth = 8362.0 ft (2548.7 m), 2000× magnification. (F) Mechanically infiltrated clay recrystallized to mixed-layer illite-smectite, lines intergranular pores in sublitharenite from the base of the middle Lance. Sample depth = 8390.7 ft (2557.5 m), 1500× magnification. (G) Recrystallized mixed-layer illite-smectite hosts overgrowths of fibrous illite. Sample depth = 8390.7 ft (2557.5 m), 3000× magnification.

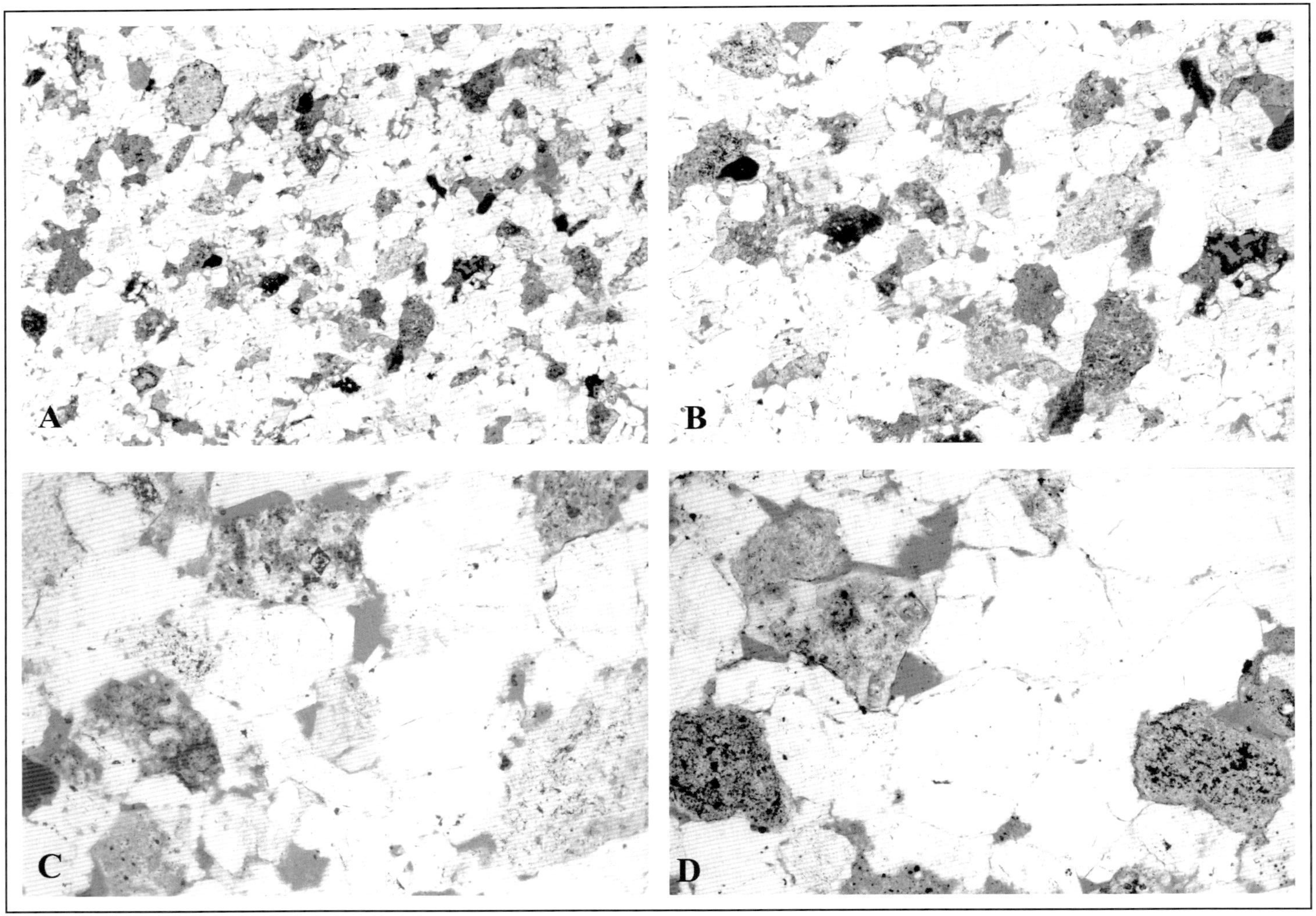

Figure 10. Photomicrographs illustrating petrographic characteristics of moderately well-sorted, medium- and fine-grained litharenite lithofacies, upper sandstone in cored interval of middle Lance sandstone, American Hunter Old Road unit 1. (A) Moderately well-sorted, upper medium-grained litharenite lithofacies, upper sandstone in cored interval of middle Lance sandstone. Lithic grains are dominated by chert with sparse altered volcanic rock fragments. Long dimension of view is 6.1 mm (0.24 in.). Plane light 20×, 8349.7 ft (2545.0 m). (B) Moderately well-sorted, upper medium-grained litharenite lithofacies, moderately well compacted, with common linear and curved intergranular contacts, and quartz overgrowth cement. Long dimension of view is 3.05 mm (0.12 in.). Plane light, 40×, 8349.7 ft (2545.0 m). (C) Intergranular pores are lined and filled by quartz overgrowths and small amounts of mixed-layer illite-smectite and local chlorite clay cement. Intergranular mesopores are poorly connected, resulting in low permeability. Long dimension of view is 1.2 mm (0.047 in.). Plane light, 100×, 8349.7 ft (2545.0 m). (D) Intragranular microporosity is present in partly dissolved chert grains (upper left and lower right). Note well-developed quartz overgrowths that line and completely fill intergranular areas. Oversized intergranular pores at the center and top center may have been formed by dissolution of prior cement. Long dimension of view is 1.2 mm (0.047 in.). Plane light, 100×, 8354.5 ft (2546.5 m).

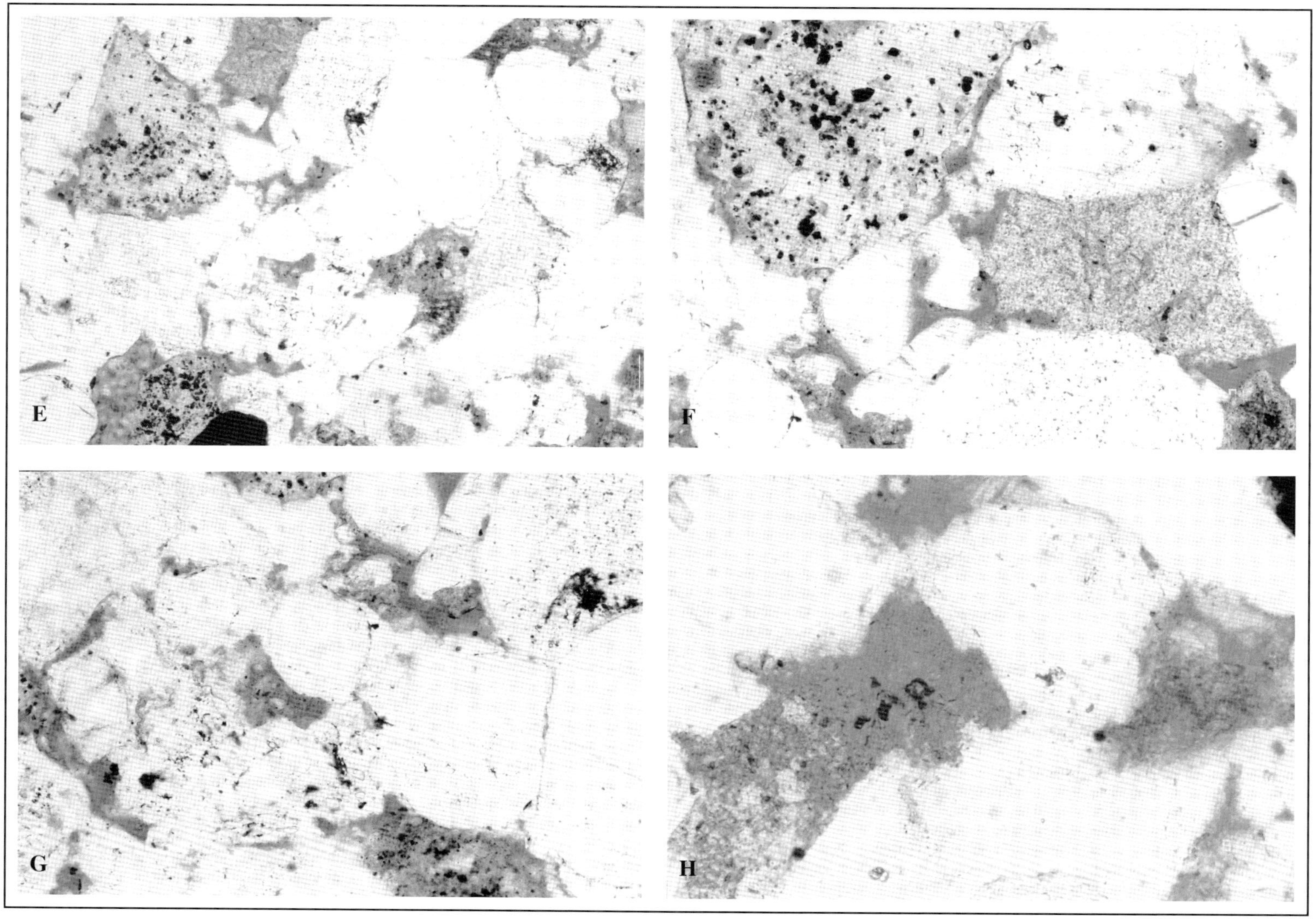

Figure 10. (cont.). (E) Framework grain-sized patches of kaolinite cement (right center and lower left) are the result of replacement of unstable grains (such as feldspar) by kaolinite. Microporosity is present in the kaolinite cement. Long dimension of view is 1.2 mm (0.047 in.). Plane light, 100×, 8349.7 ft (2545.0 m). (F) Intragranular microporosity is present in partly dissolved chert and altered volcanic rock fragments. Note crushed chert grain at right center. Long dimension of view is 0.6 mm (0.024 in.). Plane light, 200×, 8349.7 ft (2545.0 m). (G) Oversized intergranular pores were created by dissolution of prior cement. Secondary pores are moderately well connected in this portion of the sample, resulting in locally improved permeability. Long dimension of view is 0.6 mm (0.024 in.). Plane light, 200×, 8349.7 ft (2545.0 m). (H) Intergranular pores are lined and locally filled by mixed-layer illite-smectite and kaolinite cement. The clay cement reduces permeability and causes elevated irreducible water saturation. Long dimension of view is 0.3 mm (0.012 in.). Plane light, 400×, 8354.5 ft (2546.5 m).

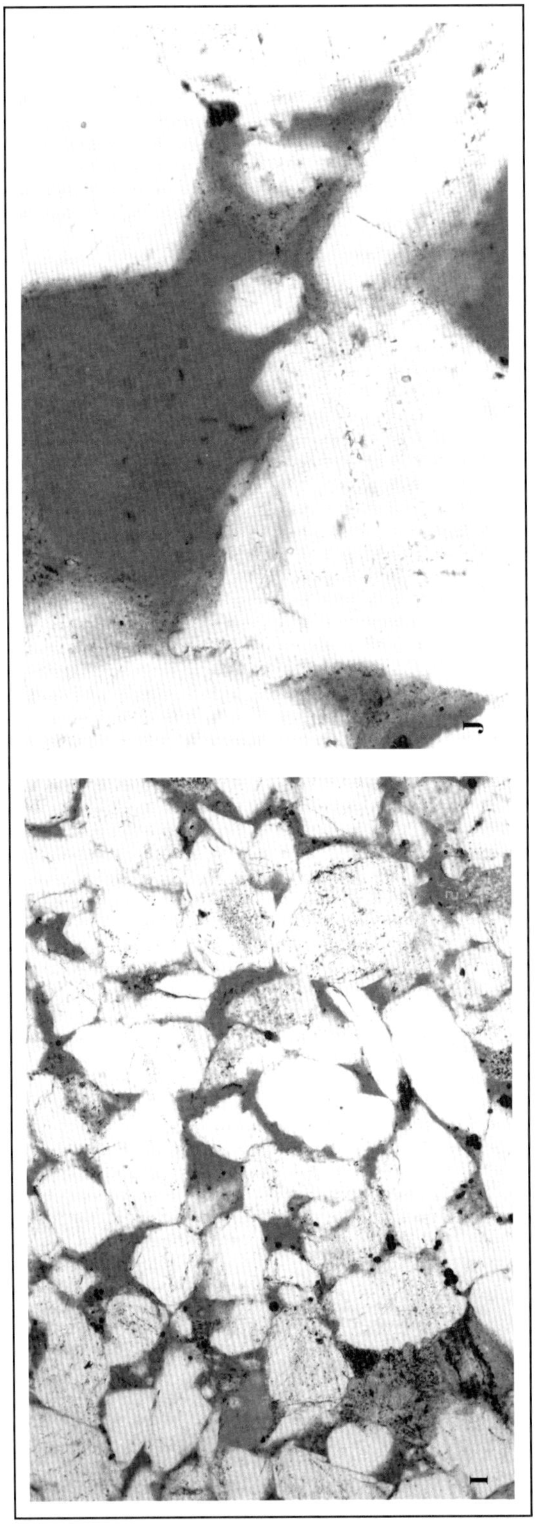

Figure 10. (cont.). (I) Open intergranular fabric of this area indicates that prior cement (probably calcite) has been removed. Long dimension of view is 1.2 mm. Plane light, 100×. (J) Mechanically infiltrated clay forms meniscus texture (right center) where the clay cement has coated intergranular pores. Thick coatings of mechanically infiltrated clay (upper left and right center) have locally inhibited the development of quartz overgrowths. Long dimension of view is 0.3 mm. Plane light, 400×.

RESERVOIR QUALITY

Total in-situ porosities range from 4 to 13% as measured from core. Correcting for water saturation, corresponding effective gas porosities range from 1.4 to 10.3%. Wire-line, log-calculated effective porosities indicate that the middle Lance interval exhibits a total porosity-height of 29.82 ϕ ft. Extensive quartz overgrowth occlusion of pore space and pore throats, the presence of pore-bridging and pore-filling mixed-layer clays, and the abundance of microporosity in many samples result in lower permeability and high irreducible water saturation. Correcting routine air permeabilities to in-situ effective gas permeability values, the estimated permeability-height for the middle Lance interval is 508.3 md ft.

Average porosity and permeability measurements for fine-grained sandstones are 9.6% and 0.339 md. Average porosity and permeability data for medium-grained sandstones (9.7% and 0.265 md, respectively) is comparable to that of the fine-grained sandstones and is lower than might be expected, given the initially larger intergranular pores in the medium sandstone. Porosity in the medium-grained sandstones has been severely occluded by quartz and clay cement, leaving very little intergranular mesoporosity (2%) (Table 2). Similarly, the average porosity and permeability in the conglomerate samples (8% and 0.763 md, respectively) are lower than might be expected, given the very large intergranular pores of the conglomerates. Poor sorting and abundant clay cement are responsible for the relatively poor values of porosity and permeability in the conglomerate samples. Moldic pores are extremely sparse in both the medium sandstone and the conglomerate, but they comprise as much as 1% of the rock volume in the fine sandstones. This amount may seem insignificant, but where moldic pores are present, they are as much as 7% of the entire porosity and 13–17% of the total mesoporosity.

Total porosity, intergranular, moldic, and clay-filled intergranular porosity are all more abundant in samples from the lower sandstone than in samples from the upper sandstone (Table 2). Most importantly, the relative proportion of microporosity in the samples of the lower sandstone ranges from 39 to 62%, compared to 69–100% for the upper sandstone and conglomerate. This relationship is largely responsible for the higher values of permeability in samples from the lower sandstone.

Microporosity in clay cement causes increased irreducible water saturation because of the high specific surface area of clay and capillary attraction of water to the clay cement (Byrnes, 1997). This is reflected in water saturation measurements in the routine core data, which can be an indicator of the relative magnitude of irreducible

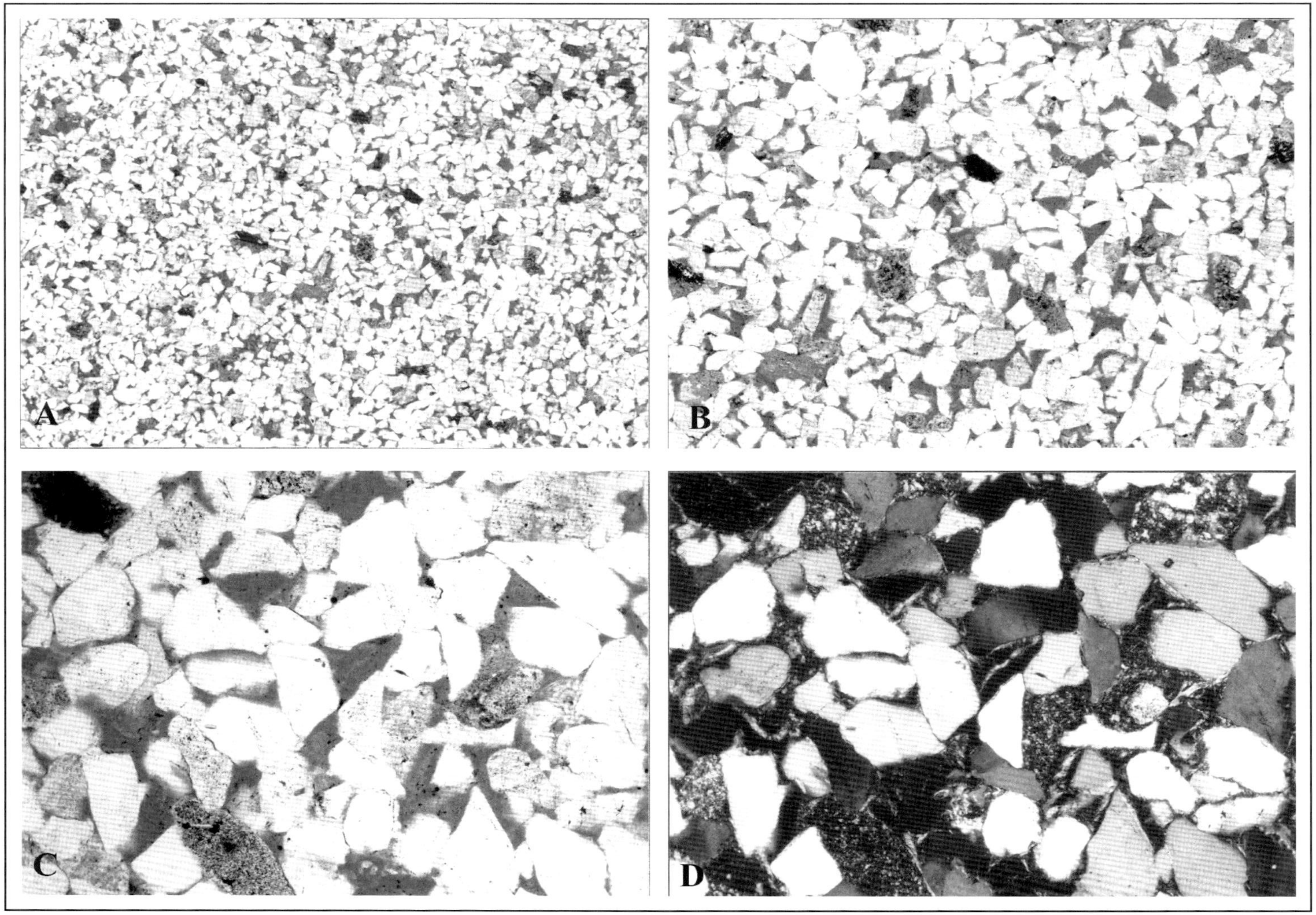

Figure 11. Photomicrographs illustrating characteristics of sublitharenite sandstones, American Hunter Old Road unit 1, 8390.7 ft (2557.5 m) core depth. (A) Well-sorted, upper fine-grained sublitharenite lithofacies, lower sandstone in cored interval of middle Lance sandstone. Lithic grains are sparse and consist of chert and altered volcanic rock fragments and traces of quartzose metamorphic rock fragments. Long dimension of view is 6.1 mm (0.24 in.). Plane light, 20×. (B) Well-sorted, upper medium-grained litharenite lithofacies, moderately compacted, with common point and linear intergranular contacts, and only traces of quartz overgrowth cement. Long dimension of view is 3.05 mm (0.12 in.). Plane light, 40×. (C). Intergranular pores are lined by mechanically infiltrated clay that has inhibited the development of quartz overgrowth cement. Inhomogeneous packing and oversized intergranular pores are indicative of secondary porosity. Intergranular mesopores are moderately connected, resulting in improved permeability over that observed in the overlying litharenite sandstone. Long dimension of view is 1.2 mm (0.047 in.). Plane light, 100×. (D) Cross-polarized light reveals the mechanically infiltrated clay that coats many intergranular areas. Note local spalling of the coating because of compaction. Long dimension of view is 1.2 mm (0.047 in.). Cross-polarized light, 100×.

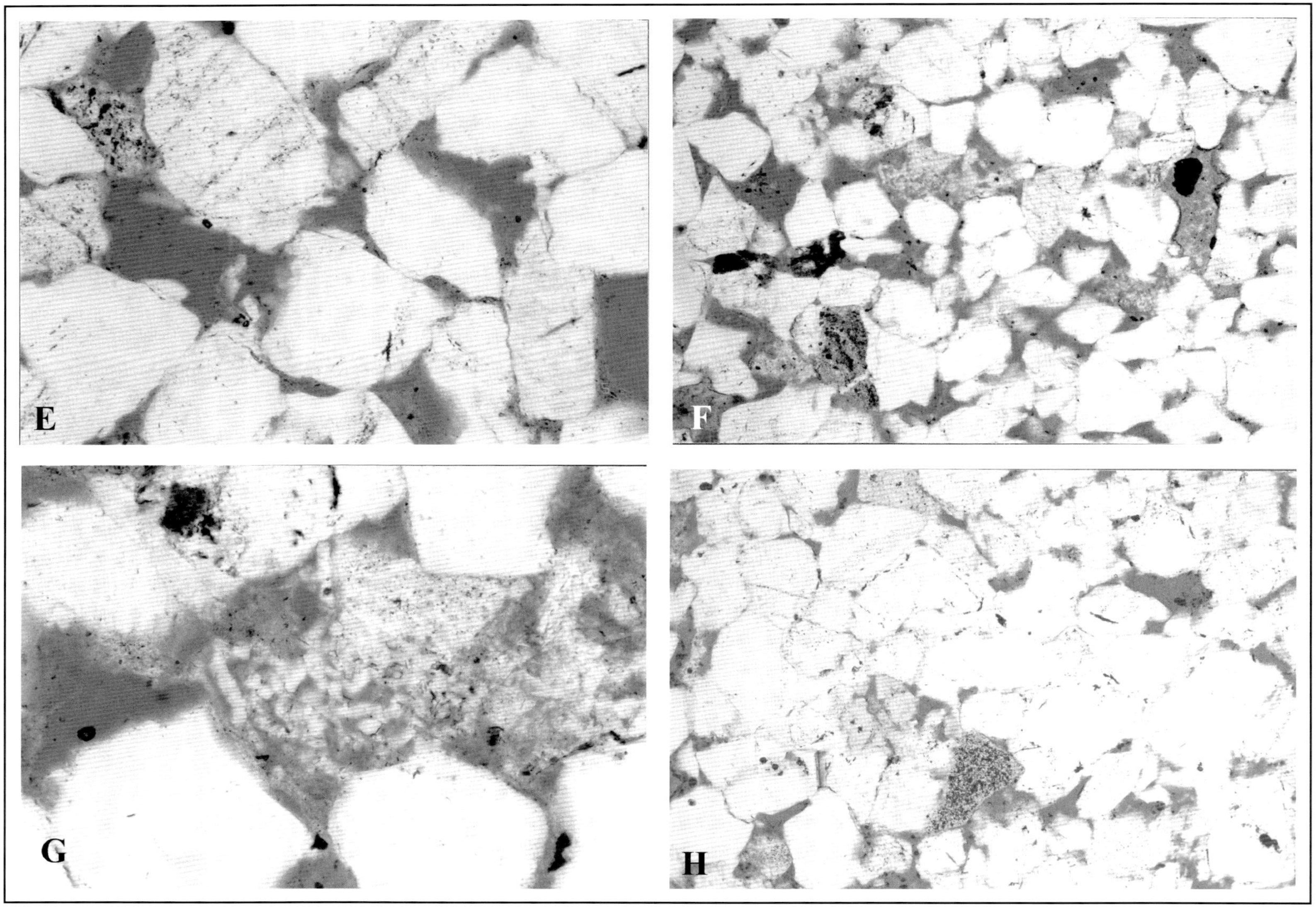

Figure 11. (cont.). (E) Oversized intergranular pores are commonly notched because of peripheral replacement of the surrounding grains by a prior cement, most likely calcite or ferroan calcite. Dissolution of unstable grains may have created moldic pores, such as the oversized void at left center. Long dimension of view is 0.6 mm (0.024 in.). Plane light, 200×. (F) Inhomogenous packing of framework grains is another criteria for recognition of secondary porosity. Partly dissolved rock fragments in this view contain intragranular microporosity. Long dimension of view is 1.2 mm (0.047 in.). Plane light, 100×. (G) Grain-sized patches of kaolinite cement are the result of replacement of an unstable grain by kaolinite. The kaolinite contains microporosity. Oversized intergranular pores were created by dissolution of prior cement. Long dimension is 0.6 mm (0.024 in.). Plane light, 200×. (H) Secondary intergranular pores are well connected and provide permeability pathways around more tightly cemented areas in this sandstone. Long dimension of view is 1.2 mm (0.047 in.). Plane light, 100×.

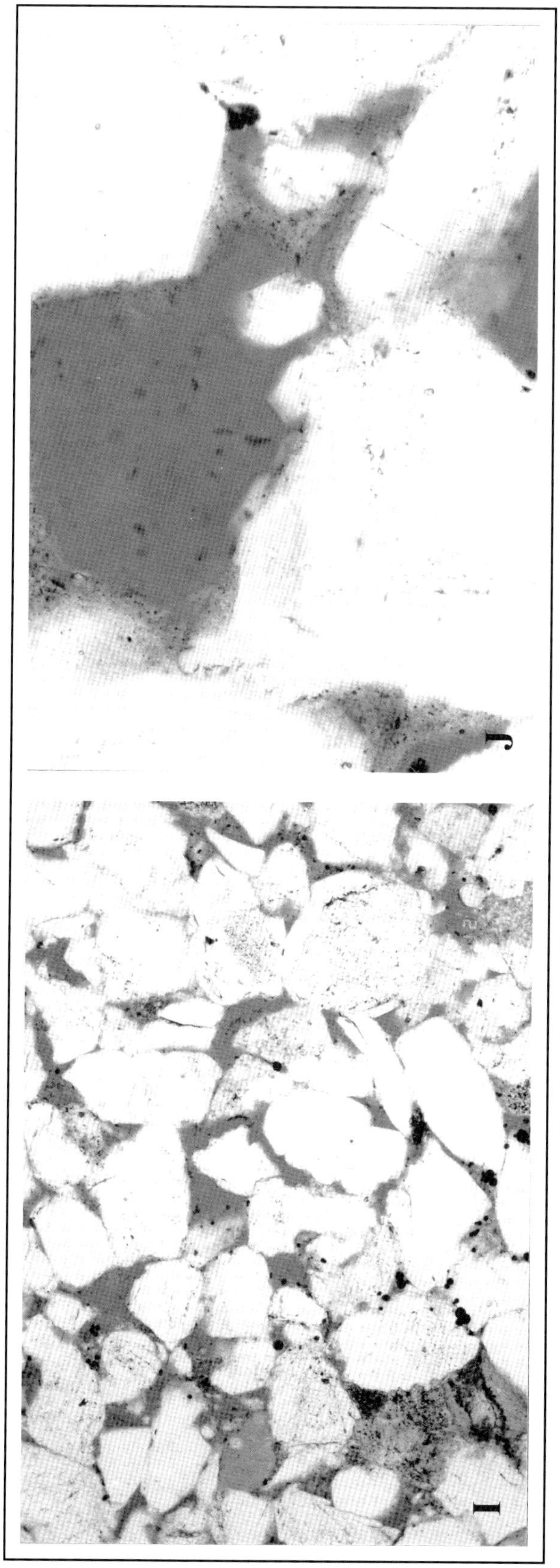

Figure 11. (cont.). (I) Open fabric of this area indicates that prior cement has been removed. Long dimension of view is 1.2 mm (0.047 in.). Plane light, 100×. (J) Mechanically infiltrated clay forms meniscus textures (right center) where the clay lines intergranular pores. The presence of a clay coating on detrital quartz grains has inhibited the development of quartz overgrowth. Long dimension of view is 0.3 mm (0.012 in.). Plane light, 400×.

water that would be quantified by more accurate methods (Table 2). A crossplot of the percentage of clay-filled microporosity observed in thin section versus water saturation from core analysis reveals the strong influence of clay cement on water saturation in the Lance sandstones (Figure 15).

Based on an examination of photomicrographs and data published for samples from the Lance sandstone at Jonah field (Yin et al., 2001), the porosity distribution in samples from the upper sandstone in the Old Road core is more typical of the Lance at Jonah field.

FORMATION DAMAGE

Changes in water saturation or chemical composition of pore fluids that occur during drilling, completion, and production can result in a significant reduction in the relative permeability of gas (Byrnes, 1997). The reduction in relative permeability to gas caused by increased water saturation in the near-wellbore or fracture face is known as "water block." Two factors influencing the susceptibility of a formation to water block are small pore-throat diameter and low permeability. Both of these factors are increased by the presence of clay cement. Because clay cement-filled porosity is a large proportion of the total porosity in the middle Lance sandstones, they are extremely susceptible to water block.

The sandstones of the middle Lance also contain swelling clay (in ML-IS) that may expand when contacted by fresh water in mud filtrate and massive hydraulic fracture fluids. Expansion of swelling clay minerals reduces pore-throat diameters and increases water saturation, resulting in reduced permeability to both gas and water. Traces of iron-bearing, acid-soluble ferroan calcite cement and chlorite are locally present, but we consider that this formation damage hazard is minimal. The formation damage reported by Schlumberger in the computer-analyzed DST report and poor fluid recovery is consistent with the reservoir characteristics of the middle Lance (Johnson Schlumberger, 1981).

LOG ANALYSIS

The open-hole logs available for this well include a dual induction log, a compensated formation

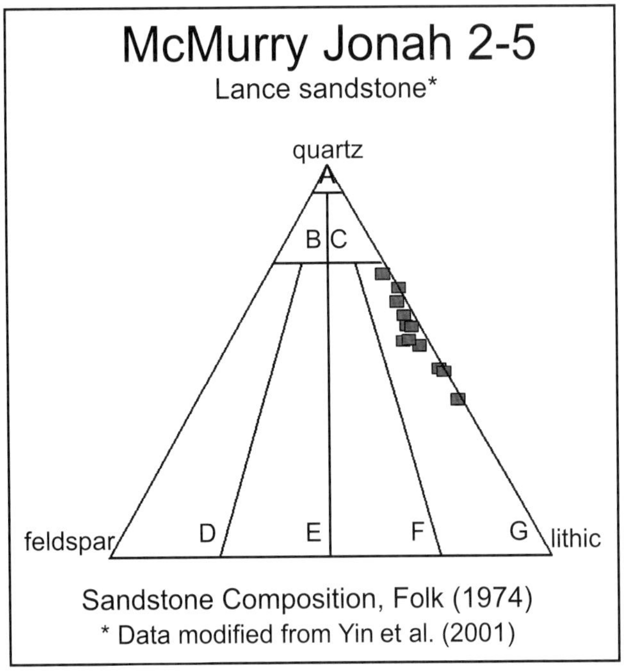

Figure 12. A ternary diagram illustrating detrital composition of sandstones from the Lance Formation, the McMurry Jonah Federal 2-5, Jonah field, Sublette County, Wyoming. Data are modified from modal analysis published by Yin et al. (2001). Sandstone clans are designated according to the classification of Folk (1974). Samples falling in field G are classified as lithorenite.

density-compensated neutron log, and a borehole-compensated sonic log. The caliper shows a rugose borehole in many intervals, and consequently, the density tool reads unreasonably high porosities in those intervals. Therefore, the shale-corrected sonic-neutron cross-plot porosity was used for effective porosity. In the cored interval, the effective porosities of the sandstones are mostly in the 8–13% range, with some shaly sandstone in the 2–6% range. The calculated log porosity and measured core porosity compare very closely (Figure 16).

The general procedures used for log analysis of the Lance sandstones are the same as those applied to Jonah field wells as described in Cluff and Cluff (2004). Water saturations were calculated using the dual water model, with 0.18 ohm m for the free-water resistivity and 0.2 ohm m for the bound-water resistivity. Calculated water saturations in the lower sandstone from 8397 to 8420 ft (2559 to 2566 m) ranged from 39 to 80%, whereas saturations in the upper sand from 8334 to 8380 ft (2540 to 2554 m) ranged from 25 to 65%. These saturations compare favorably with routine core water saturations, both in magnitude and trend, as shown in Figure 16.

Permeabilities were estimated using a modified Timur equation. Timur-calculated permeabilities ranged from less than 0.001 to 1 md, based on calibration to routine air permeabilities from core analyses (Figure 16).

CORE ANALYSIS

Both routine and advanced core analysis data are available for the Fort Union, the middle Lance interval 8346–8447 ft (2544–2575 m) (missing shaly intervals, 8372–8389 and 8418–8440 ft [2552–2557 and 2566–2573 m]), and the underlying Mesaverde Blair Formation. For the two sandstone intervals (8346–8362 and 8400–8417 ft; 2544–2549 and 2560–2566 m) cored in the middle Lance, routine core porosities (ϕ) range from 7.4 to 13.8%, averaging 10.4%. Porosities in the shaly interval (8363–8399 ft; 2549–2560 m) average 3.8 ± 2%. Routine full-diameter horizontal air permeabilities range from 0.07 to 13.0 md, with an arithmetic mean of 0.81 md and a geometric mean of 0.34 md (Figure 17). Only four samples exhibited permeability greater than 1 md and give indication of having been influenced by fracturing. Vertical permeabilities average 88 ± 70% of horizontal permeabilities. Grain densities average 2.65 ± 0.02 g/cm^3. Minor calcite cementing in the upper sandstone interval results in slightly higher grain densities (2.66–2.67 g/cm^3), whereas the lower interval exhibits lower grain densities (2.64–2.65 g/cm^3). Water saturations in the upper sandstones average 50 ± 14%, and in the lower sandstone, water saturations average 43 ± 16%, with saturations in the shaly interval averaging 76 ± 20%.

Comparison of petrophysical trends and properties of the middle Lance in the Old Road well with other western United States Upper Cretaceous low-permeability sandstones, including the Mesaverde and Frontier, illustrates that, as with most low-permeability sandstones, Old Road in-situ Klinkenberg permeabilities (k_{ik}) are significantly less than routine air permeabilities (k_a) and exhibit increasing difference with decreasing permeability (Figure 18). This difference is attributed to the influence of both confining stress on pore-throat size and Klinkenberg gas slippage effect. The Old Road Upper Cretaceous rocks, including the Lance, generally exhibit greater decrease in permeability under confining stress than the average Mesaverde rocks and can be characterized by the following relationship derived from the Old Road core plug data:

$$k_{ik} = 0.298 k_a^{1.773}$$

In-situ porosity ($\phi_{in\ situ}$) data are not available for the Old Road, but in-situ values can be approximately predicted

Table 3

Whole Rock and Clay Fraction XRD American Hunter Old Road Unit 1, Sublette County, Wyoming.

Unit	Middle Lance Sandstone	Middle Lance Sandstone	Middle Lance Sandstone
Well			
Log depth	8369	8376	8404
Sample depth	8354.5	8362.0	8390.7
Core or plug number	60	68	90
Rock number	14276	14296	14226
Whole Rock Percentage			
Quartz	89.08	89.03	95.63
Plagioclase	0.91	0.77	0.69
K-spar (Microcline)	0.91	0.96	0.92
Dolomite	1.46	1.72	0.00
Fe dolomite	0.00	0.00	0.00
Calcite	1.10	1.92	0.00
Fe calcite	0.14	1.77	0.00
Siderite	1.46	0.00	0.00
Mica or illite	1.64	1.15	1.15
Kaolinite	1.64	1.15	0.69
Chlorite	1.10	1.15	0.46
Smectite	0.00	0.00	0.00
Pyrite	0.55	0.38	0.46
Other	0.00	0.00	0.00
Total	100.00	100.00	100.00
Estimated total clay	4.39	3.45	2.30
Less than 10-μm Clay Percentage			
Kaolinite	50.6	47.3	28.0
Chlorite	12.6	16.9	17.5
Illite	28.0	27.1	29.4
ML/IS	8.8	8.7	25.1
Smectite	0.0	0.0	0.0
Total	100.0	100.0	100.0
Percent expand	20	20	20
ML/IS stacking order	Random	Random	Random

from routine porosity ($\phi_{routine}$) values using the general relationship for Mesaverde and Frontier rocks (Byrnes, 1997):

$$\phi_{in\ situ} = \phi_{routine} - 0.8$$

Using these equations to predict $\phi_{in\ situ}$ porosity and k_{ik} from the routine full-diameter $\phi_{routine}$ and k_a, respectively, and using measured k_{ik} values on selected middle Lance core plugs, Figure 19 shows that the Old Road middle Lance interval exhibits properties similar to Mesaverde and Frontier rocks. Measured core plug k_{ik} values from the middle Lance sandstones generally exhibit permeability that is as much as an order of magnitude greater for a given porosity than the average Mesaverde and Frontier rocks and represents the high end of permeability for Upper Cretaceous low-permeability sandstones. The difference between the full-diameter, predicted k_{ik} values and the measured core plug k_{ik} values may be caused by the inaccuracy of the routine to in-situ equation or may be related to lower permeabilities exhibited by full-diameter cores because of bedding and lithofacies heterogeneities that exist in full-diameter core but not in smaller, more homogeneous, core plugs.

Comparison of both middle Lance water saturations measured in core and irreducible water saturations (S_{iw}) measured on core plugs (measured at 600 psi air-brine capillary pressure, equivalent to approximately 750 ± 100 ft [230 ± 30 m] above free-water level, depending on exact pressure and water density) with general Mesaverde and Frontier S_{iw} (Figure 20) reveals that the core averages

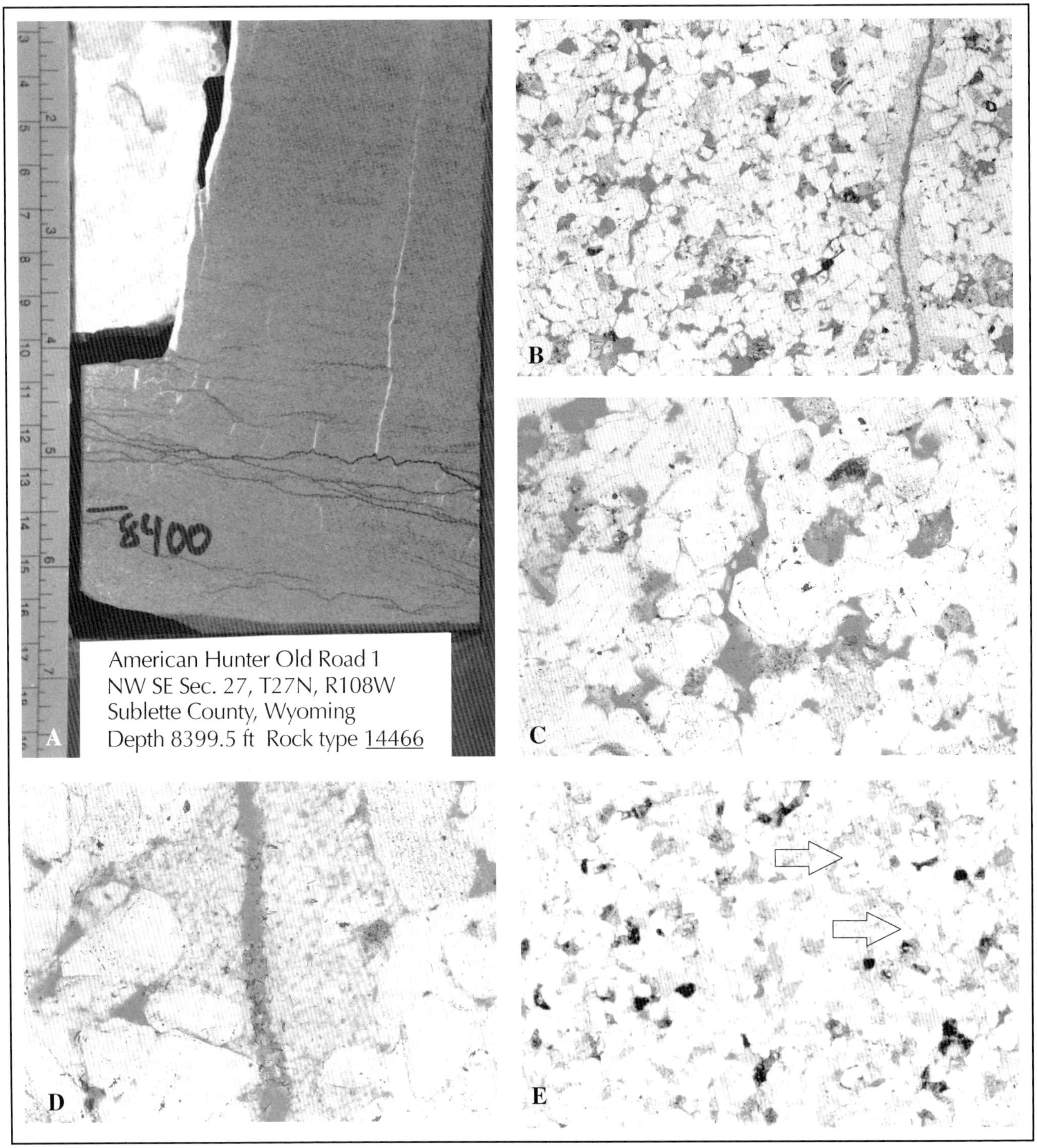

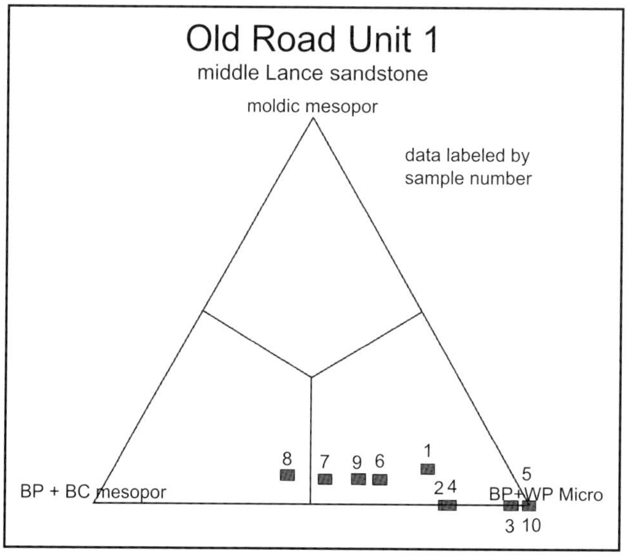

Figure 14. A ternary diagram illustrating distribution of porosity in sandstone and conglomerate samples, based on ratio of moldic (Mo) and intergranular (BP) mesoporosity and clay-filled intergranular (BP) plus intragranular microporosity (WP).

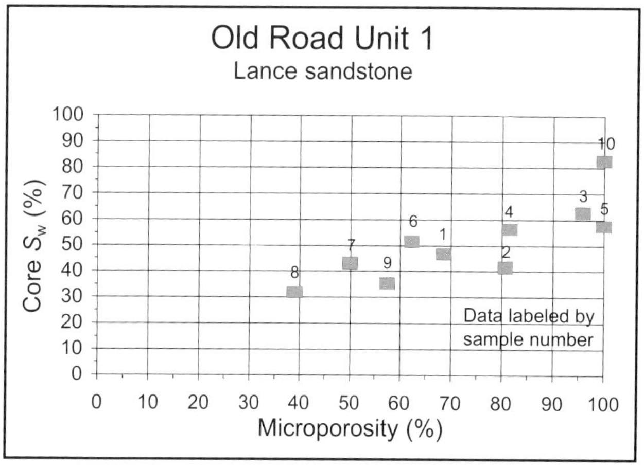

Figure 15. Crossplot of the estimated percentage of microporosity and total porosity from point-count data versus water saturation data from routine core analysis. Note the strong correlation between proportion of microporosity and the measured water saturation data.

approximately 15% higher saturations than the average Mesaverde and Frontier but are within the range of S_{iw} observed in Mesaverde and Frontier rocks. Measured S_{iw} values on middle Lance plug samples exhibit saturations similar to the full-diameter core saturations. The slightly higher water saturations in the full-diameter core would indicate the possible conditions that saturations were increased slightly during coring, or that saturations in the core are slightly greater than the S_{iw}, because the reservoir is not at the gas column height corresponding to that which was defined for the Mesaverde and Frontier population as S_{iw}. Given the good correlation between core and wire-line log saturations, it can be interpreted that water saturations in core were not significantly disturbed from those in the reservoir.

Relative gas permeability at $S_{iw}(k_{rg,S_{iw}})$ for the middle Lance is consistent with relative gas permeability properties exhibited by Mesaverde and Frontier low-permeability sandstones regionally (Figure 21). Samples with higher permeability and corresponding low S_{iw} that is less than 30% exhibit $k_{rg,S_{iw}}$ greater than 95%. With decreasing permeability, S_{iw} increases, and gas-relative permeability decreases.

CONCLUSIONS

Poor fluid recovery, low flow rates, low calculated permeability, and interpreted formation damage in the middle Lance are typical of low-permeability tight-gas

◀

Figure 13. Fractures observed in core and thin section from sandstones of the middle Lance in the American Hunter Old Road unit 1, Sublette County, Wyoming. (A) Kaolinite-filled fracture (white) in core sample of middle Lance sandstone, 8399.7 ft (2560.2 m). (B) Parallel, vertically oriented natural fractures in thin section of fine-grained sublitharenite. The fracture on the left is open and has been partly cemented by quartz overgrowths. The fracture on the right has been filled by microporous kaolinite cement (which has absorbed blue dye epoxy). Sample handling is responsible for separating the walls of the kaolinite-cemented fracture. Long dimension of view 3.05 mm (0.12 in.). Sample from 8399.5 ft (2560.2 m). Plane light, 40×. (C) Higher magnification view of quartz overgrowth-lined fracture. The fracture provides enhanced permeability and connectivity of intergranular, moldic, and clay-filled moldic pores. Core analysis data for a nearby unfractured sample are 11%, 0.30 md, and 2.64 g/cm^3. Long dimension of view is 0.6 mm (0.024 in.). Sample from 8399.5 ft (2560.2 m). Plane light, 200×. (D) Higher magnification view of kaolinite-filled fracture. Microporosity in the kaolinite cement will not hinder the movement of natural gas through this vertical fracture. Long dimension of view is 0.6 mm (0.024 in.). Sample from 8399.5 ft (2560.2 m). Plane light, 200×. (E) Linear features observed in thin section represent healed fractures (arrows). The former fractures are recemented by quartz. Long dimension of view is 3.05 mm (0.12 in.). Sample from 8354.5 ft (2546.5 m). Plane light, 40×.

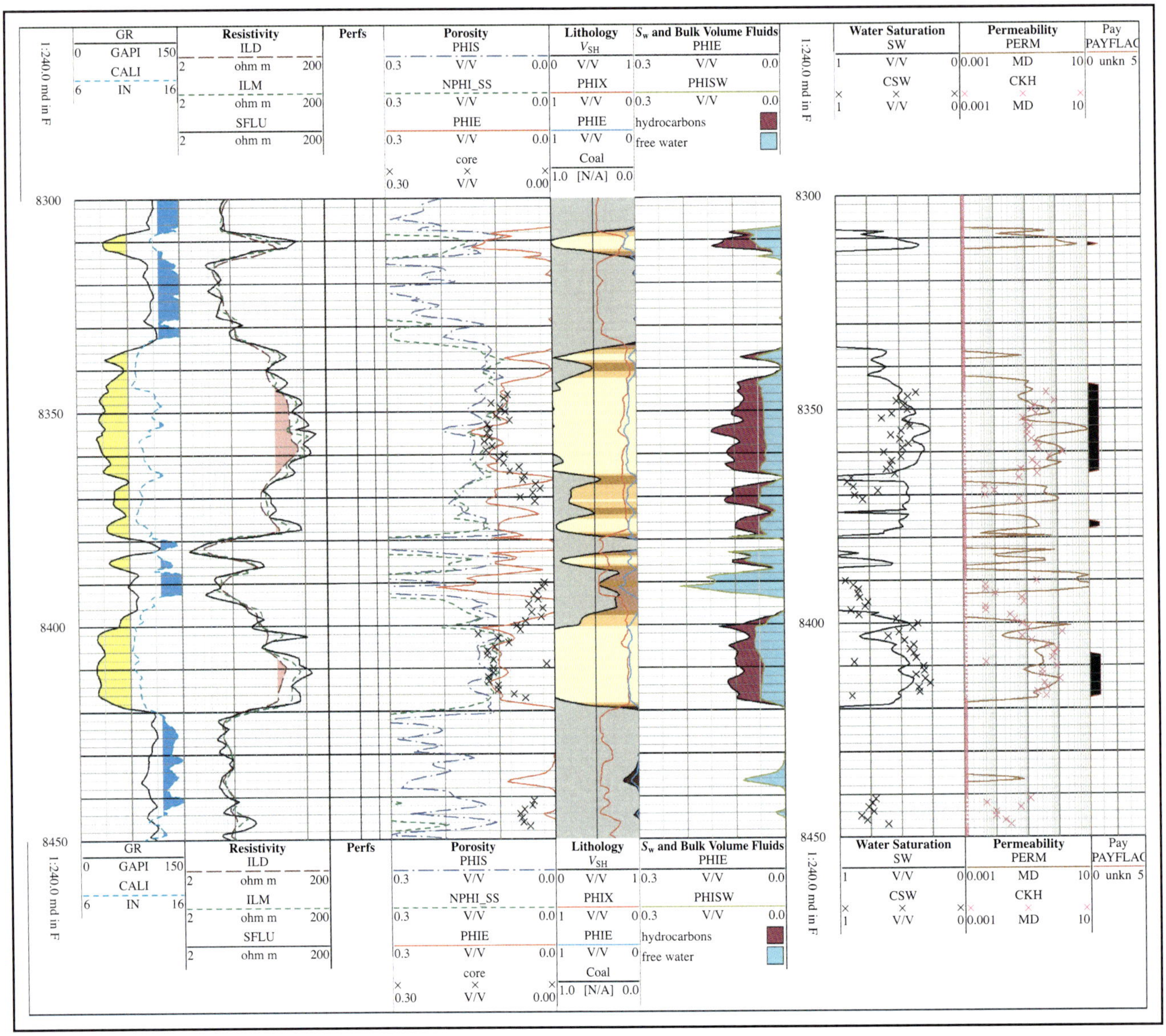

Figure 16. Gamma-ray, caliper, and resistivity log profiles are displayed in the left-hand column. Porosity log files and measured core porosity data are plotted in the second track from the left. Calculated V_{shale} and bulk volume water values are plotted in the two tracks right of center. Calculated S_w and core S_w data and calculated and core permeability data are plotted in the right-hand tracks. Pay flags are displayed in the far right track. Log-calculated porosities, water saturation, and permeability compare closely with measured core data for the cored interval in the middle and lower Lance Formation, American Hunter Old Road unit 1, Sublette County, Wyoming.

reservoirs that have been successfully exploited in the Rocky Mountains since the early 1980s.

Petrographic characterization of the distribution and mineralogy of clay cements and porosity in the middle Lance suggests that the middle Lance should be capable of economic rates of natural gas production, provided formation damage is minimized, adequate reservoir thickness is encountered, and sufficient hydrocarbon saturation and reservoir pressure are present. Massive hydraulic fracturing treatments would probably be necessary to establish economic flow rates.

The most serious potential formation damage hazard is water block, a result of reservoir pressures being insufficient to displace mud or fracture filtrate from clay-filled matrix porosity. Possible low-resistivity pay, caused by pore-lining clay cement, may have been overlooked in

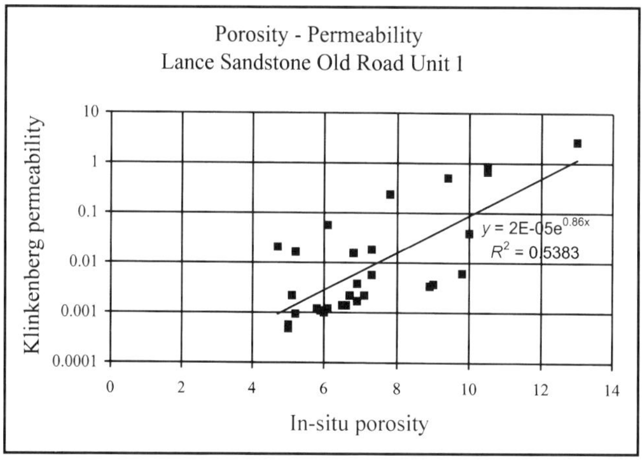

Figure 17. Porosity versus permeability crossplot for routine core analysis data, for the cored interval in the middle and lower Lance Formation, American Hunter Old Road unit 1, Sublette County, Wyoming.

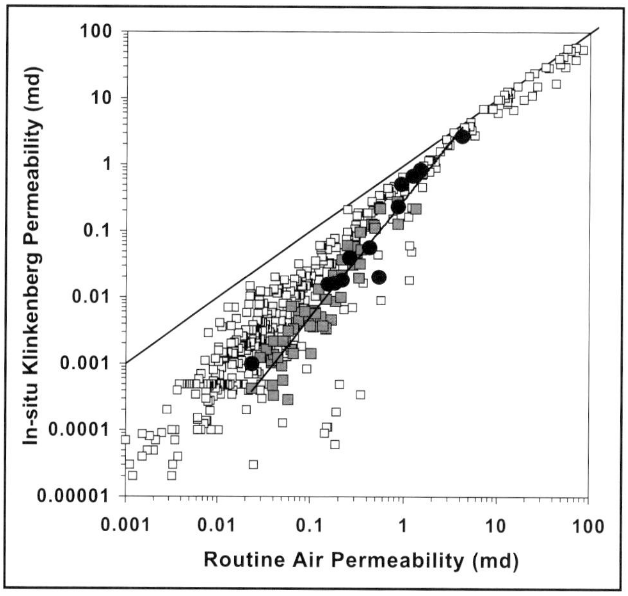

Figure 18. Crossplot of routine air permeability versus in-situ Klinkenberg gas permeability comparing trends for regional Mesaverde and Frontier samples (light-gray squares), Fort Union and Blair samples from the Old Road well (dark-gray squares), and middle Lance Old Road well samples (black circles). Old Road cores exhibit greater decrease in permeability under confining stress and with Klinkenberg gas slippage correction than general Mesaverde and Frontier rocks but lie within the overall population variance.

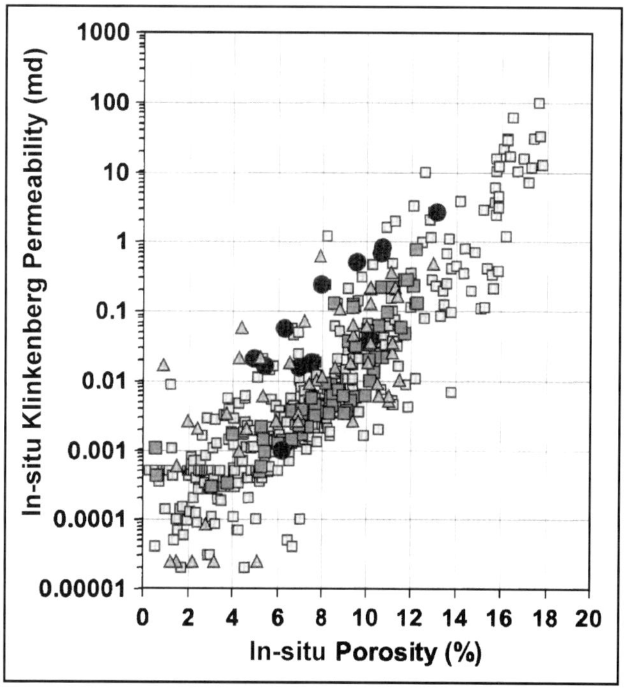

Figure 19. Crossplot of in-situ Klinkenberg gas permeability versus in-situ porosity for Mesaverde and Frontier samples (light-gray squares), Old Road Fort Union and Blair samples (dark-gray squares), Old Road middle Lance full-diameter air permeabilities and porosities transformed to in-situ values using equations (light-gray triangles), and measured middle Lance core plug in-situ Klinkenberg permeabilities and transformed in-situ porosities (black circles). Middle Lance plugs exhibit high permeability compared to that of Mesaverde and Frontier.

this cored interval. Natural fractures could enhance gas deliverability of this potential reservoir.

At the time the Old Road well was drilled, the full potential of low-permeability natural gas reservoirs was only beginning to be recognized by the petroleum industry. If this well had been drilled 10 yr later, it is likely that modern fracture-stimulation treatments may have been able to establish hydrocarbon production from the cored interval in this well.

ACKNOWLEDGMENTS

This chapter would not have been possible without the donation of core to the U.S. Geological Survey Core Research Center by American Hunter, Ltd., Calgary, Canada. The staff at the Core Research Center is responsible for archival preservation and maintenance of the core

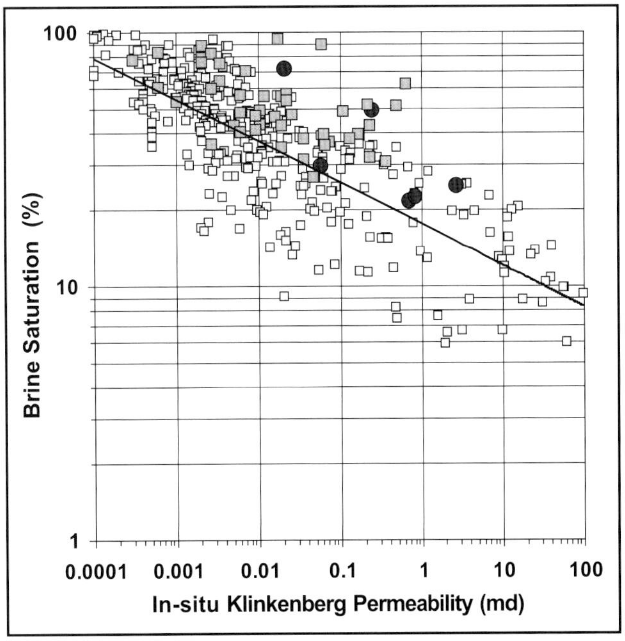

Figure 20. Crossplot of water saturation versus permeability showing irreducible water saturations for Mesaverde and Frontier samples (light-gray squares), middle Lance full-diameter routine core water saturations with in-situ permeabilities predicted from routine data using equation in text (dark-gray squares), and irreducible water saturation measured on core plugs of the middle Lance (black circles). Slightly higher water saturations in core are interpreted to reflect reservoir saturations, and indicative reservoir is close to but not at irreducible saturation.

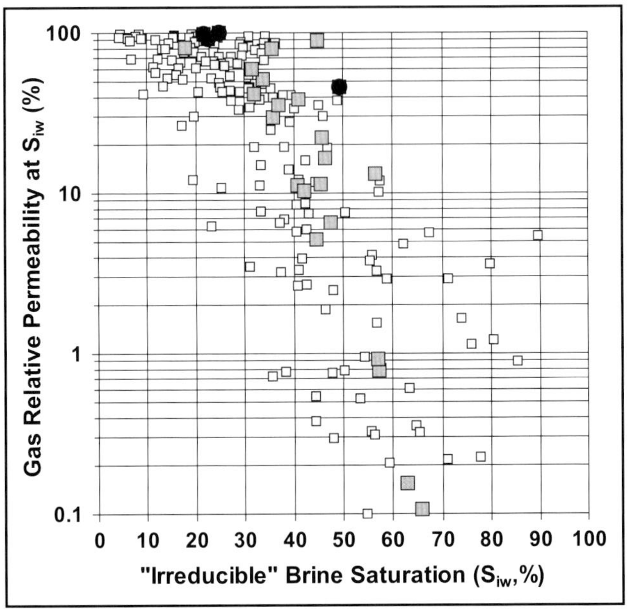

Figure 21. Gas-relative permeability at water saturation versus saturation shows that middle Lance rocks (black circles) exhibit gas-relative permeability properties consistent with Mesaverde and Frontier rocks (light-gray squares) and other Upper Cretaceous low-permeability sandstones (dark-gray squares).

samples and analytical data. Triple O. Slabbing, Denver, Colorado, slabbed cores from the middle and lower Lance. Basin Exploration and Stone Petroleum funded core slabbing. Snyder Oil Company supported special core analysis. Core analysis and DST data were provided by the operator and are on file with the Wyoming Oil and Gas Conservation Commission, Casper, Wyoming, and at the U.S. Geological Survey Core Research Center, Denver, Colorado. Drafting of the core description was completed by Ken Graff.

REFERENCES CITED

Beard, D. C., and P. K. Weyl, 1973, Influence of texture on porosity and permeability of unconsolidated sand: AAPG Bulletin, v. 57, p. 34–369.

Byrnes, A. P., 1994, Special core analysis on selected cores from the American Hunter no. 1, Aspirin 1 and Old Road unit 1, Mesaverde Group, Sublette County, Wyoming: U.S. Geological Survey Core Research Center, Denver Federal Center, Lakewood, Colorado.

Byrnes, A. P., 1997, Reservoir characteristics of low-permeability sandstones in the Rocky Mountains: The Mountain Geologist, v. 34, no. 1, p. 39–51.

Cluff, S. G., and R. M. Cluff, 2004, Petrophysics of the Lance sandstone reservoirs in Jonah field, Sublette County, Wyoming, *in* J. W. Robinson and K. W. Shanley, eds., Jonah field: Case study of a tight-gas fluvial reservoir: AAPG Studies in Geology 52, p. 215–242.

Core Laboratories, Inc., Dallas, Texas, 1981, Core analysis report for American Hunter Old Road unit 1, Sublette, County, Wyoming, File RP-2-6443: Wyoming Oil and Gas Conservation Commission, Casper, Wyoming, 36 p.

Dincau, A. R., 1989, Diagenesis and porosity evolution of Upper Cretaceous and Lower Tertiary sandstones from the Wagon Wheel # 1 well, northern Green River basin, Wyoming: M.S. thesis, University of Missouri, Columbia, 126 p.

Folk, R. L., 1974, Petrology of sedimentary rocks: Austin, Texas, Hemphill Publishing Co., unpaginated.

Jansen, G. J., 1994, Report 940507, Rocky Mountain coal petrography, archived in well data files: U. S. Geological Survey Core Research Center, Denver Federal Center, Lakewood, Colorado, 5 p.

Johnson Schlumberger Computerized Data Analysis, 1981, Field Report 28280 D, March 23, 1981: Casper, Wyoming, Wyoming Oil and Gas Conservation Commission.

Law, B. E., 1984, Relationships of source-rock, thermal maturity, and overpressuring to gas generation and occurence in low-permeability Upper Cretaceous and Lower Tertiary rocks, greater Green River basin, Wyoming, Colorado and Utah, *in* J., Woodward, F. F., Meissner, and J. L., Clayton, eds., Hydrocarbon source rocks of the greater Rocky Mountain region: Denver, Colorado, Rocky Mountain Association of Geologists, p. 469–490.

Pawlewicz, M. J., M. R. Lickus, B. E. Law, W. W. Dickinson, and C. S. V.

Barclay, 1986, Thermal maturity map showing subsurface elevation of 0.8% vitrinite reflectance in the Greater Green River basin of Wyoming, Colorado, and Utah: U.S. Geological Survey Miscellaneous Field Studies, Map MF-1890, scale 1:500,000, 1 sheet.

Pemberton, S. G., M. Spila, A. J. Pulham, T. Saunders, J. A. MacEachern, D. Robbins, and I. K. Sinclair, 2001, Ichnology and sedimentology of shallow to marginal marine systems: Ben Nevis and Avalon Reservoirs, Jeanne D'Arc basin: Geological Association of Canada Short Course, v. 15, p. 155–179.

Reynolds, R. C. Jr., and J. Hower, 1970, The nature of interlayering in mixed-layer illite-montmorillonite: Clays and Clay Minerals, v. 18, p. 25–36.

Schmidt, V., and D. A. McDonald, 1979, Texture and recognition of secondary porosity in sandstones, *in* P. A. Scholle and P. R. Schluger, eds., Aspects of diagenesis— Symposia: SEPM Special Publication 26, p. 209–225.

Shanley, K. W., 2001, Fluvial reservoir description for a tight-gas giant: Jonah field, Green River basin, Wyoming, *in* J. W. Robinson and K. W. Shanley, co-conveners, Tight-gas fluvial reservoirs: A case history from the Lance Formation, Green River basin, Wyoming: AAPG Annual Convention, Denver, Colorado: AAPG Short Course 2, 49 p.

Walker, T. R., B. Waugh, and A. J. Crone, 1978, Diagenesis in first-cycle desert alluvium of Cenozoic age, southwestern United States and northwestern Mexico: Geological Society of America Bulletin, v. 89, p. 19–32.

Yin, P., S. M. Stookey, R. C. Surdam, and T. Dunn, 2001, Tight gas sands of the Lance Formation in Jonah field— a core from the Jonah #2-5 well, *in* J. W. Robinson and K. W. Shanley, co-conveners, Tight-gas fluvial reservoirs: A case history from the Lance Formation, Green River basin, Wyoming: AAPG Annual Convention, Denver, Colorado: AAPG Short Course 2, 49 p.

CHAPTER 12

Petrophysics of the Lance Sandstone Reservoirs in Jonah Field, Sublette County, Wyoming

SUZANNE G. CLUFF
The Discovery Group, Inc., Denver, Colorado, U.S.A.

ROBERT M. CLUFF
The Discovery Group, Inc., Denver, Colorado, U.S.A.

ABSTRACT

Jonah field is a giant gas field producing from extremely low-porosity and low-permeability sandstones. Wire-line–log data from 62 wells near the center of the field were studied to characterize the porosity, permeability, and water saturation of the Lance reservoirs. The logs were environmentally corrected and normalized, shale volume and porosities were calculated, water saturations were determined by the dual water model, and net pay was calculated using field-specific pay criteria. Ultimate gas recovery per well was estimated by decline curve analysis of monthly production data.

Within the upper 2500 ft (760 m) of the Lance Formation, which includes the entire productive interval in nearly all wells, the average well has 1000 ft (30 m) of net sandstone, having an average porosity of 6.4%. The average permeability of all sandstones, estimated from core data-derived equations, is an astonishingly low 6 μd. The average water saturation of all sandstones is 45%.

Net pay criteria were determined from cumulative storage-capacity and cumulative flow-capacity plots. Although the average sandstone may have only 6% porosity, the low-porosity sandstones contribute an insignificant fraction of the reservoir flow capacity. We estimate that more than 95% of the flow capacity is from sandstones with greater than 6% porosity. A small percentage of high-porosity (>10%) and high-permeability rocks dominate the flow behavior of the reservoir and are probably critical to economic production. Using 6% porosity as an absolute net pay cutoff, the average net pay thickness at Jonah is 440 ft (130 m), with 9.3% porosity and 33% water saturation. The estimated average permeability of net pay is 25 μd. Estimated ultimate recovery per well is approximately 4 bcf gas on current 40-ac (0.16-km^2) well spacing.

INTRODUCTION

Jonah field, located in Sublette County, Wyoming, was one of the largest onshore natural gas discoveries of the 1990s in the United States. Although many aspects of this field challenge our perceptions of what is required to form a giant gas accumulation, one of the most remarkable aspects are the petrophysics of the productive sandstones. The reservoir sandstones at Jonah have average porosities and permeabilities that would be considered the sealing facies in many gas fields, which immediately raises questions of how pay is defined and what constitutes valid criteria for differentiating pay from nonpay intervals.

The purpose of this chapter is to document the petrophysical properties of the Lance sandstones at Jonah

Jonah Field: Case Study of a Giant Tight-Gas Fluvial Reservoir, J. W. Robinson and K. W. Shanley, eds.: AAPG Studies in Geology 52 and Rocky Mountain Association of Geologists 2004 Guidebook

field, evaluate criteria for defining net pay, and propose a methodology to determine net pay cutoffs.

METHODS

Data Sources

Wire-line–log data from 62 wells near the center of Jonah field were selected for study out of a total of nearly 400 wells drilled by the end of 2002 (Figure 1). Although not every well in the field was included in the summations used for this chapter, we have run comparable calculations on many other wells across the field area and consider the wells in the central field area to be a representative sample for the productive Lance Formation. The data set includes several wells in the uppermost and the lowermost quartiles of the reserve size distribution of the field.

The basic analysis procedure we used involves the following steps, each of which is described in the following sections:

(1) Acquire data from either vendor field tapes or high-accuracy digitization of paper log prints;
(2) Merge log runs and depth shift curves between logging passes, if required;
(3) Import and depth shift core data;
(4) Apply environmental corrections and normalize porosity and gamma-ray logs;
(5) Compute shale volume from the gamma ray;
(6) Compute total porosity and shale-corrected ("effective") porosity from the density, neutron, and sonic logs;

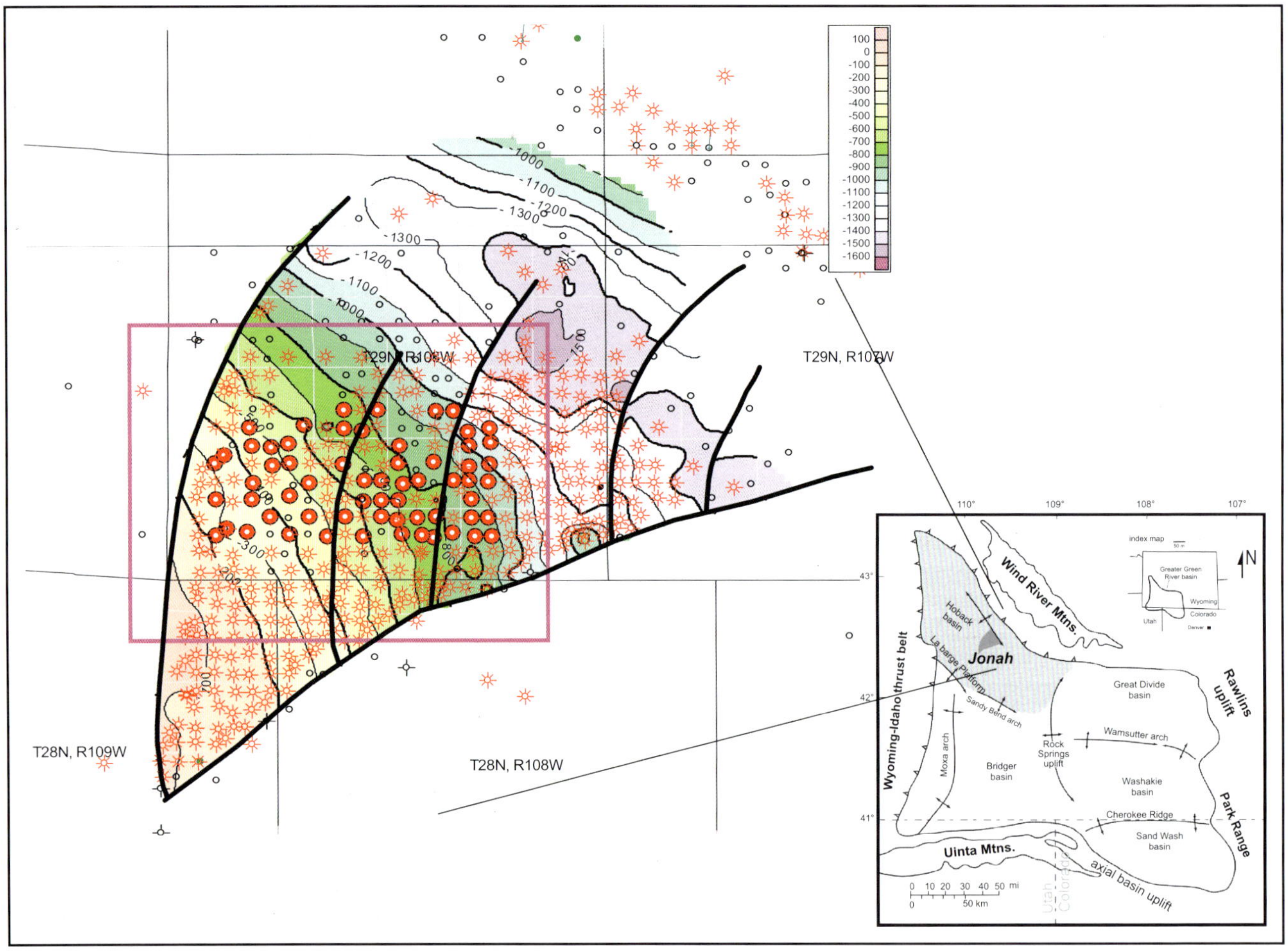

Figure 1. Structure contour map on the base of the Tertiary Fort Union Formation. Solid circles denote the location of wells used in the study. The study area is shown by the outline. Contour interval = 100 ft (30 m).

(7) Compute water saturation by dual water model;
(8) Calculate net pay using field-specific net pay cutoffs.

For this study, most of the logs were digitized from paper copies. Nearly all wells have been logged with an array induction resistivity tool and a compensated neutron-compensated density-gamma ray tool. Only a few sonic logs have been run in the field.

All logging run data were merged and depth shifted into alignment with the resistivity logs, which were assumed to be the reference depth in all cases. Because a large portion of the wells at Jonah were logged with a Schlumberger Platform Express tool, which includes the resistivity and porosity measurements all on a single tool string, relatively few wells require significant depth shifting.

Core data were available for four wells in the field, including porosity and permeability at net overburden stress conditions. The core data were imported as ASCII files, merged with the open-hole logs, then depth shifted into alignment with the logs by correlation with the gamma-ray and porosity curves. A total of 225 core sample points were used.

Environmental Corrections and Log Normalization

Most petrophysical field studies begin by normalizing the log data to minimize differences caused by random logging errors and other noise in the data set. This noise results from a variety of causes. Logging tools built by different vendors can have different responses to identical formation conditions. The study wells were drilled between 1995 and 2002 and logged by several different wire-line companies. Different generations of logging tools and different processing algorithms were used, depending on the vendor and when the well was logged. Logging tool response is also strongly affected by borehole and mud conditions. Environmental correction routines are used to compensate for most of these environmental effects, but many of the key variables are poorly known or are not measured, and all environmental correction algorithms assume ideal, commonly unrealistic conditions (e.g., a perfectly centered tool in a cylindrical borehole). The environmental corrections applied to this data included (1) borehole size and mud weight correction to the gamma-ray logs; (2) matrix corrections to the neutron porosity; and (3) invasion (or "tornado chart") corrections to the three-curve induction resistivity logs. Array induction logs were not environmentally corrected, and the deepest array curve was used as an approximation of true formation resistivity. Other common environmental corrections applied by the logging vendor were determined to be negligible for the borehole sizes and logging conditions at Jonah field or are compensated for by the normalization procedure described below.

Normalization is a process used to reduce the residual errors by comparing the log response in a zone of known and constant properties with the expected response in that lithology. For example, a clean tight limestone or an anhydrite bed might have consistent density, neutron porosity, and sonic transit time over a broad area and would be expected to have the same log response in every well. In practice, the log responses differ, because the wells were logged with different tools, borehole conditions varied, the environmental corrections did not exactly match the actual conditions during logging, a tool might have been miscalibrated, or the logging operator simply made an error. The normalization procedure to reduce these errors involves either applying an offset (bulk shift) to the log curve or applying a gain correction between two known end points. In some cases, both adjustments will be made to the observed response to achieve the expected response in the constant property intervals. In all cases, we assume the log response is linear over the range of interest and can be approximated by a simple slope-intercept method.

Unfortunately, normalization in the Green River basin is problematic because there are no zones with uniform lithology over a large geographic area. Many log analysts use statistical techniques to reduce the variation between wells toward a field or area average, but most of these methods run the risk of eliminating actual geologic variation (signal) as opposed to random noise between logging jobs. As an alternative, we make the assumption that the minimum porosity over a field size area will be constant. We believe that this assumption is reasonable, because burial depth and time-temperature history are the major controls on porosity loss, whereas mineralogical variations are secondary and tend to be minor over the area of a field. Porosity variations are minor in the Lance Formation in the immediate vicinity of Jonah field. Because the average porosity of an interval varies with the sand to shale (or net/gross) ratio, we made no attempt to normalize the average properties of the logs; only the minimum observed porosity in clean sandstones and siltstones is used.

Neutron-porosity and bulk-density logs were normalized to match a suite of eight "type" logs. Composite histograms for each measurement were constructed from the data over the top 1000 ft (300 m) of the Lance Formation in these wells (Figures 2, 3). The extreme X-axis intercepts are not significant and may be caused by a few anomalous points in one well. The minimum porosity for normalization is picked on the steep right slope of the histogram, which, in the case of Figure 2, is near a cumulative frequency

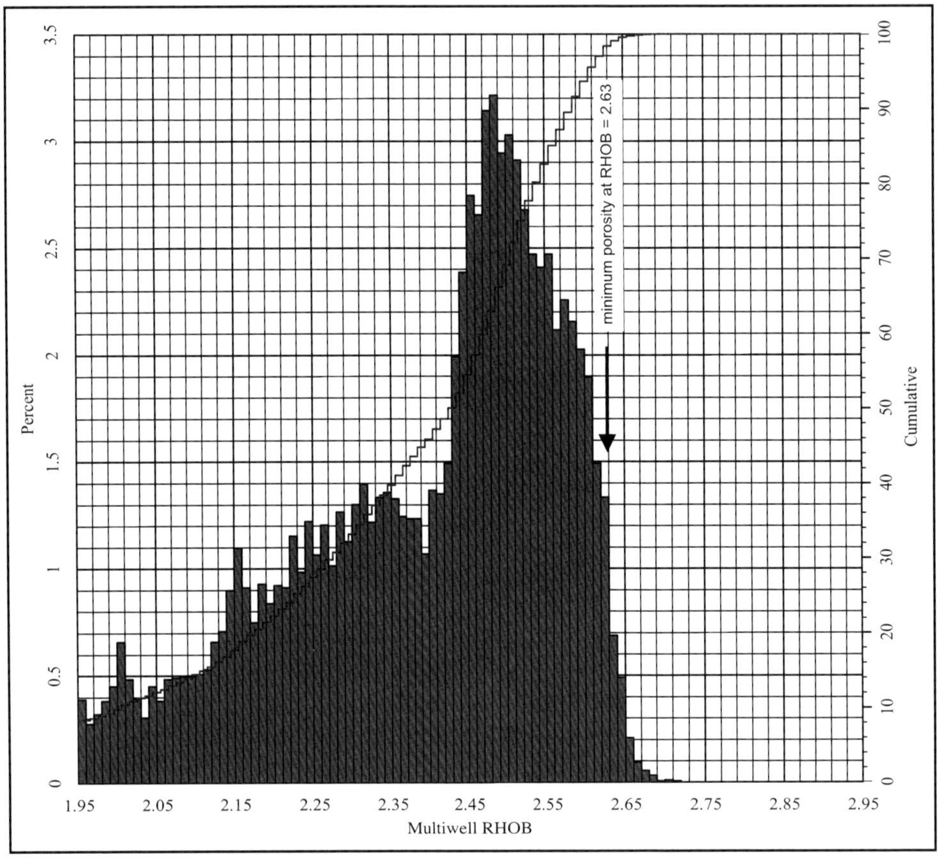

Figure 2. Type histogram for bulk-density (RHOB) log normalization. Histogram is a composite of the top 1000 ft (300 m) of the Lance Formation in eight wells. Histogram bars (frequency %) refer to left axis; cumulative frequency line (cumulative %) refers to the right axis.

of 98%. Minimum density porosity corresponds to tight siltstones with a bulk density of 2.63 g/cm^3 (1.5% porosity, Figure 2), and minimum neutron porosity in limestone units is 4.0% (Figure 3; equivalent to 7.5% on a sandstone matrix). Individual histograms of the equivalent stratigraphic interval in each well were then compared to the composite histograms to determine the normalization shift. Neutron logs show significant variation between vendors and from borehole environment effects. The differences are caused by temperature effects and variations in tool standoff. Density logs show only minor differences, on the order of ±0.02 g/cm^3, and are related to tool calibration and problems with the mud cake and borehole rugosity corrections. Because very few sonic logs were available in the study area, they were not used in the porosity analysis.

Gamma-ray logs are also normalized and require a two-point calibration procedure. A very significant and systematic difference in tool response was observed between logging vendors; we found that the published environmental correction routines only make a small reduction to these differences. In particular, two of the logging vendors in the Jonah-Pinedale area have similar tool response in the Lance Formation (after environmental corrections are applied), but one vendor reads consistently lower API values in both sandstones and shales. Reducing this variability by normalizing the gamma-ray logs greatly simplifies correlation and saturation calculations. Normalized gamma-ray logs are also essential for advanced formation evaluation methods, such as neural network modeling.

Gamma-ray logs were offset to match a consistent clean sandstone line, and a gain correction was applied to match the sandstone-to-shale span of the composite histogram in the same 8 wells used for the porosity normalization (Figure 4). The form of the equation used was

$$GRC = [(GR - GRsd) \times (GRsh\_T - GRcl\_T)/ (GRsh - GRsd)] + GRcl\_T \quad (1)$$

where GR is the environmentally corrected log value at each depth, GRcl_T is the target clean sandstone value from the type histogram (45 API units), GRsh_T is the target shale value from the type histogram (110 API units), GRsd is the observed sandstone reading on the subject well histogram, and GRsh is the observed shale reading on the subject well histogram. It should be emphasized that the choice of target values is somewhat

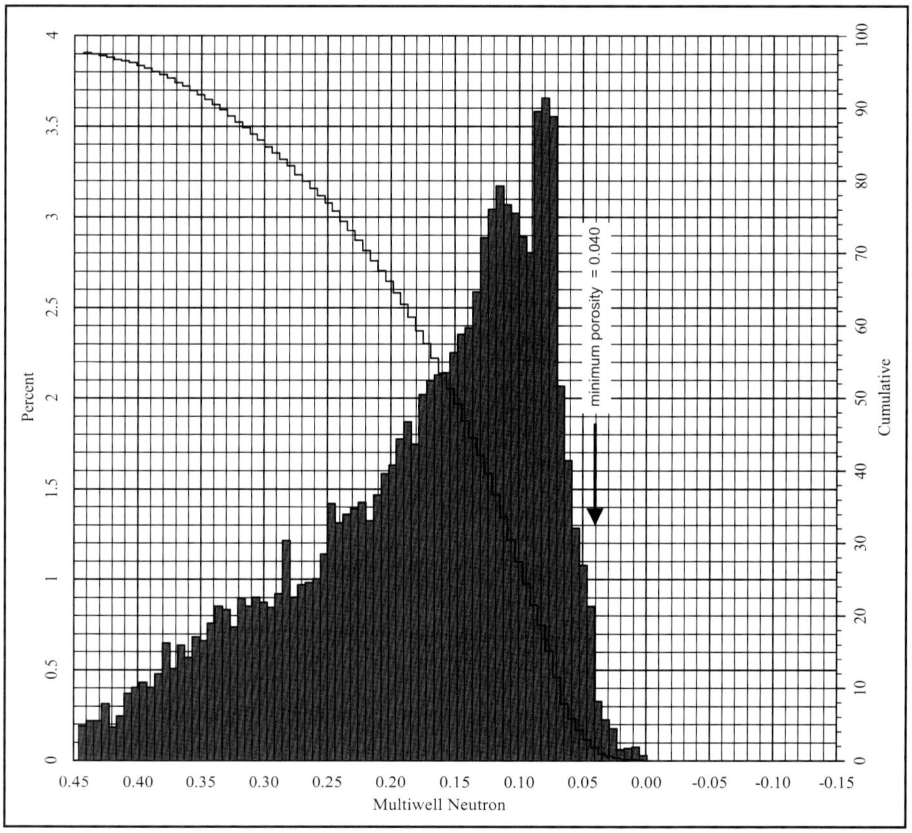

Figure 3. Type histogram for neutron-porosity normalization. Histogram is a composite of the top 1000 ft (300 m) of the Lance Formation in eight wells. Porosity was converted from field logging units (sandstone) to limestone units for this analysis. Histogram bars (frequency %) refer to left axis; cumulative frequency line (cumulative %) refers to the right axis.

arbitrary. The values selected should be distinctive features on the type histogram for which corresponding points can easily be recognized on histograms of individual logs in the area.

As a check of the normalization assumptions, the bulk shifts are mapped to see if variations form a random pattern (noise) or any kind of organized geographic distribution. If the latter is true, the assumption is that the variations are related to geology instead of logging tool response, calibration, or the borehole environment. At Jonah field, all of the normalization corrections appear either random or vendor related in most wells. With the exception of gamma-ray log shifts the corrections were relatively small.

Shale Volume, Total Porosity, and Effective Porosity

Volume of shale (V_{shale}) was computed from the gamma-ray curve with a simple linear V_{shale} equation:

$$V_{shale} = (GRC - GRsd)/(GRsh - GRsd) \qquad (2)$$

where GRC is the environmentally corrected and normalized gamma ray from equation 1, GRsd is the normalized value for clean sandstone (45 API units), and GRsh is the normalized value for shale (110 API units).

"Total porosity" is the total pore volume of the rock and includes porosity filled with hydrocarbons, moveable water, capillary-bound water, and clay-bound water (Hook, 2003). Both the density and neutron logs are considered total porosity tools, because they detect all the porosity in a region surrounding the logging tool, although they have different volumes of investigation and are responding to different physical phenomena that are indirectly related to porosity (electron density in the case of the bulk-density log and hydrogen density in the case of the neutron log). In shaly formations, they read very different values; in particular, the neutron log is strongly affected by hydrogen associated with clay-bound water and reads a much higher apparent porosity than the density log. "Effective porosity" is normally considered that part of the total porosity that is available for hydrocarbon storage and is also physically interconnected so that fluids can actually move between the pores. In vuggy carbonate rocks, for example, there may be isolated vugs that are part of the total porosity, but which have no pore throats that would allow fluids to move in or out of the vugs. These are not considered part of the effective pore system. In shaly sandstones, a

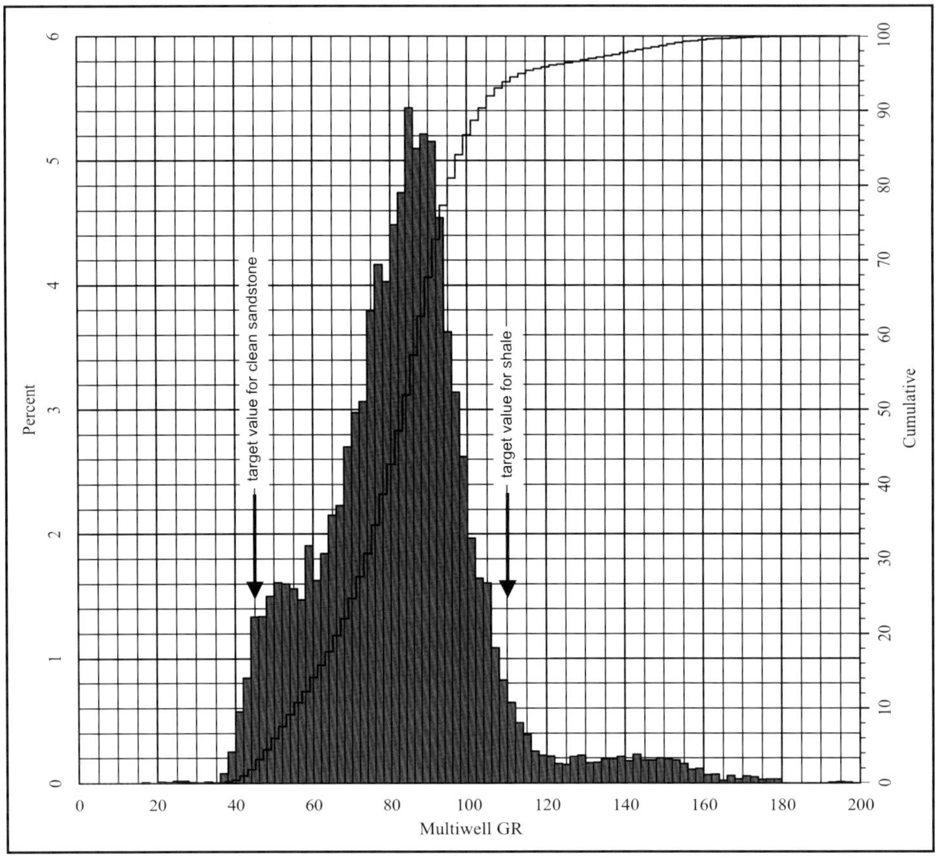

Figure 4. Type histogram for gamma-ray (GR) log normalization. Histogram is a composite of the top 1000 ft (300 m) of the Lance Formation in eight wells. Arbitrary target values for clean sandstone and shale are marked. Histogram bars (frequency %) refer to left axis; cumulative frequency line (cumulative %) refers to the right axis.

significant portion of the porosity is occupied by clay-bound water that is immobile under all conditions and consequently is not part of the effective pore space. We use effective porosity for the fraction of the total porosity that is not associated with clay minerals, thus excluding the volume of clay-bound water but including capillary-bound water on grain surfaces. By this definition, the effective porosity of shale is zero, whereas a clean sandstone has an effective porosity equal to the total porosity. Because it is unlikely that there is much truly isolated porosity in fine-grained clastic rocks, this definition of effective porosity is very close to the hydraulically connected pore volume. Our definition of effective porosity in shaly sandstones is the same as that used in several log analysis textbooks (e.g., Dewan, 1983; Asquith, 1990).

Total porosity was calculated from the normalized density and neutron logs using the appropriate (vendor-specific) crossplot. Total density porosity was calculated using 2.65 g/cm^3 grain density. The neutron porosity was converted from field-reported units (sandstone matrix) to limestone units prior to normalization and analysis. Most advanced log interpretation routines, including all crossplot porosity charts, require the neutron log to be in limestone units.

Effective porosity was computed separately for each porosity tool from the matrix-corrected total porosity, V_{shale}, and average values for the total porosity of shale by equations of the form

$$\mathrm{PHIDE} = \mathrm{PHID} - (V_{shale} \times \mathrm{PHID}_{shale}) \quad (3)$$

where PHIDE is the effective porosity from the density log, PHID is the total porosity from the density log, PHID_{shale} is the total density porosity of shale, and V_{shale} is the linear shale index from equation 2. Analogous equations were used to compute effective neutron porosity and effective sonic porosity. The average values for the total porosity of shale in these equations were 2.4% density porosity, 24% neutron porosity (sandstone matrix), and 21% sonic porosity. The calculated sum-of-squares average of the effective density and neutron porosities closely approximates a gas effect-corrected effective porosity. This value was used in all subsequent calculations and reservoir property summations.

The calculated effective porosity matches the overburden-corrected core porosity and is slightly less than the average field-measured density porosity, as would be expected if gas is present in the formation. Although routine core porosity is generally considered a total porosity measurement, it is only true if the core plugs are dried at high temperature and all clay-bound water is driven off. Many operators use density porosity as an approximation of effective porosity. This work confirmed that this is a reasonable quick-look method because the bulk density of compacted shale in the Lance is fairly close to the matrix density of sandstone. Because the density log can indicate too high a porosity in gas-bearing sandstones, the shale-corrected density-neutron effective porosity is a more accurate measure of pore volume available for hydrocarbon storage.

Water Saturation

Water saturations were computed using the dual water model (Clavier et al., 1977), with a constant R_{wf} (formation water resistivity) at formation temperature of 0.18 ohm m and an R_{wb} (bound water resistivity) of 0.20 ohm m. These values were derived from Pickett plots and crossplots of apparent water resistivity versus shale volume. Because R_{wf} and R_{wb} are similar, the clay conductivity correction is minor, and the results are similar to a simple Archie saturation calculation. The cementation exponent (m) and saturation exponent (n) were both set to 1.8 based on crossplots and our general knowledge of pore structure in tight-gas sandstones.

Substantial uncertainty is associated with both water resistivity estimates and electrical exponents. Produced water in the field is limited (average 3.1 bbl/mmcf gas), and it contains mud filtrate, completion fluids, and water of condensation from the gas stream. The R_{wb} estimate is generally consistent with the clay mineralogy of the Lance and is reasonably constrained from the log data. It is likely that R_{wf} varies through the Lance and across the field area as a function of variation in formation temperature and depth. Unfortunately, data are insufficient to determine the R_w gradient or zonation, so a constant R_w was used, which yields a simplified but easily reproducible set of calculations. The m and n exponents can be determined by special core analysis methods, but these data have not been released by any of the operators at Jonah field.

Estimated Ultimate Recovery

Estimated ultimate recovery (EUR) was determined by conventional decline curve analysis of the production histories through mid-2001 (Cluff, 2004, Appendix C). The production data were obtained from a commercial vendor. Preliminary estimates of the initial decline rates and hyperbolic decline exponents were based on the older wells in the field. A hyperbolic exponent (b) of 0.85 matched the curvature of the decline curve in most wells. Initial decline rates vary but average 70%. Recently, drilled wells with insufficient production history were compared with offset wells to estimate the decline rate and curvature. Decline rates were not limited to a constant exponential rate at the tail of the curve. All wells were projected to a constant economic rate limit (Q_{ec}) of 2 mmcf gas/month. Using a constant rate limit eliminates pricing and operating cost assumptions.

PETROPHYSICS OF THE LANCE FORMATION

Because wells in Jonah field were drilled to different total depths and few wells penetrate the entire Lance Formation, it is difficult to sum the petrophysical properties in a manner that allows meaningful well-to-well comparisons without the thickness of section drilled becoming a significant variable. No reliable stratigraphic markers in the Lance section, particularly in the lower Lance, allow us to subdivide the formation into zones that a majority of geologists would agree are correlative depositional units. As a compromise, we summed the petrophysical properties of the top 2500 ft (760 m) of the formation. The top of the interval is the base of the Tertiary Fort Union Formation (Tfu0 horizon; Cluff and Murphy, 2004), which is a more consistent and reliable correlation surface than the top of the Lance Formation. The top of the Lance can be picked at different positions within or below a 200–500-ft (60–150-m) shaly unit (averages 325 ft [100 m]), informally called the "Unnamed Tertiary shale" by several operators or the "Not Yet" by local mud-loggers (DuBois et al., 2004; Cluff and Murphy, 2004, Appendix B). Palynological data collected by Chevron U.S.A. (unpublished data shown by Bowker and Robinson, 1997) suggest that the Cretaceous–Tertiary boundary lies somewhere in the Unnamed shale section. All the wells in the study area penetrate the entire 2500-ft (760-m) interval below the base of the Fort Union. It should be noted that the upper 2500 ft (760 m) of the Lance Formation in this study includes the above-mentioned Unnamed shale unit.

Net Sandstone Thickness

Net sandstone is defined as having less than 50% V_{shale} using the linear gamma-ray index method described above. Ignoring the previously discussed vendor variability and borehole effects, the 50% V_{shale} cutoff roughly corresponds to 75 API units, which is a commonly used field definition for sandstone.

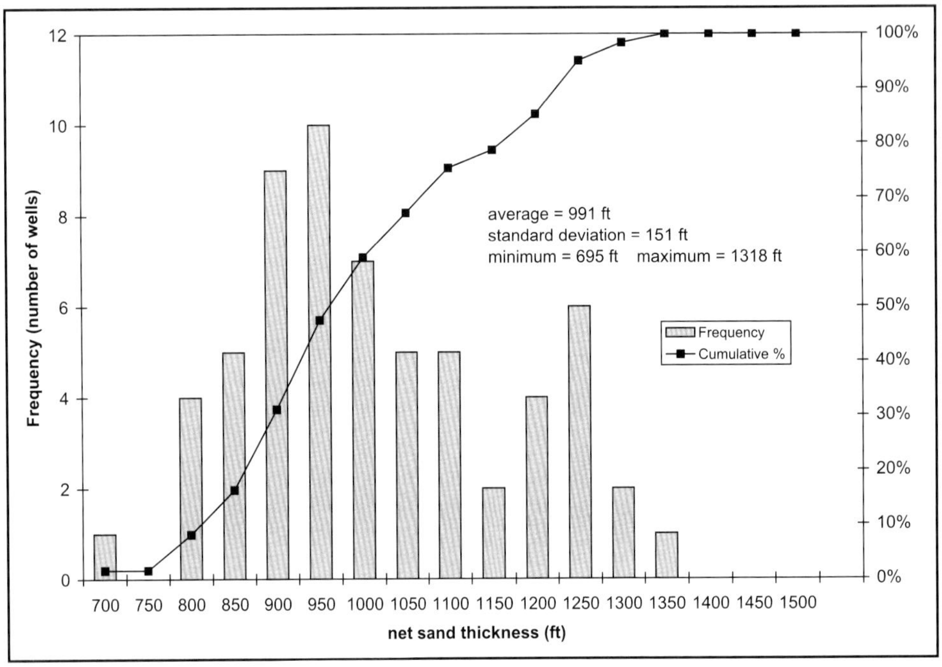

Figure 5. Distribution of total net sandstone thickness, by well, in the upper 2500 ft (760 m) of the Lance Formation. No porosity or water saturation cutoffs are applied. Histogram bars refer to left (frequency) axis; cumulative frequency line (cumulative %) refers to the right axis.

The net thickness of sandstone in the top 2500 ft (760 m) of the Lance ranges from 700 to 1300 ft (210 to 400 m) and averages a little less than 1000 ft (300 m) (Figure 5). The net/gross ratio ranges from 28 to 53% and averages 40% (Figure 6). Figures 7 and 8 show the areal distribution of these two parameters.

Porosity

The average porosity determined from the available core data at Jonah field (n = 225) is 7.7% (Figure 9). These core data are routine analyses at 800 psi confining stress and have not been corrected to net overburden

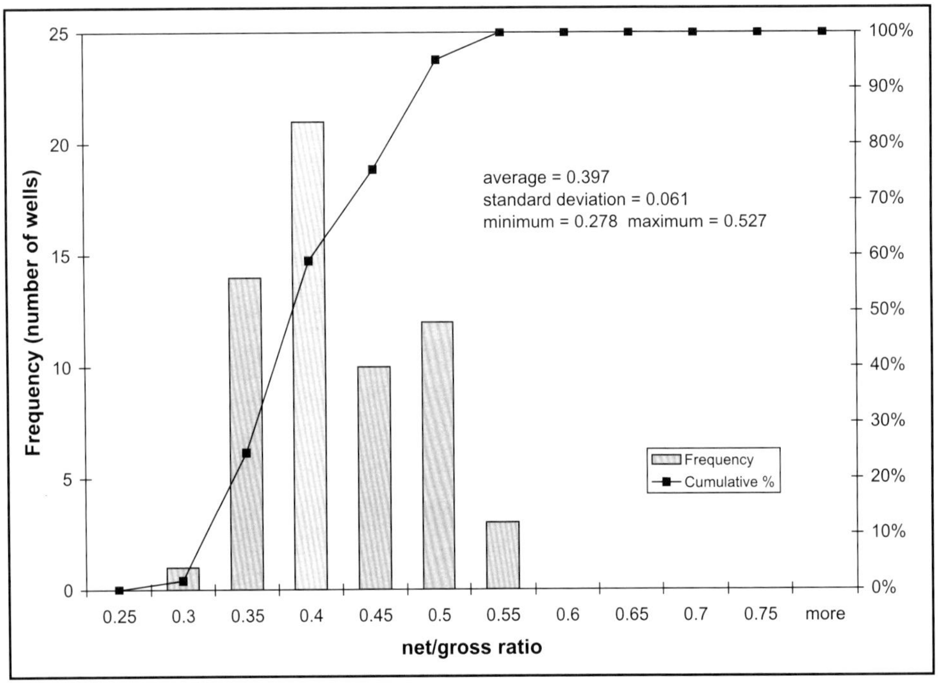

Figure 6. Distribution of the ratio of net sandstone to gross interval, by well, for the upper 2500 ft (760 m) of the Lance Formation. Histogram bars refer to left (frequency) axis; cumulative frequency line (cumulative %) refers to the right axis.

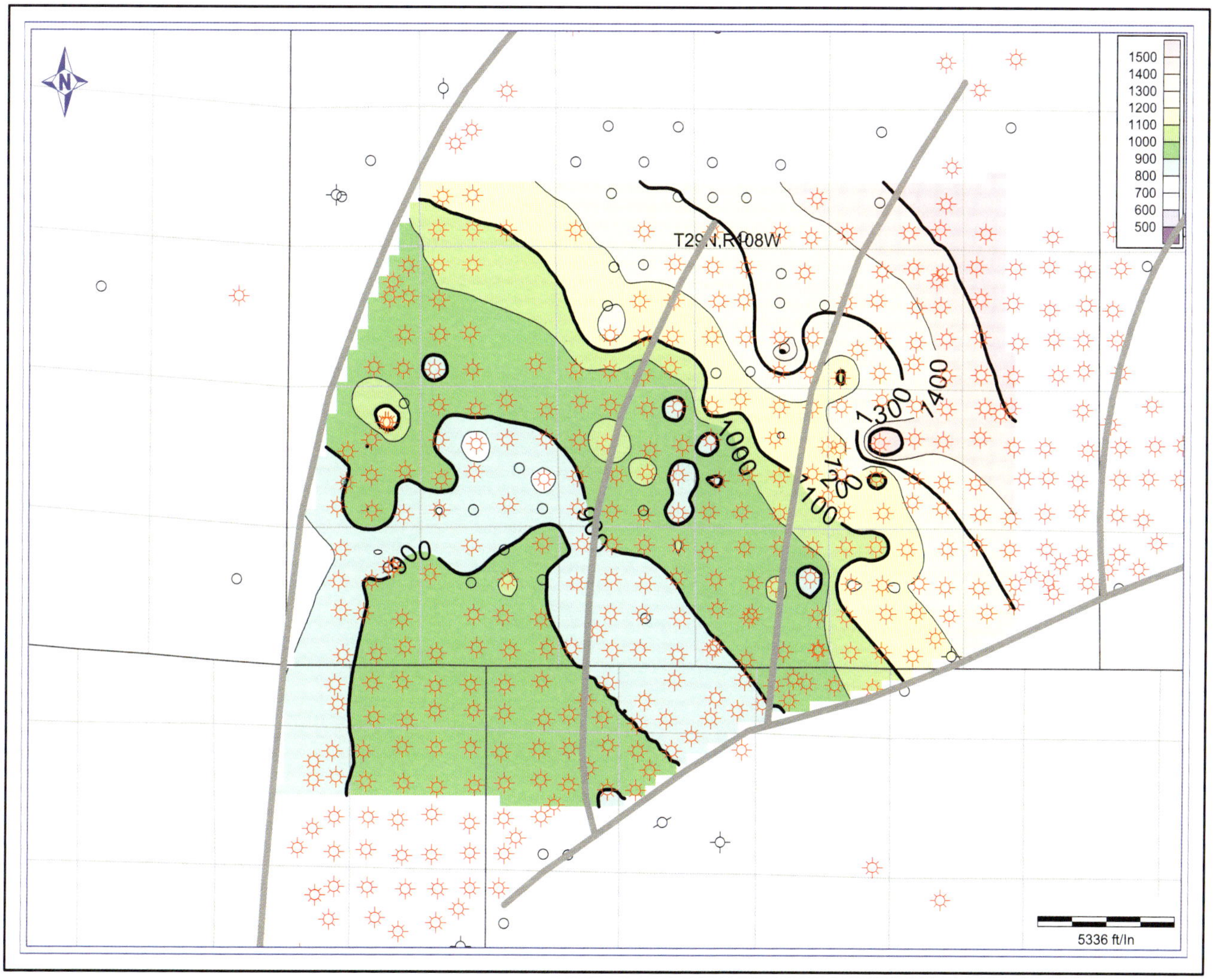

Figure 7. Net feet of sandstone in the upper 2500 ft (760 m) of the Lance Formation. Contour interval = 100 ft (30 m).

conditions. The correction from laboratory conditions to 4000 psi net effective overburden stress is given by the regression equation

$$\phi_{\text{in situ}} = 1.0077\ \phi_{\text{routine}} - 0.6482, \quad r^2 = 0.986 \qquad (4)$$

where porosity is in percent (Figure 10). By this equation, the average routine core porosity of 7.7% is equivalent to an in-situ porosity of 7.1%.

The average log-determined porosity from all wells ranges from 4 to 10%, with an average of 6.5% (Figure 11). The difference between the average log porosity and average in-situ core porosity is not significant given the small sample size of the core data compared to the log data; furthermore, core plugs are commonly selected in a manner that undersamples the lowest porosity sandstones and siltstones. Figure 12 shows the essentially random variation in average porosity over the study area. Most wells exhibit a greater range of porosity values than the range of averaged porosities shown in Figures 11 and 12. This is illustrated by the well shown in Figure 13, which has a typical average porosity of 7.4%, but the total range of porosity is from 0 to 15%.

Permeability

The average routine core permeability (800-psi net confining stress) is 145 μd (0.145 md) (Figure 14). In tight-gas sandstones, most reservoir units have average permeabilities less than 100 μd (0.1 md). To avoid the repetitive use of very small decimal fractions, we encourage using microdarcys (μd), where 1 μd = 0.001 md = 10^{-6} darcys.

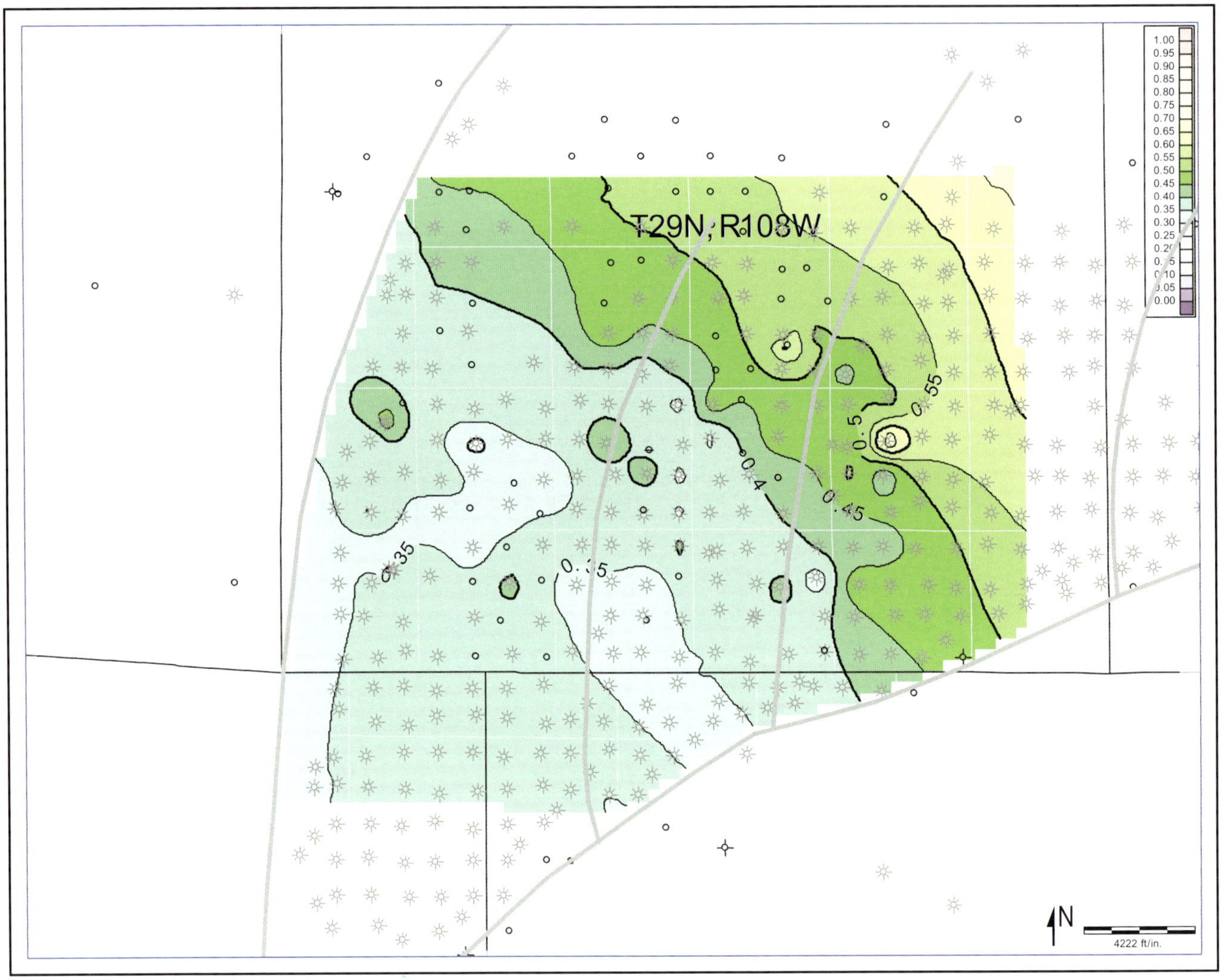

Figure 8. Ratio of net sand to gross interval in the upper 2500 ft (762 m) of the Lance Formation. Contour interval = 0.05%.

On the text figures and in equations that span to higher permeability values, we revert to the more conventional units of millidarcys.

Samples with the highest permeability are suspected to be microfractured plugs. Correction from ambient laboratory conditions to net overburden stress will reduce these values by at least 50%, and correction to effective permeability to gas at reservoir saturation will reduce them still further. Well performance data suggest that the average effective permeability to gas in typical Jonah wells are in the low tens of microdarcys (Eberhard and Mullen, 2001).

The correction from routine core analysis (P = 800 psi) to Klinkenberg corrected permeability at net effective overburden stress (P = 4000 psi) was determined by regression analysis of a subset of the core data (n = 225) with several outliers removed:

$$\log_{10} K_{\infty\ 4000\ \mathrm{psi}} = 1.4718 \times \log_{10} (K_{\mathrm{air}\ 800\ \mathrm{psi}}) - 0.4016, \quad r^2 = 0.922 \qquad (5)$$

where permeability in this equation is in millidarcys (Figure 15). Using this equation, the average routine permeability of 145 μd is equivalent to an in-situ reservoir permeability of only 23 μd, or 16% of the routine analysis. This in-situ permeability is the average single-phase (or absolute) gas permeability at 0% water saturation and will be further reduced by the effects of relative permeability, unless the reservoir sandstones are at irreducible water saturation.

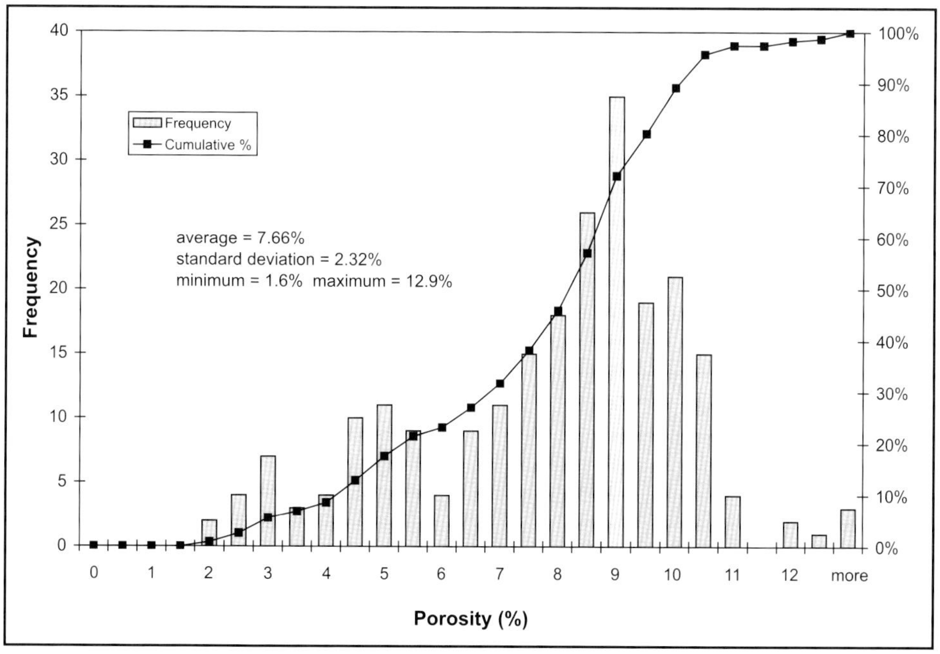

Figure 9. Distribution of routine core porosity values for the Lance Formation. Histogram bars refer to left (frequency) axis; cumulative frequency line (cumulative %) refers to the right axis.

The regression equations to estimate permeability from porosity (and vice versa) at reservoir stress condition are

$$\log_{10} K_{\infty\ 4000\ \mathrm{psi}} = 0.2135 \times \phi - 3.5861, \quad r^2 = 0.488 \tag{5.1}$$

$$\phi = 4.6838 \times (\log_{10} K_{\infty\ 4000\ \mathrm{psi}} + 3.5861) \tag{5.2}$$

where porosity is in percent, and permeability is in millidarcys (Figure 16). The poor correlation coefficient is typical of tight-gas sandstones. Equation 5.1 accurately expresses the average trend of increasing permeability with increasing porosity, but the scatter around the trend line is so great that the equation should be used with caution for quantitative prediction of flow capacity from porosity.

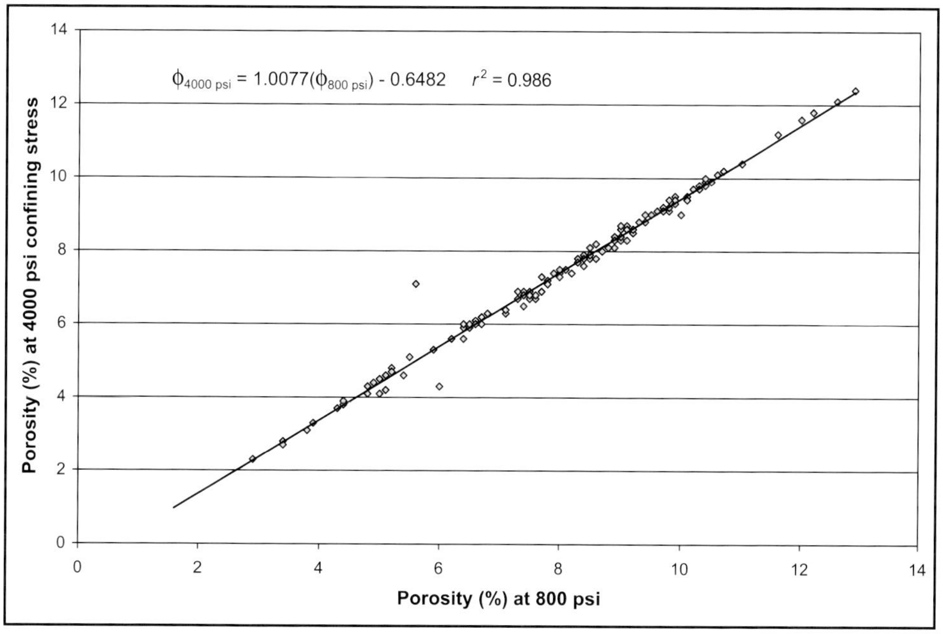

Figure 10. Correlation of routine core porosity (800 psi confining stress) and porosity at net overburden conditions (4000 psi confining stress).

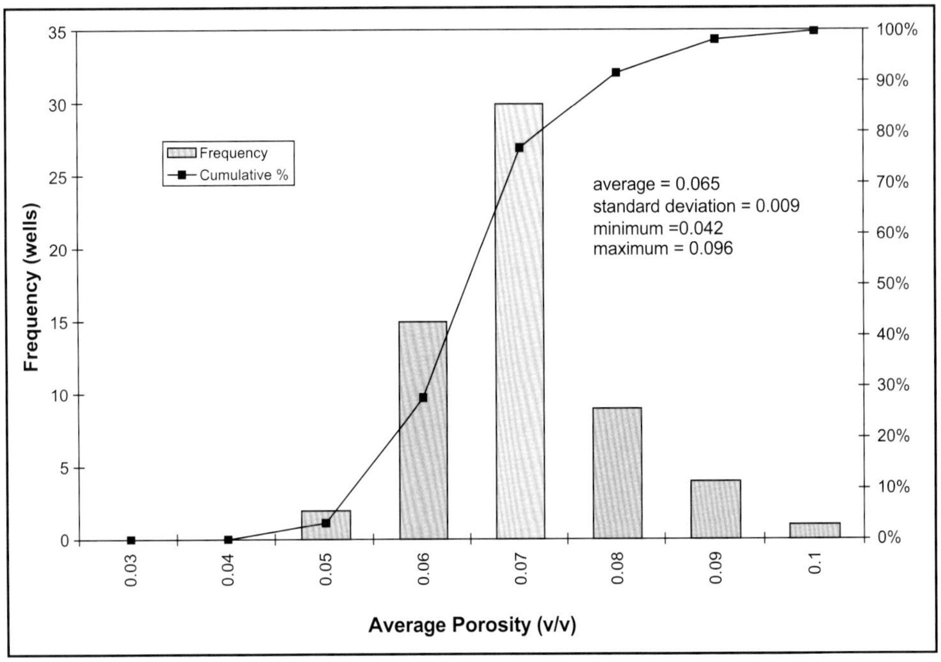

Figure 11. Distribution of average porosity, by well, in the upper 2500 ft (760 m) of the Lance Formation. Histogram bars refer to left (frequency) axis; cumulative frequency line (cumulative %) refers to the right axis.

Water Saturation

The water saturation per well by log analysis ranged from 30 to 57%, with an average of 45% (Figure 17). Eberhard and Mullen (2001) reported a similar range of water saturations (30–60%). Water saturation correlates with porosity. Lower average water saturations in higher porosity rocks are observed in the log calculations and in routine core analyses (Figure 18). The product of water saturation and porosity (bulk volume water) is nearly constant at 3 to 3.5%, which suggests that the sandstones in Jonah field are near irreducible water saturation conditions. This observation is consistent with the very low produced water/gas ratio (WGR) in the updip wells.

No oil-based cores have been cut as of mid-2003 that would be useful for testing the validity of these values. Our calculated water saturations in cored intervals are similar to reported routine core saturations in magnitude and variation. The values are reasonable based on our knowledge of capillary-pressure characteristics of similar tight-gas sandstone reservoirs in the Rocky Mountain region (e.g., Byrnes, 1997). Bowker (1995) published a gas-brine capillary-pressure curve for one sample (9781 ft [2983 m], 9.7% porosity) from the Lance in the Jonah Federal 2-5 (Sec. 5, T28N, R108W). The curve never attains an asymptotic irreducible water saturation but climbs steeply between 35 and 20% water saturation at capillary pressures equivalent to 500–2000 ft (150–610 m) of gas above a free-water level. Because the gas accumulation at Jonah is at least 2500 ft (760 m) thick and, based on pressure data, appears to act like a common pool with one pressure gradient, it is reasonable to assume that the actual saturations attained in the field will be in this range. An attempt was made to discern a trend of increasing water saturation with depth from the well-log calculations but we were unable to define a meaningful saturation-height function given the uncertainty in the R_w gradient across the field. A map of average water saturation of the Lance (Figure 19) does not show any meaningful pattern in average saturation. A map of water saturation (not included) filtered to the high-porosity rocks shows a very gradual and erratic decrease in water saturation updip toward the southwest corner of the field.

NET PAY DETERMINATION

One of the most vexing problems in tight-gas sandstone formation evaluation is the determination of net pay. Engineers and petrophysicists have never settled on a rigorous definition for net pay, much less a standard methodology to decide what is or is not pay.

In a high porosity-permeability conventional hydrocarbon accumulation, net pay is defined on flow-rate criteria and the fluids produced. Rocks with sufficient permeability to flow fluids at commercially significant rates are classified as "net sandstone" or net reservoir. If

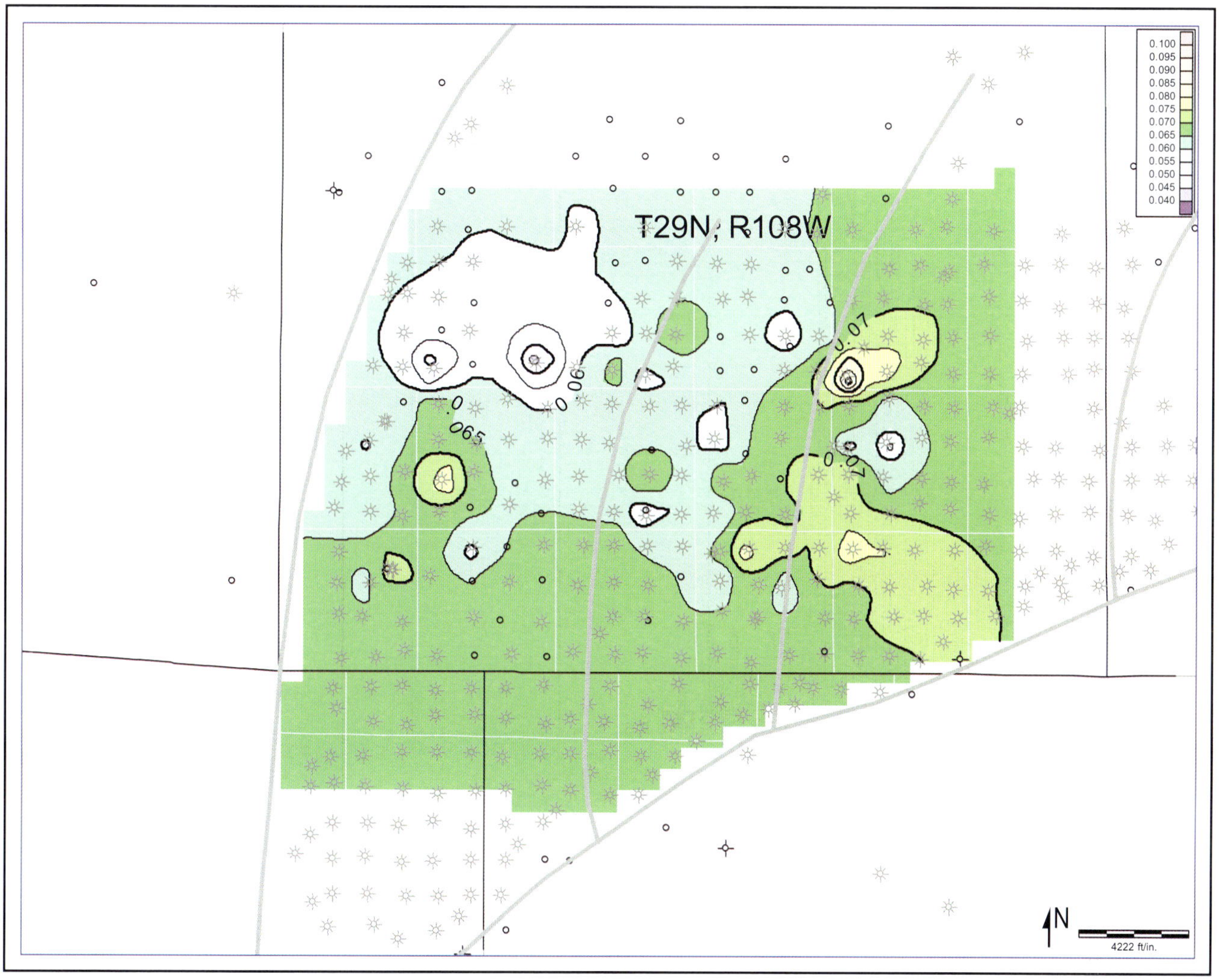

Figure 12. Contour map of average sandstone porosity in the upper 2500 ft (760 m) of the Lance Formation. Contour interval = 0.05%.

they produce hydrocarbons at a commercially acceptable hydrocarbon/water ratio, they are classified as net pay. Note the emphasis on commerciality in both cases; the word pay implies money and some kind of economic criteria. Also note that the petrophysical criteria most useful to distinguish pay from nonpay would be permeability, hydrocarbon saturation, and relative permeability to the hydrocarbon and water phases as a function of saturation. Only one of these factors (saturation) is routinely estimated from logging measurements.

This conventional notion of net pay leads to the common use of three log- or core-based petrophysical cutoffs to separate pay from nonpay:

(1) *Shaliness*: A volume of shale cutoff, commonly a gamma-ray or spontaneous potential cutoff, is used to separate potentially permeable clean formation from presumably impermeable shaly formation.

(2) *Porosity*: A porosity cutoff is used as a proxy for a permeability or flow cutoff. A porosity-permeability crossplot can be used to define a porosity equivalent to a minimum permeability considered capable of flowing hydrocarbons at commercial rates. Alternatively, a comparison of production test or drill-stem test flow rates on a large number of zones can be used to determine the minimum porosity required for a commercial completion.

(3) *Water saturation*: A calculated water saturation cutoff is used to separate wet zones from oil-bearing or gas-bearing pay intervals. In high-porosity and high-permeability sandstones, the transition zone from water wet to hydrocarbon saturation is commonly

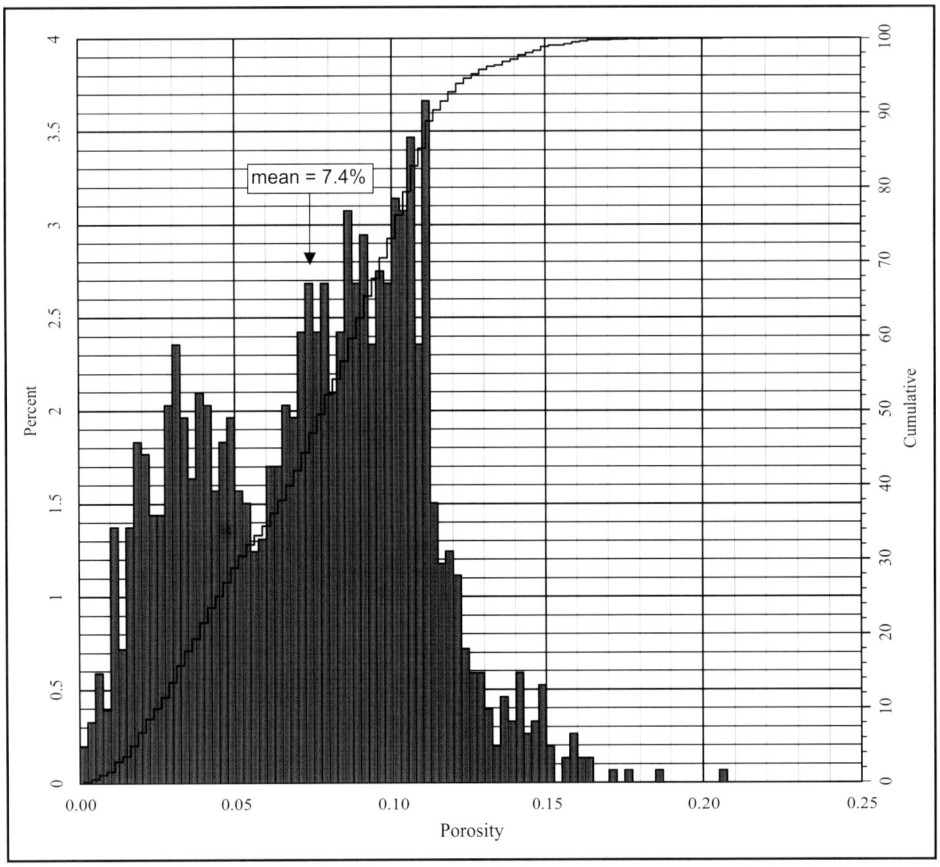

Figure 13. Sandstone porosity distribution for the upper 2500 ft (760 m) of the Lance Formation in a typical well near the center of Jonah field. Data samples greater than 12% porosity are bad log readings immediately adjacent to washed-out shale intervals. Histogram bars (frequency %) refer to left axis; cumulative frequency line (cumulative %) refers to the right axis.

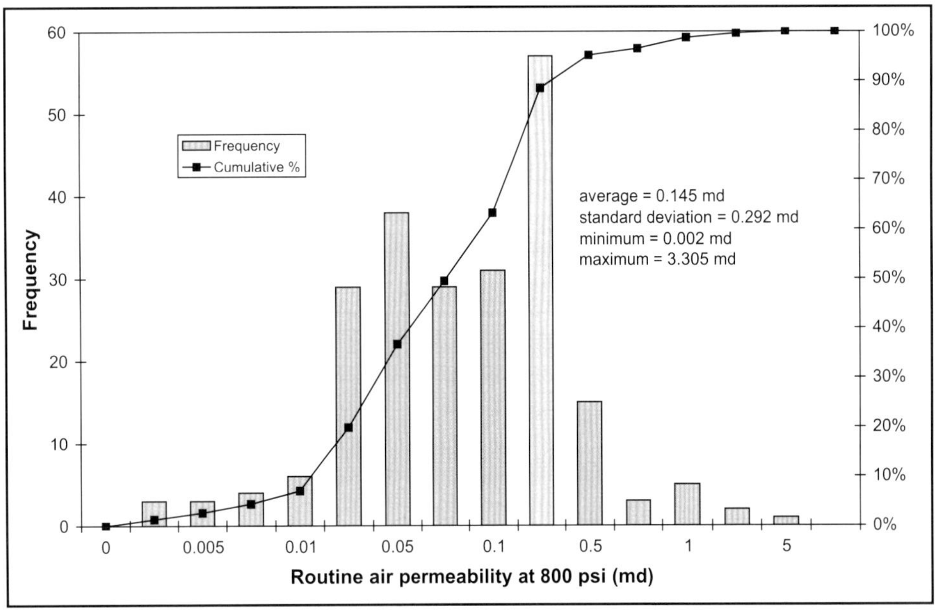

Figure 14. Distribution of routine core permeability values in the Lance Formation. Histogram bars refer to left (frequency) axis; cumulative frequency line (cumulative %) refers to the right axis.

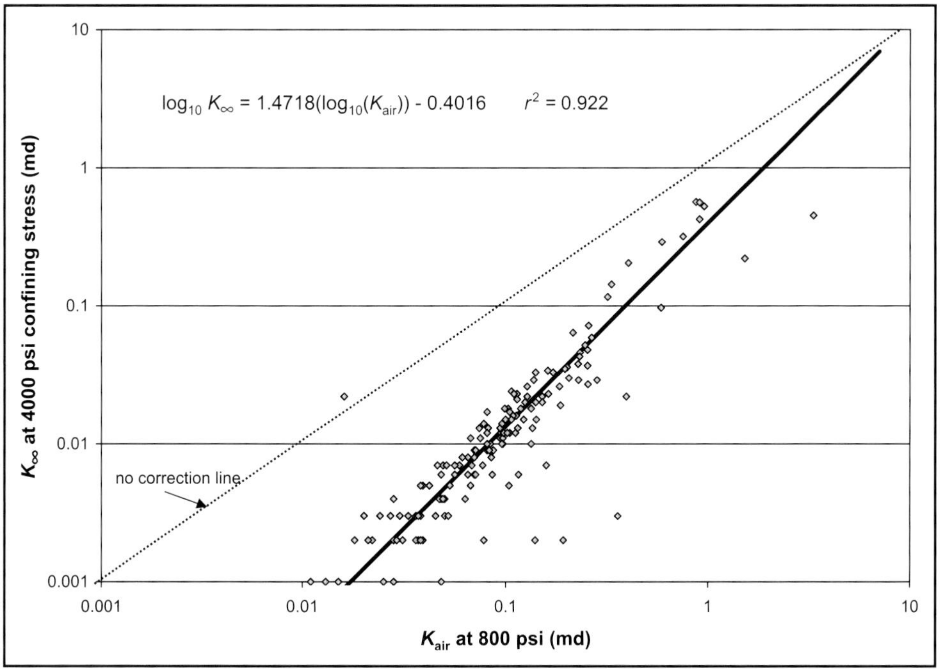

Figure 15. Correction from routine core permeability measurements to Klinkenberg corrected permeability at 4000 psi net confining stress.

very thin, with a significant difference in water saturation between high water saturation wet zones and low water saturation pay zones. Thus, the actual water saturation cutoff may not be critical, as long as it is between the average wet saturation and the typical hydrocarbon saturation.

In tight-gas sandstones, all of the above criteria become fuzzy, and decisions concerning net pay cutoffs have a significant impact on the net pay calculations. This affects both completion decisions and reserve calculations. Tight-gas sandstones include a shaly sand component, and the relationship between shaliness and permeability

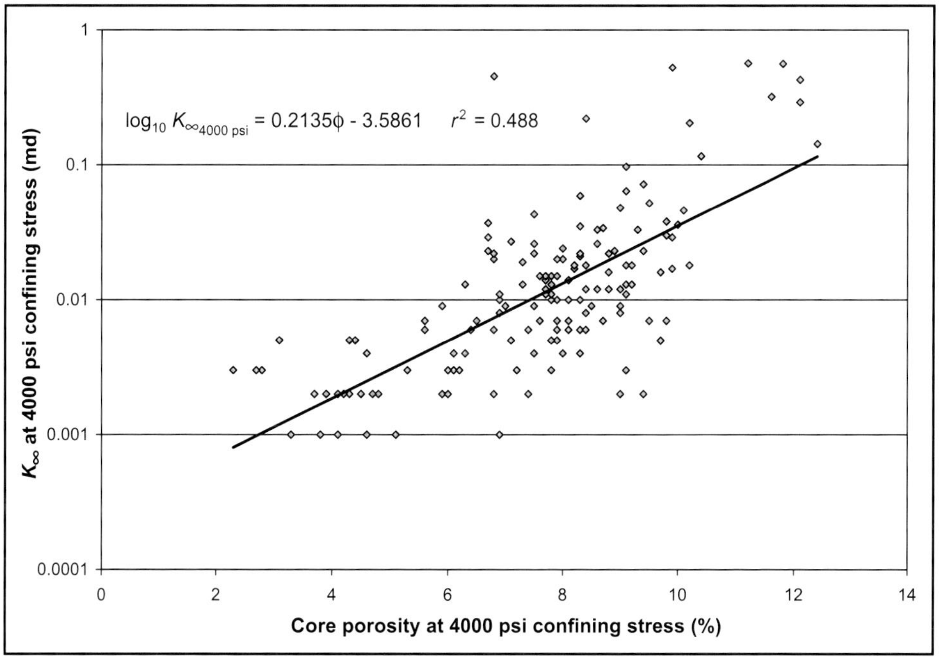

Figure 16. In-situ porosity to in-situ Klinkenberg corrected permeability correlation.

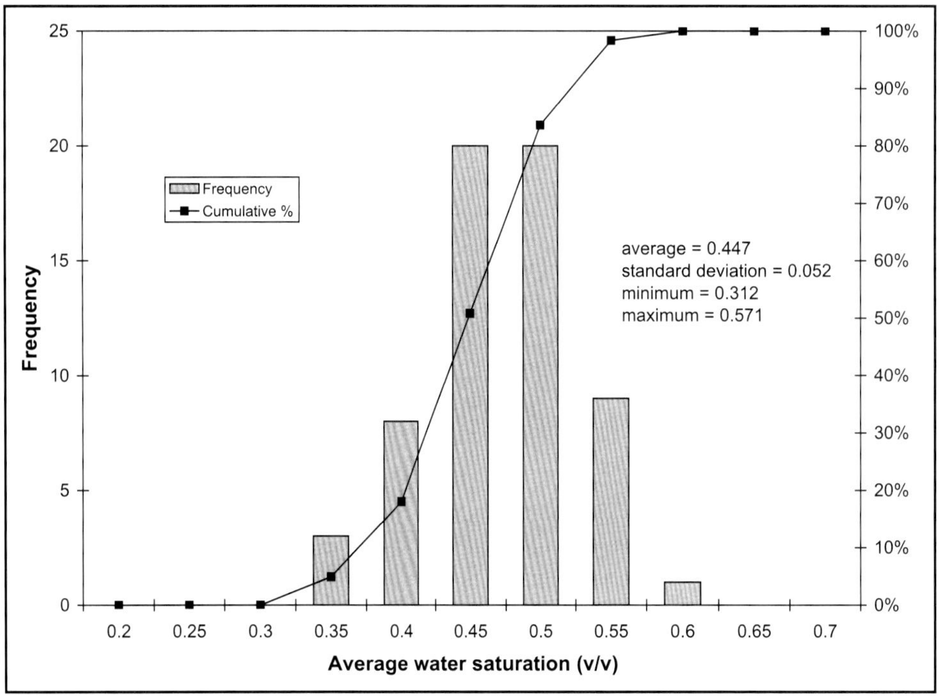

Figure 17. Distribution of average water saturation from log analysis, per well, for the upper 2500 ft (760 m) of the Lance Formation. Histogram bars refer to left (frequency) axis; cumulative frequency line (cumulative %) refers to the right axis.

is often poorly characterized. The porosity of tight-gas sandstones is also low, often times below what is considered the lower limit of pay in many other areas. The porosity range illustrated by Figures 9 and 11, for example, is entirely below what is considered net pay in the Tertiary sandstones of the offshore Gulf of Mexico. By Gulf Coast commerciality standards, none of the reservoir rock at Jonah field is net pay. Furthermore, the porosity-permeability relationship in tight-gas sandstones displays a large scatter and also has a strong confining stress

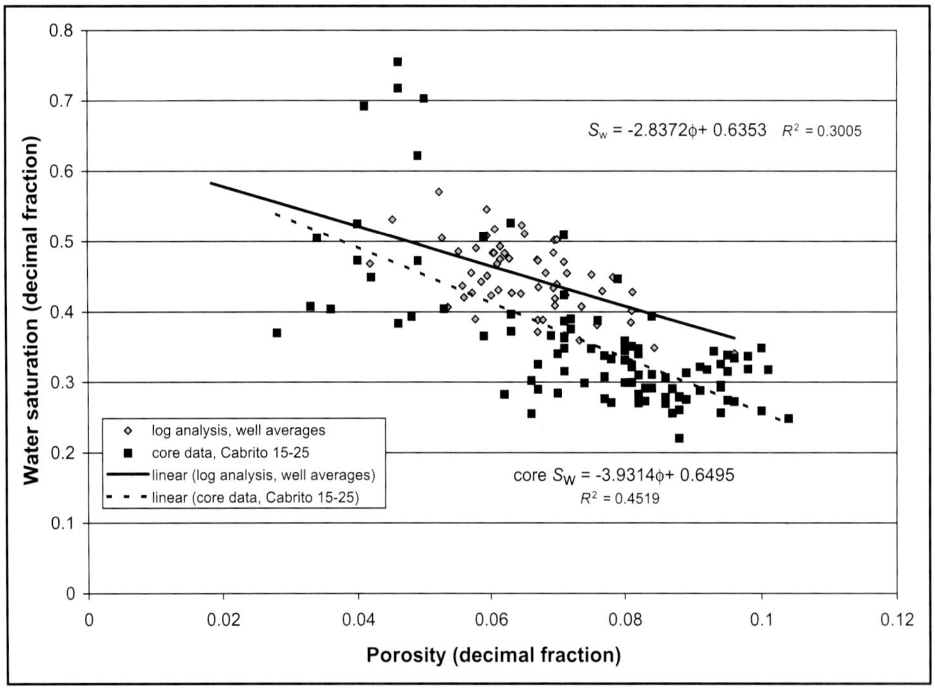

Figure 18. Trend of decreasing average water saturation with increasing average porosity in the upper 2500 ft (760 m) of the Lance Formation. Log analysis values and routine core measurements are shown.

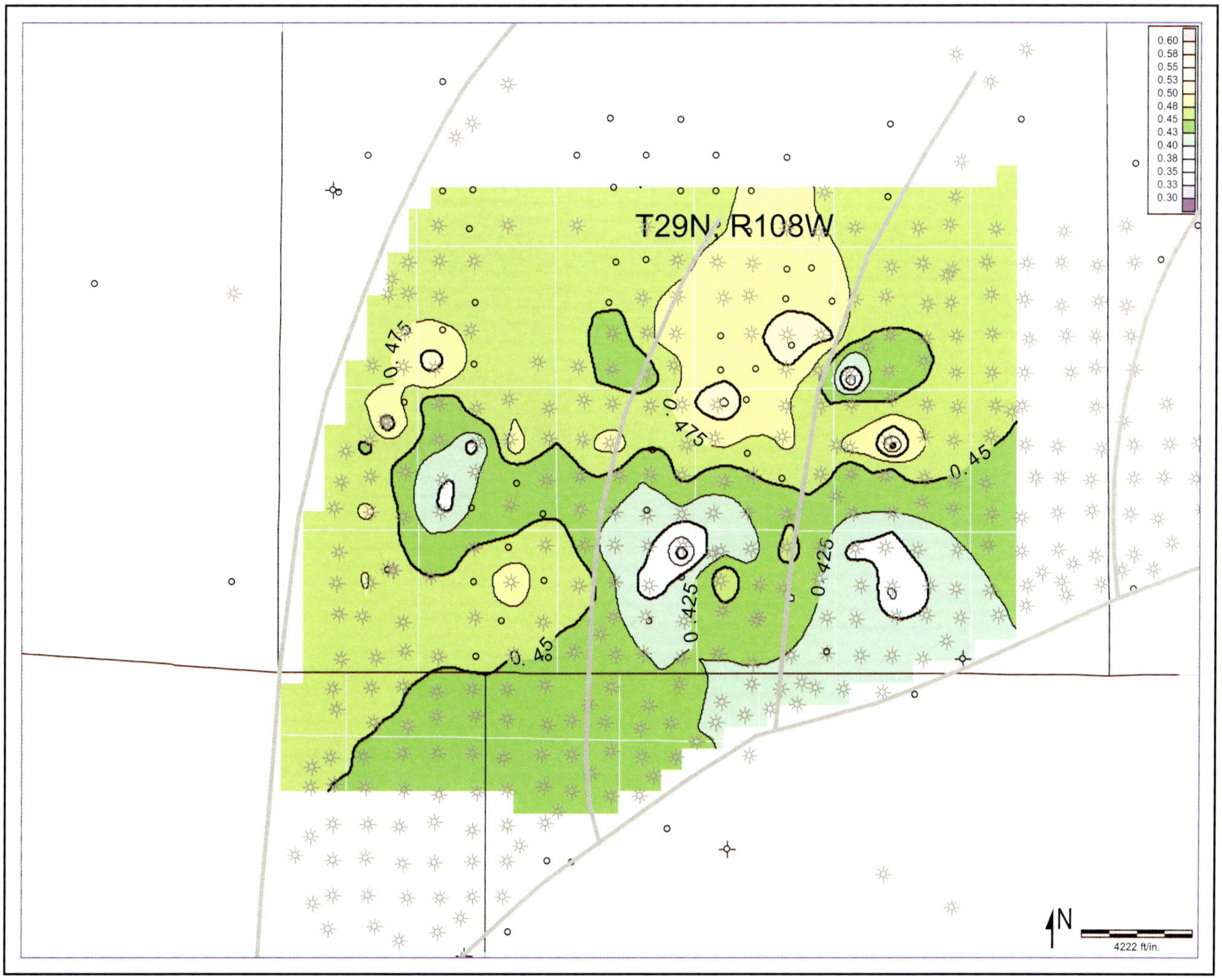

Figure 19. Contour map of average water saturation for sandstones in the upper 2500 ft (760 m) of the Lance Formation. Contour interval = 2.5%.

dependence that is not seen in higher permeability reservoir systems. Consequently, it is difficult to pick a porosity value from a porosity-permeability crossplot that correlates to a specific minimum flow criterion. To complicate matters further, tight-gas reservoirs require fracture stimulation that tends to average out differences and makes it difficult to determine which intervals are really flowing gas and at what rates. As fracture stimulation technology improves, tighter and tighter rocks can be stimulated and produced at commercial rates, so net pay criteria tends to shift with time, independent of commodity prices or development costs. Finally, tight-gas reservoirs show a wide spread in calculated water saturations. Fundamental problems in the saturation calculations exist, the most critical being the uncertainty in formation water resistivity. This uncertainty makes it difficult to pick a saturation cutoff that clearly separates wet zones from hydrocarbon-saturated rocks at irreducible water saturation.

Figure 16 illustrates the in-situ (at reservoir stress) relationship between porosity and absolute permeability. Using equation 5.2, we can estimate the average porosity at 1-μd in-situ permeability to be 2.7%. At a permeability of 5 μd, the average porosity is 6.0%. Rocks with less than 5 μd permeability at reservoir conditions produce gas at insignificant rates even after fracture stimulation. The data can also be shown as cumulative flow and cumulative storage plots (Figures 20, 21). Figure 20 shows that more than 90% of the flow capacity of the Lance reservoirs come from rocks with permeability greater than 14 μd. From equation 5.2, this equates to an average

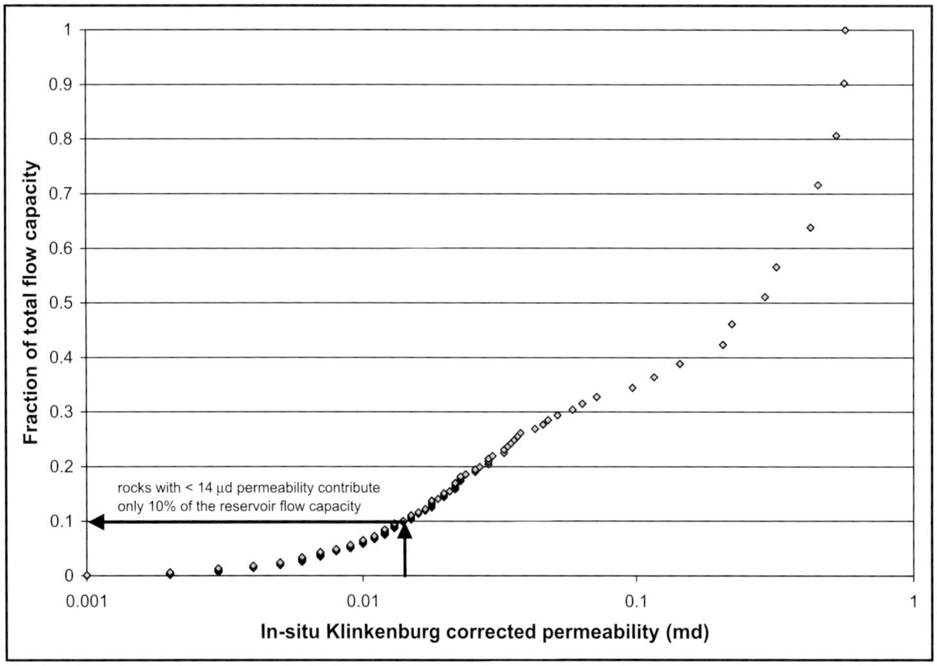

Figure 20. Cumulative flow-capacity plot showing fractional flow capacity as a function of in-situ permeability from core data sorted in permeability order.

porosity of 7.8%. Figure 21 illustrates that 38% of the total reservoir storage capacity is in rocks with 7.8% porosity or less. Therefore, although a substantial portion of the reservoir storage is in these low-porosity rocks, they contribute a relatively insignificant part of the total flow capacity.

Figure 22 is a crossplot of cumulative storage versus cumulative flow, sorted in core porosity order. The dramatic break at about 92% cumulative storage represents either a major rock type change to a significantly better porosity-permeability trend or the core plugs that contain microfractures. The top nine samples all have porosity

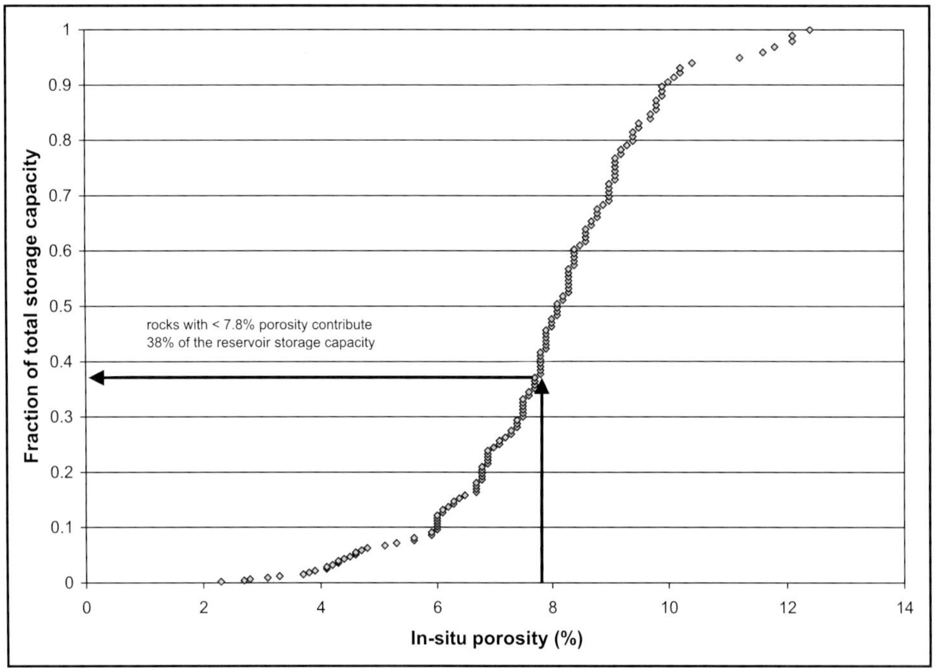

Figure 21. Cumulative storage-capacity plot showing fractional storage capacity as a function of in-situ porosity, from core data sorted in porosity order.

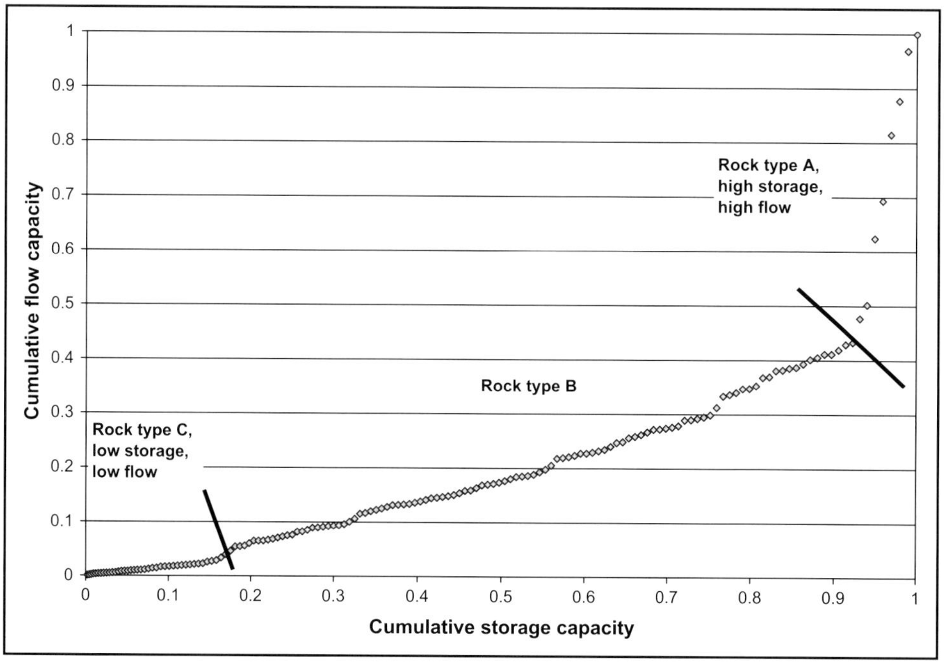

Figure 22. Cumulative storage capacity vs. cumulative flow capacity, from core data sorted in ascending porosity order.

in excess of 10.2% (Figures 16, 21). Because the high-permeability samples correlate with high porosities, we suspect that these samples all belong to a distinct, high-quality reservoir rock type (rock type A). Rock type A is so much better than the remainder of the Lance sandstones that they are likely to dominate test results and early production behavior. Based on comparison of the core depths to core descriptions shown by Shanley (2001), most of this rock type consists of trough cross-bedded and convolute-bedded sandstones.

Between 5 and 92% cumulative storage capacity (Figure 21) is a rock type that makes up the bulk of the Lance reservoir (rock type B). Rock type B is predominantly ripple cross-laminated sandstone (Shanley, 2001) but also includes the trough cross-bedded and convolute-bedded facies found in rock type A. More accurate comparisons between petrophysical rock types and core lithofacies will require spotting the precise plug locations on the cores and determining their specific lithotypes.

Rock type C samples have less than 6.7% porosity (Figure 21) and contribute only 5% to the cumulative flow capacity. It is uncertain if these rocks are capable of producing gas, but they would contribute an insignificant portion of the measured flow. If completed and tested in isolation from the better rocks, they would likely produce at noncommercial rates.

This analysis suggests two possible permeability-based pay cutoffs. One cutoff would be equivalent to an average porosity of 6.7% that represents the break between nonreservoir sandstones (rock type C) and the average reservoir rock in the Lance (rock type B). The other cutoff would be equivalent to an average porosity of 10.2%, at the break in slope from the average reservoir rock (rock type B) to the very high-quality reservoir (rock type A) that will dominate the early production behavior. Because this analysis is based on a limited core database in a very large field, and because these are average properties with substantial variance, we used 6 and 10% porosity as cutoffs between nonpay, pay, and high-quality pay.

The determination of cutoffs is actually more complicated because the data in Figure 16 represent dry rock permeability to the gas phase, but in nature, reservoirs contain both water and hydrocarbons. The effective permeability to gas in the presence of water is reduced and is dependent on saturation and the average pore-throat size of the reservoir rocks. Byrnes (1997, his figure 6) illustrates the decrease in effective permeability to gas with increasing water saturation for a large suite of tight-gas sandstones from Rocky Mountain region. Byrnes' (1997) results are generally valid for rocks of similar porosity, permeability, and grain size and are believed to be applicable to the Lance reservoirs at Jonah, although none of the samples in his data set were specifically from this field. Byrnes (1997) found that the relative permeability to gas approaches 100% at irreducible water saturations of approximately 30% or less, and it abruptly decreases to near zero at water saturations above 50–60%.

Table 1

This Table Summarizes the Average Reservoir Properties of All Sandstones in the Lance (Net of the V_{shale} Cutoff Only) Compared to the Net Pay Counts Using All Three Cutoff Values.

				Net Pay			
	Gross Thickness (ft)	Net Pay Thickness (ft)	Net/Gross (%)	Average Porosity (%)	Average Storage Capacity (ft)	Average Water Saturation (%)	Average V_{shale} (%)
Net Sand: No ϕ or S_w Cutoffs							
Minimum	2500	695	28	4.2	37.9	31	14
Maximum	2500	1318	53	9.6	102.1	57	26
Average	2500	991	40	6.5	66.6	45	17
Standard deviation		152	6	0.9	15.4	5	2
Net Pay: 6% ϕ and 50% S_w Cutoffs							
Minimum	2500	99	5	7.6	8.3	20	3
Maximum	2500	815	39	10.9	89.2	40	14
Average	2500	441	25	9.3	41.7	33	7
Standard deviation		133	6	0.7	14.5	3	2
Net Pay–Net Sandstone Ratios (%)							
Minimum		14	18	181	22	64	21
Maximum		62	73	114	87	71	54
Average		44	62	144	63	75	43

Assuming that the results of our log calculations are valid, and average water saturations in the Lance are 45% (Figure 17), over half of the sandstones will have relative permeabilities to gas of less than 10%. A full net pay cutoff analysis should incorporate relative permeability effects, which should be collected through a large number of single end-point effective permeability measurements on core. Because these data are not available, the impact of relative permeability is incorporated through a water saturation cutoff. A 50% water saturation cutoff is used as a hard net pay cutoff because the data of Byrnes (1997) indicate insignificant relative permeability to gas at higher saturations. Most sandstones with less than 6% porosity display high average water saturation (Figure 18), which reinforces the use of a 6% cutoff.

Finally, we used a 50% V_{shale} cutoff to cleanly separate the reservoir sandstones from shaly nonreservoir rocks. Interestingly, the net pay counts are not highly sensitive to the exact value used because nearly all sandstones with porosity greater than 6% are significantly cleaner than this cutoff. Some shaly siltstones with porosity in the 6–8% range may be cut out using this criterion. We have observed considerable variation in net pay counts when using an uncorrected gamma-ray cutoff, which serves to emphasize the importance of careful log normalization.

Net Pay Summations

Reservoir properties were summed over the top 2500 ft (760 m) of the Lance using the following three petrophysical cutoffs: (1) 50% V_{shale} cutoff; (2) 6% density-neutron effective porosity cutoff; and (3) 50% S_w cutoff.

Table 1 summarizes the average reservoir properties of all sandstones in the Lance (net of the V_{shale} cutoff only) compared to the net pay counts using all three cutoff values. Alternative cases were run using 4, 8, 10, and 12% effective porosity cutoffs. The latter two cases result in uncomfortably low net pay counts but are useful for mapping intervals of very high-quality pay. The net 8% case is relatively uninteresting, because it is only slightly different than the net 6% case and, based on Figure 21, is an arbitrary cut in the dominant rock type in this reservoir system.

Based on the defined cutoffs, the net pay thickness is 14–62% (average 44%) of the net sandstone counts

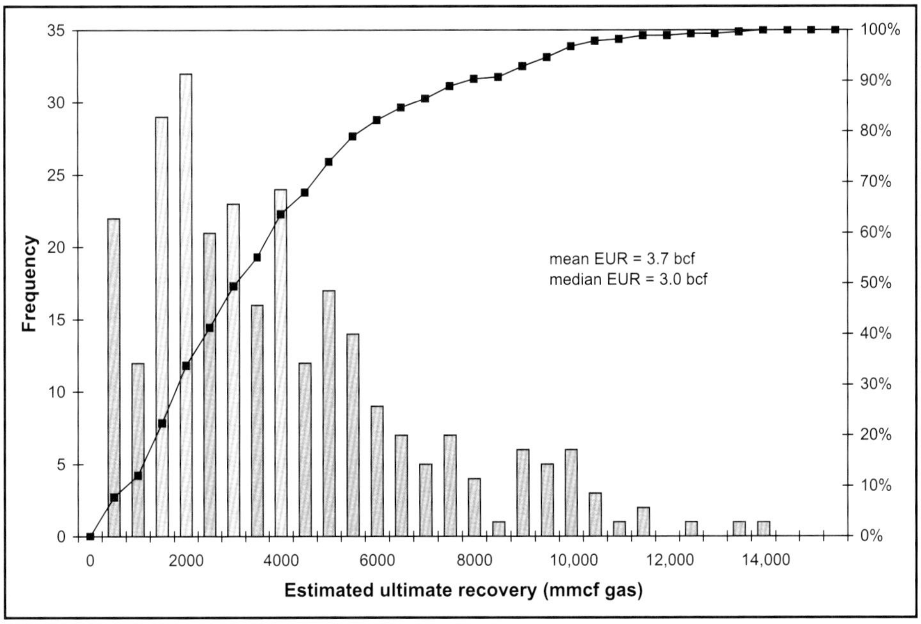

Figure 23. Distribution of estimated ultimate recoveries, in million cubic feet of gas, from decline curve analysis. Histogram bars refer to left (frequency) axis; cumulative frequency line (cumulative %) refers to the right axis.

discussed above. The average net pay thickness is 441 ft (134 m), compared to an average net sandstone thickness of almost 1000 ft (300 m). Obviously, a large portion of the sandstones passing a 50% V_{shale} cutoff are low porosity, as shown in Figures 11, 13, and 20. In particular, note the secondary low-porosity peak in Figure 13, which represents a distinct rock type of very fine-grained sandstones and siltstones with less than 10 μd in-situ permeability, which is probably not part of the productive reservoir system.

The average porosity of net pay is greater than the average porosity of all sandstones, because the low-porosity tail has been cut off. The average net pay porosity by well ranges from 7.6 to 10.9% and averages 9.3% (Table 1). Similarly, the water saturation of net pay is lower, because the high water saturation and low-porosity sandstones have been cut out, resulting in an average water saturation for net pay of 33%, as compared to 45% for all sandstones. Average water saturations are as low as 20% in a few wells.

Estimated Ultimate Recovery

The distribution of EURs based on decline curve analysis is shown in Figure 23. The mean EUR is 3.7 bcf gas, with a range from 0 to 13.7 bcf gas. The EUR distribution is log normal, so the median reserves are slightly lower at 3.0 bcf. About 20% of the wells will make more than 6 bcf gas, which is one reason why Jonah is considered such an outstanding field. The initial decline rates range from 30 to 100% per year and average 70% (Figure 24). Many Jonah wells have been recompleted during their productive life, so the initial production rate (Q_i) (in million cubic feet per month) from the decline analysis is for the most recent production history. As a result, Q_i correlates poorly with EUR. Initial production ranges from 1 to 323 mmcf/month and averages 108 mmcf/month (Figure 25).

Figure 26 is a map of EUR across Jonah field and shows the location of the major known faults (Montgomery and Robinson, 1997). The most striking feature of this map is the increased density of high-EUR wells against the updip portion of each fault block. The pattern suggests that the faults are sealing faults, and gas was structurally trapped into several pools as it migrated updip into the field area.

Net Hydrocarbon Pore Volume versus Estimated Ultimate Recovery

Net hydrocarbon pore volume was calculated from the net pay summations using a fieldwide average gas-formation volume factor. A crossplot of the calculated net feet of hydrocarbon in place versus EUR is shown in Figure 27. The regression line through the dominant cloud of points is

$$\text{EUR (mmcf gas)} = 108 \times \text{NHPV} - 873, \quad r^2 = 0.795 \tag{6}$$

where NHPV is the net hydrocarbon pore volume in feet ($Sg \times \Phi \times H$; all net of 50% V_{shale}, 50% S_w, and 6%

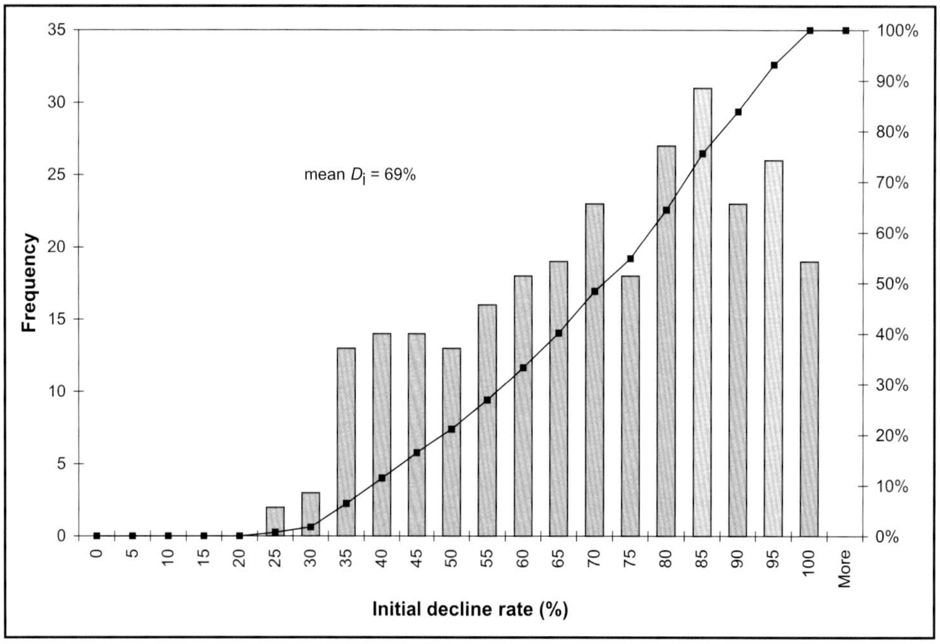

Figure 24. Distribution of initial decline rates, in %, from decline curve analysis. Histogram bars refer to left (frequency) axis; cumulative frequency line (cumulative %) refers to the right axis.

porosity cutoffs). Using the field average of 440 ft (130 m) of net pay, 9.3% porosity, and 33% water saturation, an average well will have 31.5 ft (9.6 m) net of hydrocarbon pore volume. Based on equation 6, the expected EUR is 2.5 bcf gas per well, which is close to the median reserves but is significantly less than the actual field average of 3.9 bcf gas. This difference is caused by the log-normal distribution, wherein the average is skewed toward the large wells on the high tail of the distribution (Figure 20).

The regression line in Figure 27 was determined by excluding two clouds of outliers, one representing wells that are significantly "overproducing" and another smaller group that is "underproducing." To determine if the outliers are random error, completion related, or geologically

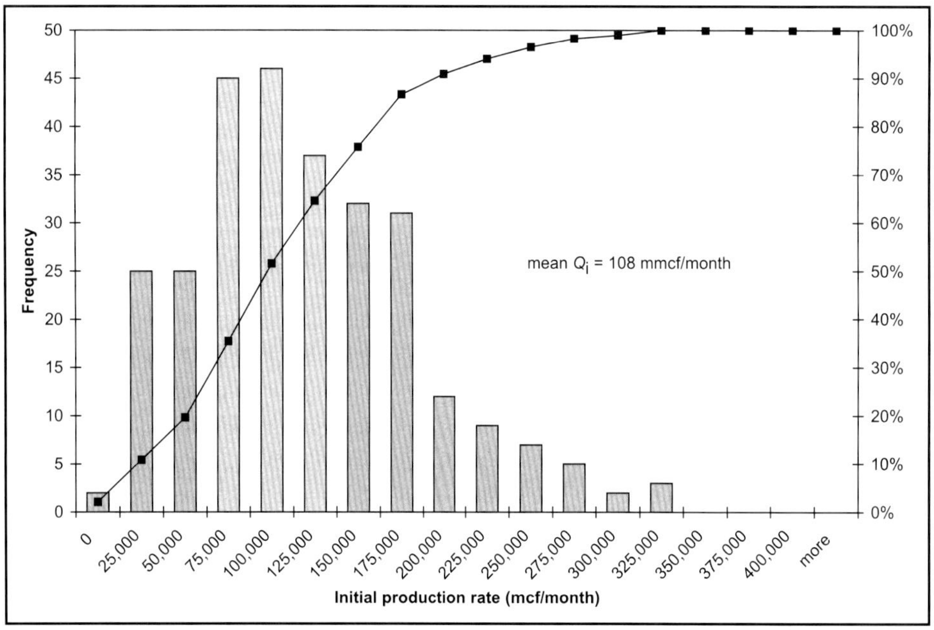

Figure 25. Distribution of average initial producing rates, in thousand cubic feet per month, from decline curve analysis. Histogram bars refer to left (frequency) axis; cumulative frequency line (cumulative %) refers to the right axis.

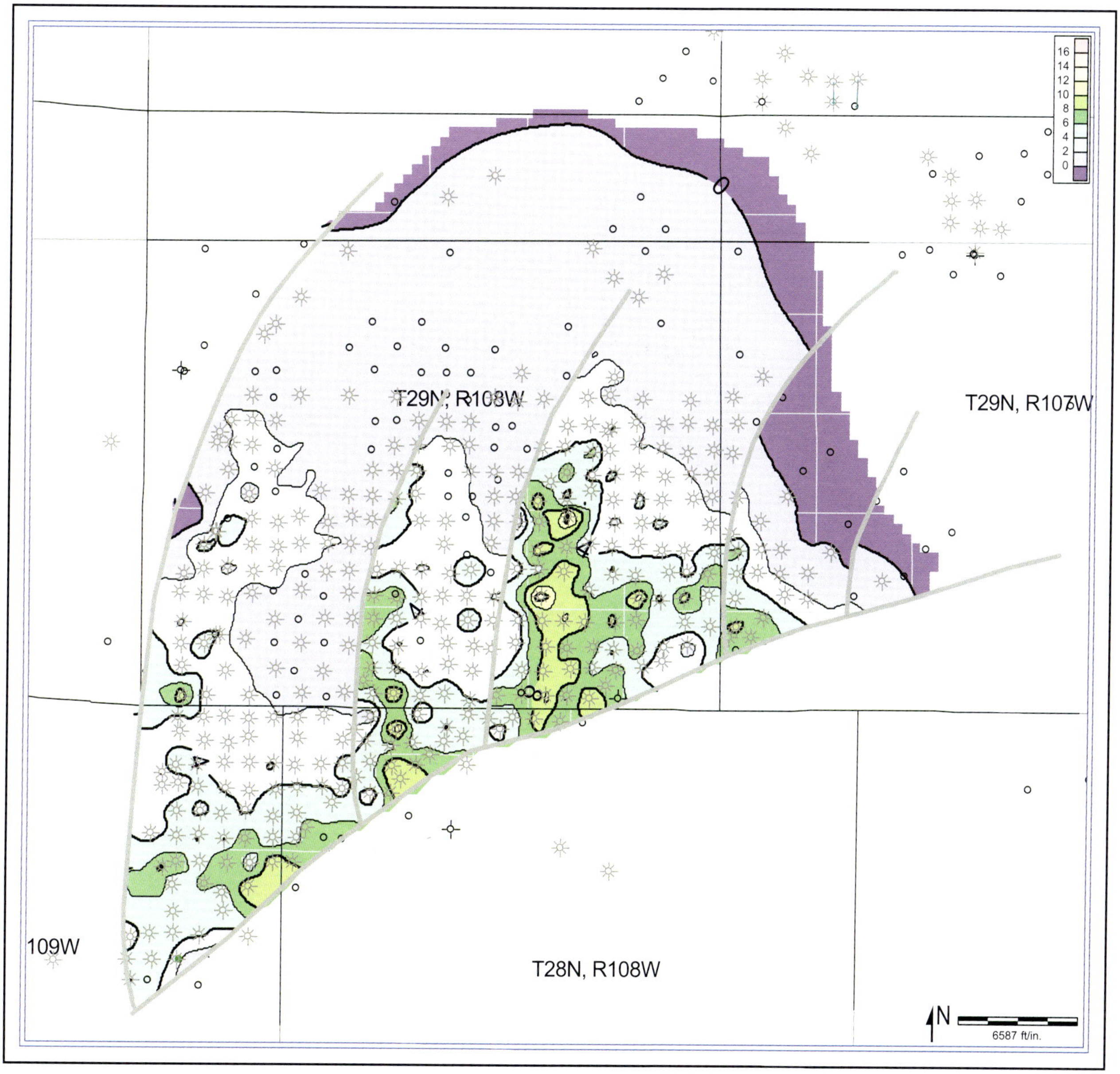

Figure 26. Estimated ultimate recovery (in bcf gas) from decline curve analysis for wells in Jonah field. Contour interval = 2.0 bcf gas.

controlled, we calculated and mapped the difference between the predicted EUR and actual EUR for all wells in the study area (Figure 28). The overproducers are clustered on the east side of the large intrafield faults, whereas the underproducers are on the west side of those faults. This pattern mimics the overall EUR pattern shown in Figure 26.

The hydrocarbon pore volume in the underproducers and the overproducers overlap, indicating that the differences across the faults are not simply caused by a greater amount of gas-charged sandstone on one side when compared to the other. The dramatic differences in productivity are entirely related to permeability and, more specifically, to relative permeability effects, instead of bulk formation permeability. Because of the steepness of the relative permeability curves in tight-gas sandstones, very small differences in saturation can have a significant impact on the recovery efficiency from the pore system. The overproducers represent high average recovery efficiency (high relative permeability to gas), whereas the underproducers represent low recovery efficiency (low relative permeability to gas). Better production in the

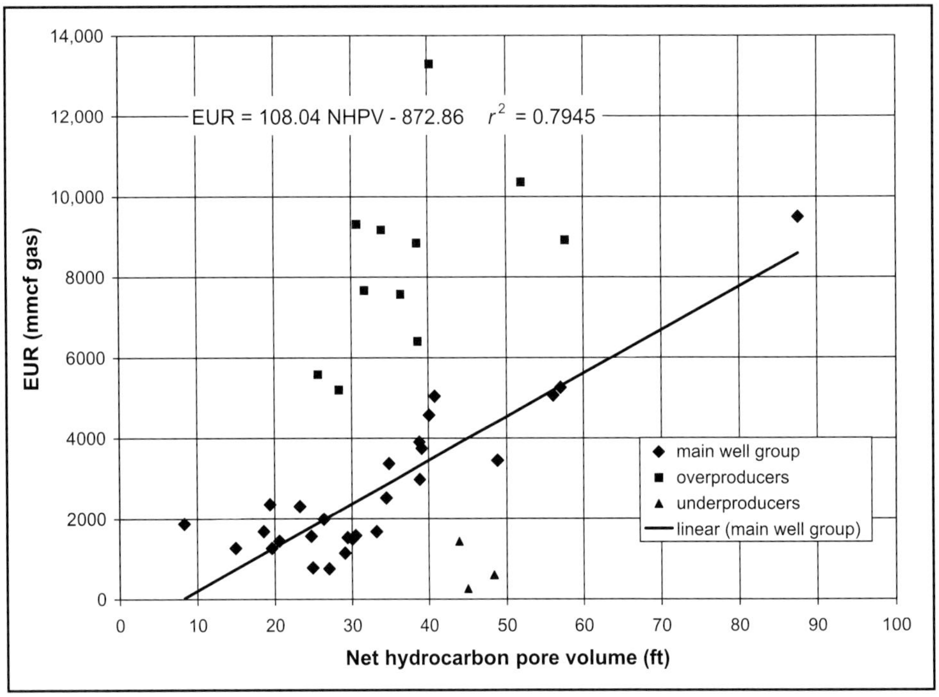

Figure 27. Net hydrocarbon pore volume in the upper 2500 ft (760 m) of the Lance Formation compared to estimated ultimate recovery (mmcf gas) from decline curve analysis.

updip regions of each fault block can be explained by the effectiveness of the sealing fault and minor changes in the water saturation from the downdip to updip portions of each fault block. The errors associated with log-derived water saturations are fairly large; a 10% change in water saturation across a fault boundary might not be detectable but could have a two to five times greater impact on recovery.

The effect of relative permeability on both gas production and water cut is dramatically illustrated by the producing water-gas ratios (WGR) of the wells at Jonah (Figure 29). The WGR at Jonah ranges from 0.1 to 70 bbl/mmcf gas, with 93% of the wells producing less than 12 bbl/mmcf gas. Until recently, it was thought that the field was making little, if any, formation water, because the average WGR for the field is a relatively insignificant 3.1 bbl/mmcf gas. Mapping revealed a smooth increase in WGR downdip across the field, with very low values (<1.0 bbl/mmcf gas) in the highest structural positions in each separate fault block. The water of condensation based on dew-point calculations should be on the order of 1 bbl/mmcf gas, so the updip wells appear to be producing dry gas free of any formation water. These wells must be near irreducible water saturation and at or below critical water saturation.

The downdip wells at Jonah are making relatively small water cuts, but the large cumulative production acts as an amplifier to reveal even small differences in relative permeability to water. As water saturation increases, the relative permeability to water also increases, and consequently, the produced WGR increases. As the relative permeability to gas decreases, the gas rates decline, and total productivity is diminished. Thus, Figures 26 and 29 are almost identical, even in their fine details, with the wells with low EURs also having significantly higher WGRs. Outside the two major bounding faults and along the downdip northeast edge of the field, the wells are producing greater than 20 bbl water/mmcf gas, and EURs are 1.0 bcf gas or less. These wells still have some relative permeability to gas, at least in the best sandstones that were perforated, and are therefore above critical gas saturation. The total spread of water saturation from the best wells in the updip areas of each fault block to the downdip limits is estimated to be 20–30%, based on our log calculations and core analysis (Figures 17, 18).

CONCLUSIONS

The upper 2500 ft (760 m) of the Lance Formation at Jonah field averages 40% net sandstone, half of which has porosity greater than 6% and is considered potential pay on the basis of porosity-permeability relationships. As a result, each well has an average of about 500 ft (150 m) of reservoir sandstone in a typical 2500-ft (760-m) section drilled. The average porosity for all sandstones at

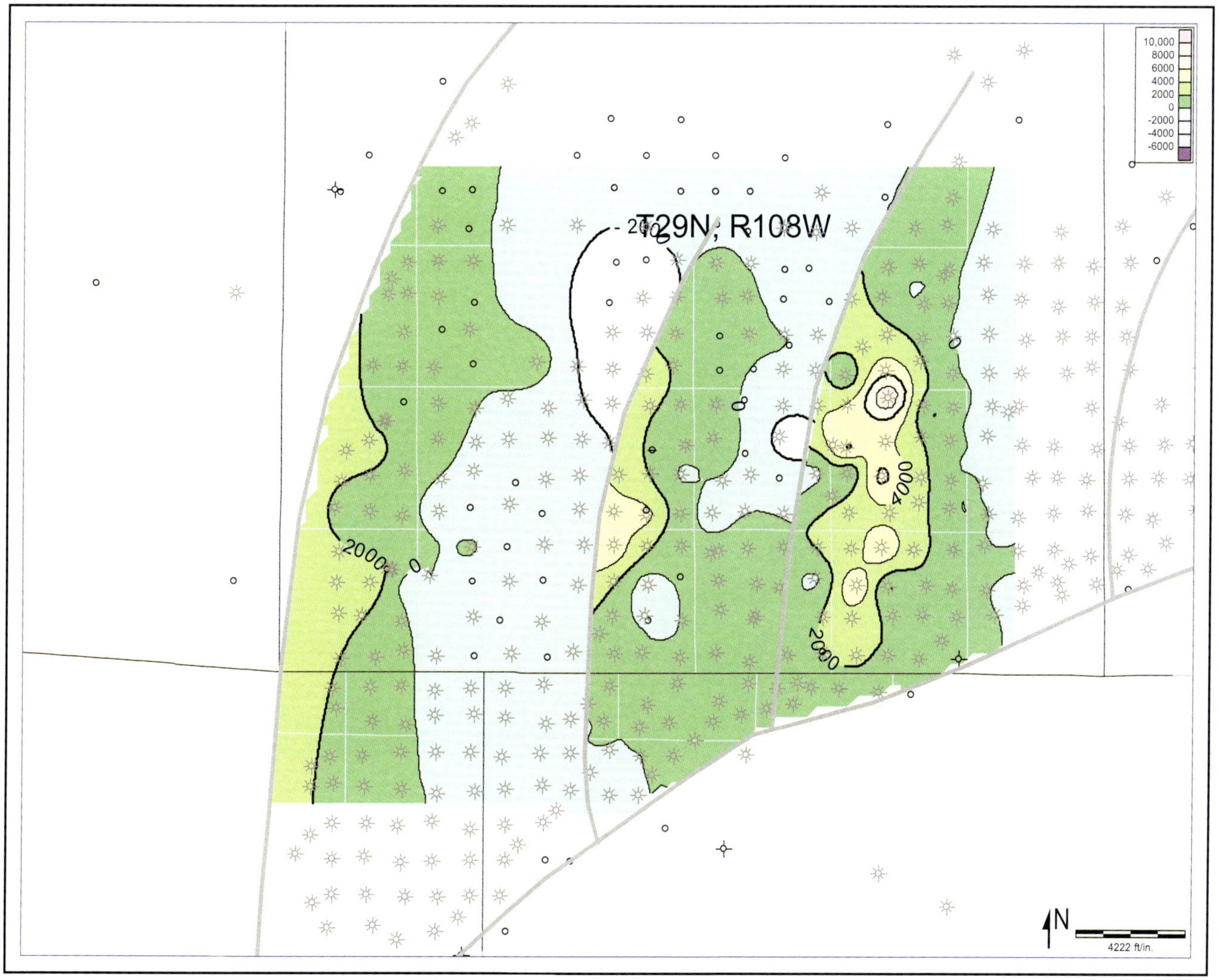

Figure 28. Map showing residual values between regression predicted EUR and actual EUR from decline curve analysis (mmcf gas). Contour interval = 2000 mmcf gas (2 bcf gas).

Jonah is 6.5% for log data and 7.1% for core data. As a subset of the sandstones meeting the 6% porosity cutoff, the average porosity of the reservoir-quality sandstones is 9.3%.

The permeability of these sandstones ranges from 1 μd to a few hundreds of microdarcys. The average in-situ reservoir permeability at the average log porosity of 6.5% is only 6 μd, whereas for the average porosity of net pay (9.3%), the average permeability improves to 25 μd. This latter value compares favorably with unpublished estimates of the average reservoir permeability from single-well modeling.

Water saturations averaged over the entire Lance range from 30 to 60%. Water saturation correlates with porosity and permeability. Higher water saturation values are associated with the lower porosity, nonreservoir rocks. Reservoir-quality sandstones have water saturations from 20 to 40% and average 33% for all wells. Subtle differences in water saturation resulting from rock-quality changes can have a dramatic impact on the effective permeability to gas because of steep relative permeability curves.

The EURs of the 40-ac (0.16-km^2) spacing wells at Jonah field range from 0 to almost 14 bcf gas, with an average of a little less than 4 bcf. Given the very low average porosity and permeability of the reservoir sandstones, it is perhaps surprising that these wells can produce gas

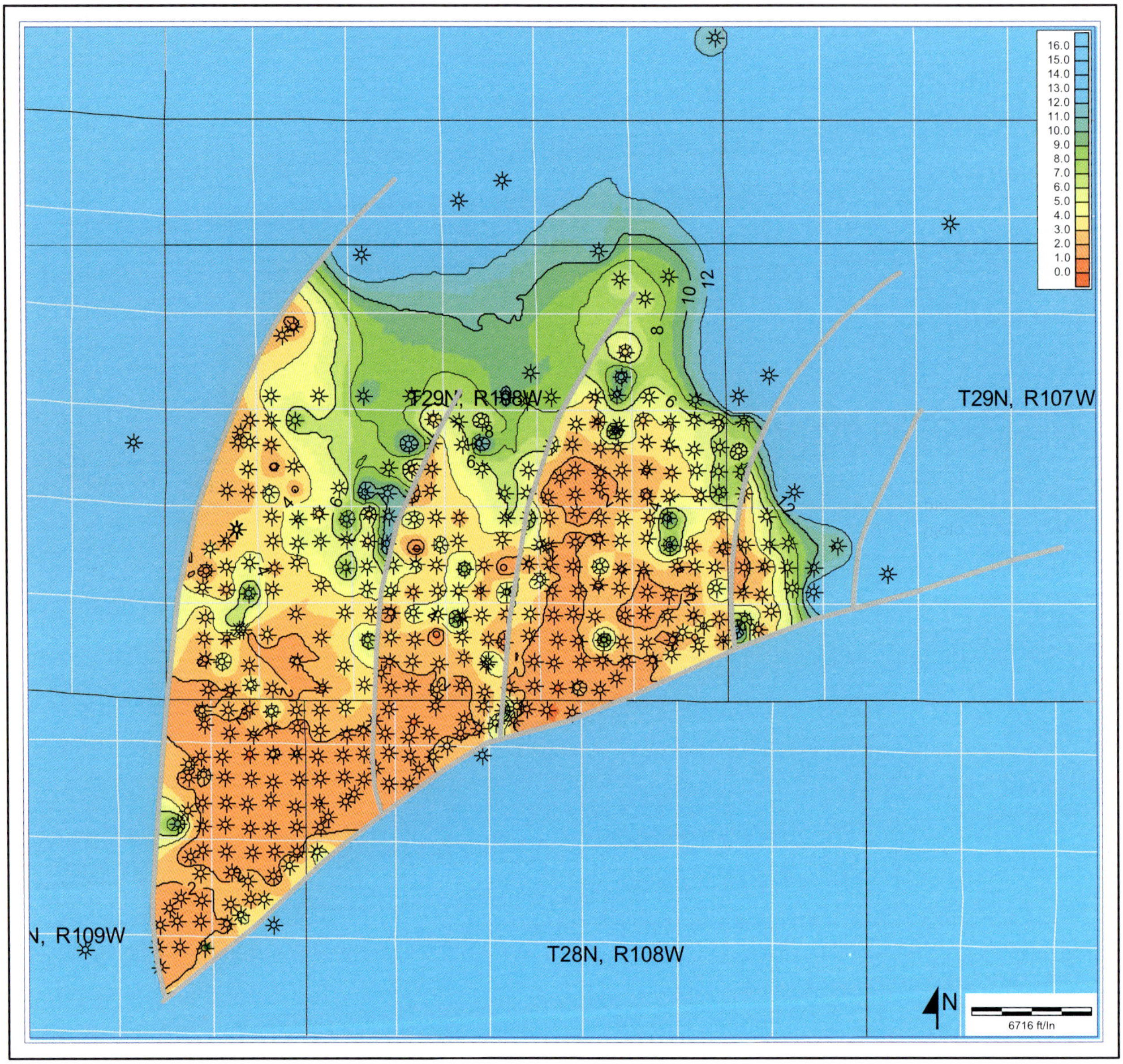

Figure 29. A Map illustrating producing water-gas ratios across Jonah Field. The water-gas ratio ranges from 0.1 bbl water/mmcf in structurally high positions of the field to 70 bbl water/mmcf in downdip portions. Contour interval is in barrels of water per million cubic feet of gas.

at all. In most sandstone reservoirs around the world, the entire Lance section at Jonah would be considered a nonreservoir rock and below the net pay cutoffs for gas production.

What separates Jonah from other fields in the Rocky Mountain region are the substantial net pay thickness of low-permeability gas sandstone in the Lance Formation and the large areal extent of the field. The application of effective stimulation methods to complete multiple-pay sandstones over the thick interval yields a world-class giant gas field in reservoirs having permeabilities in microdarcys.

REFERENCES CITED

Asquith, G. B., 1990, Log evaluation of shaly sandstones: A practical guide: AAPG Continuing Education Course Notes Series 31, 59 p.

Bowker, K. A., 1995, Lance and Mesaverde of the Hoback basin:

Preliminary results from recent drilling, *in* Geological characterization of the cretaceous gas plays in the Greater Green River basin: Gas Research Institute/Department of Energy/Independent Petroleum Association of Mountain States Greater Green River Consortium Symposium, p. 280–296.

Bowker, K. A., and J. W. Robinson, 1997, Jonah field: A shallow sweet-spot in the basin-centered gas accumulation of the northern Green River basin, Wyoming: AAPG Bulletin, v. 81, p. 1219, and Expanded Abstracts Volume, p. 25–30.

Byrnes, A. P., 1997, Reservoir characteristics of low-permeability sandstones in the Rocky Mountains: The Mountain Geologist, v. 34, p. 39–48.

Clavier, C., G. Coates, and J. Dumanoir, 1977, The theoretical and experimental bases for the "dual water" model for the interpretation of shaly sands: Society of Petroleum Engineers Paper 6859, 16 p.

Cluff, R. M., 2004, Production history plots for Jonah field, Sublette County, Wyoming, *in* J. W. Robinson and K. W. Shanley, eds., Jonah field: Case study of a tight-gas fluvial reservoir: AAPG Studies in Geology 52, CD-ROM, Appendix C.

Cluff, R. M., and C. M. Murphy, 2004, Stratigraphic cross sections through Jonah field, Sublette County, Wyoming, *in* J. W. Robinson and K. W. Shanley, eds., Jonah field: Case study of a tight-gas fluvial reservoir: AAPG Studies in Geology 52, CD-ROM, Appendix B.

Dewan, J. T., 1983, Essentials of modern open-hole log interpretation: Tulsa, PennWell Books, 361 p.

DuBois, D. P., P. J. Wynne, T. M. Smagala, J. L. Johnson, K. D. Engler, and B. C. McBride, 2004, Geology of Jonah field, Sublette County, Wyoming, *in* J. W. Robinson and K. W. Shanley, eds., Jonah field: Case study of a tight-gas fluvial reservoir: AAPG Studies in Geology 52, p. 37–60.

Eberhard, M., and M. Mullen, 2001, The effect of stimulation methodologies on production in the Jonah field: Rocky Mountain Petroleum Technology Conference, Society of Petroleum Engineers, Keystone, 21–23 May 2001, SPE paper 71048, 9 p.

Hook, J. R., 2003, An introduction to porosity: Petrophysics, v. 44, no. 3, p. 205–212.

Montgomery, S., and J. W. Robinson, 1997, Jonah field, Sublette County, Wyoming: Gas production from overpressured Upper Cretaceous Lance sandstones of the Green River basin: AAPG Bulletin, v. 81, p. 1049–1063.

Shanley, K. W., 2001, Fluvial reservoir description for a tight-gas giant: Jonah field, Green River basin, Wyoming, *in* J. W. Robinson and K. W. Shanley, eds., Tight-gas fluvial reservoirs: A case history from the Lance Formation, Green River basin, Wyoming: AAPG Annual Convention, Denver, Colorado, AAPG Short Course 2, 49 p.

CHAPTER 13

Extending the Southwest Limits of Jonah Field: Using High-quality, 3-D Seismic Data to Improve the Structural Definition

S. G. SIGUAW
Cabot Oil and Gas Corporation, Denver, Colorado, U.S.A.

D. C. FRIEND
Cabot Oil and Gas Corporation, Denver, Colorado, U.S.A.

ABSTRACT

The southwest corner of Jonah field was targeted by Cabot Oil and Gas Corporation for an exploration program to extend the limits of established production. A reprocessed aeromagnetic survey was initially used as a cost-effective method to image this area (the Yellow Point area). However, the aeromagnetic survey did not provide the necessary resolution to convince Cabot to drill a well on their acreage. Subsequently, Cabot acquired a 16.1-mi^2 (42-km^2) high-resolution, three-dimensional (3-D) survey to evaluate their acreage and define the southwestern intersection of the two faults that form the trap at Jonah field. The Yellow Point area of Jonah field is a region of complex, structural relationships and outstanding gas production. Cabot used the high-quality, 3-D seismic data in combination with available well control and aeromagnetic data to delineate the updip structural limits of Jonah field. Based on this study, a high-risk, high-reward prospect was identified as a southern extension of the Yellow Point area. A wildcat well was drilled on this high-potential prospect with negative results. The wildcat well encountered a normally pressured Lance section outside of the main Jonah fault block.

INTRODUCTION

Jonah field (Figure 1) is an overpressured, fault-bounded, structurally trapped gas accumulation (Montgomery and Robinson, 1997) in the western portion of the Green River basin, Sublette County, Wyoming (T28–29N, R108–109W). The updip southwest corner of the field is commonly called the Yellow Point area, after a topographic ridge of the same name. The Yellow Point area is structurally complex because of the intersection of a south-bounding fault zone (sometimes referred to as the "Line of Death") with a west-bounding fault zone. Both faults have a strike-slip sense of motion. Cabot Oil and Gas Corporation acquired leases in this area to test a possible extension to the field (Figure 2).

From 1996 to 1998, 3 three-dimensional (3-D) surveys were shot by Amoco and Snyder Oil Corporation over Jonah field and the adjacent Pinedale anticline. In 1999, all three surveys were merged into one large 3-D survey covering all of Jonah field (Figure 2). These 3-D surveys were shot by Veritas DGC using a diagonal source pattern, resulting in a 110 × 110-ft (33.4 × 33.4-m) bin size.

In 2000, Trace Ventures, on behalf of Forest Oil Corporation, shot a small, high-resolution, 6.15-mi^2 (16-km^2) 3-D survey (Elm 3-D) in the Yellow Point area (Figure 2), using a half-brick design, resulting in a 55 × 55-ft (16.7 × 16.7-m) bin size.

In 2001, Cabot Oil and Gas Corporation contracted Grant Geophysical to acquire the 16.1-mi^2 (42-km^2), high-resolution Yellow Point 3-D survey (Figure 2). This survey improved upon the Elm 3-D survey by enlarging the image in the Yellow Point area, by including several productive wells in Jonah field, and by extending coverage to the south, east, and west of the known production. This additional coverage not only enabled Cabot to image their existing leasehold but also provided important

Jonah Field: Case Study of a Giant Tight-Gas Fluvial Reservoir, J. W. Robinson and K. W. Shanley, eds.: AAPG Studies in Geology 52 and Rocky Mountain Association of Geologists 2004 Guidebook

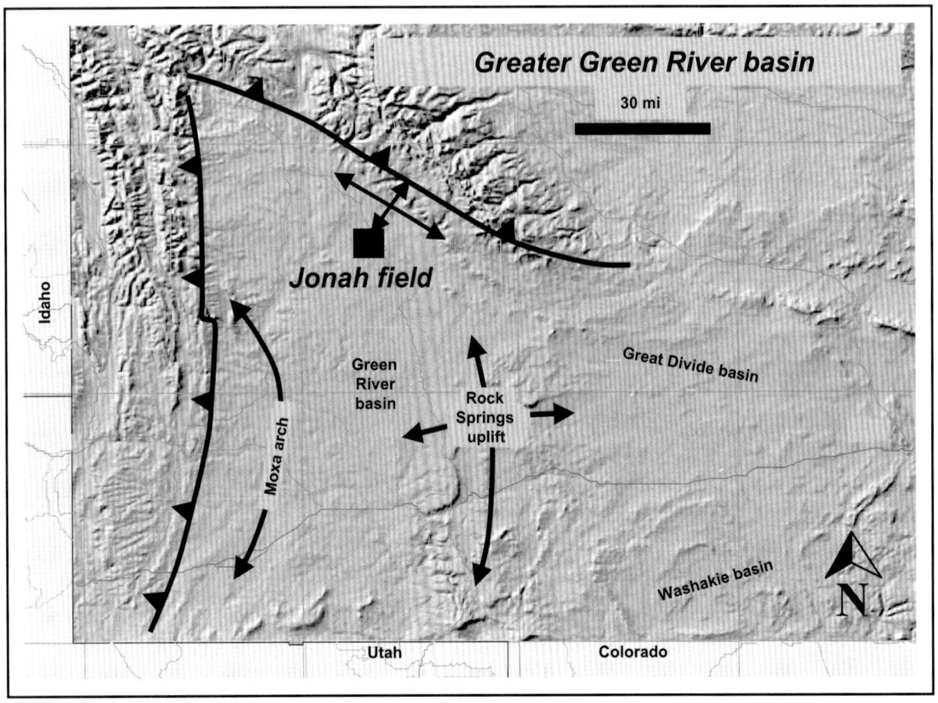

Figure 1. Generalized location map for Jonah field in southwestern Wyoming.

information about the complex fault systems in this corner of the field. Cabot also used proprietary processing and interpretation of an aeromagnetic survey in this area for the initial evaluation of the Yellow Point area.

OBJECTIVES

Previously, work on Jonah field by Warner (2000) indicated that the Yellow Point area was outside the field limits. When that study was published, there was limited subsurface control in this area and no available seismic data; hence, the interpretation was that the south-bounding fault system extended to the southwest and north of the Yellow Point area. The postulated geometry of the south-bounding fault zone (a fairly straight line) did not consider a turn in this fault system to the south to include the Yellow Point area in the field. Subsequent well information from the Forest Oil Tot unit 9 (NE NE Sec. 23, T28N, R109W) and the Elm Federal 23-14 (NE NE Sec. 23, T28N, R109W) changed this interpretation.

The Tot 9 well was drilled by Forest Oil in November and December 1998 to approximately 10,823 ft (3290 m) total depth to test an extension of the Jonah overpressured fault block to the southwest. No overpressure was encountered, based on mud-log shows, and the well was completed in the Lance with a poor initial potential of 209 mcf/day. In May and June 2000, the Elm Federal 23-14 well was drilled in a location approximately 140 ft (40 m) to the west and slightly south of the Tot 9. This slight change in location resulted in a well that had a reported initial potential of 10.7 mmcf/day from the overpressured Lance section. Based on these results, it was clear that the south-bounding fault transected the two wells. In addition, the location and proximity of these wells suggested to Cabot that the south-bounding fault changed direction to a north-northeast–south-southwest trend between the wells. Cabot's objective was to resolve the exact location of the west-bounding fault and the south-bounding fault in the Yellow Point area and determine if the area under their acreage was an extension of the field limits.

Jonah field produces gas from multiple, overpressured, fluvial-channel sandstones of the Upper Cretaceous Lance Formation (Robinson, 2000). Between June 2000 and March 2001, Forest Oil drilled and completed four additional Lance wells approximately 1 mi (1.6 km) north of Cabot's Yellow Point prospect. These wells had a combined initial potential of 48 mmcf/day (12 mmcf/day average with a range of 8.9–18.2 mmcf/day). Two of the wells had cumulative production of 1.6 and 1.2 bcf in 7 months. At the time the Yellow Point prospect was generated, there was not enough historical production information for the Forest wells to calculate estimated ultimate recoveries (EURs). However, wells immediately north of the Forest wells have EURs from 6 to more than 10 bcf per well. This indicated that the Yellow Point prospect had extremely high potential, although it encompassed only 640 ac (260 ha).

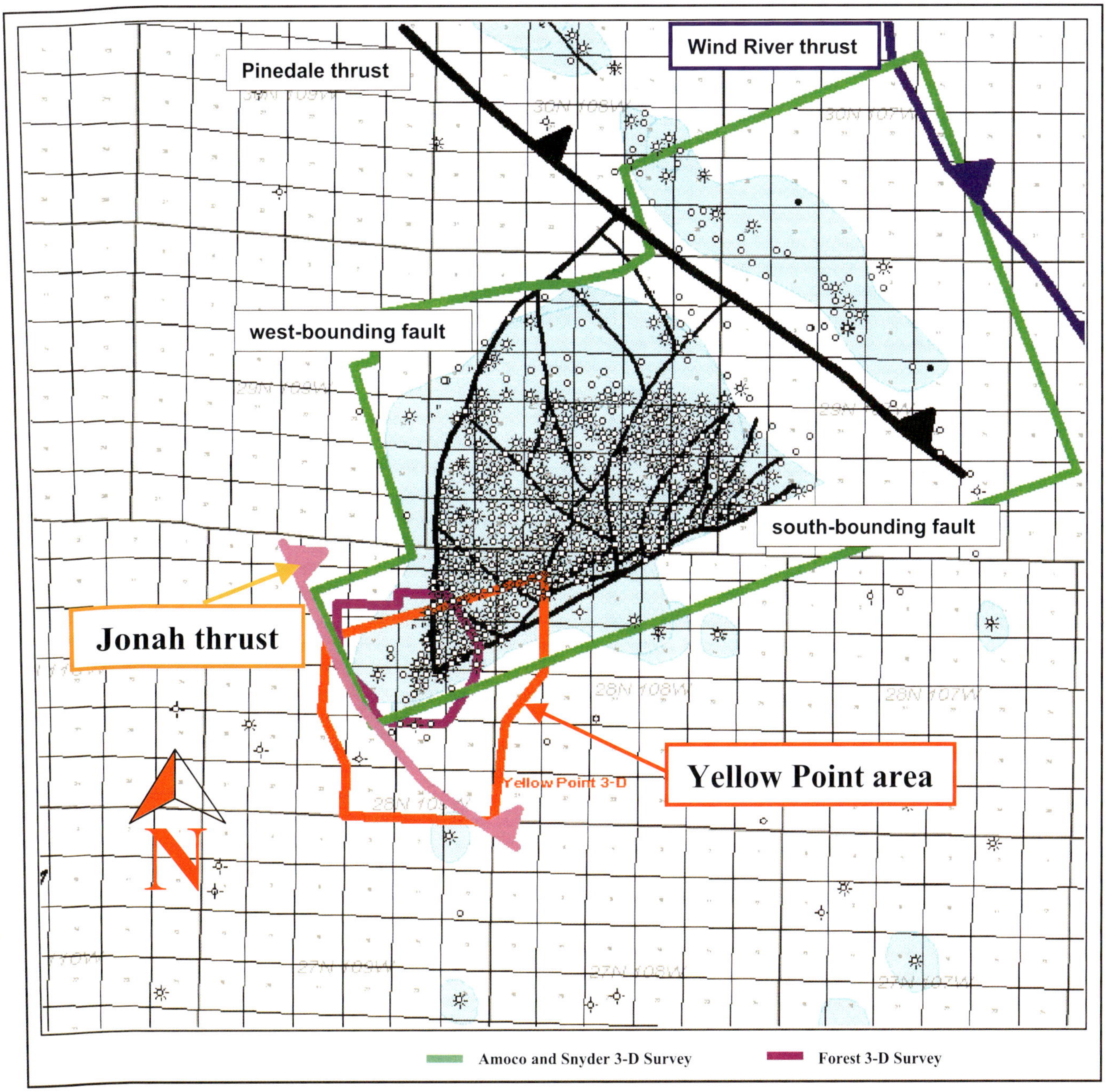

Figure 2. Map of Jonah field and the Yellow Point area showing the regional productive areas (blue), 3-D seismic survey outlines, and major fault systems. Fault pattern in Jonah field is from Warner (2000).

EXPLORATION STRATEGY AND METHODS

An aeromagnetic survey, processed with state-of-the-art technology, was used as a cost-effective method to image the subsurface structure of the Yellow Point area (Figure 3). The survey was able to detect several large-scale structural features in the Jonah area. The Pinedale thrust fault can be seen trending in a northwest–southeast direction, along with an interpreted deep-seated parallel normal fault located southwest of the Pinedale thrust. The west-bounding fault zone and the south-bounding fault zone can also be interpreted with the aeromagnetic data.

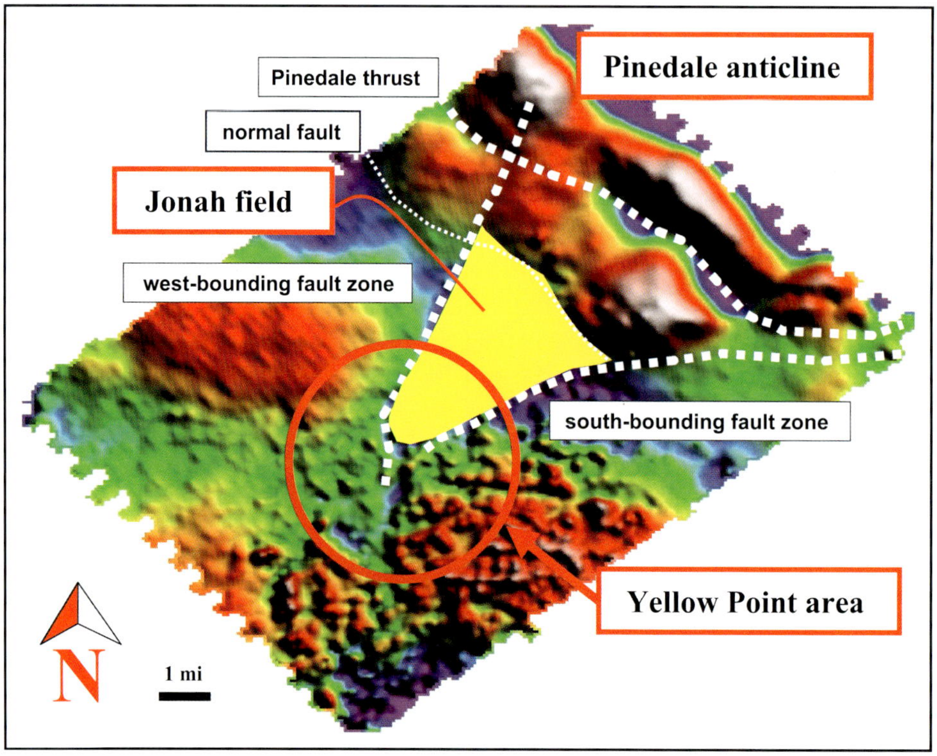

Figure 3. Proprietary aeromagnetic display of Jonah field and the Yellow Point area. The major structural features are annotated on the map.

The Yellow Point area, however, exhibits a very complex image that lacks the necessary resolution to define the structural configuration in this area.

Figure 4 shows the initial interpretation of the Yellow Point area based on well control, 1 two-dimensional (2-D) seismic line, and the reprocessed aeromagnetic data. The aeromagnetic interpretation and the published interpretation from the Jonah 3-D seismic survey (Warner, 2000) are quite good in imaging the larger structural features. However, Cabot needed better spatial resolution to allocate the necessary funds to drill a wildcat well to test their acreage holdings.

A high-resolution, 3-D survey was needed to fully define the updip structural intersection of the south-bounding fault zone with the west-bounding fault zone. The previous two 3-D surveys in this area were not available for purchase, so Cabot decided to acquire a proprietary 3-D program in the Yellow Point area. The Yellow Point survey was designed to be a larger program than the Elm 3-D to include more of Jonah field and the existing well control. Additionally, the survey used a much smaller bin size than the Jonah 3-D survey to image the fault systems more clearly.

In late 2001, Grant Geophysical acquired the proprietary Yellow Point 3-D survey for Cabot. Acquisition parameters are shown in Table 1. Vector Processing used a conventional processing flow (including 3-D DMO [dip moveout] and poststack time migration) to produce the final 3-D volume. GeoQuest's Variance Cube processing was applied to the 3-D volume to aid the final interpretation.

The Cretaceous–Tertiary (K/T) unconformity, as picked on well logs, is used as the seismic event to map the structural configuration of Jonah field and the Pinedale anticline (Figure 5). This same mapping event was used in the Yellow Point area. Previous attempts to use seismic data to map marker beds in the Lance section have been unsuccessful because of the lack of laterally continuous seismic events. The deeper Mesaverde Group can be mapped with the 3-D data throughout the Jonah and Yellow Point area.

Figure 6 shows a 3-D time slice from the variance cube seismic volume at approximately the K/T boundary level. The west-bounding fault zone and the south-bounding fault zone can be seen on this display. These two fault zones show a distinct expression of nearly merging together in this southwestern area of Jonah field. However, the complexity of the fault systems at this updip limit of the field is remarkable in that

(1) two small fault slivers are carved out at the extreme updip portion in the Yellow Point area;
(2) the south-bounding fault system becomes indistinct and curves to the southwest and possibly south–southeast from the main portion of Jonah field; and

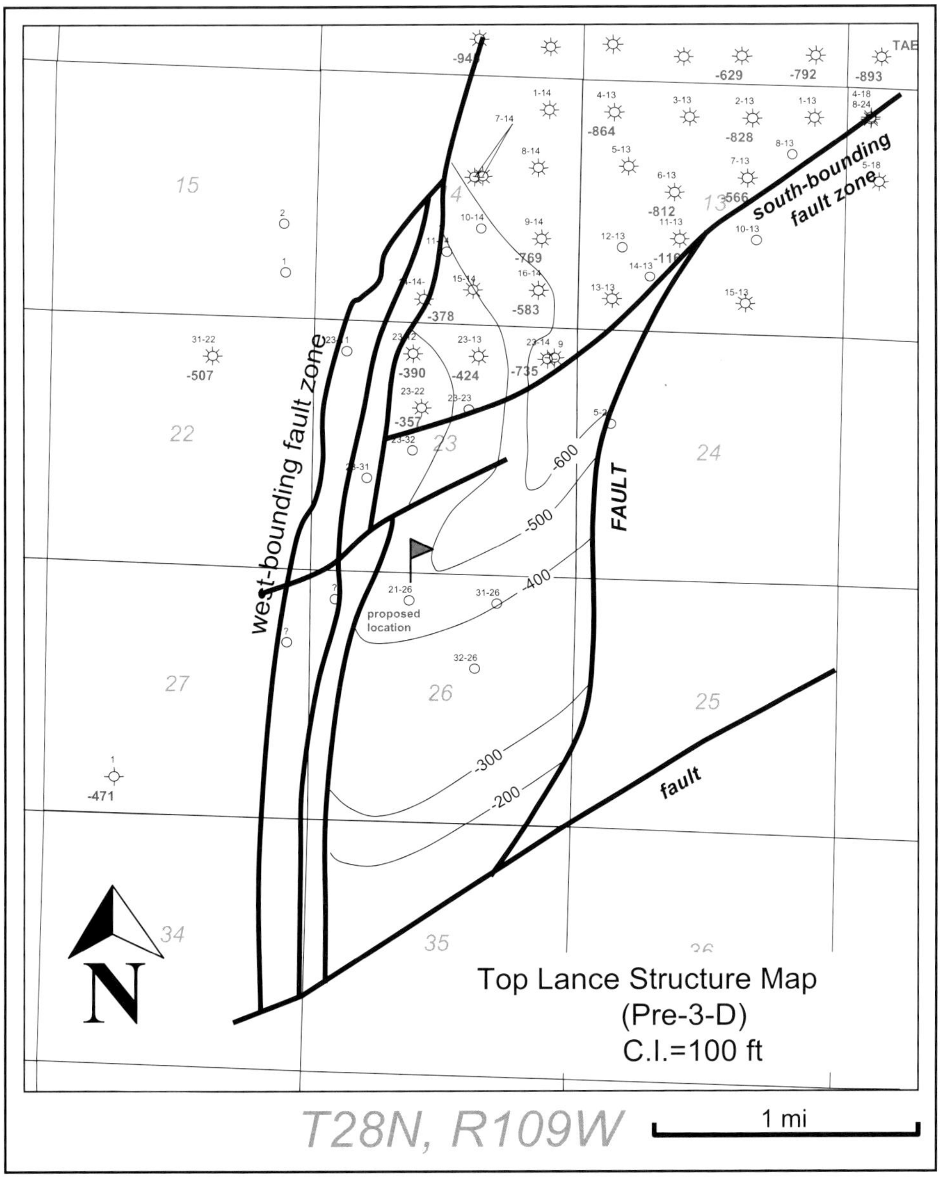

Figure 4. Initial interpretation of the structural geology of the Yellow Point area (pre-3-D).

(3) a newly recognized northwest–southeast-trending thrust fault (herein named the Jonah thrust) is interpreted from the Yellow Point 3-D; this fault was hypothesized to be the ultimate southern limit for Jonah field.

Figure 7 shows a 3-D time slice from the variance cube seismic volume near the top of the Lance interval. The west-bounding fault system is still clearly imaged, but the south-bounding fault zone seems to become diffuse as it approaches the west-bounding fault zone. Two, small south–southeast-trending faults (slivers) intersect the south-bounding fault zone and may indicate that the south-bounding fault zone does not reach the west-bounding fault zone at the Lance horizon and thus creates a larger fault block to the south as an area for exploration. Additionally, the Jonah thrust is a northwest–southeast-trending linear feature that terminates the west-bounding fault zone at the Lance horizon.

ANALYSIS AND DISCUSSION

Cabot obtained a partner (Gasco Energy, Inc.) to drill the Yellow Point 21-26 (NE NW Sec. 26, T28N, R109W) well in March 2002 (Figure 4). The well would test the

Table 1

Acquisition Parameters Used for the Yellow Point 3-D Survey.

Date acquired: December 2001
Size: 16.11 mi^2 (41.72 km^2)

Recording Instrument
I/O System Two

Spread Geometry
Bin size: 55 × 55 ft (16.7 × 16.7 m)
Patch size: 10 × 200
Line placement: Modified brick
Channels: 2000
Nominal fold: 40
Receiver lines: 1100 ft (335 m)

Source Parameters
Vibrators: 8 sweeps × 8 s
Sweep: 10–86 Hz, 3 db/oct slope
Source line: 1540 ft (470 m)

Lance Formation at a location south and west of the main Jonah field fault block. Cross sections, isopach maps, and structure maps indicated the presence of abundant sandstone at this location. As stated before, the 3-D seismic interpretation indicated that the south-bounding fault trend changed to a more southerly orientation and terminated into the Jonah thrust, creating an untested fault block. This fault configuration (Figures 4, 8) was felt to be consistent with regional stress fields set up by compressional forces from the northeast and southwest.

The location of two arbitrary seismic lines from the Yellow Point 3-D are shown overlain on a K/T structure map (Figure 9). The east–west line (Figure 10) shows the location of the west-bounding fault, the two Forest Elm Federal wells, and the Yellow Point 21-26 well in relation to the Jonah block and the west-bounding fault zone. The southwest–northeast line (Figure 11) goes through the Yellow Point 21-26 well and shows the position of the Jonah thrust that is the updip termination for the Jonah structural feature. Displacement along the thrust is evident at the K/T horizon and at the Mesaverde horizon (at approximately 1.9 s).

The Yellow Point 21-26 penetrated the Lance interval at −348 ft subsea (−106 m), which is 9 ft (2.7 m) updip to the Forest Oil Elm Federal 23-22, located 3960 ft (1200 m) due north and within the overpressured Jonah block. The well is also 159 ft (48 m) updip to the Tot 1 (map symbol 31-22) located 1.25 mi (2.0 km) to the northwest in section 22. This well was known to be normally pressured and on the west side of the west-bounding fault at Jonah field.

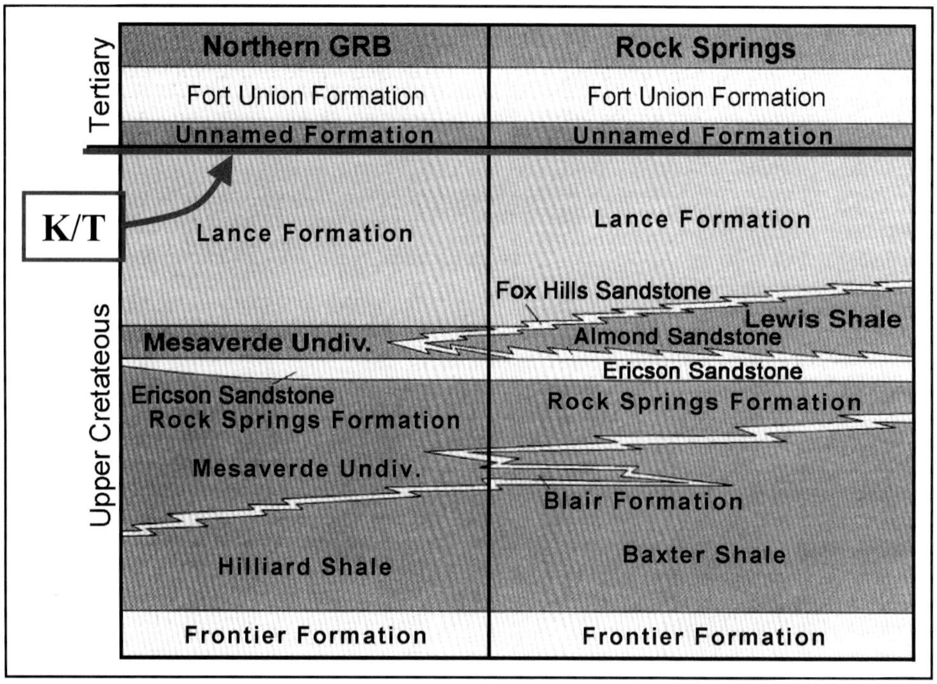

Figure 5. Stratigraphic column of Cretaceous and Tertiary rocks in the Green River basin (modified after Montgomery and Robinson, 1997). The K/T boundary and seismic marker is annotated.

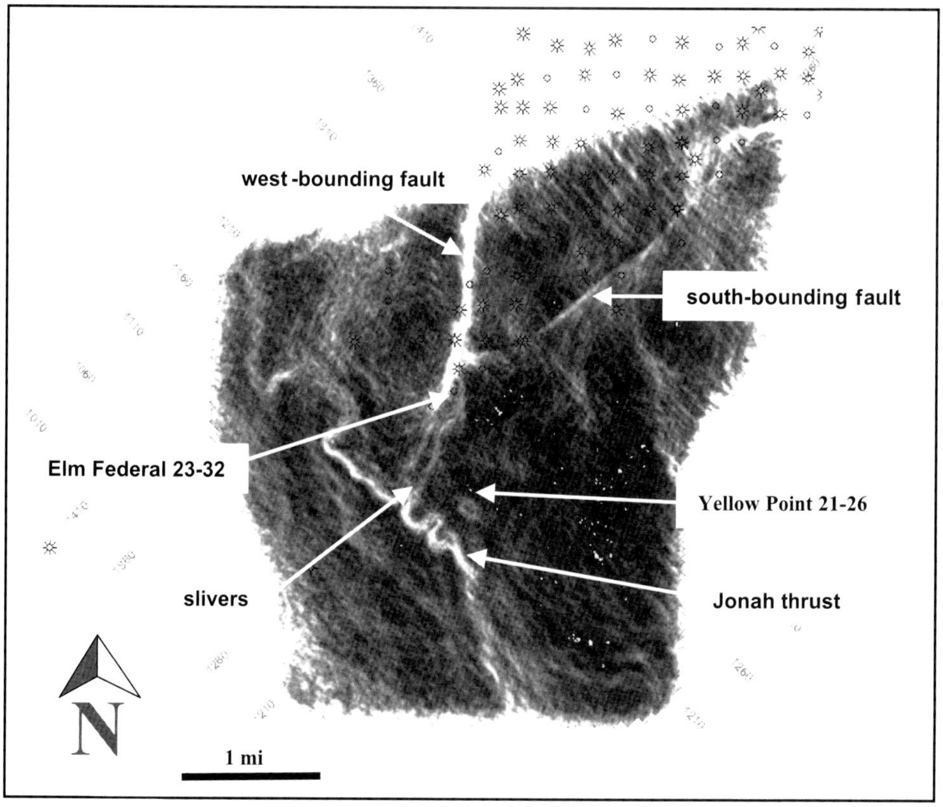

Figure 6. Variance cube time slice at the K/T boundary from the Yellow Point 3-D. The south-bounding and west-bounding faults that control the productive area of Jonah field are shown, as is the Jonah thrust fault.

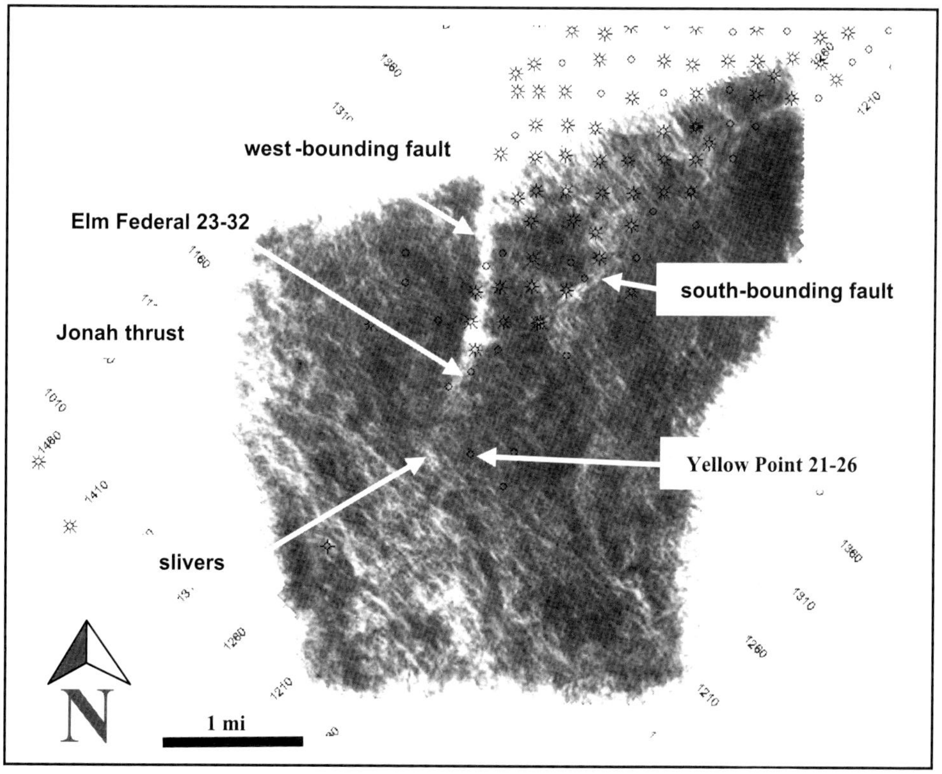

Figure 7. Variance cube time slice at the top of Lance interval from the Yellow Point 3-D. The south-bounding and west-bounding faults that control the productive area of Jonah field are shown, as is the Jonah thrust fault.

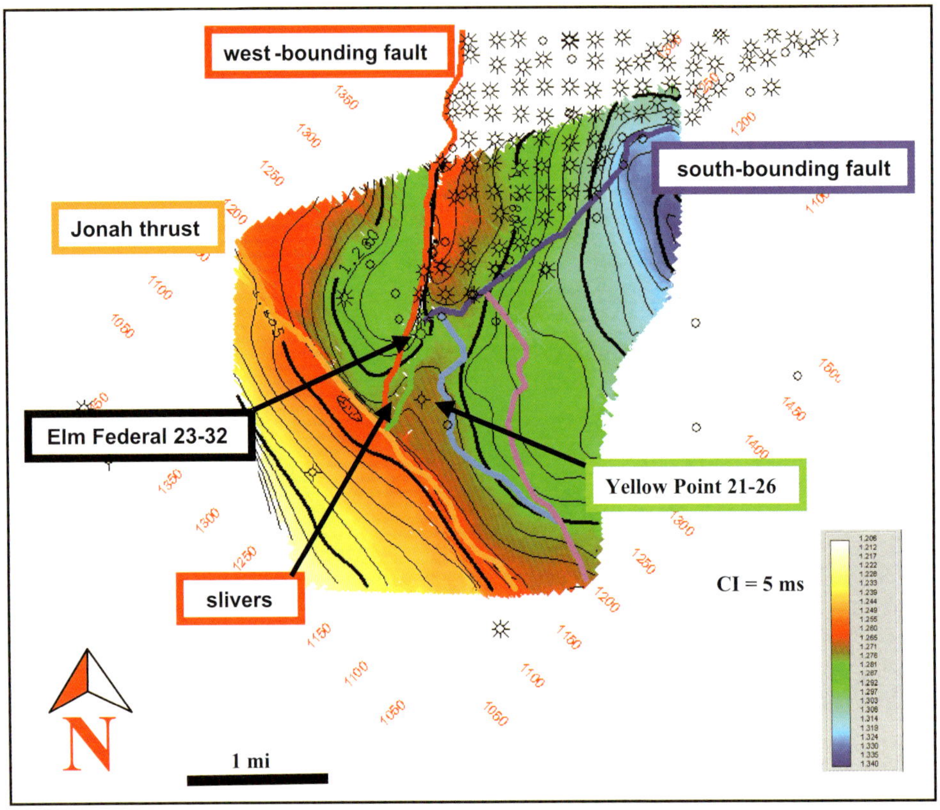

Figure 8. Time structure map on top of the K/T boundary from the Yellow Point 3-D survey. Contour interval = 5 ms.

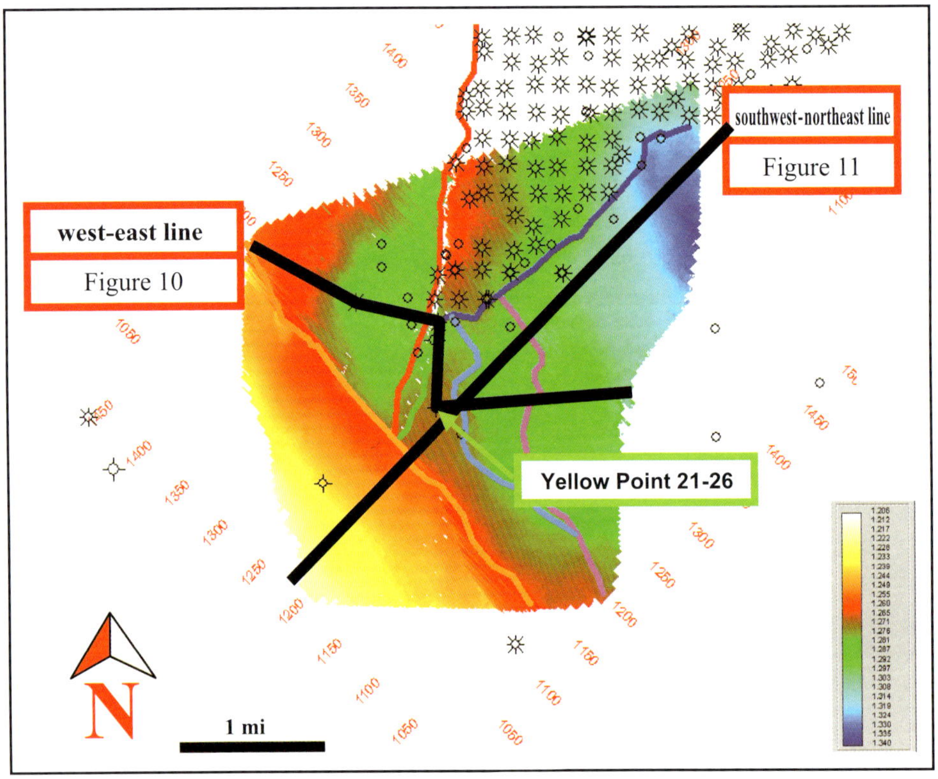

Figure 9. K/T time structure map with the two arbitrary seismic line locations. Figure 10 is east-west line, and Figure 11 is northwest-southeast line. Contour interval = 5 ms.

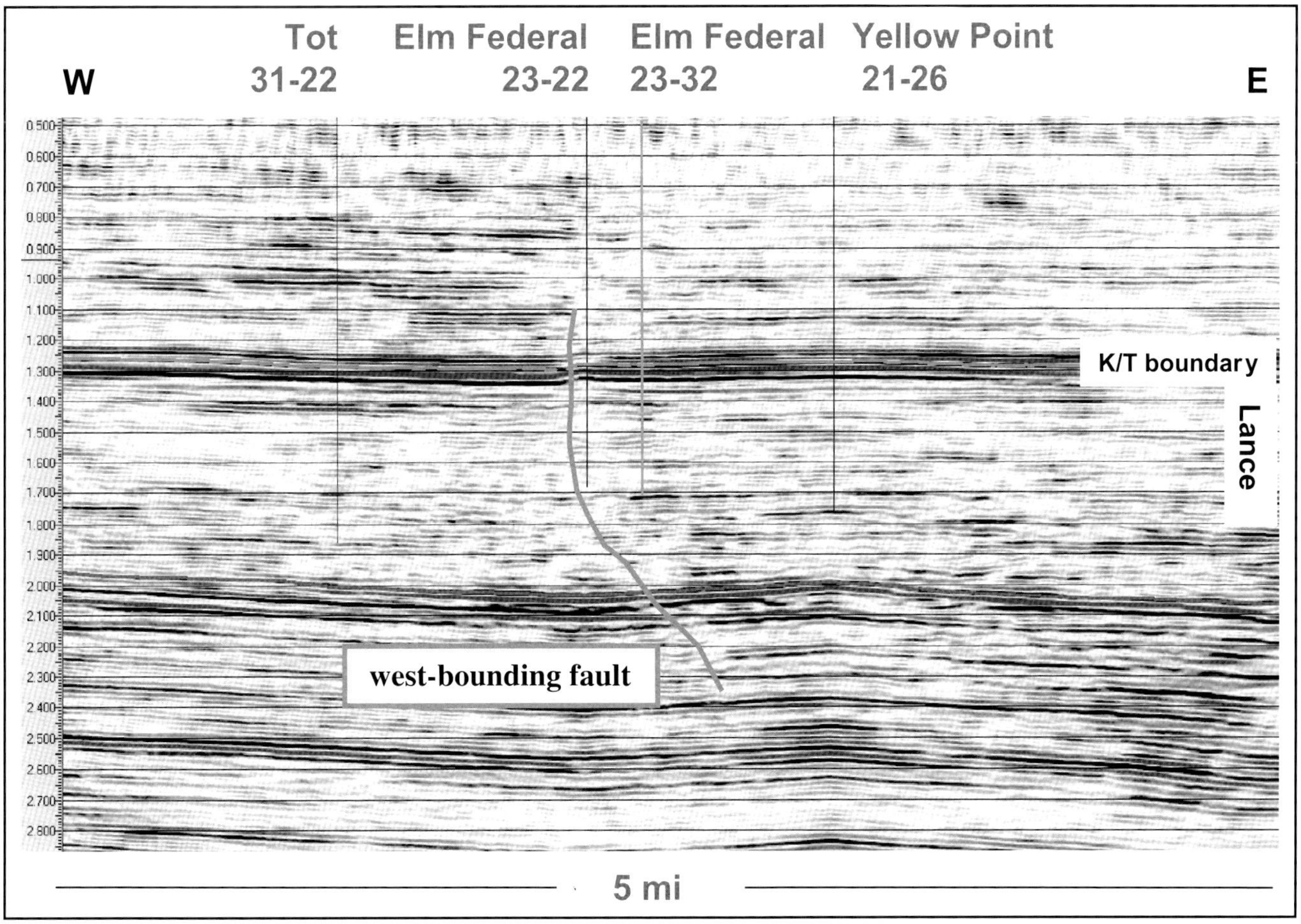

Figure 10. East–west arbitrary seismic line showing the west-bounding fault zone and its relationship to the well control. The location of the Yellow Point 21-26 is shown. The scale at left is two-way traveltime in seconds.

The Yellow Point 21-26 was drilled in accordance with the current practice of drilling at the pressure balance point to obtain the gas shows and flares necessary to identify porous and permeable Lance intervals. The top of the Lance was penetrated with a 9.3-lb/gal (1.11-kg/L) mud weight, and the well reached a total depth of 11,000 ft (3340 m) in the Ericson Sandstone with a 10.0-lb/gal (1.20-kg/L) mud weight. No significant mud-log shows were encountered while drilling the Lance and Ericson sections. The best gas show was 193 units from a 15-ft (5-m) interval in the lower Lance at 9973 ft (3032 m). Log readings from this zone indicated 14% porosity and a deep resistivity of 80–90 ohm m. Log analysis calculated high water saturation and uneconomic reserve potential. The majority of the Lance sandstones in this wellbore had porosities less than 10%.

Based on the poor mud-log gas shows, the Yellow Point 21-26 encountered a normally pressured Lance section from 7400 ft (2250 m) to the top of the Mesaverde at approximately 10,150 ft (3090 m). Figure 12 shows a section of the neutron-density porosity log and the corresponding mud log, which are representative of the entire Lance interval. Approximately 250 ft (80 m) net of Lance sandstone (gamma ray (GR) < 60° API) was encountered. This compares to 300–800 ft (90–240 m) net of sandstone in Jonah field. The Yellow Point 21-26 well was subsequently plugged and abandoned.

After the Yellow Point 21-26 was plugged, Forest Oil drilled a well in one of the small slivers of the west-bounding fault zone (Figures 4, 7, 8). The Elm Federal 23-31 (NE SW Sec. 23, T28N, R109W) did not encounter overpressure in this fault sliver, so the well was plugged back and directionally redrilled approximately 300 ft (90 m) to the north, in an attempt to encounter the overpressured Jonah block. Unfortunately, the well did not cross any significant structural boundary or reenter the Jonah block. It was plugged and abandoned in April 2002.

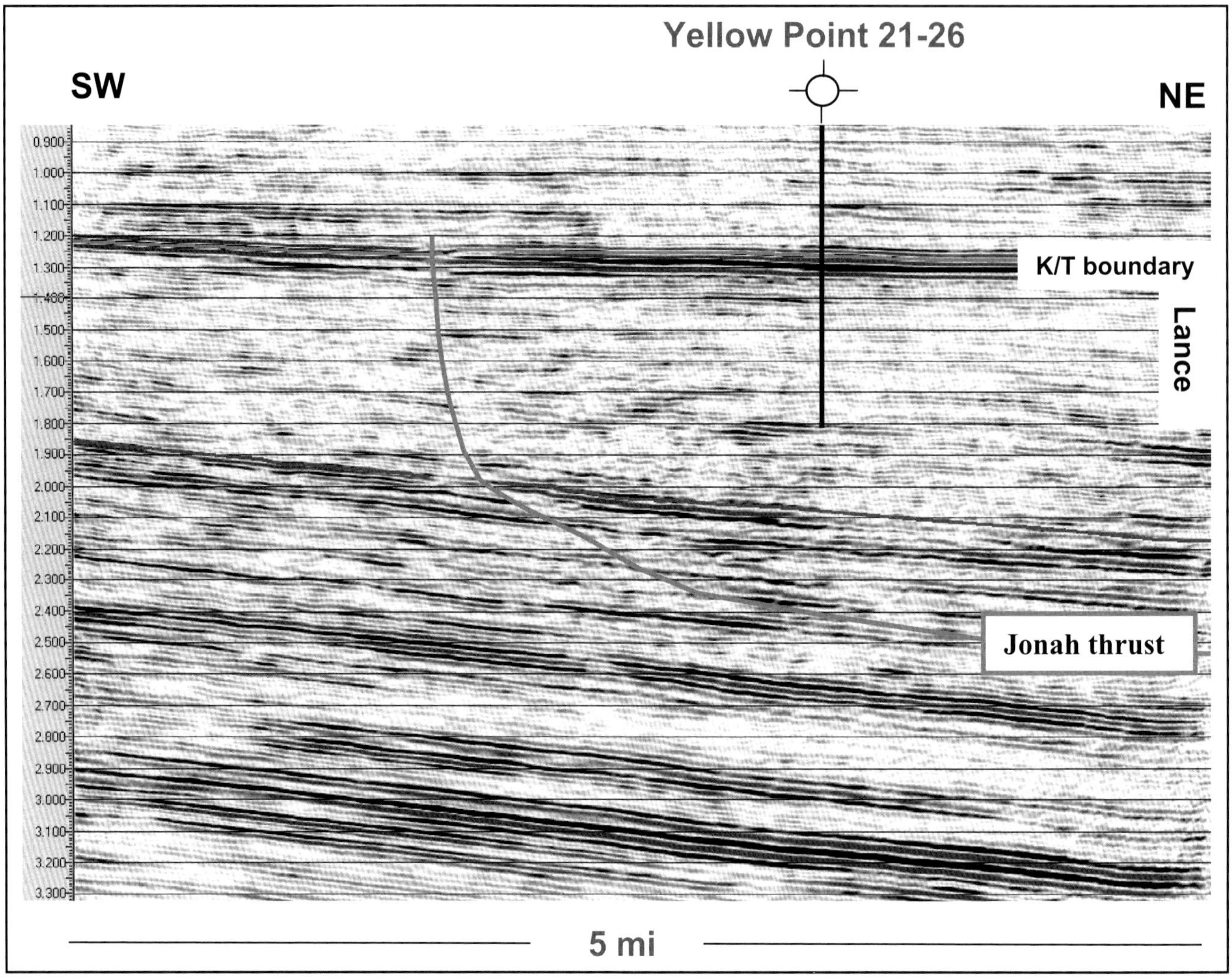

Figure 11. Northwest–southeast arbitrary seismic line through the Yellow Point 21-26 well and showing the location of the Jonah thrust. The thrust is interpreted to offset strata from the K/T boundary down to at least the top of the Mesaverde Group (at about 1.9 s). The scale at left is two-way traveltime in seconds.

SUMMARY AND CONCLUSIONS

The Yellow Point area is a region of complex structural relationships and outstanding gas production. Cabot Oil and Gas Corporation combined new, high-quality, 3-D seismic imaging with available well control and aeromagnetic data to delineate the structural limits of Jonah field. A high-risk, high-reward prospect was identified in an area adjacent to the south-bounding fault zone. A wildcat well was drilled on this high-potential prospect with

➤

Figure 12. Neutron-density porosity log and corresponding mud log from 8900 to 9156 ft (2706 to 2783 m) in the Yellow Point 21-26. Mud-log curves are rate of penetration (ROP) and total gas. Density porosity averages about 8%. Gybrn = gray brown, f gr = fine grained, arg = argillaceous, sli calc = slightly calcareous, NFSC = no fluorescence, stain or cut, gy-gygn = gray to gray green, tn-brn = tan to brown, clr-ltgy-gybrn = clear, light gray-gray brown, S&P = salt and pepper, f-md gr = fine to medium grained, num md lse grns = numerous medium-size loose grains, Coal lams = coal laminations, wht-gy = white to gray, f & sm md gr = fine and some medium grains, Qtz = Quartz, rk frags = rock fragments.

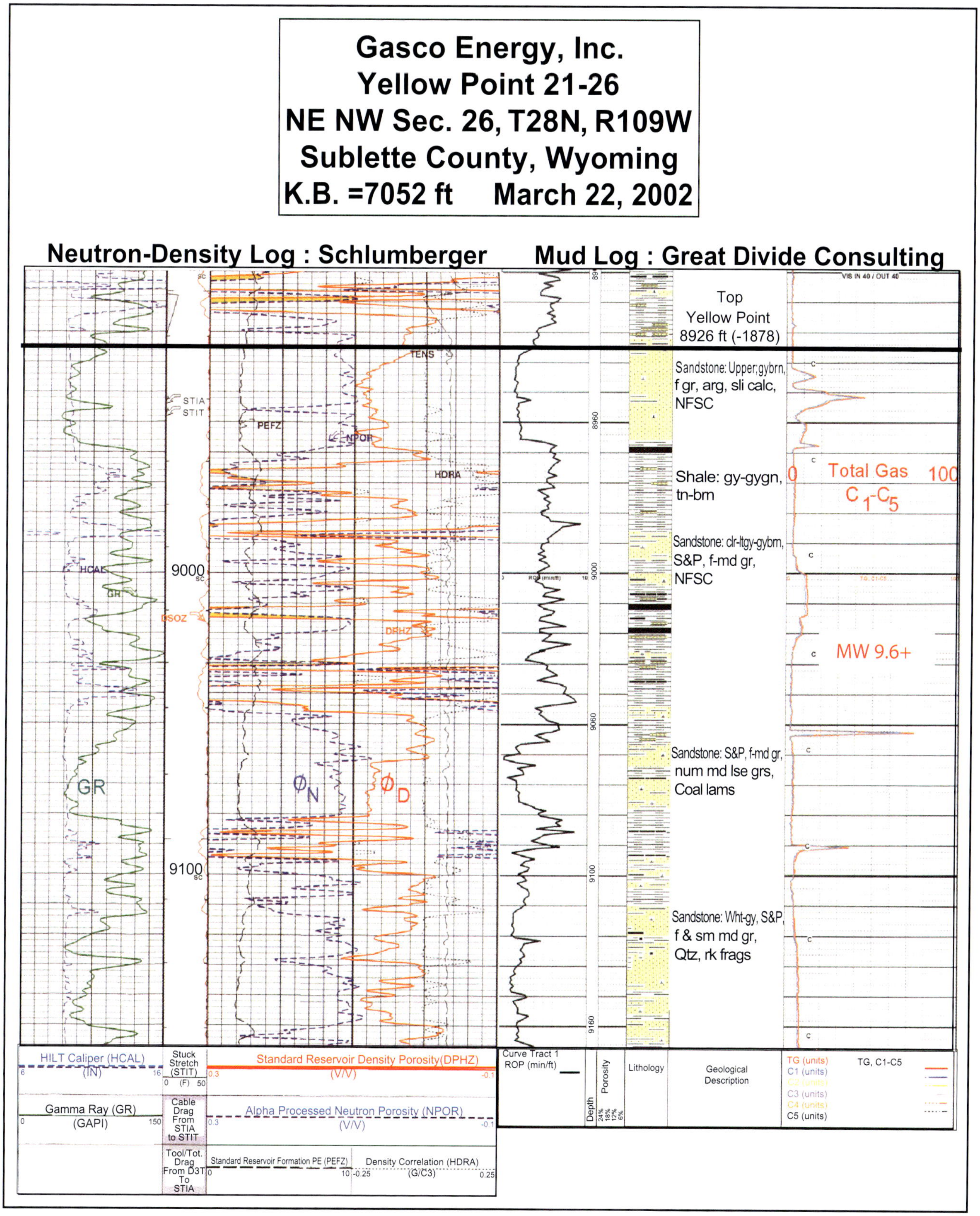
Gasco Energy, Inc.
Yellow Point 21-26
NE NW Sec. 26, T28N, R109W
Sublette County, Wyoming
K.B. =7052 ft March 22, 2002
Neutron-Density Log : Schlumberger
Mud Log : Great Divide Consulting
Top Yellow Point 8926 ft (-1878)
Sandstone: Upper;gybrn, f gr, arg, sli calc, NFSC
Shale: gy-gygn, tn-brn
Sandstone: clr-ltgy-gybrn, S&P, f-md gr, NFSC
Sandstone: S&P, f-md gr, num md lse grs, Coal lams
Sandstone: Wht-gy, S&P, f & sm md gr, Qtz, rk frags
0 Total Gas 100
C1-C5
MW 9.6+
9000
9100
GR
HCAL
STIA
STIT
PEFZ
NPOR
HDRA
DSOZ
DRHZ
TENS
ØN
ØD
HILT Caliper (HCAL) (IN) 6 16
Gamma Ray (GR) (GAPI) 0 150
Stuck Stretch (STIT) 0 (F) 50
Cable Drag From STIA to STIT
Tool/Tot. Drag From D3T To STIA
Standard Reservoir Density Porosity(DPHZ) 0.3 (V/V) -0.1
Alpha Processed Neutron Porosity (NPOR) 0.3 (V/V) -0.1
Standard Reservoir Formation PE (PEFZ) 0 10
Density Correlation (HDRA) -0.25 (G/C3) 0.25
Curve Tract 1 ROP (min/ft)
Depth
Porosity
Lithology
Geological Description
TG (units)
C1 (units)
C5 (units)
TG, C1-C5

negative results. The wildcat well encountered a normally pressured Lance section outside of the main Jonah fault block.

ACKNOWLEDGMENTS

The authors thank Cabot Oil and Gas Corporation for permission to publish information disclosed in this chapter. The Yellow Point 3-D survey was acquired by Grant Geophysical, field quality control was provided by Steve Cymbala, 3-D processing was done by Vector Seismic Data Processing, and variance cube processing was performed by Interactive Earth Sciences Corporation. The aeromagnetic survey is owned by Pearson Technologies, Inc., and the aeromagnetic processing and interpretation was done by Image Interpretation Technologies, Inc. in Calgary, Alberta, Canada. Ski Ceigerkansky of Cabot created the graphics displays.

REFERENCES CITED

Montgomery, S. L., and J. W. Robinson, 1997, Jonah field, Sublette County, Wyoming: Gas production from overpressured Upper Cretaceous Lance sandstones of the Green River basin: AAPG Bulletin, v. 81, p. 1049–1062.

Robinson, J. W., 2000, Discovery of Jonah field, Sublette County, Wyoming: The Mountain Geologist, v. 37, no. 3, p. 135–143.

Warner, E. M., 2000, Structural geology and pressure compartmentalization of Jonah field based on 3-D seismic data and subsurface geology, Sublette County, Wyoming: The Mountain Geologist, v. 37, no. 1, p. 15–30.

CHAPTER 14

The Effect of Stimulation and Completion Methodologies on Production in Jonah Field

M. EBERHARD
Halliburton Energy Services, Inc., Denver, Colorado, U.S.A.

M. MULLEN
Halliburton Energy Services, Inc., Denver Colorado, U.S.A.

ABSTRACT

The past decade has witnessed many changes in hydraulic fracture-stimulation technology and completion operation techniques used in the development of low-permeability reservoirs. Most of these changes have been effective in improving our ability to economically extract hydrocarbons from these unconventional reservoirs. By reviewing the stimulation and completion techniques used in Jonah field over the past decade, a best-practice methodology has emerged of how to treat these low-permeability gas reservoirs from both a hydraulic fracture-stimulation view and a completion practice perspective to identify damage mechanisms and to maximize productivity. The methods discussed here can be applied to low-permeability reservoirs in other basins. However, the techniques discussed here are evolutionary and will one day become dated as new innovations are developed.

The Lance Formation in Jonah field consists of several hundred feet of stacked, lenticular sandstones with reservoir permeability to gas ranging from 1 to 20 μd. The completion of multiple sandstone packages requires 5–12 stages of hydraulic fracturing. Spatial sampling and artificial neural networks, guided by an understanding of the reservoir, were used to compare specific stimulation types and completion practices in a well to its immediate offsets. The results suggest the following best practices:

(1) Never shut a well or a fracture stage in for extended periods of time.
(2) Never kill a well or fracture stage with fluid once the hydraulic fracture treatment is cleaned up.
(3) Use composite flowthrough fracture plugs instead of bridge plugs to isolate fracture stages.
(4) Fracture stages should be no longer than 350 ft (110 m).
(5) There should be no more than five entry points per fracture stage.
(6) Completion fluid should be near neutral pH and low gel loading.
(7) The stimulation treatment size should be based on the amount of net pay determined from wire-line logs.

INTRODUCTION

Modern hydraulic fracture-stimulation practices and completion techniques have been credited with making low-permeability gas reservoirs, such as those found in Jonah field and the Pinedale anticline, which are some of the most successful plays in the Rocky Mountain region in the last decade. Early wildcatters identified overpressured reservoirs in the area but were marginally successful in extracting the gas at economic rates. In this study, the completion and stimulation practices of the Jonah field were scrutinized to identify the key components of making

Jonah Field: Case Study of a Giant Tight-Gas Fluvial Reservoir, J. W. Robinson and K. W. Shanley, eds.: AAPG Studies in Geology 52 and Rocky Mountain Association of Geologists 2004 Guidebook

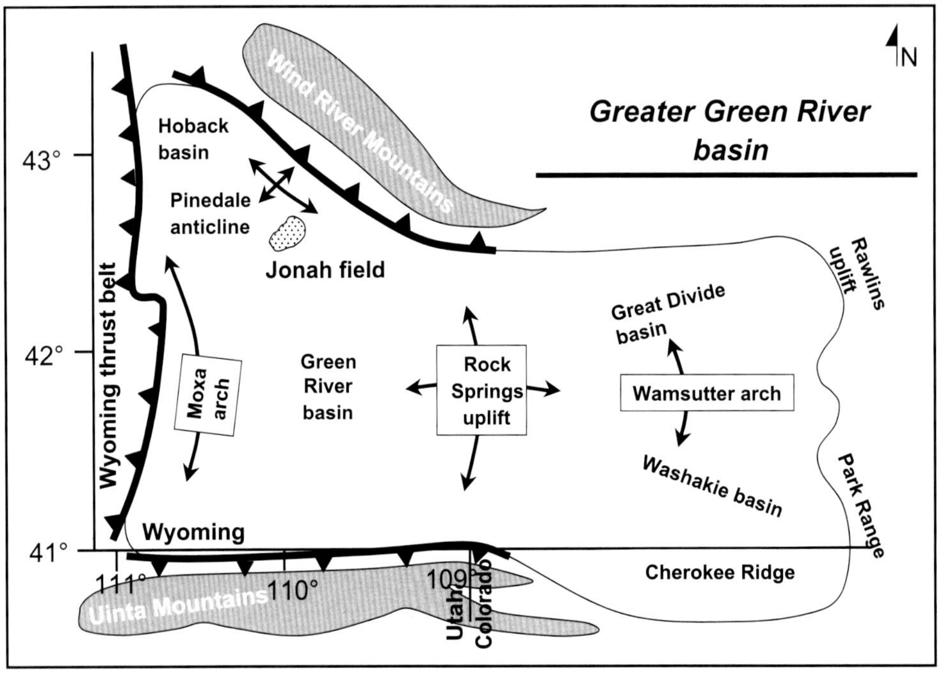

Figure 1. Location of Jonah field in the Green River basin, southwest Wyoming.

this play successful. The intent was to determine which stimulation and completion techniques have been most effective.

Jonah field is located in the northwestern corner of the Green River basin (T28–29N, R107–109W) (Figure 1) between the Wyoming thrust belt to the west and the Wind River Mountains to the east. The field is 60 mi (100 km) north of Rock Springs, Wyoming. Presently, there are more than 350 commercial wells that have been drilled on 40-ac (16-ha) spacing. The eastern (downdip) edge of the field is still being extended. The known limits of the field currently exceed 38 mi^2 (100 km^2).

Jonah field is bounded on the south and west by wrench faults (Bowker and Robinson, 1997; Warner, 1997). Wells in the field encounter overpressured conditions (0.58–0.65 psi/ft gradient) at depths of 8000–9300 ft (2400–2800 m). Overpressuring continues to the total depth of each well. To date, wells drilled outside of field-defining faults have encountered normally pressured sandstones at similar depths and have been marginally productive. Faults located in the field boundaries appear to have an effect on reservoir performance. Although they are not sealing faults, they may affect the degree of microfracturing present in the sandstones. The spatial sampling methodology used in this study was intended to consider the effect geology has on production.

The Lance Formation is Upper Cretaceous in age and consists of 2000–3000 ft (600–900 m) of interbedded fluvial sandstones and mudstones. Individual sandstone units range from 5 to more than 50 ft (1.5 to more than 15 m) in thickness and have areal extents ranging from a few acres to possibly 100 ac (40 ha). Individual sandstones are geologically heterogeneous reservoirs and are difficult to correlate because of the depositional environment. Sandstone-rich intervals used for correlation purposes are locally called the upper Lance, middle Lance, Jonah, Yellow Point, Wardell, and Mesaverde (Figure 2). The total thickness of sandstone encountered ranges from 300 to 800 ft (90 to 240 m) across the field, and drilling depth ranges from 10,000 to 12,500 ft (3040 to 3800 m). More specific geological descriptions can be found in studies by Warner (1997) and Bowker and Robinson (1997).

The reservoir sandstones have 5–14% porosity and 1–20 μd gas permeability. Water saturation ranges from 30 to 60%; currently, there is no significant water production in the field other than fracture fluid flowback. The producing condensate yield is 8–10 bbl/MMSCF with an API gravity of 52°. Although PVT fluid data are scarce, the produced fluid composition appears to be very similar throughout the productive section.

Because of the low permeability of the Lance Formation, fracture stimulation is required for economical production rates. There have been a variety of different treatment types, fracture isolation methods, and time intervals between treatments that have been tested in the field over the past decade (Integrated Petroleum Technologies, 1996; Flack et al., 1997). Until 1998, a single stage

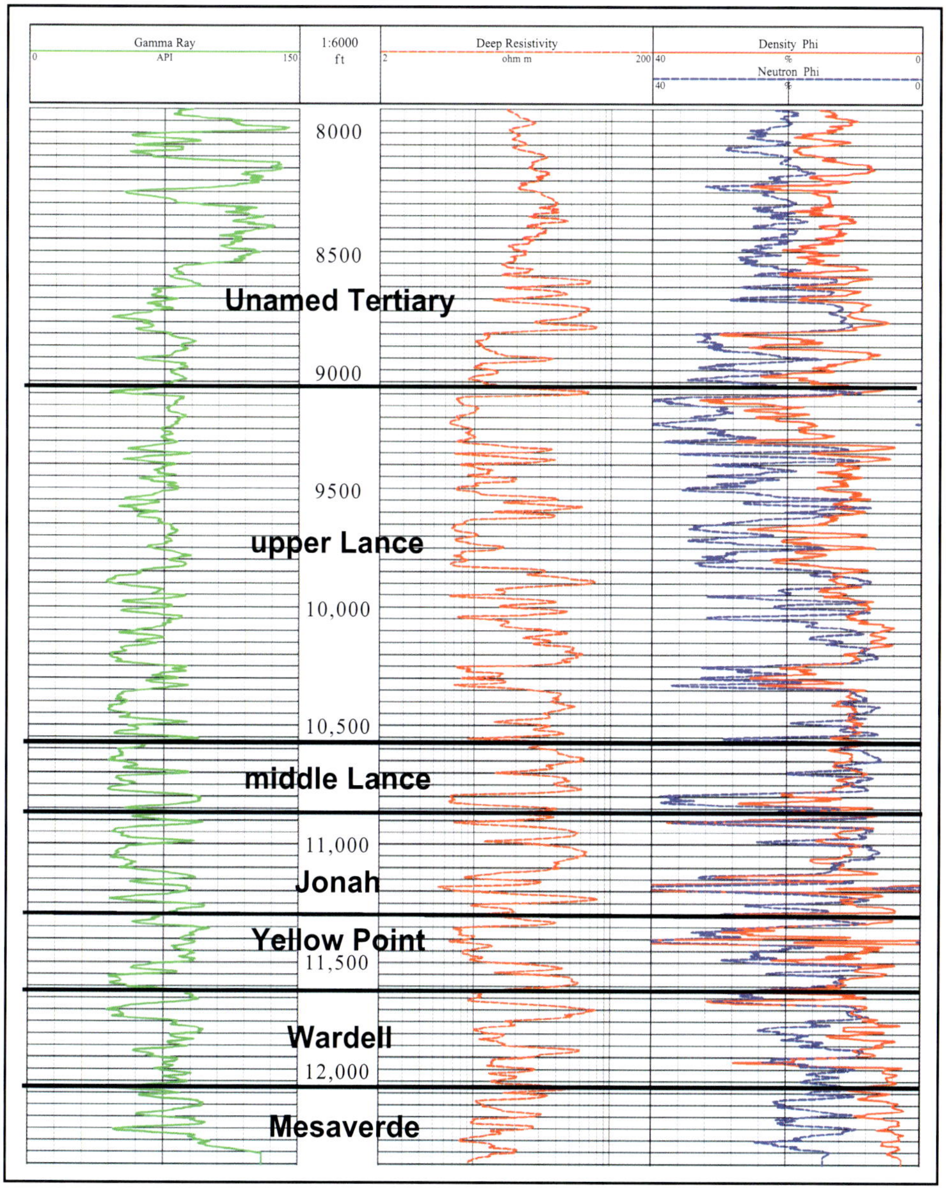

Figure 2. Type log showing the terminology of Tertiary and Cretaceous strata in the area of Jonah field. A gamma-ray (GR) curve is on the left side of the depth column. Deep resistivity and density-neutron are shown on the right side of the depth column.

consisted of fracture stimulating three to six individual sandstones using a limited entry technique. After each stage, the well would be allowed to flow back the fracture fluid and produce gas for 3–7 days to clean up the stimulated zone. Commonly, the produced gas was dry enough to send down the sales line after 24 hr. To stimulate the next stage, the producing zones would be isolated from the next treatment. Two methods were commonly used to isolate each fracture treatment, either pumping a sand plug to cover previous stages or running some type of bridge plug above the producing interval. Four to six fracture stages were performed per well.

THE DAMAGE MECHANISM

Eberhard et al. (2000) observed that 11% of the completed wells were producing at less than expected rates. Production logs were run in those wells, and it was determined that only 60–65% of the sandstones perforated and stimulated were producing. In some cases, zones that had flowed as much as several million cubic feet per day on flowback were producing little or no gas after being commingled. An integrated field study completed by Eberhard et al. (2000) cited several damage mechanisms and stimulation practices that were the cause of

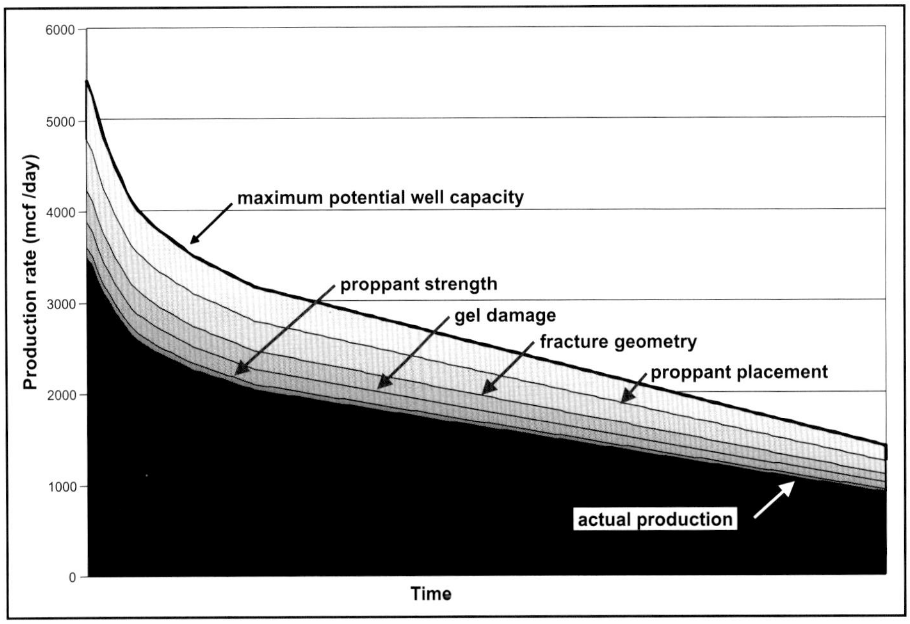

Figure 3. The adverse effects of proppant strength, gel damage, fracture geometry, and proppant placement could cause a well to produce below its maximum potential. Other stimulation considerations include the negative effects of clay damage, imbibition, embedment of proppant, turbulent flow, stress and pressure variations, and fracture fluid cleanup. The goal is to reach the maximum well capacity at an economic cost.

the poor production. Eberhard et al. (2000) discussed two factors that were thought to be the main cause of the low completion efficiency: (1) the reintroduction of fluid into a dried out hydraulic fracture when the zone was killed to shut off the flow of gas in preparation for the next stage; and (2) shutting in earlier producing zones for extended periods of time while subsequent zones in the well were being completed. Eberhard et al. (2000) suggested that fracture stage isolation methodology was the probable cause of the low completion efficiency.

Eberhard et al. (2000) investigated several components of the stimulation, including gross thickness of the stage intervals, the number of sandstones per stage, the proppant type, and the stimulation fluid type. Since 1998, many methods have been tested in Jonah to improve completion efficiency and increase well performance. Stimulation treatment optimization in a field the size of Jonah is a continuous process. The goal in stimulation optimization is to reach the maximum potential of the reservoir (Figure 3). Progress has been achieved in the use of stimulation fluid types, proppants, and breaker technologies to improve well productivity over the past decade. The most significant improvement in well productivity, however, was achieved by changing the method used to isolate each fracture stage. Two isolation methods to emerge were "induced stress diversion (ISD)" hydraulic fracturing technique and flowthrough composite bridge plugs (FTCBPs). In both cases, the number of individual sandstones treated per hydraulic fracture stage was reduced, commonly to three or less, which resulted in more treatments per well. This study compares these new completion techniques by evaluating offset wells completed using pre- and post-1998 completion practices.

COMPLETION TYPES

Prior to 1998, most operators used sand plugs and wireline-set/tubing-retrievable bridge plugs to isolate individual stages. Sand plugs are an inexpensive isolation method, in which sand is poured into the wellbore to cover the existing perforations and divert the next fracture stage into new perforations. This fracture stage isolation technique was used in the early development of the field (Flack et al., 1997). After 1998, the most common fracture isolation tool used by operators was wireline-set/tubing-retrievable bridge plugs. The drawback to this method of stage isolation was that zones were left shut in for extended periods of time during completion of the well. The challenge to the operators was to develop a method of isolating fracture stages without causing damage to the well, yet allowing the well to flow continuously once stimulated.

Induced stress diversion is a completion technique described by Hewett and Spence (1998) as an alternative to limited entry fracturing. Induced stress diversion depends on two factors: (1) Stress increases with depth, and (2) closure stress increases when the fracture closes on the proppant (Schubarth et al., 1997). To assist the stress diversion process, attempts were made to "screen out"

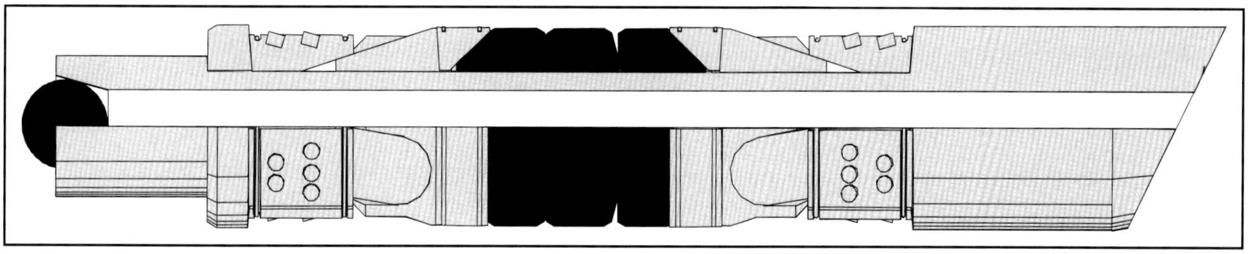

Figure 4. The flowthrough composite bridge plug acts like a conventional bridge plug when pressure is applied above the plug but differs in that it allows flow from zones below the plug during flowback or production. The composite material that composes the plug allows it to be drilled out at a later time.

a hydraulic fracturing treatment near the end of the job by pumping higher proppant concentrations. In ISD, the next stage is pumped before much of the previous fracture is flowed back. This allows the next shallower stage to be perforated and stimulated using the high pressure in the open perforations below as a virtual bridge plug. Because ISD requires increased stress from previous fracture treatments, it is a continuous process; some wells are stimulated in a continuous cycle, 24 hr a day. With 10–15 fracture stages per well, there are some practical limits to the ISD technique that make it necessary to use mechanical isolation to allow the well to flow back and clean up. This allowed time for some logistical issues like refilling water tanks, loading proppant bins, or allowing the fracture crew to get a nap. One advantage to this method is to substantially reduce completion time (2–3 days versus 4–5 weeks), which results in significant savings in surface equipment rentals and a reduction in fracture equipment charges. One drawback to this method is the uncertainty of knowing whether the fracture treatment actually went in the zone intended or was dispersed among several zones. Another problem was the negative effect of a prolonged cleanup or flowback on production after so many stages. Many early ISD wells reported similar completion inefficiency problems to using wireline bridge plugs.

The FTCBP (Figure 4) is a recent innovation in the composite bottomhole tools category. It uses composite materials to make a bottomhole tool that functions much like a conventional bridge plug in isolating hydraulic fracture-stimulation stages. The difference is the presence of a 1-in. (2.5-cm)-diameter hole through the middle of the plug body that allows flow through the plug with a ball that sits in a seat at the top of the tool. When pressure is applied to the well, such as during a hydraulic fracture treatment, the ball seals the zones below the plug. After the fracture treatment, pressure from the zones below the plug lifts the ball off the tool, and this allows gas to flow from the lower zones to assist in the cleanup of the subsequent fracture treatment(s). By using an FTCBP, a fracture stage is never killed with fluid or shut in for any extended period of time, which minimizes formation damage. After all the zones in a well have been stimulated, an FTCBP can be drilled out while landing the tubing, or they can be left in the well. Most operators drill out the plugs with a drill bit and a pump-off sub on the end of the tubing. After the plugs are drilled out, the tubing is hung off as a velocity string.

The ISD and FTCBP technologies have been used extensively over the past few years in Jonah field and have been effective in improving completion efficiency and increasing well productivity.

SPATIAL SAMPLING ANALYSIS

The remaining portion of this chapter will compare which technique (ISD versus FTCBP) has been more effective at Jonah field. Because of the complexity of the reservoir, a spatial sampling technique was applied to short-term cumulative production. Spatial sampling was developed by Voneiff and Cipolla (1996) as a method for comparing production from any given well to its direct offsets. This technique uses a computer program to investigate all wells in the study area. As the program steps from well to well, each target or center well is compared to all the offsets in a set sample radius called a domain. The primary use of this technique is to identify underperforming wells. Depending on the amount of data available, other variables, such as treatment information, log information, and production, can be compared from the center well to the offset wells. Evaluation of this information can result in a more thorough understanding of differences in well-to-well productivity. Because the location of the faults and fault blocks is a critical factor in the productivity of gas in the Jonah field, several domains were evaluated within each fault block to provide a more representative study.

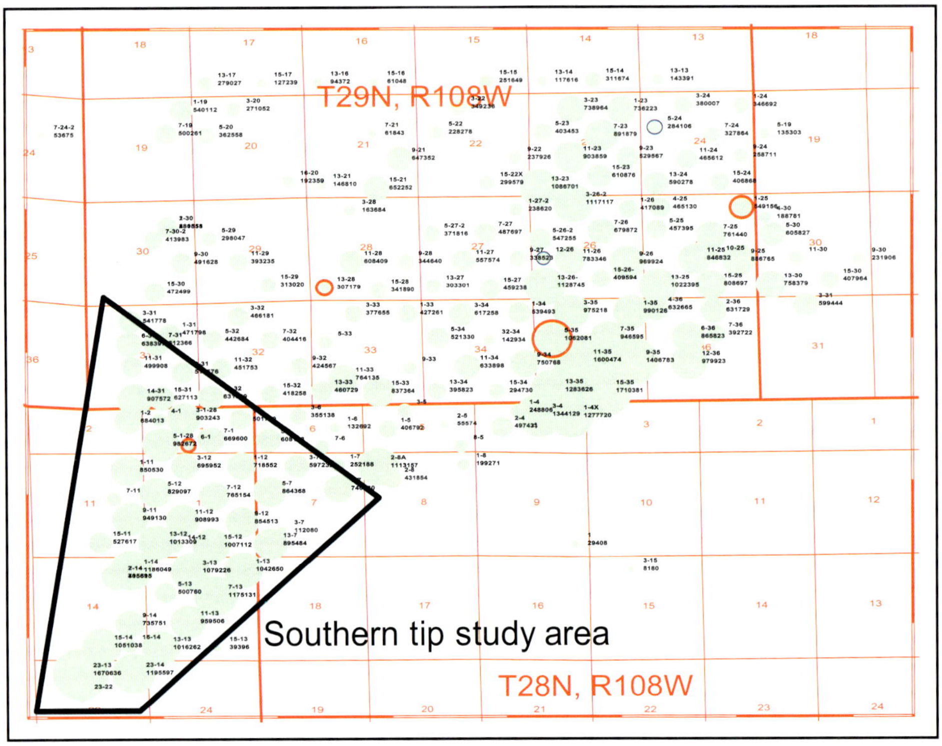

Figure 5. Wells used in the spatial sampling study. Relative bubble size represents 6-month cumulative gas production. The boxed area is the southern tip "fault block" used in a separate spatial sampling study.

WELL INFORMATION

Production data for wells in the study area were obtained from several public sources. Combining production data from several sources provided a more complete and up-to-date data set than any single data source. The study area includes only those wells inside the boundaries of the field (Figure 5). A total of 177 wells with at least 1-month production, 162 wells with 6-month production, and more than 1000 individual hydraulic stimulation treatments were used in the study. The data had to be screened carefully for completeness and accuracy to remove any production artifacts that were the result of curtailed production, bridge plug isolation, etc.

Treatment data for these wells were obtained from both public and private sources. Proppant volumes were available for a majority of the treatments, but fluid volumes needed to be carefully scrutinized when nitrogen was used. Wells in the study area were separated into five categories based on completion type:

- Nonfoam: treatments that used mechanical stage isolation and no nitrogen
- 30%: treatments with mechanical stage isolation and 30% nitrogen assist
- 70%: treatments with mechanical stage isolation and 70% nitrogen assist
- ISD: treatments using mostly or all ISD
- FTCBP: treatments using FTCBP for stage isolation.

BEST PERIOD OF PRODUCTION

The start of production for each well was determined by identifying the month of highest production. This practice was used to minimize the effects of fracture cleanup and the differences in how operators bring wells on production after completion. This method makes allowances for the time some operators take to complete wells and compensates for partial month production. Figure 5 shows 6-month cumulative gas production for each well represented by the size of the bubble around the well. The best-month production was used to start the clock by which 1-, 3-, 6-, 9-, and 12-month cumulative gas production summaries were calculated.

ANALYSIS

The first step for this analysis was to determine the minimum production period that would provide good correlation to ultimate production. A graph of 3-, 6-, 9-, and best-month cumulative gas production was compared to

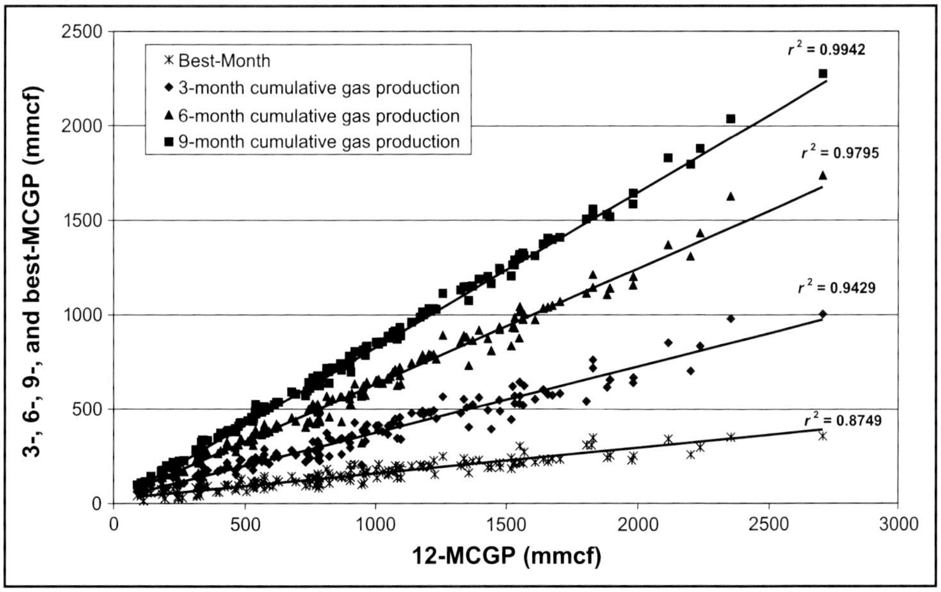

Figure 6. A graph of 3-, 6-, 9-, and best-month cumulative gas production versus 12-month cumulative gas production (12-MCGP) for Jonah field. From this graph, it was determined that 6-month cumulative gas production would be used for analysis. It was also determined that because of the good correlation (r^2 = 0.87) between best-month production and 12-month cumulative gas production, best-month production could also be used for evaluating the most recently completed wells by extrapolating early time cumulative production data.

12-month cumulative gas production (Figure 6). From this graph, it was determined that 6-month cumulative gas production would be used for analysis. It was also determined that because of the good correlation (r^2 = 0.87) between best-month production and 12-month cumulative gas production, best-month production could also be used for evaluating the most recently completed wells by extrapolating early time cumulative production data.

Production versus completion date was evaluated to determine if depletion is a factor with the infill wells drilled since 1998. As Figure 7 shows, there is no apparent indication of depletion with wells completed since 1998. In fact, some of the best wells in the field have been treated with post-1998 completion techniques.

A cumulative frequency graph of 6-month cumulative gas production (Figure 8) shows a median production value of 558,000 mcf and an average of 623,000 mcf. For comparison, the ISD and FTCBP wells are highlighted. This analysis shows that half of the wells completed with ISD produce less than the field average, and five out of seven of the FTCBP wells are in the top third of the producers in the field. This comparison, however, does not consider the complexity of the reservoir. The spatial sampling study was used to compare and account for variations in sandstone thickness, as well as to reflect the influence of faulting in the field.

The domain area for the spatial sampling analysis was 640 ac (260 ha). In the Jonah field study, each sample contained between four and eight wells in which the center well was compared. Other domain sizes, both larger and smaller, were also investigated with no significant changes in the results. Because detailed log data were not available for the majority of the wells, the analysis was limited to operator, completion type, and proppant volumes using best-month production and 6-month cumulative gas production. For all spatial sampling analyses, the comparison was based on the value of the center well minus the average value for the offsets. A positive value indicates that the center well has a higher value than the average of the offsets, and a negative value indicates that the center well has a lower value than its offsets.

Spatial sampling study results compare the center well variance of offsets by date and also by completion fluid type (Figure 9). The interpretation of this spatial sampling study compares more than just ISD and FTCBP completions. It also incorporates completion fluid types that have been reported by Eberhard et al. (2000). In terms of comparison to offsets, the first ISD well was, by far, the best ISD completion and was also the best producing ISD well. Overall, 13 of 21 ISD completions have less production than their offsets. Induced stress diversion wells produced an average of 18% less gas in the first 6 months as compared to their offsets, completed using FTCBP fracture stage isolation methods. Conversely, six of seven wells completed with FTCBP have higher best-month production than their offsets. On average, the

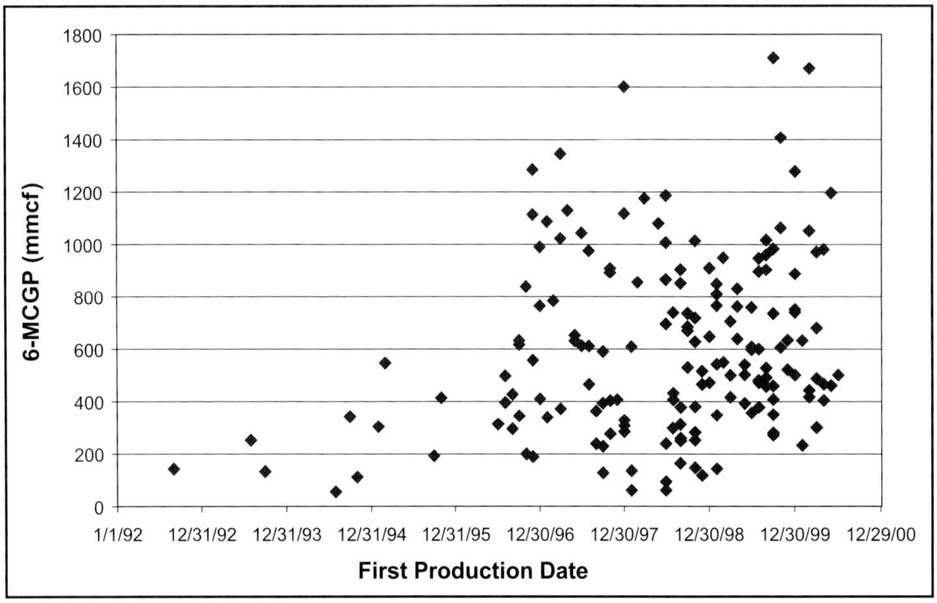

Figure 7. Graph of 6-month cumulative gas production (6-MCGP) over time. The figure shows that there is no indication of depletion in wells completed since 1998 that were drilled as part of an infill program.

wells that were completed using FTCBP produced 50% better in the first 6 months than their offsets using conventional fracture stage isolation methods.

Figure 10 shows a bar chart of the average difference between the center well and offsets for each completion type. The chart shows that before the introduction of FTCBP, there were not any large differences in production between different completion types (±50,000 mcf in 6 months). However, when FTCBP was used, a step change occurred in the productivity of these wells, resulting in an average increase of 250,000 mcf for 6-month cumulative gas production when compared to their offsets.

SOUTHERN TIP AREA SPATIAL SAMPLING STUDY

Because of the structural and stratigraphic complexity of Jonah field, there may be a bias in the fieldwide

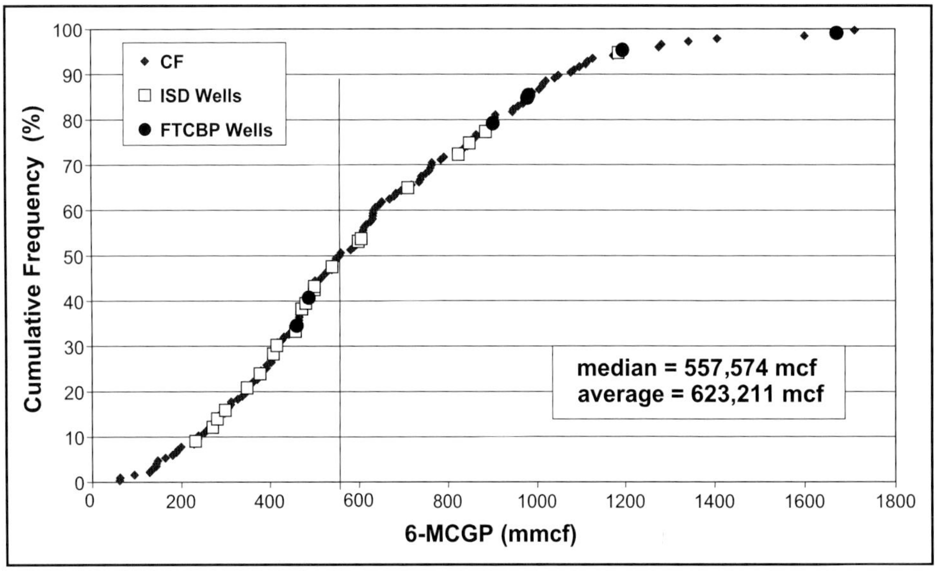

Figure 8. Cumulative frequency chart of 6-month cumulative gas production (6-MCGP) for Jonah field. Median production is 558,000 mcf, and average production is 623,000 mcf. Half the wells completed with ISD methods produce less than the field average. Five out of seven wells completed with FTCBP are in the top 30% of this category.

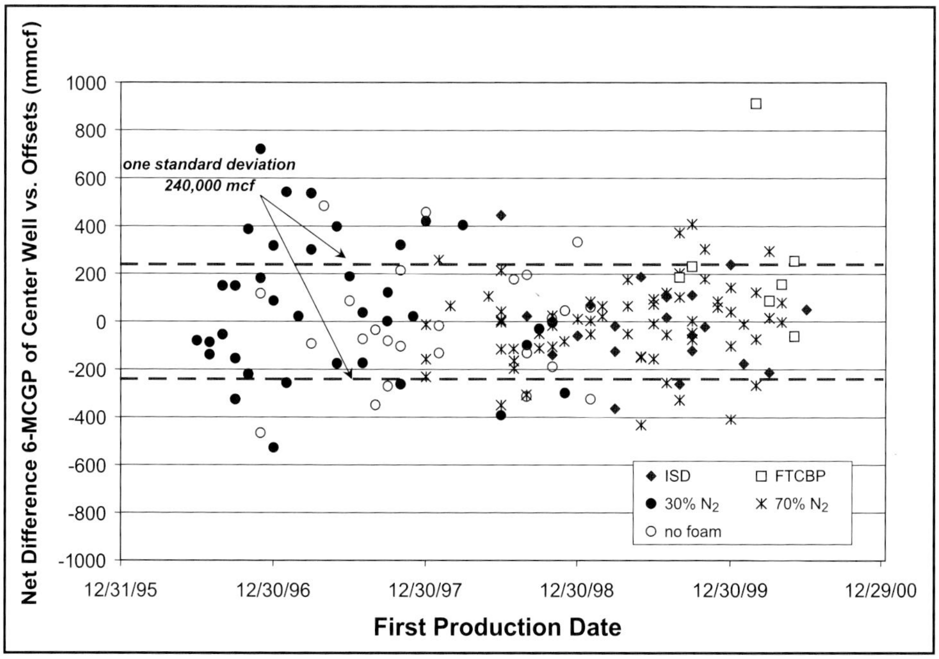

Figure 9. Comparison of center well 6-month cumulative gas production (6-MCGP) to the average of the offset wells' 6-month cumulative gas production by completion date and type for Jonah field. A positive value indicates that the center well has a higher value than the average of the offsets, and a negative value indicates that the center well has a lower value than its offsets.

spatial sampling study. The study was rerun on an isolated fault block in the southwestern tip of the field (Figure 5) that contains 60 wells. Half the wells in this area were completed after July 1999; therefore, the best-month production was used to maintain the largest sample size.

A cumulative frequency chart of best-month production for the southern tip fault block (Figure 11) shows a median best-month production of 171,622 mcf and an average of 174,196 mcf. These values compare to 130,436 and 144,806 mcf for the field. Induced stress diversion wells and wells completed with FTCBP are highlighted in Figure 11.

A graph of spatial sampling results shows the variation in production for the five completion fluid types and fracture stage isolation methods in this single fault block (Figure 12). These results show that in this area, an average

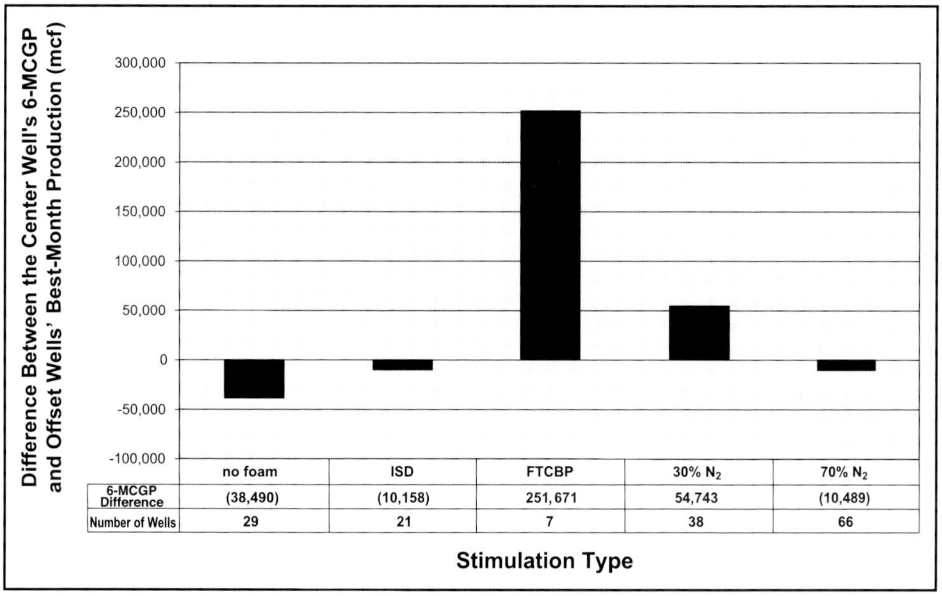

Figure 10. Average of the difference between the center well's 6-month cumulative gas production (6-MCGP) to the offset wells' 6-month cumulative gas production based on completion type. The chart shows that before the introduction of FTCBP, there were not any large differences in production between different completion types (±50,000 mcf in 6 months). However, with FTCBP, an improvement in the productivity of these wells resulted in an average increase of 250,000 mcf for 6-month cumulative gas production when compared to their offsets.

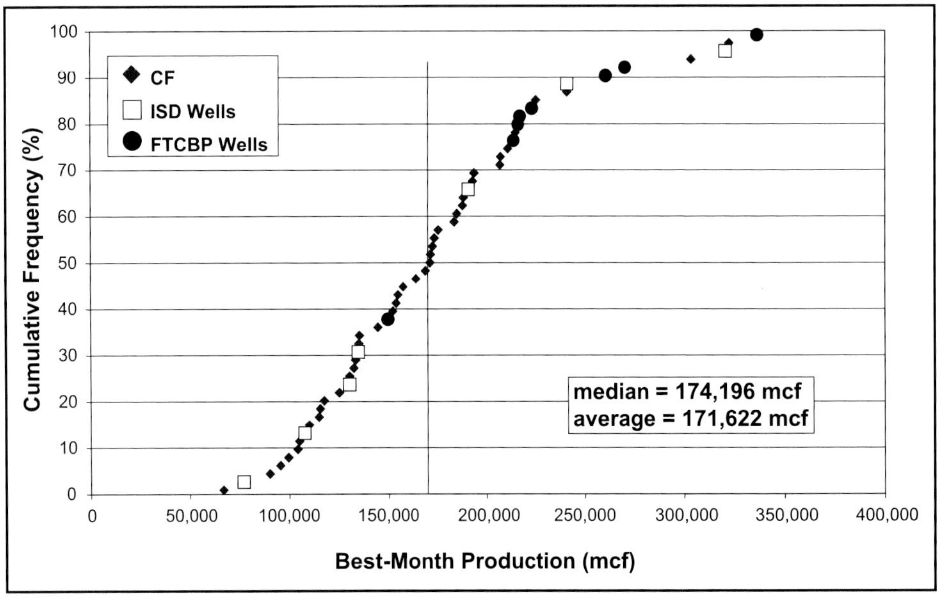

Figure 11. Cumulative frequency chart of best-month gas production for the southern tip study area in which the spatial sampling was limited to a single fault block. The study area had a median best-month production of 174,196 mcf and an average of 171,622 mcf. These values compare to 130,436 and 144,806 mcf for the field.

ISD well produced less gas than its offsets, whereas an average FTCBP well produced 40% better than its offsets. A bar chart shows similar results in this fault block to the fieldwide analysis (Figure 13). It should be noted that there are small sample sizes in some groups, but there remains a significant difference in FTCBP-completed wells and the others.

Flowthrough composite bridge plugs may improve completion efficiency, although there are very few production logs to confirm this observation. Figure 14 shows a typical production curve for a well completed using FTCBP. With conventional fracture stage isolation methods (i.e., bridge plugs), commingled production to a sales line uncommonly added up to the sum of the individual well tests. When FTCBPs are used, the final flow rate is close to what is ultimately produced to the sales line.

COMPLETION FLUIDS AND PROPPANT AMOUNTS

This study analyzed the stimulation fluid type and total proppant amounts pumped to determine which of

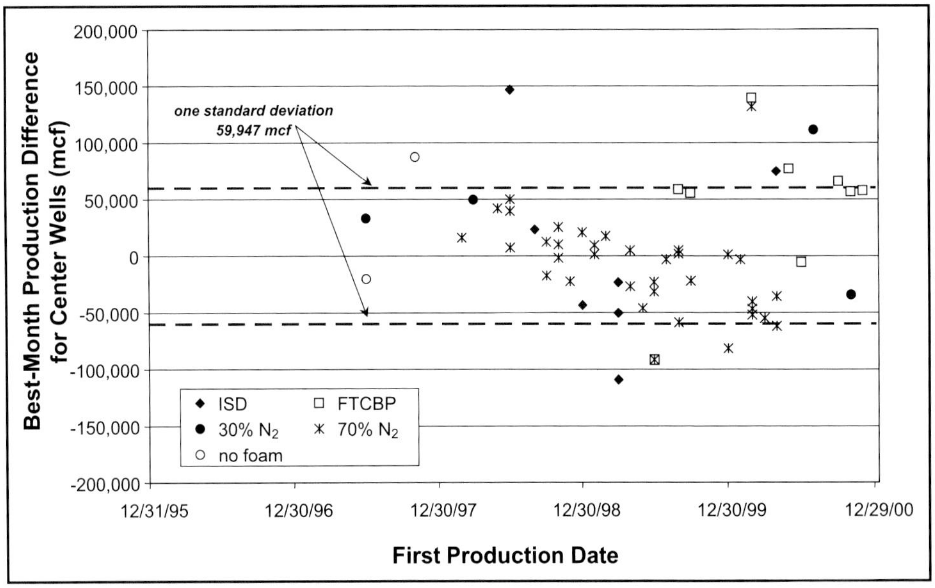

Figure 12. Comparison of center well's best-month gas production to the average of the offset wells' best-month gas production by completion date and type for southern tip study area. These results show that in this area, an average ISD well produced less gas than its offsets, whereas an average FTCBP well produced 40% better than its offsets.

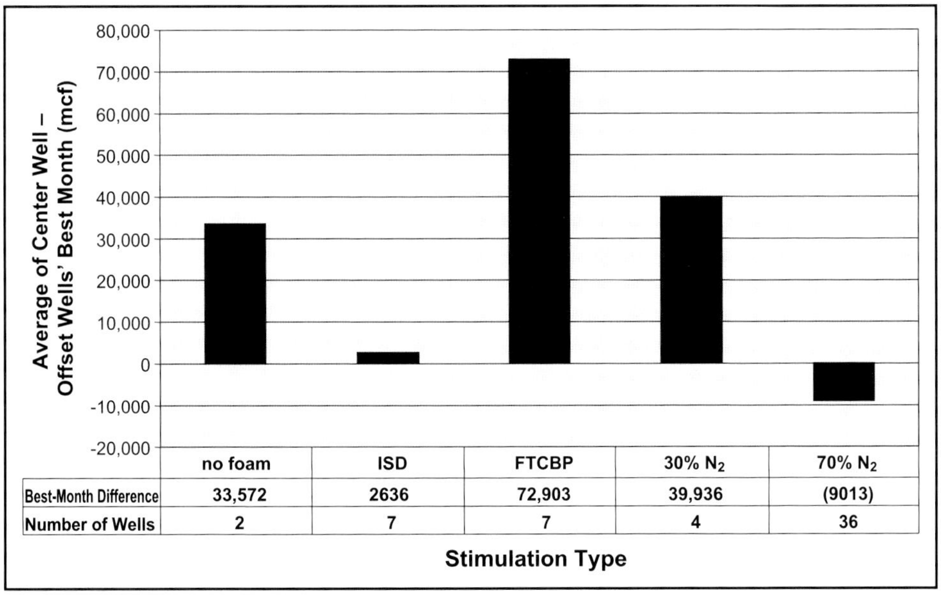

Figure 13. Overall average of the difference between the center well's best-month gas production to the offset wells' best-month gas production based on completion type for the southern tip study area.

these combinations resulted in the best well performance. A graph comparing proppant treatment volumes in the southern tip study area (Figure 15) shows that ISD wells used less proppant volume than the other completion types. The ISD and FTCBP wells average eight and nine treatments per well, respectively. Foam wells average five treatments per well.

ECONOMIC CONSIDERATIONS

Economic evaluation of a best-practice methodology is a complicated process. Every operator has different economic drivers and objectives. So the "best practice" by operators in the field may not always be the same. At the time of this study, ISD completions were

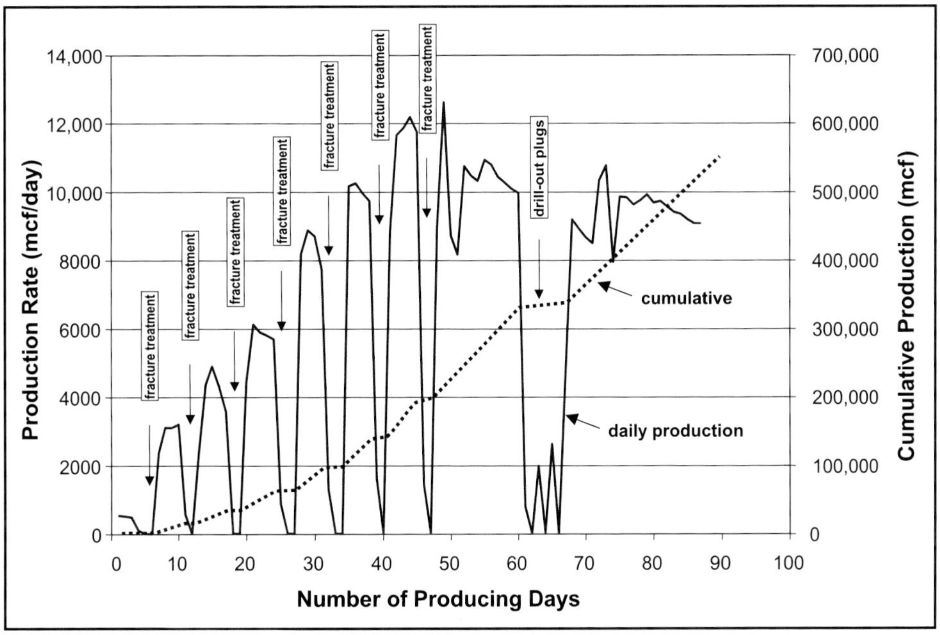

Figure 14. Typical production chart for a completion using an FTCBP where the well produced nearly 550 mmcf during completion. Note the increase in production between each fracture stage. With conventional fracture stage isolation methods (i.e., bridge plugs), commingled production to a sales line uncommonly added up to the sum of the individual well tests. When FTCBPs are used, the final flow rate is close to what ultimately is produced to the sales line. The production from this well was choked back after the tubing was landed because of production facility limitations.

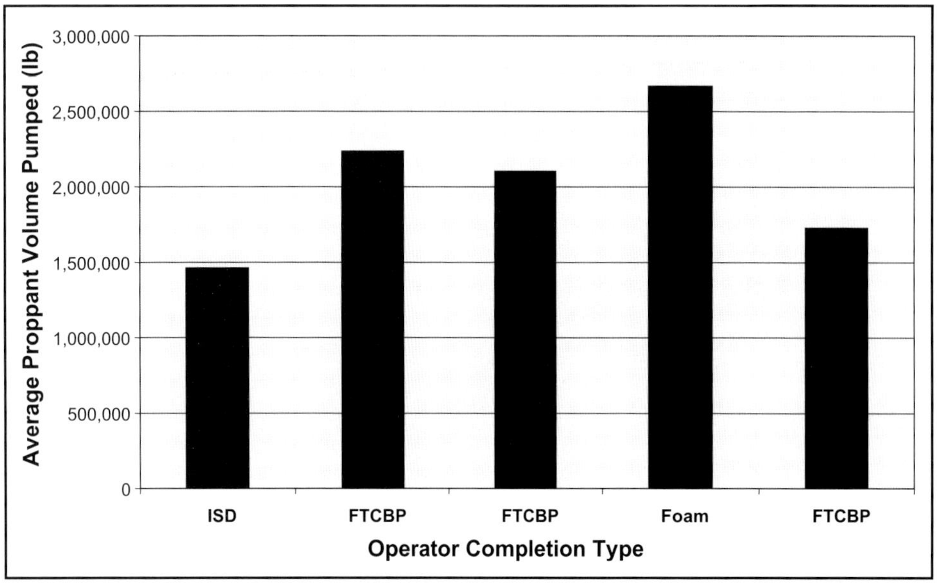

Figure 15. Average proppant volume pumped based on treatment type for southern tip. Induced stress diversion wells used less proppant volume than the other completion types.

\$200,000–\$300,000 less expensive than FTCBP completions. If the first 6 months of cumulative gas production is considered, FCTBP wells produce 250,000 mcf more gas than the ISD wells. At a net gas price of \$1.00/mcf, ISD is more economic because of its low cost. However, as gas prices rise, FTCBPs become progressively more economic.

Operators should also consider as much additional reservoir information as possible when considering completion best practices. Stimulation design criteria are highly dependent on the wire-line–log evaluation and pore pressure. It may be necessary to use positive isolation methods, such as composite bridge plugs, to obtain pore pressure and permeability information on individual sandstone units to aid in formation modeling. Although there may be some risk involved, this information is very valuable in reservoir modeling, formation evaluation, and stimulation design.

CONCLUSIONS

The study of stimulation and completion practices in Jonah field has identified several sources of posttreatment formation damage. Two new technologies, ISD and FTCBP, in conjunction with other changes in hydraulic fracture fluids and proppants, have been used to help overcome these obstacles to economic gas production. The ISD approach reduced the completion cost, commonly at the expense of cumulative production, when compared to the offsets completed using conventional fracture stage isolation methods. The FTCBP completions, on average, showed as much as a 50% increase in the first 6 months of gas production. In today's gas price market, this step change in production characteristics has a significant economic impact.

Based on this study and gleaning the recommendations for completion strategy from other authors, the following recommendations are provided as a best-practice guide for completing wells that require multiple hydraulic fracture stages to stimulate stacked, low-permeability sandstone formations:

(1) Never shut a well or a fracture stage in for extended periods of time.
(2) Never kill a well or fracture stage with fluid once the fracture is cleaned up.
(3) Use composite flowthrough fracture plugs to isolate fracture stages instead of bridge plugs.
(4) Fracture stages should be no longer than 350 ft (110 m).
(5) There should be no more than five entry points per fracture stage.
(6) Completion fluid should be near neutral pH and low gel loading.
(7) The stimulation treatment size should reflect the net pay observed in the wire-line logs.

FUTURE CONSIDERATIONS

As with any study of this type, these results will one day become dated, as newer technologies are tested and applied to multistage completions in low-permeability reservoirs. The keys to future improvements in multistage completions are in developing an understanding of the damage caused by our operational practices, being open to new innovations and technologies, and maintaining good databases, so the effect of these innovations can be quantified. Although the results of this study may not be repeated in other fields, the application of these best-practice guidelines in other comparable areas will give the producers the best chance at unlocking the gas trapped in low-permeability reservoirs.

ACKNOWLEDGMENTS

The authors thank everyone who helped compile data and provided insight on the analysis. Thanks are also extended to Halliburton Energy Services, Inc., for permission to publish this chapter.

REFERENCES CITED

Bowker, K. A., and J. W. Robinson, 1997, Jonah field: A shallow sweetspot in the basin-centered gas accumulation of the northern Green River basin: AAPG Bulletin, v. 81, p. 1219, and Expanded Abstracts Volume, p. 25–30.

Eberhard, M. J., M. J. Mullen, C. A. Seal, and B. P. Ault, 2000, Integrated field study for production optimization: Jonah field, Sublette County, Wyoming, *in* Society of Petroleum Engineers/Canadian Energy Research Institute Gas Technology Symposium: SPE paper 59790, p. 683–695.

Flack, D., J. Shaw, J. Martin, and J. Oriet, 1997, Modified drilling and completion procedures enhance recovery in southwest Wyoming field, *in* Society of Petroleum Engineers Rocky Mountain Regional Meeting: SPE paper 38344, p. 85–98.

Hewett, T. W., and C. J. Spence, 1998, Induced stress diversion: A novel approach to fracturing multiple pay sands of the NBU field, Uintah Co., Utah, *in* Society of Petroleum Engineers Rocky Mountain Regional/Low-Permeability Reservoirs Symposium: SPE paper 39945.

Integrated Petroleum Technologies, Inc., 1996, Advanced stimulation technology deployment program, Snyder Oil Corporation, Lance Formation, Jonah field, Sublette County, Wyoming: Topical Report, GRI-96/0371, 126 p.

Schubarth, S. K. et al., 1997, Understanding proppant closure stress, *in* Society of Petroleum Engineers Production Operations Symposium: SPE paper 37489.

Voneiff, G. W., and C. Cipolla, 1996, A new approach to large-scale infill evaluations applied to the OZONA (Canyon) gas sands, *in* Society of Petroleum Engineers Permian Basin Oil & Gas Recovery Conference: SPE paper 35203.

Warner, E. M., 1997, Geology of Jonah field, a major gas accumulation in the Upper Cretaceous Lance Formation, Sublette County, Wyoming, *in* E. B. Coalson, J. C. Osmond, and E. T. Williams, eds., Innovative applications of petroleum technology in the Rocky Mountain area: Rocky Mountain Association of Geologists Guidebook, p. 1–12.

CHAPTER 15

Jonah Field Completions: An Integrated Approach to Stimulation Optimization with an Enhanced Economic Value

J. L. JENNINGS
Schlumberger, Denver, Colorado, U.S.A.

B. P. AULT
Ultra Resources Inc., Denver, Colorado, U.S.A.

ABSTRACT

Gas production in Jonah field is derived from more than 100 lenticular sandstone units that are interbedded with mudstones and siltstones over a 3000-ft (900-m) section of the Cretaceous Lance Formation. Productive sandstones average 12 ft (3.7 m) thick and are dispersed unevenly throughout the stratigraphic section. Historically, completion methods in Jonah have been costly and inefficient. Efforts to reduce completion costs while maintaining reserves in Jonah field required a new approach in evaluating and completing such thick productive zones. Analysis of fracture-simulation results for selected wells in the field suggests that stimulation of smaller, selected intervals would result in enhanced recovery by exploiting previously bypassed sandstone intervals. To obtain an optimal fracture-stimulation design, an enhanced petrophysical model was developed to provide reliable permeability, mechanical rock properties, fluid saturation, net pay, and net stress predictions. To further optimize the fracture stimulation, induced stress diversion was implemented. Induced stress diversion is a new technology that allows for multiple-stage stimulation without the use of a plug or mechanical device between stages. Using induced stress diversion, completion time was reduced by 4 weeks, and the percentage of producing sandstones in a treated interval increased from 60 to 90%. The net result was a 40% reduction in time and costs without negative impact on reserves. These optimization methods reduced development costs in Jonah field by 22%, from \$2.8 million to \$2.2 million per well, while achieving similar production and reserves.

INTRODUCTION

In the 1980s, pumping large amounts of sand proppant in hydraulic fracture treatments was thought to be the key to unlocking commercial success. Many subsequent operators realized that there was a point of diminishing returns, and large hydraulic fracture treatments were replaced by optimization. Although optimization of many well simulators has increased in the past 20 yr, fracture treatments in a particular field today are still based primarily on historical treatments. Stimulations using comprehensive data and detailed designs are not yet common practice.

Because of the low permeability of many gas reservoirs in the Rocky Mountain region, well stimulation is mandatory to achieve an economic well. Even for reservoirs with higher permeability, stimulation is an effective way to improve production or accelerate recovery, especially during periods of rising gas prices or when well economics require a quick return on investment. Movement of formation fines, scale deposition, and organic deposits are productivity-related problems that stimulation technology can delay or help avoid.

Reservoir characterization has become more effective because of advances in three-dimensional simulation. Geologists, petrophysicists, and engineers who design drilling, completion, or stimulation programs are challenged to acquire data for these simulation models. A vast amount of data must be gathered, some of which are

Jonah Field: Case Study of a Giant Tight-Gas Fluvial Reservoir, J. W. Robinson and K. W. Shanley, eds.: AAPG Studies in Geology 52 and Rocky Mountain Association of Geologists 2004 Guidebook

critical for predicting potential production rates, reserves, and recovery factors, to determine well completion or stimulation strategies.

Key reservoir characteristics such as permeability, porosity, fluid saturation, magnitude and direction of tectonic stress, and other properties impact field development decisions. The accuracy of these data is crucial in designing optimal stimulations. In reality, operators still rely on limited data sets, rock catalogs, prior experience, or local practices, which may lead to inaccuracies, miscalculations, and completion inefficiencies. The key parameters influencing completion and stimulation designs are commonly based on averages, estimates, standard correlations, or even assumptions.

To quantify important reservoir parameters and improve predictive modeling, the use of advanced formation evaluation tools is essential. These tools provide estimates of in-situ reservoir conditions and yield high-resolution, continuous data acquisition across potential zones of interest. Other data sources, such as core, pressure and production data, minifracture treatments, and advanced formation measurements, can provide correlations to complement and verify empirically derived values. Enhanced interpretation techniques and innovative formats for processing data are paving the way for improved models and simulations by providing values that were previously unknown or difficult to determine accurately, especially for low-permeability and heterogeneous reservoirs. Evaluating data at this level of detail can aid in identifying productive zones that might otherwise be overlooked.

Collaboration is necessary between operating and service companies to select and apply the best stimulation technologies and complete solutions. Together, this team brings a full range of skills and expertise on the targeted reservoir in addition to experience derived from work in a variety of fields and basins. This partnership provides improved stimulation results and enhanced well performance in a timely, cost-effective manner. For this reason, established data-processing services and a knowledge-management infrastructure are indispensable for real-time data exchange between widely dispersed office and field locations.

Improvements made to the reservoir characterization based on comprehensive data acquisition, combined with Web-based information technology, aid in stimulation optimization. Proper knowledge management allows previous well completion information to be readily accessible. This efficient distribution of information allows multidisciplinary teams to work together, creating a more responsive and reliable reservoir characterization. Stimulation optimization takes many forms, ranging from slight modification of fracture designs to application of new techniques or a complete revision of field-development schemes.

This chapter will focus on production improvement using enhanced reservoir stimulation and completion techniques. This chapter describes a field model that can be used to customize well completions. Improved engineering methodology, unique Web-based workflow, and aggressive field development are the essence of the initiative. Case histories will illustrate how this approach capitalizes on stimulation opportunities and improves financial results by fully using data captured during the drilling, evaluation, completion, and production phases. Al-Qarni et al. (2000) identify fields in south Texas and New Mexico, where this technique has been successful.

GEOLOGICAL OVERVIEW

Jonah field is a structurally trapped, tight-gas sandstone accumulation in the foredeep syncline of the northern Green River basin. It is located between the Wind River uplift on the east and the La Barge platform on the west, along the east-dipping, western flank of the Hoback basin. Faults are present on the west and south margins of the Jonah field (Warner, 2000). Production is from sandstones in the Upper Cretaceous (Maastrichtian) Lance Formation and (Campanian) Mesaverde Formation. Elevated gas pressures and improved porosity and permeability in the Lance and Mesaverde, relative to the surrounding area, characterize this "sweet spot" in the basin (Bowker and Robinson, 1997). Gross thickness of the Lance in the field ranges from 2000 ft (600 m) in the west to more than 3000 ft (900 m) in the east. Pressure gradient at the top of the Lance is typically 0.56 psi/ft and increases with depth to more than 0.71 psi/ft below 12,000 ft (3660 m).

The reservoir sandstones in the field are litharenites that are rich in quartz and rock fragments and poor in feldspar. Rock fragments are dominated by chert, siltstone, and shale, with lesser amounts of carbonates and volcanics. The only compositional difference between individual sandstones in the Lance is the relative percentages of quartz, rock fragments, and clay. Physical compaction decreases intergranular volume (i.e., the combined volume of pore space and cement in a given sample). Intergranular volume is also reduced by the presence of abundant lithic rock fragments. Cementation is comprised primarily of quartz overgrowths and ferroan calcite. The occurrence of ferroan minerals, including ferroan-calcite, is generally greatest in finer grained sediments. A more detailed review of the Jonah field geology is presented in Law and Spencer (1986), Robinson et al. (1996), Bowker and Robinson (1997), Robinson and Evans (1997), and other chapters in this volume.

PETROPHYSICS

Pay intervals completed within the field range from several hundred feet to more than 1000 ft (300 m). Pay is typically identified as those sandstones exhibiting the following wireline-derived characteristics:

- gamma-ray readings less than 65° API
- density porosity of 8% or greater (using a 2.65 g/cm^3 grain density)
- a deep resistivity of 40 ohm m or greater

In general, these parameters identify sandstones that have been successfully completed and have proven productive. Low-permeability sandstones at Jonah field are the reason why large hydraulic fracturing treatments were used in the early history of the field (Robinson, 2000).

Permeability measurements gathered using traditional methods, such as core, sidewall cores, and pressure-volume-temperature (PVT), may not give a representative value. In addition to the possible variations from true permeability, traditional methods can be costly and time consuming. Valuable information is provided by core, but only a statistically small portion of the zone of interest is sampled. Pressure buildup test and production-history matching can provide average permeability over a zone, but they lack information on mudstones in the adjacent bedding. In cases of extremely low permeability, some pay intervals may need to be stimulated to provide any data to determine permeability.

Predicting fracture geometry, selecting treatment systems (proppant and fluids), and determining job execution parameters (pumping rates and pressure) are dependent on the accuracy of permeability measurements from individual layers in sections with multiple-pay zones. Continuous wireline measurements to estimate permeability are available from several service companies. This study used Combinable Magnetic Resonance (CMR™; mark of Schlumberger Technology Corporation) log information to obtain permeability data in wells with sidewall core data. These data provided the necessary information to calibrate the log response to known core values. Four reservoir rock types were identified, and correlations were developed to compute permeability using the CMR logs, Platform Express (mark of Schlumberger Technology Corporation) logs and Reservoir Saturation Tool (RST™; mark of Schlumberger Technology Corporation). Measurements from these logs were used in wells, where adverse borehole conditions prevent acquisition of open-hole log data. Similar correlations were developed to compute rock mechanical properties. A unique zone-by-zone layering routine was developed to identify and evaluate each of the hundreds of distinct layers in the Lance Formation. To create a Jonah field model that would predict permeability, quantify reservoir stress profiles, and identify net pay, especially in low-productivity zones, it was necessary to calibrate log- and core-permeability data. The objective was to make the final logging program more cost efficient than alternate methods of acquiring permeability data. The production optimization team relied on a formation evaluation summary and independent data to obtain mechanical rock properties and stress profiles (Figure 1). Wells were perforated, and bottomhole pressures were obtained.

Net pay calculations were determined using a formation evaluation summary that included effective porosity greater than 6% over a 4-ft (1.2-m) interval. This technique increased pay by using the more complete data to compute different porosities and shale volumes. Upon completion of the reservoir characterization, Ultra Resources and Schlumberger Technology Corporation were able to create a model for Jonah field that used more accurate and consistent permeability, fluid saturation, net pay, and stress predictions.

PRODUCTION OPTIMIZATION WORKFLOW

The production optimization process combines the available data to understand wells, fields, and basins to present specific recommendations in the form of a comprehensive montage (Figure 2). This montage is an effective presentation of the formation evaluation, well analysis, reservoir characterization, and stimulation design. It displays well locations, relevant background data, mud log, open-hole and/or cased-hole logs, core analysis, and other test data (Figure 3). In addition to the geological and petrophysical data, the montage includes a wellbore configuration that shows completion design and perforation records, stimulation designs, fluid designs, and production forecasts. As the production optimization process continues, actual fluid treatments, production log data, production analysis, and actual production data are added to the display.

A Web-based intranet tool provides a framework for collaboration among the various technical disciplines as well as between the operator and the service company. The internet is used throughout the process, including

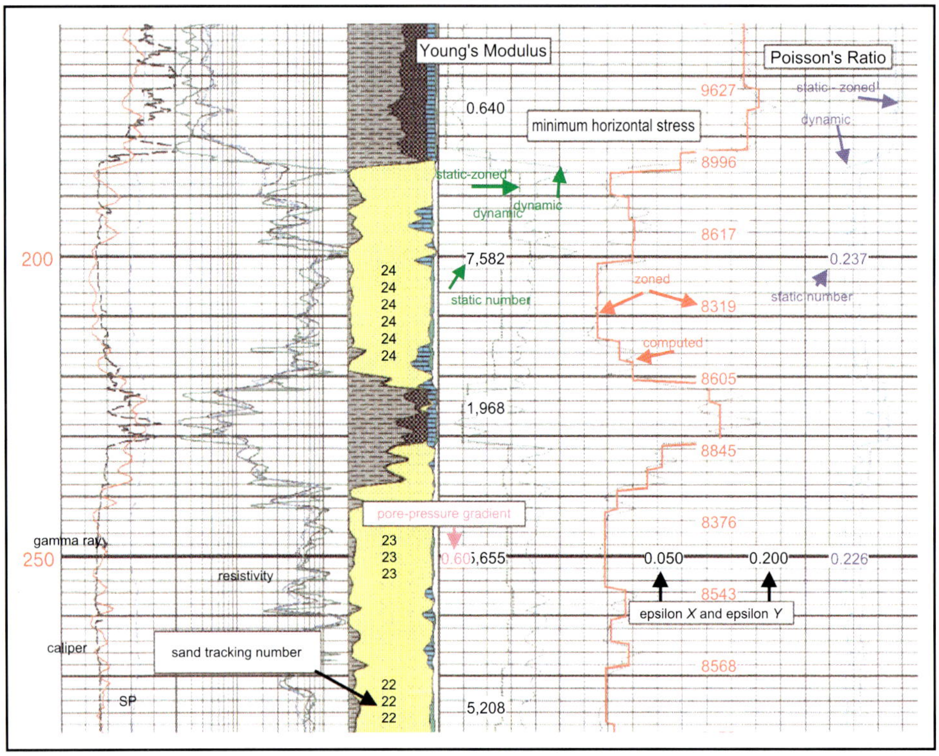

Figure 1. A profile of mechanical rock properties and stress profiles used for fracture simulation.

uploading open-hole data in the field, moving data between disciplines, providing perforation intervals and job design data to the field, and real-time stimulation reporting. The rapid integration of knowledge from several sources facilitates teamwork. The ability to quickly deliver completion recommendations is the main benefit of this workflow. Applications of the production optimization methodology are as follows:

(1) optimization of profitability in heterogeneous reservoirs where conditions vary
(2) improvement of marginal field performance
(3) solutions to completion problems or failures in areas where others are succeeding
(4) reengineering of completion methods to maintain or exceed past production at costs that are economically justifiable (Al-Qarni et al., 2000).

METHODOLOGY

The integration of open-hole and cased-hole reservoir characterization, drilling, completion, and stimulation technologies provides a workflow and the tools for

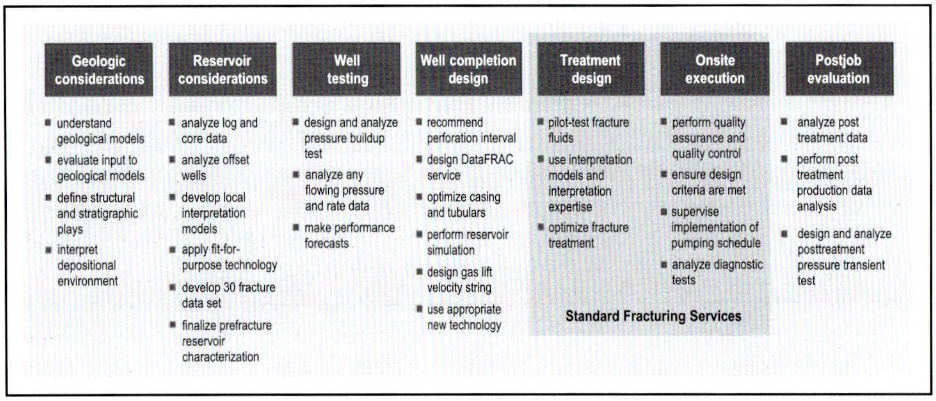

Figure 2. The production optimization methodology used in this chapter encompasses geologic and reservoir considerations and can be used to develop optimized stimulation designs. Critical to the success of this methodology is the ability to provide a feedback loop for continual improvement of well completions, stimulation, and field development (modified from Al-Qarni et al., 2000).

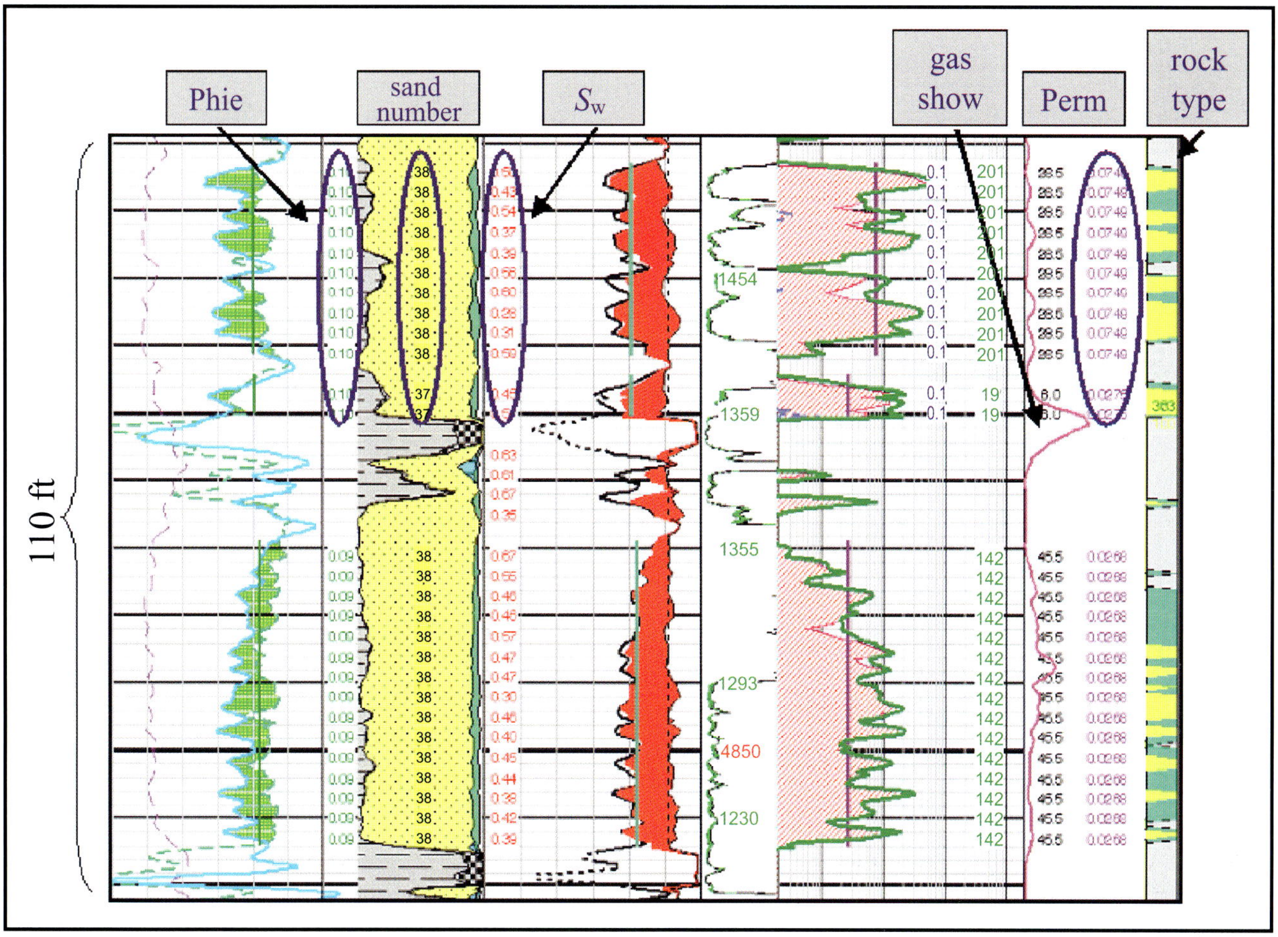

Figure 3. This formation evaluation summary includes data from the petrophysical model, including porosity, permeability, water saturation, and rock type. It also shows mud-log data, formation tops, and perforation intervals. Phie = effective porosity. S_w = water saturation. Perm = calculated permeability.

reengineering well completions and stimulation treatments in the production optimization framework. In addition, combining the improved assessment of water saturation, porosity, permeability, rock mechanics, stress profiles, and net pay forms the basis of specific solutions for field development. The production optimization workflow consists of two stages. The first stage concentrates on three to five wells. Once an initial model is established, emphasis shifts to identifying technologies for improving well performance. Geologists, petrophysicists, and reservoir or production engineers use this model to make completion and stimulation recommendations during the productive life of Jonah field.

In the second stage, geoscientists and engineers refine the reservoir model and completion designs to provide integrated stimulation recommendations for future wells. The team can deliver updated models within hours of logging operations. This "in-time" approach makes the production optimization methodology an integral part of completion planning instead of an afterthought. Knowledge gained from acquiring, interpreting, and formatting comprehensive data sets using the latest logging suites, cores, in-situ tests, and state-of-the-art stimulation technology is crucial to the success of the project.

One tool that is used in the process is the Dipole Shear Sonic Imager (DSI™; mark of Schlumberger Technology Corporation), which operates by bombarding the formation with acoustic waves then measures the resulting compressional and shear transit times (Brie et al., 1998). Transit times are converted into rock properties, such as shear modulus, Young's modulus, and Poisson's ratio. These measurements can be improved by correlation with measurements from core and in-situ formation tests. Radial or azimuthal sonic data can also be measured to derive the preferred fracture plane direction. A Fullbore Formation MicroImager (FMI™; mark of Schlumberger Technology Corporation) tool can be used to identify faults, natural fractures, secondary porosity, and borehole breakout. Borehole breakouts identified on the FMI log are perpendicular to the principal stress direction and help confirm DSI data (Al-Qarni et al., 2000).

After the well is completed, a production log can be run to correctly allocate production characteristics by zone in commingled, multistage reservoirs. The production log data indicate that only 60–65% of the sandstones stimulated contribute to production. Production optimization and induced stress diversion techniques have increased production to 90–95%.

COMPLETION AND STIMULATION METHODS

Jonah field has a thick pay interval consisting of individual and stacked channel sequences. To produce at economic rates, each sandstone unit requires stimulation. A typical completion interval consists of 50–200 ft (15–60 m) of gross interval. Stimulating too many stages will result in a significant increase in cost and a reduced return on investment. Stimulating too few stages will result in a reduction in production potential. A plan has to be developed to determine which sandstones to complete and how to group them into stages, and then a choice of isolation methods must be implemented.

In the early work on the field, limited-entry diversion treatments over longer intervals were used to minimize the number of fracturing treatments and to control costs (Flack et al., 1997). Treatment stages ranged from 100 to 450 ft (30 to 140 m), with 20–28 perforations per stage. Once an interval was stimulated, the well was flowed to recover fracturing fluids and to test productivity. After 1–2 weeks of flow, the fractured intervals were isolated with mechanical plugs (retrievable or drillable) or sand plugs, and the next interval was perforated and fractured. This process continued, until the well was fully completed. Three to six intervals were fractured per well over a 5-week period. These completion methods could take more time than drilling the well.

Investigation of traditional completion techniques developed by other operators (e.g., Finch et al., 1996; Flack et al., 1997; Eberhard et al., 2000) revealed several concerns. Radioactive tracer surveys run on early wells showed that many of the perforated sandstones were not effectively stimulated, and production logs indicated that only about 60–65% of the perforated intervals produced gas. Sand plugs used to isolate zones were difficult to place and were commonly set too low in the section and resulted in costly procedures to reset them. In some wells, proppant from the near-wellbore region was displaced when sand plugs were set before fracturing subsequent stages. Tracer surveys indicated near-wellbore fracture containment, but net-pressure plots corrected for perforation tunnel erosion show considerable fracture-height growth (Figure 4). Even with limited-entry perforating for diversion, there may be connectivity between perforated and unperforated sandstones. The observation that fracture jobs rarely screened out and sand plugs for zonal isolation commonly ended up being displaced into the formation suggested that there was potential to increase the treatment size.

A production optimization team evaluated these traditional fracture methods and undertook a program to gather additional data, including stress gradients, proppant selection, Young's modulus, fluid leakoff, and reservoir pressure, in an attempt to modify the stimulation strategy. Minifracture treatments were acquired to determine fracture gradient, Young's modulus, and leakoff parameters for the fracture fluids. Dipole Shear Sonic Imager data were used to build near-wellbore stress models, determine preferred fracture direction, and calibrate stress profiles for sand-shale sequences. These studies confirmed the stress values from the minifractures.

Fracturing models using more reliable stress measurement in sandstones and bounding layers, coupled with Young's modulus values, provided improved fracture height and width estimates. Core data were used to verify mechanical rock properties and understand fluid compatibility. Previous geologic interpretations assumed that there were natural fractures throughout the field, but image logs (FMI) do not indicate significant fracturing. Lower-than-expected leakoff indicated that pad volumes could be decreased, which would reduce costs.

The use of induced stress diversion techniques was developed by Amoco (now BP) and Schlumberger and has been reported on by Christensen et al. (2000). It was developed to be an improved method for completing multiple horizons without sand plugs or mechanical isolation. Some Jonah operators decided to test the method by fracturing smaller vertical intervals and by performing multiple treatments in a single day. The ideal technique would allow cleanup of fracture intervals without the need to wash out sand or retrieve packers/plugs. Net pressure generated during stimulation would be used to divert subsequent fracturing treatments into the next interval.

After each zone is fractured, it is flowed back long enough to recover at least 1.5 times the casing volume of fluid and to allow the fracture near the wellbore to close. The next interval is then perforated, and fracture is stimulated. This process is repeated until the entire producing

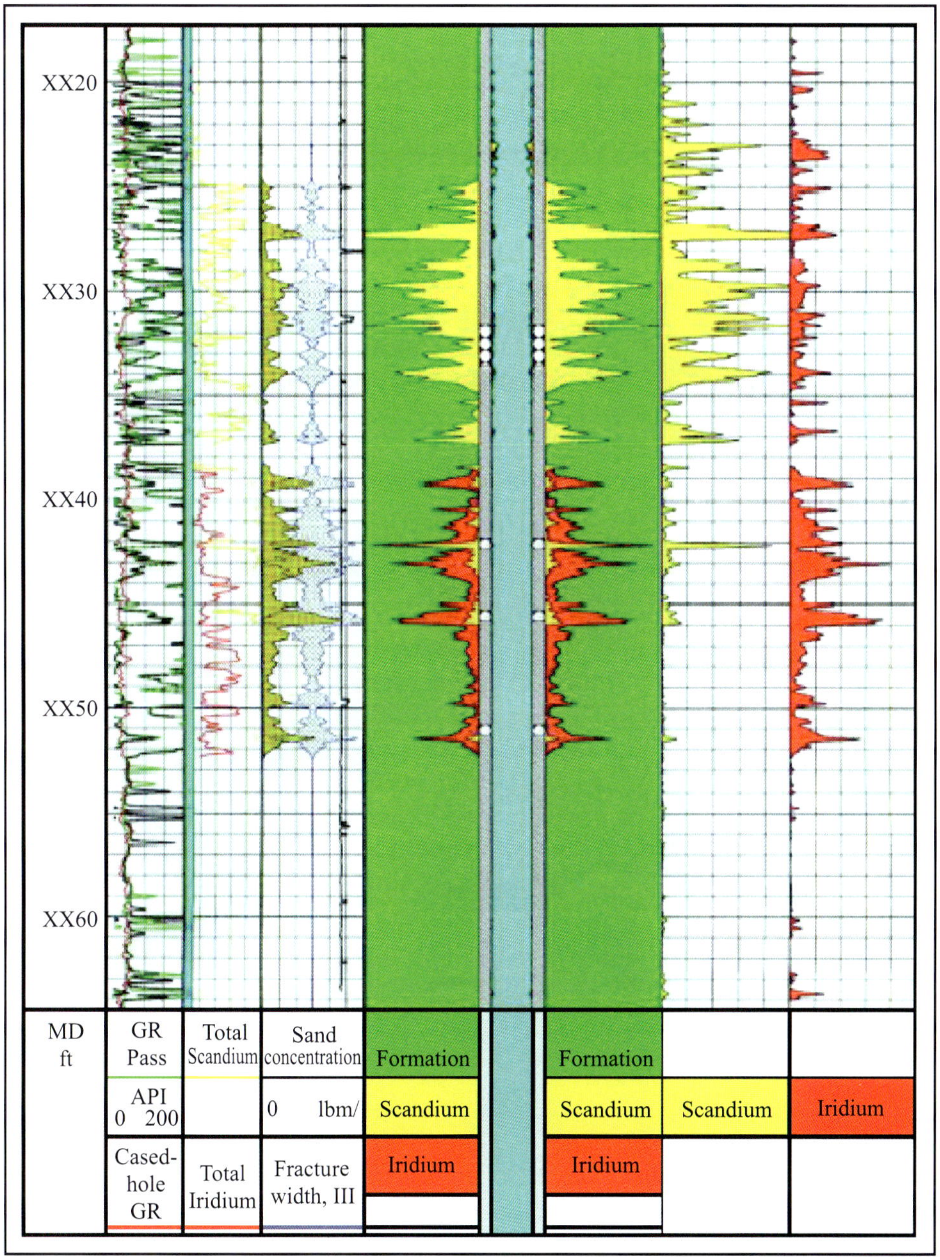

Figure 4. This radioactive tracer survey indicates successful placement of proppant into separate fractures less than 100 ft (30 m) apart, without sand plugs or mechanical isolation (modified from Al-Qarni et al., 2000).

interval is treated. As many as 18 fracture stages can be performed in 36 hr, reducing the time required to completely stimulate a well from as long as 5 weeks to less than 4 days.

This completion technique provides for a single fracture interval to have as much as 120 perforations, whereas in past methods, an entire wellbore might contain 120 perforations. The additional perforations reduce the probability of leaving any pay sandstones untreated by the stimulation. Sturm et al. (2000) have shown that smaller treated intervals (i.e., more stages) will generate more effective fracture half-length.

Wellbore tubulars were the final aspect to be analyzed. Early operators used 4.5- or 5.5-in. (11- or 14-cm) casing to accommodate the high pump rates and pressures that are necessary to fracture large intervals. After all zones were treated, the operators would run 2.375- or 2.875-in. (6- or 7.3-cm) tubing using a snubbing unit. Jonah wells typically produce water and condensate, so it was important that tubing sizes allowed fluids to unload to sustain production. Induced stress diversion methods use more and shorter treatment intervals, so each stimulation can be placed at lower pump rates, making 3.5-in. (8.9-cm) casing feasible.

Schlumberger has developed a technique called PowerSTIM that combines formation evaluation and stimulation design to predict production. This technique allows an investigative team to quickly evaluate multiple completion scenarios to determine which combination yields the maximum production at the lowest cost. Using PowerSTIM, it takes about 4 hr to generate predictions for production rates for all sandstones after receiving open-hole logs and other well data. Schlumberger constructs a montage, including as many as 18 stages, showing individual stimulation designs, production forecasts, and economic evaluations. More effective use of the internet may reduce this turnaround time in the future.

After 3–6 months, the acquisition of production logs on key wells provides data to calibrate the production contribution from each of the stimulated intervals. These data can be used to confirm that fracture-model conductivity and geometry are actually attained. By constantly evaluating the production optimization process and refining the model, this methodology can achieve remarkable predictive accuracy.

INDUCED STRESS DIVERSION

Induced stress diversion is a completion technique that uses staged hydraulic fracture treatments without the aid of sand plugs, bridges, baffles, or any type of positive or mechanical isolation. The process relies on diversion created from both natural and induced stress from each previous hydraulic treatment. Figure 5 is pore-pressure profile from data collected using induced stress diversion. Stress diversion results from a combination of the normal increase in stress with depth, the temporary supercharge from fracturing, the incremental closure stress on proppant, and pipe friction (Hewett and

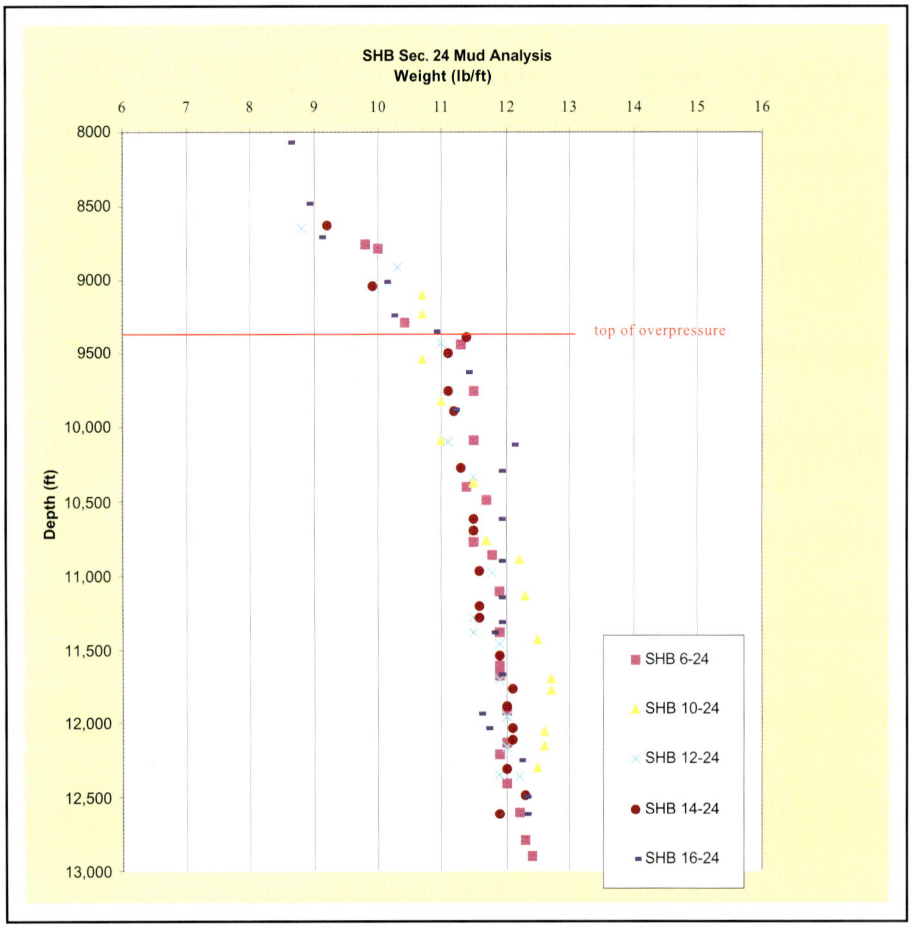

Figure 5. Mud weight versus drilled depth for five wells in Sec. 24, T29N, R108W, Sublette County, Wyoming. Increasing mud weights at about 9400 ft (2900 m) indicate increasing pore pressure with depth. This is an important requirement for employing the induced stress diversion technique. SHB = Stud Horse Butte.

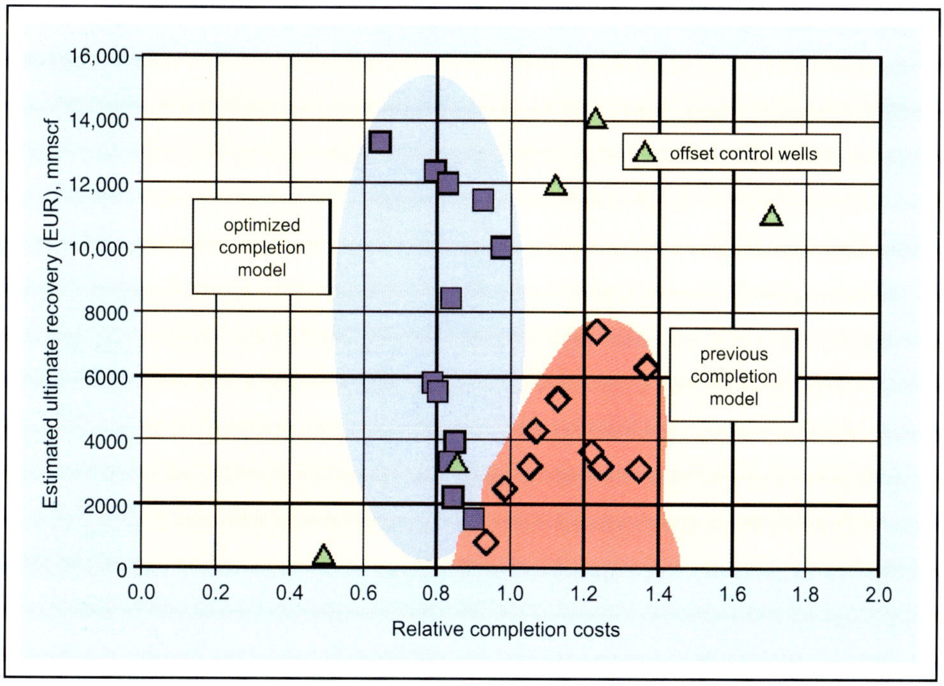

Figure 6. Estimated ultimate recovery (EUR) versus completion costs for wells investigated in this study. Using induced stress diversion, stimulation methods increases EUR per completion dollar spent over the conventional completion techniques applied in Jonah field (modified from Al-Qarni et al., 2000).

Spence, 1998). Once the initial stage has been perforated and hydraulically fractured, it is allowed to flow back for approximately 1.5 times the casing volume. The flow-back period allows the wellbore to begin to clean up, the fracture to close in the near-wellbore region, and the supercharged pressure from the stimulation to begin to dissipate. At this point, the next stage is perforated, and the second fracture treatment is initiated. The buildup of induced stress from the first stimulation forces the hydraulic fracture to take the path of least resistance into the newly perforated zone. In Jonah, the matrix properties, in-situ stresses, and fracture density are heterogeneously distributed within the gross productive interval and create difficulty in achieving proper stimulation (Sturm et al., 2000).

ECONOMICS

Average completion costs for Ultra Resources in 1998 were approximately $2.5 million per well. This cost included stimulating eight stages at $250,000 per stage, used the limited entry perforating technique, and placed large proppant volumes (Figure 6). Production optimization described in this chapter reduced this cost to less than $1.0 million per well in 2001 (Figure 6). These wells used the induced stress diversion method and had an average of 11 fracture stages at a cost of $60,000 each. A significant reduction in cost and completion time was realized in these wells. Original completion procedures that range from 45 to 90 days per well were reduced to 6–10 days per well (Figure 7). Induced stress diversion yielded a 50% reduction in cost and was achieved while sustaining an equivalent or greater reserve per well (Figure 8). For example, using an average total well cost of $2.2 million and an average estimated ultimate recovery of 6 bcf at a wellhead price of $2.00/mcf, the rate of return is 46% (Figure 8). Based on this scenario, the finding and development cost is less than $0.40/mcf, and payout is approximately 24 months. Decline curve analysis on wells using induced stress diversion indicates an increase in estimated ultimate recovery (Figure 9).

CONCLUSIONS

Production optimization methods and induced stress diversion reservoir stimulation have been used to create models for generating improved well-completion strategies. A better understanding of reservoir characteristics across several pay zones in a well, field, or a basin will result in optimized stimulation treatments and completion techniques that can reduce costs and maximize production. Reservoir modeling can accelerate this understanding.

The Lance Formation in Jonah field provides an excellent opportunity to test this production optimization

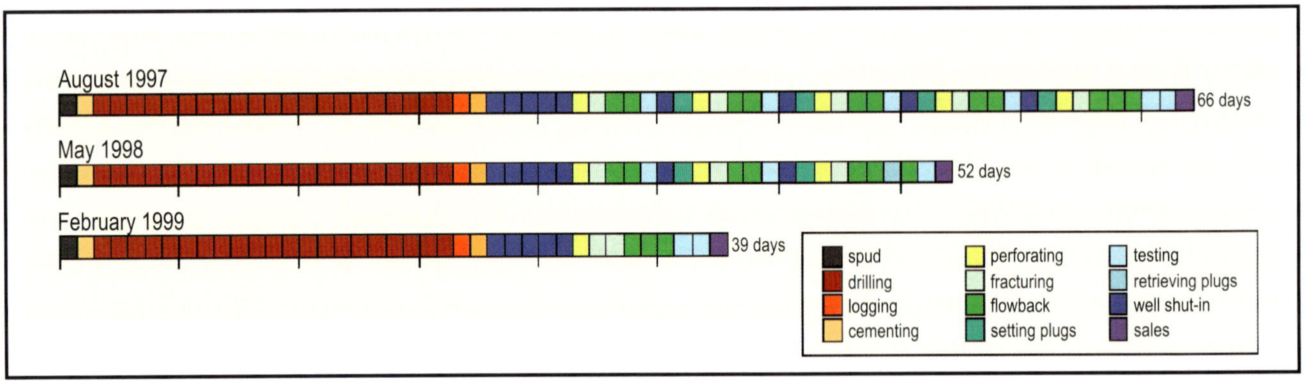

Figure 7. Over an 18-month period, the average time to first production was reduced by 4 weeks, primarily because of completing Jonah field wells without sand plugs or other forms of mechanical isolation (modified from Al-Qarni et al., 2000).

and characterization methodology. A comprehensive data set was developed to improve completion strategies and designs. Mechanical rock properties, hydrocarbon resource profiles, and sand-prediction models were either developed or improved. Optimal fracture design was achieved by stimulating smaller gross intervals and by evaluating each interval zone by zone. In February 2001, the first well was stimulated using recommendations based on reservoir and completion models developed by the production optimization team. The more precise model aided in design optimization for fracture stimulations that reduced well completion costs from \$2.8 million to \$1.0 million per well while delivering equal or greater gas production and reserves per well. Other cost savings were achieved through optimized pad volumes, proper fluid and proppant selection, and optimized tubular design. Overall completion costs in the Jonah field were cut by 50%. Finding costs decreased from \$0.45 to \$0.23/MSCF.

ACKNOWLEDGMENTS

We thank Schlumberger Technology Corporation and Ultra Resources, Inc., for allowing us access to the data used in this chapter. We also thank the following individuals

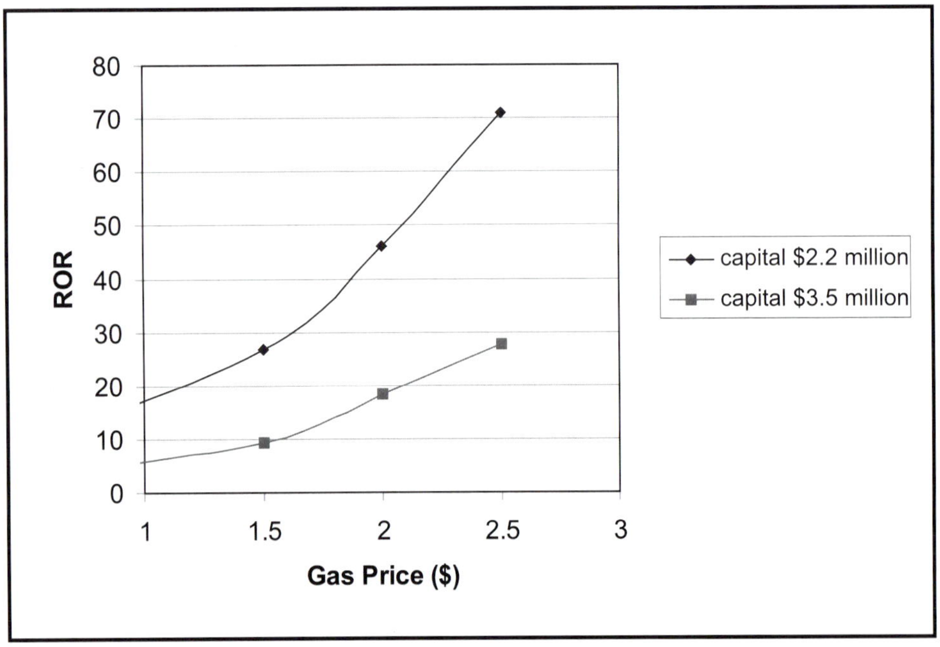

Figure 8. Using induced stress diversion completion methods reduces capital costs (\$2.2 million) and achieves an improved rate of return (ROR) at gas prices between \$1.50 and \$2.50/mcf.

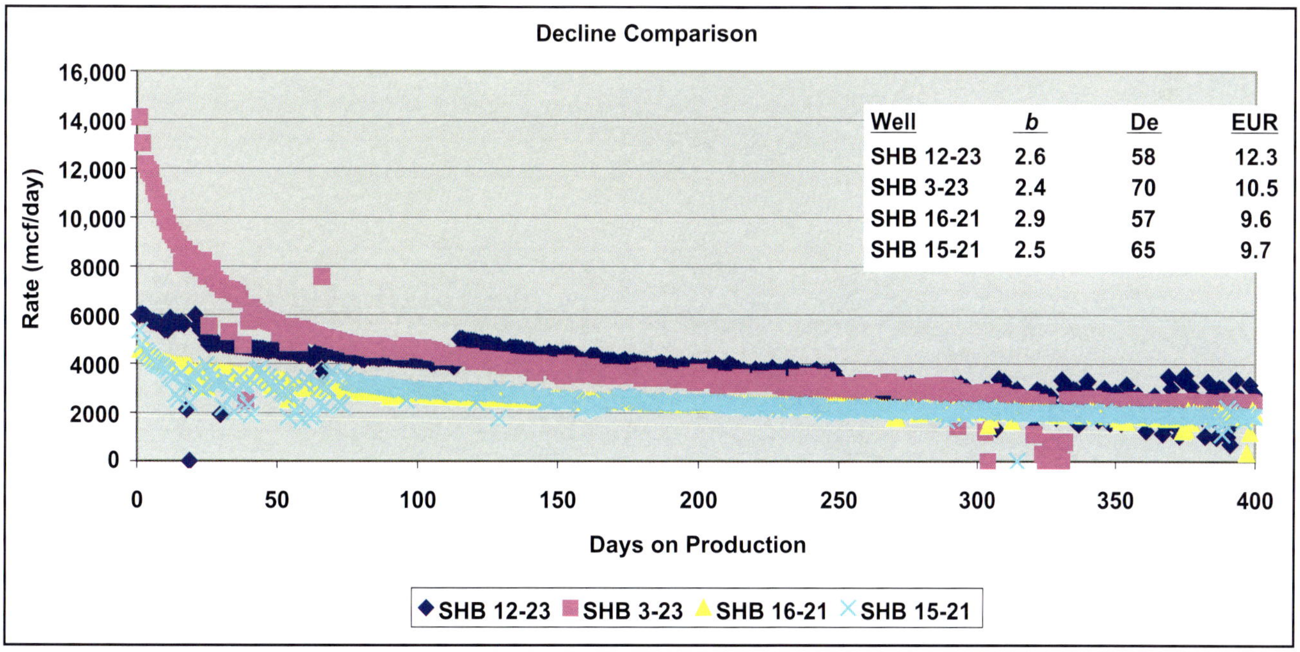

Figure 9. Decline curve analysis comparing an early well completion (SHB 3-23, SHB 15-21) versus an optimized (SHB 12-23, SHB 16-21) well completion method for two sections in Jonah field. The significant difference in initial decline rates (De) and the sustained or greater estimated reserves result in improved economic return.

for their support and contributions, without which this chapter would not have been possible: Bob McDermott, Sam McClure, Lesley Evans, Wayne Rowe, Stan Denoo, and Tim Waite.

REFERENCES CITED

Al-Qarni, A. et al., 2000, From reservoir specifics to stimulation solutions: Oilfield Review, v. 12, no. 4, p. 42–61.

Bowker, K. A., and J. W. Robinson, 1997, Jonah field: A shallow sweet spot in the basin-centered gas accumulation of the northern Green River basin: AAPG Bulletin, v. 81, p. 1219, and Expanded Abstracts Volume, p. 25–30.

Brie, A. et al., 1998, New directions in sonic logging: Oilfield Review, v. 10, no. 1, p. 40–55.

Christensen, C. J., D. L. Cox, E. A. Lake, V. B. Dolan, J. D. Crisler, and J. P. Lima, 2000, Optimized completion techniques in Jonah: Society of Petroleum Engineers (Western Region) and AAPG (Pacific Section) Regional Meeting, Long Beach, California, U.S.A., 19–23 June, and SPE Paper 62853.

Eberhard, M. J., M. J. Mullen, C. A. Seal, and B. P. Ault, 2000, Integrated field study for production optimization, Jonah field, Sublette County, Wyoming: Society of Petroleum Engineers/Canadian Energy Research Institute Gas Technology Symposium Proceedings, p. 683–695, and SPE Paper 59790.

Finch, R. W., J. L. Skees, W. W. Aud, J. L. Johnson, C. Esphahanian, and J. T. Hansen, 1996, Evolution of completion and fracture stimulation practices in the Jonah field, Sublette County, Wyoming: Society of Petroleum Engineers Annual Conference Proceedings, p. 521–531, and SPE Paper 36734, 11 p.

Flack, D., J. Shaw, J. Martin, J. Oriet, and C. Koenen, 1997, Modified drilling and completion procedures enhance recovery in southwest Wyoming field: Society of Petroleum Engineers Rocky Mountain Regional Meeting Proceedings, p. 85–98, and SPE paper 38344.

Hewett, T. W., and C. J. Spence, 1998, Induced stress diversion: A novel approach to fracturing multiple pay sands of the NBU field, Uintah County, Utah: Society of Petroleum Engineers Rocky Mountain Regional Meeting, Low Permeability Reservoirs Symposium and Exhibition, Denver, Colorado, U.S.A., April 5–8, and SPE paper 39945.

Law, B. E., and C. W. Spencer, 1986, Geology of tight gas reservoirs in the Pinedale anticline area, Wyoming, and at the multiwell experiment site, Colorado: U.S. Geological Survey Bulletin, v. 1886, p. 1–260.

Robinson, J. W., 2000, Discovery of Jonah field, Sublette County, Wyoming: The Mountain Geologist, v. 37, p. 135–143.

Robinson, J. W., and L. W. Evans, 1997, Reservoir lithofacies in the Lance Formation, Jonah field, Sublette County, Wyoming: AAPG Bulletin, v. 81, p. 1232.

Robinson, J. W., D. Delozier, and R. Finch, 1996, Integrated reservoir description and analysis of the Lance Formation at Jonah field, Sublette County Wyoming: AAPG Bulletin, v. 80, p. 297–308.

Sturm, S. et al., 2000, Multi-disciplinary analysis of tight gas sandstone reservoirs, Almond Formation, Siberia Ridge field, Greater Green River basin: Gas Research Institute Topical Report, GRI-00/0026, 339 p.

Warner, E. W., 2000, Structural geology and pressure compartmentalization of Jonah field based on 3-D seismic data and subsurface geology, Sublette County, Wyoming: The Mountain Geologist, v. 37, p. 15–30.

CHAPTER 16

From Prospect to Giant Gas Field: History of the Environmental Analyses of Jonah Field

ROBIN M. SMITH
Mountaintop Consulting, Casper, Wyoming, U.S.A.

THE NATIONAL ENVIRONMENTAL POLICY ACT AND ITS APPLICATION TO JONAH FIELD DEVELOPMENT

Jonah field and the area around it have been the subject of an immense amount of environmental study since the field was discovered. Between 1993 and 2001, the Bureau of Land Management (BLM) completed one Environmental Impact Statement (EIS) and four Environmental Assessments (EAs) (Figure 1). In addition, another EIS for downspacing in Jonah field is currently in preparation by the BLM. Before any surface disturbance is authorized, each project component is also subject to an additional, site-specific EA prior to final site approval and construction.

Each of these studies was conducted to comply with the requirements of the National Environmental Policy Act (NEPA), passed by Congress in 1969. NEPA-mandated environmental impact analysis and assessment is a discovery function for ascertaining the range of risks and benefits of proposed actions on public lands. Where practicable and technically feasible, the resulting decision recommends and requires various mitigation measures and alternatives designed to reduce project impacts. The NEPA was also written as a means to require the various agencies in the federal government to include the public in identifying issues and concerns and to disclose the estimated impacts of the proposed action under consideration.

PROPOSED ACTIONS AND COSTS

Project components analyzed and approved by previous Jonah field environmental analyses include the following:

- 497 well locations, developed at a maximum of 16 well pads per 640-ac (260-ha) section over an area of approximately 29,200 ac (11,800 ha) or at a maximum of eight well pads in an area encompassing an additional 30,400 ac (12,300 ha)
- gas compression and associated facilities
- two produced water disposal facilities
- 16 water-supply wells
- a condensate stabilization facility pilot project
- equipment yards, office sites, sand pit, roads, and numerous pipelines (for both gathering and sales)

The EIS currently underway is analyzing a range of alternatives that are listed below.

No Action: No additional well pads beyond the 497 existing authorized well pads. BLM will consider new development only to alleviate drainage conditions or as identified in alternate proposals;

Proposed Action Scenario: As many as 1250, 2200, and/or 3100 new wells on as many as 850, 1500, and/or 2100 new well pads, respectively; as many as 40 well pads per 640-ac (260-ha) section;

Minimum Surface Disturbance Scenario: As many as 1250, 2200, and/or 3100 wells, all from the existing 497 well pads; 16 well pads per 640-ac (260-ha) section maximum, as currently authorized;

Maximum Surface Disturbance Scenario: As many as 1250, 2200, and/or 3100 wells on a maximum of as many as 1250, 2200, and/or 3100 new well pads; no well pad density restrictions;

Surface Disturbance Density Restriction Scenario: As many as 1250, 2200, and/or 3100 wells on as many as 755 well pads (16 pads/section) or 1510 well pads (32 pads/section). It is anticipated that the new EIS will be completed in late 2004.

The estimated cost of the completed and ongoing analyses, borne entirely by industry, is more than \$1,500,000. Wildlife monitoring (a required mitigation) and investigations have resulted in an additional cost of more than \$175,000, also paid for by industry. The cost of

Jonah Field: Case Study of a Giant Tight-Gas Fluvial Reservoir, J. W. Robinson and K. W. Shanley, eds.: AAPG Studies in Geology 52 and Rocky Mountain Association of Geologists 2004 Guidebook

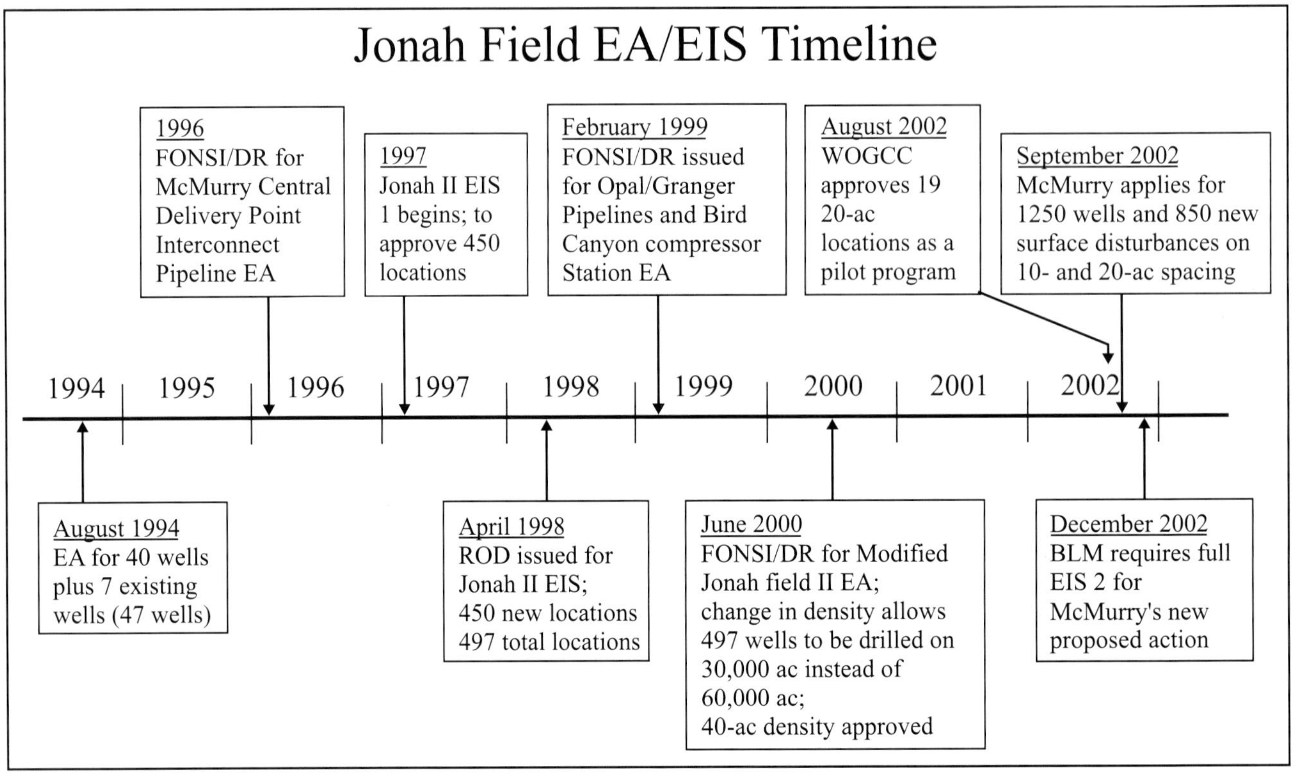

Figure 1. Timeline illustrating the sequence of environmental studies at Jonah field. WOGCC = Wyoming Oil and Gas Conservation Commission; FONSI/DR = Finding of No Significant Impact/Decision Record; ROD = Record of Decision.

implementing other mitigation required by the various decisions is difficult to calculate, but it is estimated that those costs far exceed the total cost of the analyses and wildlife monitoring.

RESOURCE CONCERNS

Oil and gas development activities can affect other resources in a positive or negative way. Affected areas can include many species of wildlife and their habitat, air quality, viewsheds, local socioeconomic conditions, water quality and supply, paleontology, other mineral resources, soils, surface water and wetlands, cultural resources, and other land uses, such as livestock grazing and recreation.

The purpose of environmental analysis is to determine the magnitude of the potential positive and negative impacts of development activities on other resources in the project area. The impacts are analyzed and estimated in both an immediate (project-specific impacts) and a cumulative sense (including impacts from other nearby projects). The NEPA process is a means to ascertain whether the impacts are considered significant by various affected parties and to publicly develop workable alternatives and mitigation strategies that are technically and economically feasible. Ultimately, the goal is to estimate and reveal the anticipated impacts of the project, as approved, so an informed decision is reached with public participation.

ALTERNATIVES AND MITIGATION

Many alternatives to the proposed actions have been analyzed and include the following:

- reduced scope of drilling activity, i.e., allow fewer wells
- directional drilling
- staged development
- no action, i.e., not allow any further development of Jonah field

Mitigation actions consist of avoidance of sensitive areas, timing restrictions to limit activities during specific time frames, seasonal construction limitations, reclamation guidelines, noise restrictions, various measures to limit emissions, and numerous measures to minimize visual impacts to other users of the public lands.

CONCLUSIONS

Jonah field development continues to demonstrate that oil and gas activities can be conducted with minimal negative environmental impacts and can provide significant, positive socioeconomic impacts if mitigation measures are carefully considered and front-end loaded into project planning. By using a multiagency, interdisciplinary approach with emphasis on public involvement, a thorough mitigation plan has been developed and implemented during the development of Jonah field. As development of the field has proceeded, monitoring of impacts to other resources has shown that impact levels are consistently below the predicted level, except for positive socioeconomic impacts, which have far exceeded predications in the various analyses. Ultimately, land management goals, the need for reduced dependency on foreign energy supplies, the public's concerns, and the operators' proposed actions have all been well balanced and considered.